Linear Algebra

Linear Algebra

John B. Fraleigh
UNIVERSITY OF RHODE ISLAND

Raymond A. Beauregard
UNIVERSITY OF RHODE ISLAND

Addison-Wesley Publishing Company
READING, MASSACHUSETTS   MENLO PARK, CALIFORNIA   DON MILLS, ONTARIO
WOKINGHAM, ENGLAND   AMSTERDAM   SYDNEY   SINGAPORE   TOKYO
MADRID   BOGOTÁ   SANTIAGO   SAN JUAN

Sponsoring Editor   *Jeffrey M. Pepper*
Production Supervisor   *Nancy Lasker Behler/Herbert Nolan*
Editorial and Production Services   *Quadrata, Inc.*
Designer   *Nancy Sugihara/Quadrata, Inc.*
Illustrator   *Scientific Illustrators*
Manufacturing Supervisor   *Ann DeLacey*
Cover Designer   *Marshall Henrichs*

IBM is a registered trademark of the International Business Machines Corporation.

Library of Congress Cataloging-in-Publication Data

Fraleigh, John B.
    Linear algebra.

    Bibliography: p.
    Includes index.
    1. Algebra, Linear.   I. Beauregard, Raymond A.
II. Title.
QA184.F73   1986   512′.5   85-30647
ISBN 0-201-15459-5

Reprinted with corrections, May 1987

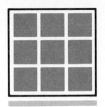

To The Instructor

Because linear algebra provides the tools to deal with many problems in fields ranging from forestry to nuclear physics, it is desirable to make the subject accessible to a variety of students. Recognizing this, we have attempted to achieve an appropriate blend of intuition and rigor in our presentation. Applications, which appear throughout, serve to emphasize the practicality of the subject.

Our text is designed for a first undergraduate course in linear algebra. However, more material is included than can be covered in a single semester. In class testing, we found that Chapters 1, 3, 4, 5, and 6 alone required one semester to cover.

We outline some distinctive features of our book.

MOTIVATING LINEAR ALGEBRA

In a course in linear algebra, considerable time is devoted to solving linear systems and studying algebraic structure. All too often, applications showing the great utility of the subject do not appear until the course is almost over. We believe that students have a greater appreciation of the subject if its importance is made clear at the outset. To that end, Chapter 1 begins with an indication of the role that linear algebra plays in the sciences. Three of the seven sections in Chapter 1 are devoted to applications. Section 1.1 illustrates in an intuitive way the approximation of tough nonlinear problems by more easily solved linear ones. A brief introduction to linear programming is given in Section 1.2 using a geometric treatment of two-variable linear optimization. Both of these sections are designed to emphasize the importance of linear algebra rather than to develop particular techniques. Markov chains appear in Section 1.7, where we study the distribution of a population among states. This section provides a satisfying application of matrix multiplication. Any or all of these three application sections of Chapter 1 may be omitted if an instructor desires to make the usual dive directly into the matrix algebra and solutions of linear systems that comprise Sections 1.3–1.6. If that is done, we urge that the three application sections be assigned for independent reading to provide motivation. In class testing this material, we covered these three sections and found them a very satisfying way to start the course. Our students never asked, "What is linear algebra good for?"

EIGENVALUES AND EIGENVECTORS

This important topic appears in several places. It is encountered briefly when dealing with Markov chains in Section 1.7, although eigenvalues and eigenvectors are not formally defined or developed there. With Section 1.7 as motivation, they are treated in detail in Section 4.5 using determinants. Section 4.6 is devoted to diagonalization of a square matrix A and to the study of vectors $A^k\mathbf{x}$ for large values of k. In Section 6.4 we examine eigenvalues and eigenvectors from the point of view of linear transformations. Chapter 7 is devoted to further applications and to methods for computation of eigenvalues.

COMPUTATIONS IN LINEAR ALGEBRA

Chapter 2 is devoted to problems encountered in the solution of large linear systems by a computer. (This chapter title is followed by an asterisk, indicating that the material is not needed in subsequent unstarred sections.) Counting flops, the LU-factorization, partial pivoting, and ill-conditioned matrices are treated here. The chapter is of special interest to computer science majors. In class testing our manuscript, one of the authors included this chapter and his students found it one of the most interesting in the text.

TREATMENT OF GENERAL VECTOR SPACES

We deal chiefly with the vector spaces $\mathbb{R}^n$. However, many definitions and theorems concerning vectors in $\mathbb{R}^n$ are identical with those for general vector spaces. General vector spaces are defined in Section 3.2 and whenever such definitions or theorems and proofs occur, they are phrased in terms of general vector spaces. A few examples of function spaces are given, but on the whole, we stick to finite-dimensional vector spaces, in particular, $\mathbb{R}^n$. An instructor who so desires can easily give a quick sketch of the material in Chapter 3, making sure that students grasp the essential notions of subspace, independence, and bases, and continue on to other topics.

CALCULUS AND LINEAR ALGEBRA

Recognizing that some students of linear algebra may not have studied calculus, we have chosen not to scatter distracting examples from calculus throughout the text. Rather, we devote Chapter 8 (starred) to an explanation of the relationship between linear algebra and calculus.

LINEAR PROGRAMMING

Linear programming is so widely used today that we would like all linear algebra students to know the nature of the problems and have a little experience with them. As stated above, Section 1.2 gives a geometric introduction to two-variable linear programming. Chapter 9 gives an intuitive explanation of the simplex method using tableaux, and can be presented any time after Chapter 1.

APPLICATIONS

At the close of a linear algebra course, the instructor often regrets that there was not time to present more applications. We tried to take care of that by placing three applications in Chapter 1. Section 5.2 is devoted to the method of least squares and the following two sections continue to treat this application. Applications of eigenvalues and eigenvectors are presented in Section 4.6 and the first three sections of Chapter 7. Many applications of a geometric nature appear throughout the text as well.

EDUCATIONAL AIDS

Exercises. The exercises are class-tested, graded, and varied. Typically, each section concludes with from 20 to 30 exercises. Optional computer exercises are provided.

Summaries. Each section concludes with a summary, which we have found to be very useful for both students and instructors. Students find it a handy reference for formulas and theorems. Instructors can see by a glance at the summary what is covered in a section and what notation is used.

Historical Notes. A number of historical notes by Victor Katz are scattered throughout the text. A bit of personal and professional information about the people who contributed to a discipline always adds a dash of spice to its study.

Microcomputer Software. A diskette of software developed especially for this text by one of the authors is available from the publisher. It is designed for the IBM PC, and also runs acceptably on several compatibles. Use of the software is, of course, optional. We believe it is helpful both for lecture demonstrations and for use by students in drill and problem solving. Upon adoption of the text, a license to use the software in the classroom and to copy it as desired for use by students can be obtained without charge. Optional exercises for its use appear throughout the text. We believe it is important for students to experience the fact that linear algebra computations that may appear formidable to do by hand can be executed without difficulty these days. In our class testing, students were encouraged to check handwritten assignments using the software. A list of the programs with a brief description of each appears in Appendix B.

ACKNOWLEDGMENTS

We wish to thank the following reviewers for their helpful suggestions: Ross A. Beaumont, University of Washington; Lawrence O. Cannon, Utah State University; Daniel Drucker, Wayne State University; Bruce Edwards, University of Florida; Christopher Ennis, University of Minnesota; W. A. McWorter, Jr., Ohio State University; John Morrill, DePauw University; and James M. Sobota, University of Wisconsin–LaCrosse.

We also thank Geri Davis and Martha Morong of Quadrata, Inc., for their excellent work in design and editing. Our editor, Jeffrey Pepper, worked closely with us all the way, and we deeply appreciate his enthusiasm and guidance. We are grateful to Don S. Balka and Mike Kasper for preparing a solutions manual.

Historical notes in a text are often limited to a line or two giving the dates, country, father's occupation, and the professional position of a mathematician mentioned in the text. When we saw the first historical note (on page 2) by Victor Katz, telling how problems involving linear systems were written down more than 2000 years ago, we realized that students of our text would get a special treat. Those who read these notes will acquire some insight into the evolution of linear algebra to the form in which it appears today, and can only wonder how it will look 100 years from now. We are greatly indebted to Victor Katz for these fascinating contributions to our text.

Kingston, R.I.   J.B.F.
                                                                                R.A.B.

To The Student

If you look through our text, you will get some idea of the wide variety of fields in which linear algebra is useful. Many different types of problems can be attacked using linear methods. Quite a bit of our study will be devoted to developing efficient techniques for solving problems that at first glance might seem to require long and laborious computations. Many of these techniques are easily implemented on a computer. Not surprisingly, the present accessibility of computers is responsible for a surge in the study of linear algebra as well as in its application.

Our text is organized so that you can see the utility of the subject at the outset and throughout your study. Most courses in linear algebra begin with the basic techniques in Sections 3–6 of Chapter 1. It may be that your instructor will not have time in class to discuss Sections 1, 2, and 7, which present applications. We encourage you to at least leaf through these sections and read the paragraphs describing the nature of the applications, even if you don't have time to follow the mathematical explanations.

Chapter 2 is especially appropriate if you are interested in computers. Once again, a half hour looking through the chapter to see what it is about is time well spent if it is not covered in class. If you follow this policy throughout the text as your course unfolds, you will acquire an intuitive understanding of the nature of the subject and of the scope of its applications. Then too, if in the future you run into some problem where our book could be of help, you may recognize that fact and know where to look.

If you have had calculus, we especially recommend that you look at Chapter 8 at the end of your course, if you did not study it in class. A large part of calculus deals with linear approximations of functions. Chapter 8 may well give you a whole new perspective on calculus. In any case, it will provide some fine reading during summer vacation.

Software is available to your instructor. After the text has been adopted, the software can be copied without charge for your use. We encourage its use so that you can experience firsthand how feasible the techniques of linear algebra are using a computer. You may also wish to check some of your handwritten computations. However, you should not attempt to substitute software for all handwritten computations. A certain amount of pencil-and-paper work is necessary for most of us to ensure that we really assimilate concepts and techniques.

We close with one warning. The basic techniques in Sections 3–6 of Chapter 1 are quite easily acquired, and our students have sometimes become too complacent early in the course. Beginning with Chapter 3, you will frequently have to decide what technique to use and how to arrange data for that technique. Often this requires a good deal more thought than executing the technique. You will probably find that the course requires more time and effort when you get to Chapter 3.

Kingston, R.I.   J.B.F.
                                                                            R.A.B.

Contents

4
Determinants and Eigenvalues   183

46

5
Applications of Vector Geometry and of Determinants   245

x (Q R-factor) →

43.

6
Linear Transformations   309

48

7
Eigenvalues: Applications and Computations   363

8
The Role of Linear Algebra in Calculus* 407

9
Linear Programming 443

Linear Algebra

Introduction to Linear Systems

A great deal of linear algebra involves finding simultaneous solutions of linear equations, like the system

$$x - y = 8$$
$$3x + 2y = -6.$$

Solving such a linear system is a fundamental numerical technique in linear algebra. In this chapter, we will study the Gauss and Gauss–Jordan methods for finding all solutions of a linear system, even when the number of equations is different from the number of unknowns. The Gauss and Gauss–Jordan methods are practical when the number of equations and the number of unknowns are not too large. Chapter 2 will discuss solving large linear systems.

Matrices, which are just rectangular arrays of numbers, arise naturally in an attempt to work efficiently with a linear system. We will also develop matrix arithmetic in this chapter.

When we taught undergraduate linear algebra before, students sometimes asked "But what is linear algebra good for?" Sections 1.1, 1.2, and 1.7 will address this question. It is not too easy to answer this in an introductory course since many applications of linear algebra are quite sophisticated. Linear algebra is a useful tool for solving quite complicated problems, as we will show. Sections 1.1, 1.2, and 1.7 can be considered optional or deferred for later study, but we recommend at least reading through them as motivation for your work. The emphasis in these three sections is on appreciating the importance of linear equalities and inequalities, rather than on computational expertise. Pencil-and-paper illustrations of Section 1.1, in particular, can be quite tedious. Of course, a computer can handle the computations easily. Computer software to accomplish this is available to users of our text.

1.1
Systems of Linear Equalities: An Overview

We have all solved systems of two equations in two unknowns, but the "applications" we saw were often artificial and not of practical importance. For instance, we have all had to solve contrived problems like the following one.

Example 1 Alice is now twice as old as Paul. Three years ago, she was two years younger than three times Paul's age. How old are Paul and Alice now?

Solution Let x be Alice's age now and let y be Paul's age now. Then $x = 2y$ and $x - 3 = 3(y - 3) - 2$. This gives us the linear system

$$\begin{aligned} x - 2y &= 0 \\ x - 3y &= -8. \end{aligned}$$

Subtracting, we obtain $y = 8$, so $x = 2y = 16$. Thus Alice's age is 16 and Paul's is 8. $\triangleleft$

Here is a somewhat more realistic application.

Example 2 The Hadleys have just inherited \$350,000, which they wish to use to increase their annual income. Since they are already in a high income-tax bracket, they decide to invest in tax-free bonds. They want the bonds to generate \$32,500 income annually. They can invest in highly rated "safe" bonds that pay 8% annually or lower rated "risky" bonds that pay 11%. How much should they invest in each type of bond in order to achieve their income goal?

SYSTEMS OF LINEAR EQUATIONS are found in ancient Babylonian and Chinese texts dating back well over 2000 years. The problems are generally stated in "real-life" terms, but it is clear that they are artificial and designed simply to train students in mathematical procedures. As an example of a Babylonian problem, consider the following, which has been slightly modified from the original found on a clay tablet of about 300 B.C.: There are two fields whose total area is 1800 square yards. One produces grain at a rate of $\frac{2}{3}$ bushel per square yard, the other at a rate of $\frac{1}{2}$ bushel per square yard. The total yield of the two fields is 1100 bushels. What is the size of each field? This problem leads to the system

$$x + y = 1800$$
$$(\tfrac{2}{3})x + (\tfrac{1}{2})y = 1100.$$

A typical Chinese problem, taken from the Han dynasty text "Nine Chapters of the Mathematical Art" (about 200 B.C.) reads: There are three classes of corn, of which three bundles of the first class, two of the second, and one of the third make 39 measures. Two of the first, three of the second, and one of the third make 34 measures. And one of the first, two of the second, and three of the third make 26 measures. How many measures of grain are contained in one bundle of each class? The system of equations here is

$$3x + 2y + z = 39$$
$$2x + 3y + z = 34$$
$$x + 2y + 3z = 26.$$

Solution   Let x be the amount invested in the safe bonds and y the amount in the risky bonds. Then we have the linear system

$$0.08x + 0.11y = 32{,}500$$
$$x + y = 350{,}000.$$

We multiply the second equation by 0.08 and subtract, obtaining

$$0.08x + 0.11y = 32{,}500$$
$$\underline{0.08x + 0.08y = 28{,}000} \qquad \textbf{Subtract}$$
$$0.03y = 4{,}500.$$

We find that $y = 4{,}500/0.03 = \$150{,}000$ and $x = \$200{,}000$. The Hadleys should invest \$200,000 in the safe bonds and \$150,000 in the risky bonds. ◁

In a beginning linear algebra text it is difficult to illustrate really practical problems that correspond to systems of linear equations. There are some situations in which linear problems do occur, as in some electric circuit computations, but the class of realistic problems that are themselves linear is not a very large one. Why, then, is it important to study linear systems?

Reason 1 Solutions of linear systems can be found by algebraic methods, unlike many nonlinear problems.

Reason 2 An adequate approximate solution of a nonlinear problem can often be found by solving a sequence of linear problems.

To illustrate reason 1, consider just one equation in one unknown. A *linear* equation in one unknown, like $3x = 8$, is very easy to solve. The solution of every such linear problem $ax = b$ is simple to determine. But the nonlinear equations $x^5 + 3x = 1$, $x^x = 100$, and $x - \sin x = 1$ are all tough to solve algebraically.

The remainder of this section is devoted to illustrating reason 2. One technique for dealing with a nonlinear problem consists of *approximating* the problem by a linear one and solving the easy linear problem to get at least a rough solution of the nonlinear one. Sometimes this rough solution can be used to generate a new and better linear approximation of the original problem. Solving the new linear problem may give a still better approximate solution of the original problem. In a moment we will illustrate such a technique, called Newton's method, for solving systems of nonlinear equations.

Calculus is based on approximating problems by linear problems. Linear algebra and calculus are closely allied, a fact that is all too often not apparent to the student of either topic. No knowledge of calculus is assumed in this text except in Chapter 8, which demonstrates part of the connection between calculus and linear algebra.

APPROXIMATING FUNCTIONS BY LINEAR FUNCTIONS

We start with a function $f(x)$ having a nice, smooth graph as shown in Fig. 1.1. We want to find the equation of a linear function—that is, one having a straight-line graph—that is a good approximation to the graph of $f(x)$ near the point $(x_0, f(x_0))$ on the graph. In order to find the equation of a line, we need to know a point on the line and the slope of the line. We know the point $(x_0, f(x_0))$. It remains to determine what to use for the slope.

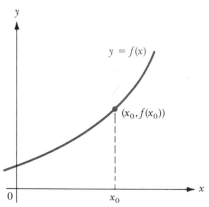

Figure 1.1
A smooth graph $y = f(x)$.

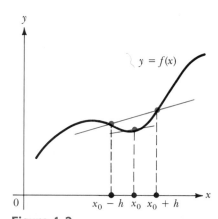

Figure 1.2
The color lines are almost parallel.

Referring to Fig. 1.2, we see that if h is a small positive number, the line through the points

$$(x_0 - h, f(x_0 - h)) \quad \text{and} \quad (x_0 + h, f(x_0 + h))$$

has roughly the same steepness as the graph of $f(x)$ at the point $(x_0, f(x_0))$. We denote the slope of this line by m, so

$$m = \frac{f(x_0 + h) - f(x_0 - h)}{2h}. \tag{1}$$

Figure 1.3 shows why we should make h fairly small. However, we must not make h too small, even using a computer, for then $f(x_0 + h)$ and $f(x_0 - h)$ may be so nearly equal that their *difference*, which appears in the numerator of Eq. (1), will have very few significant figures. For example, if we compute using 16 significant figures, we find

$$2.173492653124869 - 2.173492653124842 = 0.000000000000027$$

and we have an accuracy of only two significant figures in this difference. In practice, $h = 0.0001$ works quite well when a computer is used. For pencil-and-paper illustrations, we prefer to take a larger value, like $h = 0.1$, to save labor in computation.

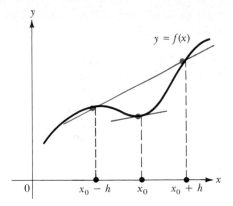

Figure 1.3
The color lines are not almost parallel.

In summary, if $f(x)$ has a smooth graph as in Fig. 1.1, we may take as approximation to the graph of $f(x)$ near $(x_0, f(x_0))$ the line found as follows:

Point: $(x_0, f(x_0))$

Slope: $m = \dfrac{f(x_0 + h) - f(x_0 - h)}{2h}$

Equation: $y - f(x_0) = m(x - x_0)$ or $y = mx + [f(x_0) - mx_0]$

Our linear approximation for $f(x)$ near $x = x_0$ is therefore

$$mx + [f(x_0) - mx_0], \tag{2}$$

which is of the form $ax + b$ with $u - m$ and $b = f(x_0) - mx_0$.

Example 3   Find the linear approximation (2) to $f(x) = 2x^2 - x$ at $x_0 = 2$, taking $h = 0.1$.

Solution   Point: $(x_0, f(x_0)) = (2, 6)$

Slope: $m = \dfrac{f(2.1) - f(1.9)}{2(0.1)} = \dfrac{6.72 - 5.32}{0.2} = \dfrac{1.4}{0.2} = 7.$

Our approximating linear formula (2) becomes

$$7x + [6 - 7(2)] = 7x - 8. \;\; \lhd$$

Now suppose that we have a two-variable function $f(x, y)$ with a nice, smooth graph, and we want to approximate it by a linear function near a point $(x, y) = (x_0, y_0)$. We will not attempt to sketch the function. A linear function in x and y is of the form

$$ax + by + c. \tag{3}$$

We must find suitable values for a, b, and c.

If we set $y = y_0$, then $f(x, y_0)$ becomes a function of the one variable x. For example, if $f(x, y) = x^2 + xy + 2y^3$ and we set $y = y_0 = 2$, we obtain the function $x^2 + 2x + 16$. Also, the formula (3) becomes $ax + by_0 + c$, which should be a single-variable linear approximation of the single-variable function $f(x, y_0)$. From Eq. (1) above, we see that we can take as coefficient a of x

$$a = m_1 = \frac{f(x_0 + h, y_0) - f(x_0 - h, y_0)}{2h} \tag{4}$$

for small positive h. A similar argument setting $x = x_0$ and approximating $f(x_0, y)$ by the linear formula $ax_0 + by + c$ shows that we can take

$$b = m_2 = \frac{f(x_0, y_0 + h) - f(x_0, y_0 - h)}{2h}. \tag{5}$$

Therefore, Eq. (3) becomes

$$m_1x + m_2y + c. \tag{6}$$

It remains to determine c. We want our approximation (6) to have the value $f(x_0, y_0)$ at (x_0, y_0), so we must have

$$m_1x_0 + m_2y_0 + c = f(x_0, y_0) \tag{7}$$

and

$$c = f(x_0, y_0) - m_1x_0 - m_2y_0. \tag{8}$$

Therefore, (6) becomes

$$\boxed{m_1x + m_2y + [f(x_0, y_0) - m_1x_0 - m_2y_0].} \tag{9}$$

Compare approximation (9) with approximation (2).

Analogous reasoning shows that for a function $f(x, y, z)$ of three variables, a linear approximating function at (x_0, y_0, z_0) should be given by

$$\boxed{m_1x + m_2y + m_3z + [f(x_0, y_0, z_0) - m_1x_0 - m_2y_0 - m_3z_0],} \tag{10}$$

where

$$m_1 = \frac{f(x_0 + h, y_0, z_0) - f(x_0 - h, y_0, z_0)}{2h}$$

and m_2 and m_3 are similarly defined.

Example 4   Find a linear function approximating

$$f(x, y) = x^3 - 3xy + y^2 + 3x,$$

where $(x, y) = (1, 1)$, taking $h = 0.1$.

Solution   Using pencil and paper or using a calculator, we find that

$$m_1 = \frac{f(1.1, 1) - f(0.9, 1)}{0.2} = \frac{2.331 - 1.729}{0.2} = 3.01$$

and

$$m_2 = \frac{f(1, 1.1) - f(1, 0.9)}{0.2} = \frac{1.91 - 2.11}{0.2} = -1.$$

Thus our approximation (9) becomes

$$(3.01)x - y + [2 - 3.01 - (-1)] = (3.01)x - y - 0.01. \quad \triangleleft$$

NEWTON'S METHOD

The approximation of functions by linear functions just presented is the basis for *Newton's method* for solving a system of nonlinear equations. Figure 1.4 illustrates the idea behind Newton's method for the case of a nonlinear equation $f(x) = 0$ in one variable. Here is an outline of the method.

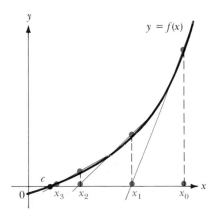

Figure 1.4
Newton's method to estimate a solution c to the equation $f(x) = 0$.

Step 1   Make a reasonable estimate x_0 of a solution.

Step 2   Replace $f(x)$ in the equation $f(x) = 0$ by a linear approximating function at x_0, and solve the resulting linear equation. Call the solution x_1.

NEWTON'S METHOD first appears in his manuscript "De Analysi per Aequationes Numero Terminorum Infinitas" (On Analysis by Equations of Infinitely Many Terms), which was written about 1669 and which circulated through the English mathematical community over the next decades. It is interesting that Newton does not present a justification by either geometry or calculus, but merely presents the method as an algorithm for solving polynomial equations. Similar methods for solving such equations were known to Chinese mathematicians at least several hundred years earlier.

Step 3 Compute $f(x_1)$.

Step 4 Continue the above procedure, starting this time with step 2 and using x_1 in place of x_0. Call the solution obtained x_2 and compute $f(x_2)$ in step 3. Keep going back to step 2, and compute x_3, x_4, and so on, until the values $f(x_3)$, $f(x_4)$, and so on, obtained in step 3, become so close to zero that the approximate solution is judged satisfactory.

A few comments on this outline are in order. Concerning step 1, in a particular problem we often have a rough idea of an approximate solution x_0. Otherwise, we can use a computer to graph the function or a calculator to compute some values and see where the function values are near zero. As a matter of fact, the first approximation may not need to be a good one. Often Newton's method will converge to a solution if a random first estimate is tried. In a practical problem, such a random first estimate might not lead to the solution desired, however.

Steps 2–4 may seem like a lot of work, and they often are using pencil and paper or a calculator. However, it is exactly the kind of iterative computation at which a computer excels. The method is very practical using today's technology.

Step 3 is very important. We really must compute these function values as a check. It is possible that the solutions of the linear equations obtained in step 2 do not approach a solution of the original equation $f(x) = 0$. Figure 1.5 illustrates one possible problem if the initial estimate in step 1 is a poor one, on the wrong side of a hump in the curve.

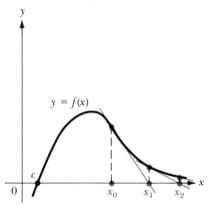

Figure 1.5
x_0, x_1, x_2 . . . do not get close to c.

Example 5   Use a calculator and Newton's method to find an approximate solution of $x^x = 100$.

Solution   We write our equation as $f(x) = x^x - 100 = 0$. For any x, the formula for m in

(1) then becomes

$$m = \frac{f(x + h) - f(x - h)}{2h} = \frac{[(x + h)^{x+h} - 100] - [(x - h)^{x-h} - 100]}{2h}$$

$$= \frac{(x + h)^{x+h} - (x - h)^{x-h}}{2h}$$

We find that $f(3) = 3^3 - 100 = 27 - 100 = -73$ and $f(4) = 4^4 - 100 = 256 - 100 = 156$. This sign change from -73 to 156 indicates that $f(x) = 0$ has a solution between $x = 3$ and $x = 4$. Using a scientific calculator, starting with $x_0 = 4$ and $h = 0.00001$, we obtained the data given in Table 1.1. Thus we see that $x^x = 100$ for approximately $x = 3.597285024$. ◁

Table 1.1

i	x_i	$f(x_i)$
0	4	156
1	3.744635441	40.3491215
2	3.620734188	5.500400569
3	3.597934159	0.1481299455
4	3.597285529	0.00011535
5	3.597285024	0.0000000004
6	3.597285024	0.0000000004

Newton's method is also effective with a nonlinear system

$$f(x, y) = 0,$$
$$g(x, y) = 0$$

of two equations in two variables, or with

$$f(x, y, z) = 0,$$
$$g(x, y, z) = 0,$$
$$h(x, y, z) = 0$$

as well as with even larger nonlinear systems of n equations in n variables. The method is essentially the same as for one variable. The general procedure is summarized below and then illustrated with an example.

Newton's Method

Step 1 Make an estimate of a solution. Often the estimate need not be very accurate.

Step 2 Approximate the nonlinear equations in the system by linear equations at the estimated solution, and solve the resulting linear system to obtain the next estimate of a solution.

> **Step 3** Compute the values of the functions at the approximation obtained.
>
> **Step 4** Return to step 2 to generate the next approximation to a solution. Continue until the significant figures in your estimates stabilize and the values obtained in step 3 are all close to zero. If this does not occur, start again with a new first estimate.

Example 6   Perform the first iteration of Newton's method to approximate a solution of the nonlinear system

$$f(x, y) = x^2 + 2y^2 - 8 = 0$$
$$g(x, y) = 3xy + 4 = 0$$

starting with $(x_0, y_0) = (-1, 2)$ and taking $h = 0.1$.

Solution   Using a calculator, we obtain m_1, m_2, and c for the linear approximation $m_1x + m_2y + c$ to $f(x, y)$:

$$m_1 = \frac{f(-0.9, 2) - f(-1.1, 2)}{0.2} = \frac{-0.4}{0.2} = -2,$$

$$m_2 = \frac{f(-1, 2.1) - f(-1, 1.9)}{0.2} = \frac{1.6}{0.2} = 8,$$

$$c = f(-1, 2) - m_1(-1) - m_2(2) = 1 - 2 - 16 = -17.$$

Proceeding similarly for $g(x, y)$, we obtain m_1', m_2', and c' for the linear approximation $m_1'x + m_2'y + c'$:

$$m_1' = \frac{g(-0.9, 2) - g(-1.1, 2)}{0.2} = \frac{1.2}{0.2} = 6,$$

$$m_2' = \frac{g(-1, 2.1) - g(-1, 1.9)}{0.2} = \frac{-0.6}{0.2} = -3,$$

$$c' = g(-1, 2) - m_1'(-1) - m_2'(2) = -2 + 6 + 6 = 10.$$

We now solve simultaneously these two linear approximations:

$$\begin{aligned}
-2x + 8y - 17 &= 0 \quad &\textbf{Multiply by 3} \\
\underline{6x - 3y + 10} &= 0 \quad &\textbf{Add} \\
21y - 41 &= 0 &
\end{aligned}$$

Thus $y = \frac{41}{21}$. Substituting into the second equation, we find that

$$6x - \frac{41}{7} + 10 = 0$$

$$6x = -\frac{29}{7}$$

$$x = -\frac{29}{42},$$

so we obtain as an estimated solution $(x_1, y_1) = (-\frac{29}{42}, \frac{41}{21})$. Computation shows that

$$f(-1, 2) = 1 \quad \text{while} \quad f\left(-\frac{29}{42}, \frac{41}{21}\right) \approx 0.1$$

and

$$g(-1, 2) = -2 \quad \text{while} \quad g\left(-\frac{29}{42}, \frac{41}{21}\right) \approx -0.04.$$

Thus this first iteration of Newton's method produced a decrease in the magnitude of the values of the functions. ◁

In contrast to Example 6, we may not always decrease the function values as an iteration of Newton's method is performed. Our next example will illustrate this.

The computer software available to schools using this text contains a program NEWT123, which makes these computations for nonlinear systems containing one, two, or three equations. The program proceeds precisely as we have described here, taking $h = 0.0001$. Our final example shows data obtained using NEWT123. We used as a starting estimate $(1, 1, 1)$, where the value of each function in the example is fairly small. Again we mention that this initial estimate is often not critical; Newton's method frequently achieves a solution if we start with a bad estimate. Of course, if there are several different solutions, and nonlinear systems often have more than one solution, the one we obtain starting with a bad estimate may not be the one we want. Note in the following example that the first iteration of Newton's method actually increased the magnitude of all function values.

Example 7   Use the program NEWT123 or similar software to find an approximate solution of the nonlinear system

$$f(x, y, z) = z^3 - 2^{xy} = 0$$
$$g(x, y, z) = x^2 + y^2 + z^2 - 6 = 0$$
$$h(x, y, z) = y^2 + z^4 - 3 = 0.$$

Solution   Using NEWT123 we obtained the data given in Table 1.2. The first group of two lines shows the organization of the data. The next group of two lines shows the data for the initial estimate. Each subsequent group of two lines gives data for an iteration of Newton's method. Note that the last two estimates generated agree to all significant figures the computer prints and that the values of the function there are small. We are convinced that we are very close to an actual solution of this nonlinear problem. When used in conjunction with a computer, linear algebra is a very powerful tool! ◁

Table 1.2

x-estimate f(estimate)	y-estimate g(estimate)	z-estimate h(estimate)
1	1	1
−1	−3	−1
2.702059949380162	9.588011528095497D-02	1.702059935338883
3.734171094511983	4.207328990036415	5.401848484647414
2.07832154619196	0.6993138426911154	1.422314655906721
0.1387876979376031	0.8314392803522045	1.581483805748174
2.004647186012171	0.5717462687632218	1.300407130756668
−1.415931615434413D-02	3.656284195397075D-02	0.1865733420455031
2.016390081965199	0.5363859968797477	1.28379334859411
−4.758614558474372D-04	1.66426219078053D-03	4.027146168480766D-03
2.016428122094907	0.5355885681523603	1.2834185957312
−6.955309482248318D-07	7.777793346042117D-07	2.024415043988537D-06
2.016428181052007	0.5355880305933138	1.283418424421588
−4.06286115861576D-13	3.217426325363704D-13	5.880296249927142D-13
2.016428181052011	0.5355880305931036	1.283418424421545
−1.110223024625157D-16	1.110223024625157D-16	0
2.016428181052011	0.5355880305931036	1.283418424421545
0	1.110223024625157D-16	−5.551115123125783D-17

SUMMARY

1. The class of practical problems that correspond to systems of linear equalities is not very large. However, systems of linear equalities are important since:

   a) they are comparatively easy to solve, and

   b) they can often be used to give information about the solution of nonlinear problems that are difficult to solve otherwise.

2. Let $y = f(x)$ be a function with a smooth graph, and let $y_0 = f(x_0)$. A linear function approximating $f(x)$ near $x = x_0$ is given by

$$mx + [f(x_0) - mx_0],$$

where

$$m = \frac{f(x_0 + h) - f(x_0 - h)}{2h}$$

for a small positive value of h. Typically, $h = 0.0001$ works well using a computer and double-precision arithmetic.

3. A linear function approximating a nonlinear function $f(x,y)$ near $(x, y) = (x_0, y_0)$ is given by

$$m_1 x + m_2 y + [f(x_0, y_0) - m_1 x_0 - m_2 y_0],$$

where

$$m_1 = \frac{f(x_0 + h, y_0) - f(x_0 - h, y_0)}{2h} \quad \text{and}$$

$$m_2 = \frac{f(x_0, y_0 + h) - f(x_0, y_0 - h)}{2h}$$

for a small positive value of h. Similar approximations can be made for a function $f(x, y, z)$, as in Eq. (10).

4. Newton's method for a system of nonlinear equations consists of repeatedly approximating the nonlinear system by a linear system and solving the linear system. See the boxed outline on pp. 9–10.

EXERCISES

Exercises 1–15 can be done with pencil and paper.

1. Let $f(x) = ax^2 + bx + c$ be a quadratic function, so a is nonzero.
   a) Compute the value of m in Eq. (1) at $x = x_0$ in terms of h, and observe that the result is independent of the value of nonzero h.
   b) Give the formula for the approximation of $f(x)$ in Eq. (2) in terms of a, b, c, x, and x_0.

2. Find a linear approximation of the function $f(x) = 3x^2 - 4x + 5$ near $x_0 = 2$. (Exercise 1 shows that we can use any nonzero value of h.)

3. Repeat Exercise 2 for the function $f(x) = -5x^2 + 3x - 8$ with $x_0 = -2$.

4. Repeat Exercise 2 for the function $f(x) = -x^2 + 10x - 7$ at $x_0 = 1$.

5. Consider the equation $f(x) = x^2 - 3x + 1 = 0$. We have $f(0) = 1$ and $f(1) = -1$. Approximate $f(x)$ at $x_0 = 0$ by a linear function, and replace the equation $f(x) = 0$ by its linear approximation. Solve this linear equation, calling the solution x_1. (Exercise 1 shows that we can use any nonzero value for h.) Is $f(x_1)$ of smaller magnitude than $f(x_0)$?

6. Repeat Exercise 5 for $f(x) = 2x^2 + 5x + 1 = 0$, taking $x_0 = 0$.

7. Repeat Exercise 5 for $f(x) = x^3 + x - 1 = 0$, taking $x_0 = 1$ and $h = 0.1$.

8. Repeat Exercise 5 for $f(x) = -2x^3 + x^2 - 4 = 0$ for $x_0 = -1$, taking $h = 0.1$.

9. Repeat Exercise 5 for $f(x) = x^5 + 3x - 1 = 0$ for $x_0 = 1$, taking $h = 0.1$.

10. a) Conclude from Exercise 1 that the values m_1 and m_2 of Eqs. (4) and (5) are independent of the nonzero value taken for h for any quadratic function $f(x, y) = ax^2 + bxy + cy^2 + dx + ey + k$.
    b) Is the analogous statement true for quadratic functions $f(x, y, z)$ of three variables and quadratic functions of even more variables? Why?

11. Find the linear approximation in Eq. (9) of the function $f(x, y) = 3x^2 - 4xy + 5y^2$ at (x_0, y_0). (See Exercise 10.)

12. Repeat Exercise 11 for $f(x, y) = x^2 - 7xy + 3y^2 - 4x + 5y - 2$ at $(x_0, y_0) = (1, 1)$.

13. Repeat Exercise 11 for $f(x, y) = x^3 - 2xy^2$ at $(x_0, y_0) = (-1, 1)$, taking $h = 0.1$.

14. Consider the linear system

$$f(x, y) = x^2 + y^2 - 6 = 0$$
$$g(x, y) = xy - 1 = 0.$$

Apply one iteration of Newton's method, starting with $(x_0, y_0) = (2, 0)$, and find the point (x_1, y_1). Are $f(x_1, y_1)$ and $g(x_1, y_1)$ of smaller magnitude than $f(x_0, y_0)$ and $g(x_0, y_0)$, respectively?

15. Repeat Exercise 14 for the nonlinear system

$$f(x, y) = 2x^2 - y^2 + 3 = 0$$
$$g(x, y) = -3xy - 2y^2 + 1 = 0,$$

starting with $(x_0, y_0) = (1, -2)$.

Exercises 16–20 can be solved with a scientific calculator as well as with a computer. In these exercises, use Newton's method to estimate a solution of the given equation as accurately as you can, starting with the given x_0. Use radian measure for all trigonometric functions.

16. $f(x) = x^5 + 3x - 1 = 0, x_0 = 1$

17. $f(x) = x - 2 \sin x = 0, x_0 = 2$

18. $f(x) = \cos x - x = 0, x_0 = 1$

19. $f(x) = x^x - 5 = 0$, you choose x_0

20. $f(x) = 2^x + 3^x - 50 = 0$, you choose x_0

█ *For Exercises 21–29, use the program NEWT123 available to schools using this text, or similar software, to find approximate solutions to the given problems.*

21. Solve $x^x = 1000$.

22. Solve $2^x + 3^x = 85$.

23. Find both solutions of $x^2 - 2x + \sin x - 8 = 0$.

24. Find both solutions of $x^4 + 3^x - 8 = 0$.

25. The nonlinear system $x^2 + y^2 = 4$, $xy^2 = 1$ has four solutions. Try to find them all.

26. Find both solutions of the system $x^2 - 4x + y^2 = 0$, $xy = 1$.

27. Find both solutions of the system $x^2 + y^2 - 8y + 2x = 0$, $y^2 - x^2 = 2$.

28. The system of three nonlinear equations in Example 6 has three more solutions. Find them.

29. Find both solutions of the system $xy + x = -1$, $z^3 = \sin xy$, $y^2 - x^2 = 4$.

1.2
Systems of Linear Inequalities: An Overview

We turn now to a discussion of a system of linear inequalities, such as

$$2x + 3y \geq 6$$
$$2x + y \geq 2 \tag{1}$$
$$x - y \leq 3.$$

We said in Section 1.1 that the class of those problems that lead directly to a system of linear equalities is not a very large one. However, there are a great many problems that lead directly to a system of linear inequalities. Oddly enough, many of these problems involving linear inequalities were not formulated and studied until the 1940s.

The *linear programming* problem we will discuss arose partly from our modern technology in medicine, transportation, forestry, agriculture, fuels, and so on. Linear programming has contributed to the preservation of environmental resources. It has also saved industries, and eventually the consumer, millions of dollars. New applications arise constantly. George Dantzig developed the simplex method to solve linear programming problems in 1947. By contrast, Newton's method for estimating solutions of systems of nonlinear equalities has been around for more than 200 years.

Real-world linear programming problems may involve hundreds or even thousands of variables. However, we will discuss only the two-variable case of linear programming at this time, and our treatment will be very geometric and intuitive. This section, like the preceding one, is designed to illustrate the scope and importance of linear algebra before we delve into the nuts and bolts of the subject.

STATEMENT OF A LINEAR PROGRAMMING PROBLEM

To illustrate the type of problem we wish to study, we state a simple linear programming problem in two variables and then analyze it mathematically.

Problem   A lumber company owns two mills that produce hardwood veneer sheets of plywood. Each mill produces the same three types of plywood. Table 1.3 shows the daily production and the daily cost of operation of each mill. The rightmost column shows the amount of plywood required by the lumber company for a period of six months. Find the number of days that each mill should operate during the six months to supply the required sheets in the most economical way.

Table 1.3

Plywood Type	Mill 1 Per Day	Mill 2 Per Day	Six-month Demand
A	100 sheets	20 sheets	2000 sheets
B	40 sheets	80 sheets	3200 sheets
C	60 sheets	60 sheets	3600 sheets
Daily Costs	$3000	$2000	

Analysis   If mill 1 operates x days and mill 2 operates y days, then the cost is $3000x + 2000y$. To fulfill the company's requirements, these inequalities must be satisfied simultaneously:

$$100x + 20y \geq 2000$$
$$40x + 80y \geq 3200$$
$$60x + 60y \geq 3600$$
$$x \geq 0, y \geq 0.$$

Formulated mathematically, our problem is to minimize the linear function $3000x + 2000y$ subject to these inequalities. ◁

THE SOLUTION SET OF A SYSTEM OF LINEAR INEQUALITIES

Every linear programming problem is concerned with maximizing or minimizing a linear function on the solution set of a system comprised of linear equations or inequalities. To solve a problem like the one stated above, we consider the nature of such solution sets.

We know that a single linear equation in x and y has a line in the plane as its graph. The equation $y = mx + b$ describes a line of slope m and y-intercept b, as shown in Fig. 1.6.

The inequality $y < mx_0 + b$ describes the points (x_0, y) on $x = x_0$ lying under the line $y = mx + b$. (See Fig. 1.7.) Since x_0 can be any value, we see that $y \leq mx + b$ describes the half-plane consisting of the points on or below the line $y = mx + b$. This half-plane is shaded in Fig. 1.7. Similarly, the inequality $y \geq mx + b$ describes all points on or above the line $y = mx + b$, which gives the half-plane shaded in Fig. 1.8.

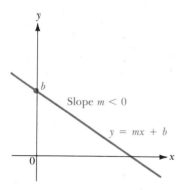

Figure 1.6
The line $y = mx + b$.

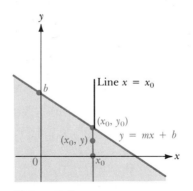

Figure 1.7
The half-plane $y \leq mx + b$.

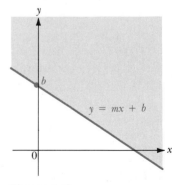

Figure 1.8
The half-plane $y \geq mx + b$.

As the preceding arguments indicate, any linear inequality $ax + by \leq c$, where not both a and b are zero, has as its solution set a half-plane consisting of the line $ax + by = c$ together with the part of the plane on one side of the line.

Given a system of linear inequalities in x and y such as the system (1), we want to find all (x, y) for which all the inequalities hold simultaneously. Then

(x, y) must lie in all the half-planes defined by the inequalities, that is, in the intersection of these half-planes.

> The solution set of a system of linear inequalities consists of an intersection of half-planes.

Example 1   In the plane, shade the solution set of the system (1) given at the beginning of this section:

$$2x + 3y \geq 6$$
$$2x + y \geq 2$$
$$x - y \leq 3.$$

Solution   The three half-planes corresponding to these inequalities are shown in Fig. 1.9. Once the line $2x + 3y = 6$ has been drawn as in Fig. 1.9(a), an easy way to determine which side of the line to shade is simply to pick a point on one side and see if it satisfies the inequality. For example, if we pick $(0, 0)$, we see that it does *not* satisfy $2x + 3y \geq 6$, so we shade the half-plane on the opposite side of the line in Fig. 1.9(a).

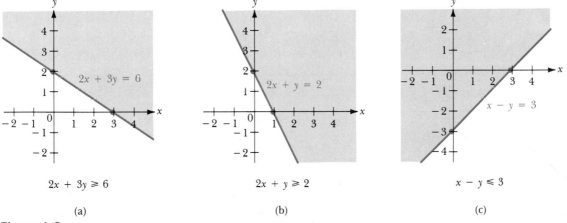

Figure 1.9
(a) $2x + 3y \geq 6$; (b) $2x + y \geq 2$; (c) $x - y \leq 3$.

In Fig. 1.10, the intersection of the half-planes in Fig. 1.9 is shaded. Points in this shaded region are precisely those in the solution set of the system (1). ◁

Example 2   Shade the solution set of the system $3x + 4y \leq 12, x \geq 0, y \geq 0$.

Solution   We easily see that the solution set is the shaded region in Fig. 1.11. ◁

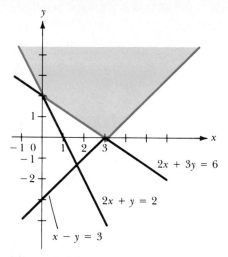

Figure 1.10
The set $2x + 3y \geq 6$, $2x + y \geq 2$, $x - y \leq 3$.

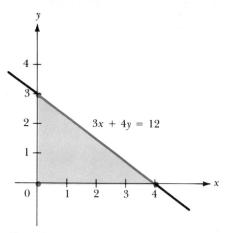

Figure 1.11
The set $3x + 4y \leq 12$, $x \geq 0$, $y \geq 0$.

Example 3   Describe the solution set of the system $3x + 4y \leq -12$, $x \geq 0$, $y \geq 0$.

Solution   The half-plane described by $3x + 4y \leq -12$ is shaded in Fig. 1.12. Since none of the points in that half-plane satisfies both $x \geq 0$ and $y \geq 0$, we see that the solution set is empty. ◁

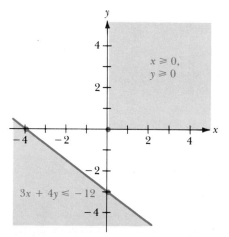

Figure 1.12
The shaded regions do not intersect.

The preceding examples make it apparent that a nonempty solution set of a system of linear inequalities is bounded by lines or line segments and that all the corners point outward. We call such a set a *convex polygonal set*. Let us define the notion of convexity, and then prove the convexity of such a solution set as a theorem. Figure 1.13 illustrates the definition.

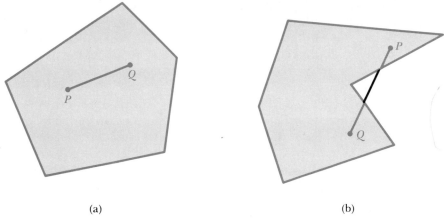

(a)   (b)

Figure 1.13
(a) A convex set; (b) not a convex set.

Definition 1.1 Convex Set

A subset of the plane is *convex* if for each two points P and Q in the set, the line segment joining them also lies in the set.

Theorem 1.1 Convexity of Solution Sets

Any nonempty intersection of half-planes is a convex set.

Proof Let P and Q be points in the intersection. Then P and Q lie in each of the half-planes. A half-plane is obviously convex, so the line segment joining P and Q lies in each half-plane. Therefore, the line segment lies in the intersection of the half-planes. This completes the proof. ■

MAXIMIZING OR MINIMIZING $ax + by$ ON AN INTERSECTION OF HALF-PLANES

Consider the linear programming problem of maximizing or minimizing a linear function $ax + by$ on the solution set of a system of linear inequalities. The inequalities are called the **constraints** and the function $ax + by$ is the **objective function**. With economic applications in mind, the objective function is called the **profit function** if it is to be maximized and the **cost function** if it is to be minimized. In applied problems, two of the constraints are always taken to be $x \geq 0$ and $y \geq 0$, but we will not assume this for the moment. The convex solution set of the constraints is called the **feasible set**, and any point in this set is a **feasible solution**. Our task is thus to determine from among all the feasible solutions those that maximize or minimize $ax + by$.

Suppose that for a feasible solution (x_0, y_0) and cost function $ax + by$, we have $ax_0 + by_0 = c_0$. Then every feasible solution lying on the line $ax + by = c_0$ also gives the same value c_0 for this cost function. For a feasible solution (x_1, y_1) not on this line, we obtain some cost c_1, where $c_1 = ax_1 + by_1$. Again, all feasible solutions (x, y) on the line $ax + by = c_1$ give this same value c_1. Now the lines $ax + by = c_0$ and $ax + by = c_1$ are *parallel,* for the slope of a line is completely determined by the coefficients of x and y. This is illustrated in Fig. 1.14. If we hold a and b fixed and increase c, the parallel lines $ax + by = c$ move upward or downward in the plane, depending on the sign of b. They move upward if $b > 0$ and downward if $b < 0$, as we see easily by considering the y-intercepts c/b of the lines. Of course, if $b = 0$, the lines $ax = c$ for fixed a move either left or right as c increases.

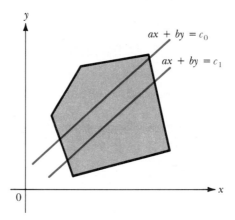

Figure 1.14
Different values of c give parallel lines.

Suppose that the feasible set is as shown in Fig. 1.15(a). Then if $b > 0$, the maximum value $c_{\max}$ is the value of $ax + by$ at the last point where the lines $ax + by = c$ touch the feasible set as c increases and the lines move upward. The minimum value $c_{\min}$ is assumed at the last point where the lines touch the set as c decreases and the lines move downward. The case $b < 0$ where the lines move downward as c increases is shown in Fig. 1.15(b). The case for vertical lines where $b = 0$ is handled in a similar way.

It may be that the final position of a line $ax + by = c$ before leaving a feasible set falls on a whole boundary segment of the set. Figure 1.15(b) shows the case where $c_{\max}$ is assumed at every point on such a line segment.

If the feasible set is unbounded, as in Fig. 1.10, then there may be no maximum or minimum attained by the cost function. For example, if $b > 0$, then $ax + by$ attains no maximum value on the set in Fig. 1.10. Similarly, if $b < 0$, no minimum value is attained.

From considering all these cases, we obtain a theorem.

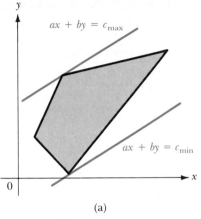

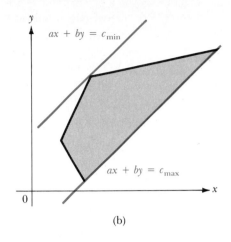

(a)   (b)

Figure 1.15
(a) The case $b > 0$; (b) the case $b < 0$.

Theorem 1.2   Assumption of Optimal Cost or Profit

If a linear objective function $ax + by$ assumes a maximum value on a set defined by linear constraints, then this maximum value is assumed at a corner point of the set. The same is true if an objective function assumes a minimum value on the feasible set.

Theorem 1.2 and the discussion preceding it suggest the following procedure for finding the maximum or minimum attained by an objective function $ax + by$.

**Finding the Maximum (Minimum) Attained by $ax + by$
Subject to a System of Linear Constraints**

Step 1 Sketch the feasible set defined by the constraints.

Step 2 Determine whether a maximum (minimum) is attainable.

Step 3 If so, find the coordinates of all corner points of the set.

Step 4 Compute the value of $ax + by$ at each corner point of the set. The largest (smallest) value obtained is the maximum (minimum) of the objective function on the feasible set.

Concerning step 2, it is usually clear in a practical problem that the desired maximum or minimum is assumed. If the feasible set is bounded, then both a maximum and a minimum are assumed. If the set is unbounded, one or both may fail to be assumed, as shown in Example 5.

Example 4   Find the maximum and minimum values attained by $x + 2y$ subject to the constraints $3x + 4y \leq 12$, $x \geq 0$, $y \geq 0$ given in Example 2.

Solution   Referring to Fig. 1.11, we see that the feasible set is bounded, so both a maximum and a minimum are attained by $x + 2y$. The figure shows that the corner points are $(0, 0)$, $(4, 0)$, and $(0, 3)$. Thus we have the following.

Corner point	Value of $x + 2y$
$(0, 0)$	0
$(0, 3)$	6
$(4, 0)$	4

Therefore, the maximum value is 6, attained at $(0, 3)$, and the minimum value is 0, attained at $(0, 0)$.   ◁

Example 5   Discuss the values of a and b for which $ax + by$ attains a maximum or minimum subject to the constraints $2x + 3y \geq 6$, $2x + y \geq 2$, $x - y \leq 3$ of Example 1.

Solution   The feasible set was found in Example 1, and is shown again in Fig. 1.16. In order for $ax + by$ to attain either a maximum or a minimum value on this set, the parallel lines $ax + by = c$ must eventually fail to intersect the feasible set as c increases or decreases. Referring to Fig. 1.16, we see that the rightmost boundary line of the feasible set has slope 1, whereas the leftmost boundary line has slope -2. As illustrated in Fig. 1.16(a), every line of slope greater than 1 intersects the feasible set, so we see that an objective function $ax + by$ attains no maximum or minimum on this set if the line $ax + by = c$ has slope greater than 1. A similar argument shows that this is also true if the line is steeper than the left boundary line, that is, if it has slope less than -2. However, as shown in Fig. 1.16(b), if $ax + by = c$ has slope m where $-2 \leq m \leq 1$, then the cost function $ax + by$ attains a minimum value if $b > 0$ and a maximum value if $b < 0$. Thus we see that $ax + by$ attains a maximum or a minimum if $-2 \leq -a/b \leq 1$.   ◁

Example 6   As a continuation of Example 5, find the maximum or minimum attained by $x + 3y$ and by $x - 2y$ on the feasible set of Fig. 1.16.

Solution   The slope of $x + 3y = c$ is $-\frac{1}{3}$, whereas the slope of $x - 2y = c$ is $\frac{1}{2}$, both of which are between -2 and 1. Example 5 shows that we will have a maximum or minimum attained in each case. We see from the figure that the corner points are $(0, 2)$ and $(3, 0)$.

The value of $x + 3y$ is 6 at $(0, 2)$ and 3 at $(3, 0)$. Clearly, $x + 3y$ takes on very large values in this unbounded feasible set. Thus 3 is the minimum value attained.

The value of $x - 2y$ is -4 at $(0, 2)$ and 3 at $(3, 0)$. Clearly, $x - 2y$ attains negative values of great magnitude at some points $(0, y)$ in the feasible set, so 3 is the maximum value attained by $x - 2y$.   ◁

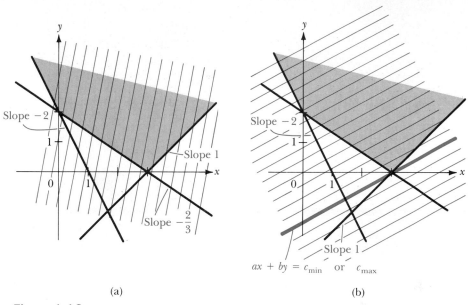

(a)   (b)

Figure 1.16
(a) Neither a minimum nor a maximum; (b) either a minimum or a maximum.

THE SOLUTION TO OUR OPENING PROBLEM

We restate and solve the problem we posed at the beginning of this section.

Example 7 Each of two mills produces the same three types of plywood. Table 1.4 gives the production, demand, and cost data. Find the number of days that each mill should operate during the six months in order to supply the required sheets in the most economical way.

Table 1.4

Plywood Type	Mill 1 Per Day	Mill 2 Per Day	Six-month Demand
A	100 sheets	20 sheets	2000 sheets
B	40 sheets	80 sheets	3200 sheets
C	60 sheets	60 sheets	3600 sheets
Daily Costs	$3000	$2000	

Solution If mill 1 operates x days and mill 2 operates y days, then the cost is $3000x + 2000y$. In order to fulfill the company's requirements, we must have the following constraints:

$$100x + 20y \geq 2000$$
$$40x + 80y \geq 3200$$
$$60x \overset{.}{+} 60y \geq 3600$$
$$x \geq 0, y \geq 0.$$

The feasible set is sketched in Fig. 1.17. For this practical problem, it is clear that a minimum cost must exist, and we need only find the corner points and compute $3000x + 2000y$ there.

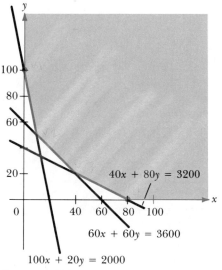

Figure 1.17
The feasible set.

We can see from the figure that two corner points are (0, 100) and (80, 0). To find another corner point, we solve the equations for two consecutive boundary lines of the feasible set simultaneously. We have

$$100x + 20y = 2000 \quad \text{Multiply by 3 and subtract.}$$
$$\underline{60x + 60y = 3600}$$
$$240x \qquad = 2400, \ x = 10, \ y = 50.$$

For our final corner point, we solve:

$$40x + \ 80y = 3200 \quad \text{Multiply by 3.}$$
$$\underline{60x + \ 60y = 3600} \quad \text{Multiply by 2 and subtract.}$$
$$120y = 2400, \ y = 20, \ x = 40.$$

Thus we have the following.

Corner point (x, y)	$3000x + 2000y$
(0, 100)	\$200,000
(80, 0)	\$240,000
(10, 50)	\$130,000
(40, 20)	\$160,000

We see that it is best to run mill 1 for 10 days and mill 2 for 50 days. This will produce the necessary 2000 sheets of type A plywood and the 3600 sheets of type C. Note that it will actually produce 4400 sheets of type B, for a surplus of 1200 sheets over the 3200 required of that type! ◁

Problems in linear inequalities involving just two variables serve to illustrate the concepts. Chapter 9 presents the *simplex method,* developed by Dantzig in 1947, which is presently the main algebraic algorithm for solving similar problems involving more than two variables. It is simply not feasible to graph and find all corner points in spaces of more than two dimensions. The simplex method has been implemented on computers to solve a great variety of important applications.

At the present time, there is great interest in another algorithm as an alternative to the simplex method, one that was developed by Narendra Karmarkar at AT&T Bell Labs. Dantzig's simplex method finds a corner point of the convex set and then travels along edges to adjacent corner points until an optimal solution is found. Karmarkar's method attempts to take shortcuts through the interior of the convex set. The AT&T report to stockholders for the quarter ending December 31, 1984, has a paragraph on Karmarkar's work. The report mentions that finding the most efficient way to route an overseas call to the Pacific area might involve 20,000 variables. People at Bell Labs think it possible that a computer could solve the problem using Karmarkar's algorithm in less than one minute. Clearly, we can't give problems involving 20,000 variables in this text! We have tried to present a concept, and provide a little insight into its importance.

SUMMARY

1. The solution set of a linear inequality $ax + by + c \leq 0$ is a half-plane.

2. A subset of the plane is convex if for every two points in the set, the line segment joining them is in the set.

3. The solution set of a system of linear inequalities in two variables is either empty or a convex polygonal set in the plane. The solution set may or may not be bounded.

4. A linear programming problem consists of maximizing or minimizing a linear function on the solution set of a system containing linear equations and/or inequalities. The conditions in the system are called constraints, their solution set is called the feasible set, and the linear function to be optimized is called the objective function.

5. The box on page 21 lists a four-step procedure for solving linear programming problems in two variables.

EXERCISES

In Exercises 1–6, shade the solution set in the plane of the system of linear inequalities.

1. $x \geq 0, y \leq 0$

2. $x + y \geq -3$

3. $x - y \leq 5, x + y \geq 2$

4. $x \geq 2, y \leq 4, x + y \leq 8$

5. $x \geq 0, y \geq 0, x + 3y \geq 12, y + 2x \geq 6$

6. $x \geq 0, y \geq 0, 4x + 3y \geq 24, 5 + 2y \geq 20,$ $6x + y \geq 12$

In Exercises 7–12, find the maximum and minimum values, if they exist, of the objective functions on the feasible set defined by the constraints.

7. Constraints: $x \geq 2, y \geq 3$
Objective functions: (a) $2x + y$; (b) $y - 2x$; (c) $-2x - 3y$

8. Constraints: $x \geq 0, y \geq 0, x + y \geq 1$
Objective functions: (a) $x - 2y$; (b) $4x + 5y$; (c) $3x + 2y$

9. Constraints: $x - y \leq 0, x + y \geq 0$
Objective functions: (a) $2x - 3y$; (b) $3y + x$; (c) $y + 2x$

10. Same as Exercise 9 with the additional constraint $x + 9y \leq 80$.

11. Constraints: $x \geq 0, y \geq 0, x + 2y \geq 5,$ $2x + y \geq 4, x + 4y \geq 7$
Objective functions: (a) $2x + 9y$; (b) $x + 3y$; (c) $x + y$

12. Same as Exercise 11 with the additional constraint $7x + 4y \leq 28$.

13. Referring to Example 7 in the text, suppose that production and requirements remain the same, but daily costs are subject to change. Find the condition on the ratio

$$\frac{\text{(Daily cost for mill 1)}}{\text{(Daily cost for mill 2)}}$$

in order that:

a) it is best to run only mill 1;
b) it is best to run only mill 2.

14. Solve the problem in Example 7 if the data for the two mills is changed to that given in Table 1.5.

Table 1.5

Plywood Type	Mill 1 Per Day	Mill 2 Per Day	Six-month Demand
A	200 sheets	100 sheets	5000 sheets
B	40 sheets	10 sheets	1100 sheets
C	20 sheets	80 sheets	2800 sheets
Daily Costs	$4000	$2000	

15. A food packaging house receives daily 1500 lb of type C coffee and 2000 lb of type K coffee. It can sell blend A, which consists of 2 parts C to 1 part K, at a profit of 10 cents per pound, and blend B, which consists of 2 parts K to 1 part C, at a profit of 12 cents per pound. Find the amount of each blend that the house should prepare daily to maximize the profit for sales of just these two blends.

16. Referring to Exercise 15, find the amount of each blend necessary for a maximum profit if the house is also able to sell the leftover types C and K for a profit of:
a) 6 cents per pound each;
b) 9 cents per pound each;
c) 12 cents per pound each.

17. A dietician wants to use food 1 and food 2 to supply certain minimum amounts of vitamins A, B, and C. The data for the amounts of the vitamins, the requirements, and the costs are given in Table 1.6. How much of each food should be used to meet the vitamin requirements in the most economical way?

Table 1.6

Vitamin	Food 1	Food 2	Requirements
A	30 units/oz	20 units/oz	120 units
B	40 units/oz	10 units/oz	80 units
C	20 units/oz	40 units/oz	100 units
Cost	10 cents/oz	15 cents/oz	

18. Answer Exercise 17 if other data remains the same but:
a) Food 1 costs 5 cents/oz and food 2 costs 15 cents/oz;
b) Food 1 costs 25 cents/oz and food 2 costs 15 cents/oz;
c) Food 1 costs 25 cents/oz and food 2 costs 5 cents/oz.

19. A gardener has two bags of fertilizer, each weighing 80 lb. Bag 1 contains 10% nitrogen, 5% phosphate, and 10% potash, while bag 2 contains 5% nitrogen, 10% phosphate, and 10% potash. Bag 1 costs $8.00 and bag 2 costs $10.00. The gardener wishes to put at least 5 lb of nitrogen, 4 lb of phosphate, and 7 lb of potash on his garden. How many pounds from each bag should he mix together to fertilize most economically?

20. Answer Exercise 19 if the cost of the fertilizer is:
a) $10.00 for bag 1 and $8.00 for bag 2;
b) $4.00 for bag 1 and $10.00 for bag 2.

⊙ *The program LINPROG in the software available to schools using this text can be used to graph the feasible set of a system of linear inequalities in two variables where the system includes $x \geq 0$ and $y \geq 0$. The user may then specify an objective function, and estimate graphically the maximum and minimum values of this function on the feasible set, if they exist. Information on using the program is given at its beginning. In the remaining exercises, use this program, or similar software, to find as accurately as you can from the graph the maximum and minimum of the given objective function on the feasible set defined by the constraints.*

21. Constraints: $x \geq 0, y \geq 0, x + y \geq 10$, $x \leq 15, x - y \leq 10, y \leq 7$
Objective function: (a) $x + 2y$; (b) $2x + y$

22. Constraints: $x \geq 0, y \geq 0, x + y \geq 4$, $2x + y \leq 8, 5x + 8y \leq 50$
Objective function: (a) $2x + 3y$; (b) $3x + 2y$; (c) $3x + y$; (d) $x - 2y$

23. Same as Exercise 22, but with the additional constraint $2x - y \leq 4$.

24. Constraints: $x \geq 0, y \geq 0, x \leq 8, 5x + y \geq 10, 3x - 2y \leq 15$
Objective function: (a) $x + 6y$; (b) $2x - y$; (c) $x + 4y$; (d) $y - x$

25. Same as Exercise 24 with the additional constraint $6x + 5y \leq 60$.

1.3
Matrices and Their Algebra

THE NOTION OF A MATRIX

In Section 1.1, we indicated one way in which systems of linear equations arise. The linear inequalities in Section 1.2 also give rise to linear systems. As we will see in Section 1.4, solving a system of linear equations is handled efficiently by ignoring the letters and equals signs. The result of this is an array of numbers known as a *matrix* (pl., matrices). Matrices are a fundamental tool for efficient computations in linear algebra.

A **matrix** is an ordered rectangular array of numbers, usually enclosed in parentheses or square brackets. For example

$$A = \begin{pmatrix} 2 & -1 & 3 \\ 4 & 2 & 1 \end{pmatrix} \quad \text{and} \quad B = \begin{pmatrix} -1 & 0 & 3 \\ 2 & 1 & 4 \\ 4 & 5 & -6 \\ -3 & -1 & -1 \end{pmatrix}$$

are matrices. We will generally use upper-case letters to denote matrices.

The size of a matrix is specified by the number of rows and the number of columns that it contains. The matrix A above contains 2 rows and 3 columns and is called a 2×3 (read "2 by 3") matrix. Similarly, B is a 4×3 matrix. In writing the notation $m \times n$ to describe the shape of a matrix, we always write the number of rows first.

The location of any particular entry in a matrix can be described by the number of the row and the number of the column in which it lies. For example, the entry in row 2 and column 1 of matrix A above is 4.

THE TERM "MATRIX" is first mentioned in the mathematical literature in an 1850 paper of James Joseph Sylvester (1814–1897). The standard nontechnical meaning of that term is "a place in which something is bred, produced or developed." For Sylvester, then, a matrix, which was an "oblong arrangement of terms," was an entity out of which one could form various square pieces to produce determinants. These latter quantities, formed from square matrices, were quite well known by this time.

James Sylvester (his original name was James Joseph) was born into a Jewish family in London, and was to become one of the supreme algebraists of the nineteenth century. Despite having studied for several years at Cambridge University, he was not permitted to take his degree there because he "professed the faith in which the founder of Christianity was educated." Therefore, he received his degrees from Trinity College, Dublin. In 1841 he accepted a professorship at the University of Virginia; he remained there only a short time, however, his horror of slavery preventing him from fitting into the academic community. In 1871 he returned to the United States to accept the chair of mathematics at the newly opened Johns Hopkins University. In between these sojourns, he spent about 10 years as an attorney, during which time he met Arthur Cayley (see the note on p. 55), and 15 years as Professor of Mathematics at the Royal Military Academy, Woolwich. Sylvester was an avid poet, prefacing many of his mathematical papers with examples of his work. His most renowned example was the "Rosalind" poem, a 400-line epic, each line of which rhymed with "Rosalind."

Of special interest are matrices with only one row or only one column; these are called **vectors**. The entries in a vector are called its **components**. For example, a $1 \times n$ matrix is a **row vector** with n components, and an $m \times 1$ matrix is a **column vector** with m components. This text uses lower-case bold-face letters to denote vectors. We follow the common practice of separating the components of a row vector with commas. For example, we might have

$$\mathbf{v} = \begin{pmatrix} 2 \\ 3 \\ 4 \end{pmatrix}, \qquad \mathbf{w} = (4, 2), \quad \text{and} \quad \mathbf{x} = \begin{pmatrix} x_1 \\ x_2 \\ x_3 \end{pmatrix}.$$

Here $\mathbf{v}$ and $\mathbf{x}$ are column vectors with 3 components, and $\mathbf{w}$ is a row vector with 2 components. In written work, it is customary to place an arrow over a letter to denote a vector, as in $\vec{v}$, $\vec{w}$, and $\vec{x}$.

Note that when we use subscripts in columns to denote position, we number from top to bottom, as in the vector $\mathbf{x}$ above. For a row vector, we number from left to right:

$$\mathbf{a} = (a_1, a_2, a_3, a_4).$$

Double subscripts are commonly used to indicate the location of an entry in a matrix. The first subscript gives the number of the row in which the entry appears, counting from the top, and the second subscript gives the number of the column, counting from the left. Thus an $m \times n$ matrix A can be written as

$$A = (a_{ij}) = \begin{pmatrix} a_{11} & a_{12} & a_{13} & \cdots & a_{1n} \\ a_{21} & a_{22} & a_{23} & \cdots & a_{2n} \\ a_{31} & a_{32} & a_{33} & \cdots & a_{3n} \\ & & \vdots & & \\ a_{m1} & a_{m2} & a_{m3} & \cdots & a_{mn} \end{pmatrix}.$$

MATRIX MULTIPLICATION

Matrices are used in this chapter to simplify work with systems of linear equations, saving a lot of writing. Consider, for example, the linear system

$$\begin{aligned} 2x_1 - 3x_2 &= 7 \\ 3x_1 - 4x_2 &= 2. \end{aligned} \tag{1}$$

The important data for this linear system are contained in the **coefficient matrix**

$$A = \begin{pmatrix} 2 & -3 \\ 3 & -4 \end{pmatrix} \tag{2}$$

and the column vector

$$\mathbf{b} = \begin{pmatrix} 7 \\ 2 \end{pmatrix} \tag{3}$$

of those constants that appear after the equals signs. We let

$$\mathbf{x} = \begin{pmatrix} x_1 \\ x_2 \end{pmatrix} \tag{4}$$

be the column vector of unknowns in the system (1). It is customary to abbreviate the system (1) by

$$A\mathbf{x} = \mathbf{b}, \tag{5}$$

or, written out,

$$\begin{pmatrix} 2 & -3 \\ 3 & -4 \end{pmatrix} \begin{pmatrix} x_1 \\ x_2 \end{pmatrix} = \begin{pmatrix} 7 \\ 2 \end{pmatrix}. \tag{6}$$

Equation (6) suggests the notion of *multiplication* of a matrix A times a column vector $\mathbf{x}$. As preparation for this, we now define the *dot product* of any two vectors with the same number of components.

Definition 1.2   Dot Product (Scalar Product)

Let $\mathbf{a}$ and $\mathbf{b}$ be any two vectors with the same number of components. Each may be either a row vector or a column vector. Let the ordered components of $\mathbf{a}$ be given by $a_1, a_2, \ldots, a_n$ and those of $\mathbf{b}$ be given by $b_1, b_2, \ldots, b_n$. The **dot product** (or **scalar product**) of $\mathbf{a}$ and $\mathbf{b}$ is

$$\mathbf{a} \cdot \mathbf{b} = a_1 b_1 + a_2 b_2 + \cdots + a_n b_n.$$

It is essential to keep the notions of *vector* and *number* distinct. The term **scalar** is commonly used in linear algebra in place of *number*. The dot product is also called the scalar product, since $\mathbf{a} \cdot \mathbf{b}$ is a scalar quantity, rather than a vector quantity.

Example 1   Let

$$\mathbf{a} = (1, -3, 4) \quad \text{and} \quad \mathbf{b} = \begin{pmatrix} 2 \\ 0 \\ 8 \end{pmatrix}.$$

Find $\mathbf{a} \cdot \mathbf{b}$.

Solution   We have

$$\mathbf{a} \cdot \mathbf{b} = (1)(2) + (-3)(0) + (4)(8) = 2 + 0 + 32 = 34. \quad \triangleleft$$

Example 2   Let $\mathbf{a} = (1, 5, 6)$ and $\mathbf{b} = (2, 4)$. Find $\mathbf{a} \cdot \mathbf{b}$.

Solution   The dot product is not defined since $\mathbf{a}$ and $\mathbf{b}$ have different numbers of components. $\triangleleft$

Let us return to the matrix expression

$$\begin{pmatrix} 2 & -3 \\ 3 & -4 \end{pmatrix} \begin{pmatrix} x_1 \\ x_2 \end{pmatrix} = \begin{pmatrix} 7 \\ 2 \end{pmatrix} \tag{7}$$

for the linear system

$$\begin{aligned} 2x_1 - 3x_2 &= 7 \\ 3x_1 - 4x_2 &= 2. \end{aligned} \tag{8}$$

We see that the number 7 in this system appears as the dot product of the vectors

$$(2, -3) \quad \text{and} \quad \begin{pmatrix} x_1 \\ x_2 \end{pmatrix}.$$

in the first equation of the system, whereas 2 appears as the dot product of

$$(3, -4) \quad \text{and} \quad \begin{pmatrix} x_1 \\ x_2 \end{pmatrix}$$

in the second equation.

The product AB of two matrices A and B is formed by taking all possible dot products of *row vectors* of A times *column vectors* of B. In order for such dot products to be defined, the number of components in each row of A must equal the number of components in each column of B. If AB is defined and A is $m \times n$ so the rows have n components, then B must be $n \times s$ so the columns of B also have n components.

Definition 1.3   Matrix Multiplication

Let $A = (a_{ik})$ be an $m \times n$ matrix and let $B = (b_{kj})$ be an $n \times s$ matrix. The **matrix product** AB is the $m \times s$ matrix $C = (c_{ij})$, where c_{ij} is the dot product of the ith row vector of A and the jth column vector of B.

We illustrate the choice of row from A and column from B to find the element c_{ij} in AB according to Definition 1.3 by the equation

$$AB = (c_{ij}) = \begin{pmatrix} a_{11} & \cdots & a_{1n} \\ \vdots & & \vdots \\ a_{i1} & \cdots & a_{in} \\ \vdots & & \vdots \\ a_{m1} & \cdots & a_{mn} \end{pmatrix} \begin{pmatrix} b_{11} & \cdots & b_{1j} & \cdots & b_{1s} \\ \vdots & & \vdots & & \vdots \\ & & \vdots & & \\ & & \vdots & & \\ b_{n1} & \cdots & b_{nj} & \cdots & b_{ns} \end{pmatrix}.$$

In summation notation, we have

$$c_{ij} = a_{i1}b_{1j} + a_{i2}b_{2j} + \cdots + a_{in}b_{nj} = \sum_{k=1}^{n} a_{ik}b_{kj}. \tag{9}$$

Note again that AB is defined only when the second size-number (the number of columns) of A is the same as the first size-number (the number of rows) of B. The product matrix has size

(first size-number of A) $\times$ (second size-number of B).

Example 3   Let A be a 2×3 matrix and B be a 3×5 matrix. Find the size of AB and BA, if they are defined.

Solution   Since the second size-number 3 of A equals the first size-number 3 of B, we know that AB is defined; it is a 2×5 matrix. However, BA is not defined, since the second size-number 5 of B is not the same as the first size-number 2 of A. $\triangleleft$

Example 4   Compute the product

$$\begin{pmatrix} -2 & 3 & 2 \\ 4 & 6 & -2 \end{pmatrix} \begin{pmatrix} 4 & -1 & 2 & 5 \\ 3 & 0 & 1 & 1 \\ -2 & 3 & 5 & -3 \end{pmatrix}.$$

Solution   The product is defined since the left-hand matrix is 2×3 and the right-hand one is 3×4; the product will have size 2×4. The entry in the product in the

---

MATRIX MULTIPLICATION originated in the composition of linear substitutions. Though there are hints of such ideas in the work of Euler and Lagrange, they are fully discussed by Karl Friedrich Gauss (1777–1855) in his 1801 work *Disquisitiones Arithmeticae* in connection with quadratic forms (functions in two variables of the form $Ax^2 + 2Bxy + Cy^2$). In particular, a linear substitution of the form

$$x = ax' + by' \qquad y = cx' + dy' \tag{1}$$

transforms one such form F in x and y into another form F' in x',y'. If a second substitution

$$x' = ex'' + fy'' \qquad y' = gx'' + hy'' \tag{2}$$

transforms F' into a form F'' in x'',y'', then the composition of the substitutions, found by replacing x',y' in (1) by their values in (2), gives a substitution transforming F into F'':

$$x = (ae + bg)x'' + (af + bh)y'' \qquad y = (ce + dg)x'' + (cf + dh)y''. \tag{3}$$

The coefficient matrix of substitution (3) is the product of the coefficient matrices of substitutions (1) and (2). Gauss performed an analogous composition of substitutions for forms in 3 variables, which gives the multiplication of 3×3 matrices.

Gauss is often called the "prince of mathematicians," since over a long scientific career he contributed greatly to such varied fields as number theory, algebra, geometry, complex analysis, astronomy, geodesy, and mechanics.

first row and first column position is obtained by taking the dot product of the first row vector of the left-hand matrix with the first column vector of the right-hand one, as follows:

$$(-2)(4) + (3)(3) + (2)(-2) = -8 + 9 - 4 = -3.$$

The entry in the second row and third column of the product is the dot product of the second row vector of the left-hand matrix with the third column vector of the right-hand one, that is,

$$(4)(2) + (6)(1) + (-2)(5) = 8 + 6 - 10 = 4,$$

and so on. Eight such computations show that

$$\begin{pmatrix} -2 & 3 & 2 \\ 4 & 6 & -2 \end{pmatrix} \begin{pmatrix} 4 & -1 & 2 & 5 \\ 3 & 0 & 1 & 1 \\ -2 & 3 & 5 & -3 \end{pmatrix} = \begin{pmatrix} -3 & 8 & 9 & -13 \\ 38 & -10 & 4 & 32 \end{pmatrix}. \quad \triangleleft$$

Example 3 shows that AB may be defined while BA may not be. Even if both AB and BA are defined, it need not be true that $AB = BA$. As the following example shows:

$$\boxed{\text{Matrix multiplication is not commutative.}}$$

Example 5   Let

$$A = \begin{pmatrix} -1 & 2 \\ 3 & -4 \end{pmatrix} \quad \text{and} \quad B = \begin{pmatrix} 0 & 1 \\ 2 & 5 \end{pmatrix}.$$

Compute AB and BA.

Solution   We easily compute that

$$AB = \begin{pmatrix} 4 & 9 \\ -8 & -17 \end{pmatrix}, \quad \text{whereas} \quad BA = \begin{pmatrix} 3 & -4 \\ 13 & -16 \end{pmatrix}. \quad \triangleleft$$

It can be shown that matrix multiplication is associative, that is,

$$A(BC) = (AB)C$$

whenever the product is defined. This is a bit difficult to prove from the definition. We will see an easy proof in Section 6.3.

It is too bad that matrix multiplication is not easier to execute. However, the definition we gave for this multiplication is precisely what is needed to make the expression for the linear system in (7) a valid multiplication. The historical note on page 32 indicates in another way why we multiply matrices in the manner given in Definition 1.3. This historical motivation is still relevant and appears in Theorem 6.10.

THE $n \times n$ IDENTITY MATRIX

Let I be the $n \times n$ matrix (a_{ij}) such that $a_{ii} = 1$ for $i = 1, \ldots, n$ while $a_{ij} = 0$ for $i \neq j$. That is,

$$I = \begin{pmatrix} 1 & 0 & 0 & \cdots & 0 \\ 0 & 1 & 0 & \cdots & 0 \\ 0 & 0 & 1 & \cdots & 0 \\ \vdots & \vdots & \vdots & & \vdots \\ 0 & 0 & 0 & \cdots & 1 \end{pmatrix} = \begin{pmatrix} 1 & & & & \\ & 1 & & & \\ & & 1 & & \\ & & & \ddots & \\ & & & & 1 \end{pmatrix},$$

where the circles above and below the diagonal in the last matrix are regarded as large zeros indicating that each entry of the matrix in those positions is 0. If A is any $m \times n$ matrix and B is any $n \times s$ matrix, then it is easy to show that

$$AI = A \quad \text{and} \quad IB = B.$$

For instance, we can easily check that

$$\begin{pmatrix} 2 & 3 \\ -1 & 7 \end{pmatrix} \begin{pmatrix} 1 & 0 \\ 0 & 1 \end{pmatrix} = \begin{pmatrix} 2 & 3 \\ -1 & 7 \end{pmatrix} = \begin{pmatrix} 1 & 0 \\ 0 & 1 \end{pmatrix} \begin{pmatrix} 2 & 3 \\ -1 & 7 \end{pmatrix}.$$

Because of these relations $AI = A$ and $IB = B$, the matrix I is called the **$n \times n$ identity matrix**. It behaves for $n \times n$ matrices as the scalar 1 behaves for scalar arithmetic. We have one such square identity matrix for each integer $n = 1, 2, 3, \ldots$. To keep notation simple, we denote them all by I, rather than by $I_1, I_2, I_3, \ldots$. The dimensions of I will be clear from the context.

OTHER MATRIX OPERATIONS

Although multiplication is the most important matrix operation for our work, other operations with matrices do arise. We define matrix addition, the product of a scalar with a matrix, and the transpose of a matrix.

Definition 1.4 Matrix Addition

Let $A = (a_{ij})$ and $B = (b_{ij})$ be two matrices of the same size $m \times n$. The **sum** $A + B$ of these two matrices is the $m \times n$ matrix $C = (c_{ij})$, where

$$c_{ij} = a_{ij} + b_{ij}.$$

That is, the sum of two matrices of the same size is the matrix of that size obtained by adding corresponding entries.

Example 6 Find

$$\begin{pmatrix} 1 & 2 & -4 \\ 0 & 3 & -1 \end{pmatrix} + \begin{pmatrix} -1 & 0 & 2 \\ 1 & -5 & 3 \end{pmatrix}.$$

Solution   The sum is the matrix

$$\begin{pmatrix} 0 & 2 & -2 \\ 1 & -2 & 2 \end{pmatrix}. \quad \triangleleft$$

Example 7   Find

$$\begin{pmatrix} 1 & -3 \\ 2 & 4 \end{pmatrix} + \begin{pmatrix} -5 & 4 & 6 \\ 3 & 7 & -1 \end{pmatrix}.$$

Solution   The sum is undefined, since the matrices are not the same size. $\triangleleft$

Let A be an $m \times n$ matrix and let O be the $m \times n$ matrix all of whose entries are zero. Clearly

$$A + O = O + A = A.$$

The matrix O is called the $m \times n$ **zero matrix**; the size of such a zero matrix is always clear from the context.

Definition 1.5   Scalar Multiplication

Let $A = (a_{ij})$ and let r be a scalar. The **product** rA of the scalar r and the matrix A is the matrix $B = (b_{ij})$ having the same size as A, where

$$b_{ij} = ra_{ij}.$$

Example 8   Find

$$2\begin{pmatrix} -2 & 1 \\ 3 & -5 \end{pmatrix}.$$

Solution   Multiplying each entry of the matrix by 2, we obtain the matrix

$$\begin{pmatrix} -4 & 2 \\ 6 & -10 \end{pmatrix}. \quad \triangleleft$$

For matrices A and B of the same size, we define the **difference** $A - B$ to be

$$A - B = A + (-1)B.$$

Example 9   If

$$A = \begin{pmatrix} 3 & -1 & 4 \\ 0 & 2 & -5 \end{pmatrix} \quad \text{and} \quad B = \begin{pmatrix} -1 & 0 & 5 \\ 4 & -2 & 1 \end{pmatrix},$$

find $2A - 3B$.

Solution   We easily find that

$$2A - 3B = \begin{pmatrix} 9 & -2 & -7 \\ -12 & 10 & -13 \end{pmatrix}. \quad \triangleleft$$

The transpose of a matrix interchanges the roles of row and column.

Definition 1.6 Transpose of a Matrix; Symmetric Matrix

The matrix B is the **transpose** of the matrix A, written $B = A^T$, if each entry b_{ij} in B is the same as the entry a_{ji} in A, and conversely. If A is a matrix and if $A = A^T$, then the matrix A is **symmetric.**

Example 10   Find A^T if

$$A = \begin{pmatrix} 1 & 4 & 5 \\ -3 & 2 & 7 \end{pmatrix}.$$

Solution   We have

$$A^T = \begin{pmatrix} 1 & -3 \\ 4 & 2 \\ 5 & 7 \end{pmatrix}.$$

Note that the rows of A became the columns of A^T.   ◁

PROPERTIES OF MATRIX OPERATIONS

For handy reference we have boxed the algebraic properties of the dot product, matrix arithmetic, and the transpose operation. These properties are valid for all vectors, scalars, and matrices for which the indicated quantities are defined.

Properties of the Dot Product

$\mathbf{u} \cdot \mathbf{v} = \mathbf{v} \cdot \mathbf{u}$	Commutative property
$\mathbf{u} \cdot (\mathbf{v} + \mathbf{w}) = \mathbf{u} \cdot \mathbf{v} + \mathbf{u} \cdot \mathbf{w}$	Distributive property
$(r\mathbf{u}) \cdot \mathbf{v} = \mathbf{u} \cdot (r\mathbf{v}) = r(\mathbf{u} \cdot \mathbf{v})$	Homogeneous property

---

Properties of Matrix Arithmetic

$A + B = B + A$	Commutativity of addition
$(A + B) + C = A + (B + C)$	Associativity of addition
$A + O = O + A = A$	Identity for addition
$r(A + B) = rA + rB$	A left distributive law
$(r + s)A = rA + sA$	A right distributive law
$(rs)A = r(sA)$	Associativity of scalar multiplication
$A(BC) = (AB)C$	Associativity of matrix multiplication

$$IA = A \text{ and } BI = B \qquad \text{Identity for matrix}$$
multiplication

$$A(B + C) = AB + AC \qquad \text{A left distributive law}$$

$$(A + B)C = AC + BC \qquad \text{A right distributive law}$$

Properties of the Transpose Operation

$$(A^T)^T = A \qquad \text{Transpose of the transpose}$$

$$(A + B)^T = A^T + B^T \qquad \text{Transpose of a sum}$$

$$(AB)^T = B^T A^T \qquad \text{Transpose of a product}$$

All these properties are easy to check, although writing out the proof of the associativity of matrix multiplication is a bit tedious. Proofs are just routine computations with components of vectors. We illustrate this with an example.

Example 11 Prove that $r(A + B) = rA + rB$ for any two $m \times n$ matrices A and B and any scalar r.

Solution Let $A = (a_{ij})$ and $B = (b_{ij})$. The entry in the ith row and jth column position of $r(A + B)$ is

$$r(a_{ij} + b_{ij}) = ra_{ij} + rb_{ij},$$

which is the sum of the entries in the ith row and jth column position of rA and the corresponding position of rB. But this is by definition the entry in the ith row and jth column position of the matrix $rA + rB$. ◁

SUMMARY

1. An $m \times n$ matrix is an ordered rectangular array of numbers containing m rows and n columns.

2. An $m \times 1$ matrix is a column vector with m components and a $1 \times n$ matrix is a row vector with n components.

3. The dot product of vector **a** with components $a_1, a_2, \ldots, a_n$ and vector **b** with components $b_1, b_2, \ldots, b_n$ is the scalar $\mathbf{a} \cdot \mathbf{b} = a_1 b_1 + a_2 b_2 + \cdots + a_n b_n$.

4. The product AB of an $m \times n$ matrix A and an $n \times s$ matrix B is the $m \times s$ matrix C whose entry c_{ij} in the ith row and jth column is the dot product of the ith row vector of A with the jth column vector of B. In general, $AB \neq BA$.

5. If $A = (a_{ij})$ and $B = (b_{ij})$ are matrices of the same size, then $A + B$ is the matrix of that size with entry $a_{ij} + b_{ij}$ in the ith row and jth column.

6. For any matrix A and scalar r, the matrix rA is found by multiplying each entry in A by r.

7. The transpose of an $m \times n$ matrix A is the $n \times m$ matrix A^T having as kth row vector the kth column vector of A.

8. Properties of the dot product and matrix operations are listed in the boxed displays on pages 36 and 37.

EXERCISES

In Exercises 1–24, let

$$A = \begin{pmatrix} -2 & 1 & 3 \\ 4 & 0 & -1 \end{pmatrix}, \quad B = \begin{pmatrix} 4 & 1 & -2 \\ 5 & -1 & 3 \end{pmatrix}, \quad C = \begin{pmatrix} 2 & -1 \\ 0 & 6 \\ -3 & 2 \end{pmatrix}, \quad and \quad D = \begin{pmatrix} -4 & 2 \\ 3 & 5 \\ -1 & -3 \end{pmatrix}.$$

Compute the indicated quantity, if it is defined.

1. $3A$
2. $0B$
3. $A + B$
4. $B + C$
5. $C - D$
6. $3A + 4D$
7. $4A - 2B$
8. $2D - 3C$
9. AB
10. CD
11. AC
12. DB
13. A^2
14. $(AC)^2$
15. $(2A - B)D$
16. $(A - 2B)(3C + D)$
17. ADB
18. DBA
19. $BDAC$
20. $(A^T)A$
21. $A(A^T)$
22. $C(D^T)$
23. $(C^T)D$
24. $(CD)^T$

25. Consider the row and column vectors

$$\mathbf{x} = (-2, 3, -1) \quad and \quad \mathbf{y} = \begin{pmatrix} 4 \\ -1 \\ 3 \end{pmatrix}.$$

Compute the matrix products $\mathbf{xy}$ and $\mathbf{yx}$.

26. Show that if A is a matrix and $\mathbf{x}$ is a row vector, then $\mathbf{x}A$, if defined, is again a row vector.

27. Show that if A is a matrix and $\mathbf{y}$ is a column vector, then $A\mathbf{y}$, if defined, is again a column vector.

In Exercises 28–37, show that the given relation holds for all vectors, matrices, and scalars for which the expressions are defined.

28. $\mathbf{u} \cdot \mathbf{v} = \mathbf{v} \cdot \mathbf{u}$
29. $\mathbf{u} \cdot (\mathbf{v} + \mathbf{w}) = \mathbf{u} \cdot \mathbf{v} + \mathbf{u} \cdot \mathbf{w}$
30. $A + B = B + A$
31. $(A + B) + C = A + (B + C)$
32. $(r + s)A = rA + sA$
33. $(rs)A = r(sA)$
34. $A(B + C) = AB + AC$
35. $(A^T)^T = A$
36. $(A + B)^T = A^T + B^T$
37. $(AB)^T = B^TA^T$
38. If B is an $m \times n$ matrix and $B = A^T$, what is the size of A?
39. Show that every symmetric matrix must be square.
40. Show that the $n \times n$ identity matrix I is symmetric.

41. Fill in the missing entries in the 4×4 matrix

$$\begin{pmatrix} 1 & -1 & 2 & 5 \\ -1 & 4 & -7 & 8 \\ 2 & -7 & -1 & 6 \\ 5 & 8 & 6 & 3 \end{pmatrix}$$

so that the matrix is symmetric.

42. Show that if A is a square matrix, then $(A^2)^T = (A^T)^2$ and $(A^3)^T = (A^T)^3$.

43. State the generalization for $(A^n)^T$ of Exercise 42.

> *The software available to schools using this text includes a program MATCOMP, which performs the matrix operations described in this section. Let*

$$A = \begin{pmatrix} 4 & 6 & 0 & 1 & -9 \\ 2 & 11 & 5 & 2 & -5 \\ -1 & 2 & -4 & 5 & 7 \\ 0 & 12 & -8 & 4 & 3 \\ 10 & 4 & 6 & 2 & -5 \end{pmatrix} \quad and \quad B = \begin{pmatrix} -8 & 15 & 4 & -11 \\ 3 & 5 & 6 & -2 \\ 0 & -1 & 12 & 5 \\ 1 & 13 & -15 & 7 \\ 6 & -8 & 0 & -5 \end{pmatrix}.$$

Use MATCOMP, or similar software, to enter and store these matrices, and then compute the matrices in Exercises 44–51, if they are defined.

44. $A^4 + A$

45. $A^2 B$

46. $A^3 (A^T)^2$

47. BA^2

48. $B^T A$

49. $AB(AB)^T$

50. $A^3 - A^5$

51. $(A^T)^5$

1.4
Square Linear Systems with Unique Solutions

We turn now to techniques for solving systems of linear equations. As a first step, we discuss systems with the same number of equations as unknowns; such systems occur often. In a moment, we will see that a system of n equations in n unknowns can be written in matrix form as

$$A\mathbf{x} = \mathbf{b},$$

where the *coefficient matrix* A is a square $n \times n$ matrix. Usually such a square system has a unique solution; we restrict our work to these systems in this section. However, it is also possible that there is no solution or an infinite number of solutions. Although we will study these cases in detail in Section 1.6, we will present the geometry of general systems now.

THE GEOMETRY OF LINEAR SYSTEMS

Frequently, students have the impression that a linear system containing the same number of equations as unknowns always has a unique solution, whereas a system with more equations than unknowns never has a solution. The geometrical interpretation of the problem quickly shows that these statements are not true.

We know that a single linear equation in two unknowns has a *line* in the plane as its solution set. Similarly, a single linear equation in three unknowns has a *plane* in space as its solution set. The solution set of the equation $x + y + z = 1$ is the plane shown in Fig. 1.18. This geometric analysis can be extended to an equation with more than three variables, but is difficult for us to represent graphically.

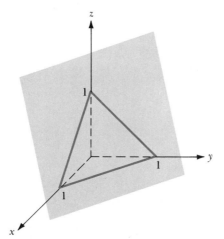

Figure 1.18
The plane $x + y + z = 1$.

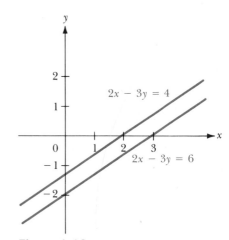

Figure 1.19
$2x - 3y = 4$ and $2x - 3y = 6$ represent parallel lines.

Two lines in the plane usually intersect at a single point; here "usually" means that if the lines are selected in some random way, the chance that they are either parallel (have empty intersection) or coincide (have an infinite number of points in their intersection) is very small. Thus we see that a system of two randomly selected equations in two unknowns can be expected to have a unique solution. However, it is *possible* for the system to have no solution or an infinite number of solutions. For example, the equations

$$2x - 3y = 4$$
$$2x - 3y = 6$$

correspond to distinct parallel lines as shown in Fig. 1.19, and the system consisting of these equations has no solution. Moreover, the equations

$$2x - 3y = 4$$
$$-4x + 6y = -8$$

correspond to the same line, as shown in Fig. 1.20. All points on this line are solutions of this system of two equations.

It is surely possible to have any number of lines in the plane, say fifty lines, pass through a single point. Therefore, it is possible for a system of fifty equations in only two unknowns to have a unique solution.

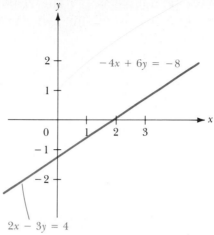

Figure 1.20
$2x - 3y = 4$ and $-4x + 6y = -8$
represent the same line.

Similar illustrations can be made in space, where a linear equation has a plane as its solution set. Three randomly chosen planes can be expected to have a unique point in common: Two of them can be expected to intersect in a line, which in turn can be expected to meet the third plane at a single point. However, it is *possible* for three planes to have no point in common, creating a linear system with no solution. It is also possible for all three planes to contain a common line, and the corresponding linear system would then have an infinite number of solutions.

THE GAUSS REDUCTION METHOD

We discuss an important method for solving a linear system of n equations in n unknowns having a unique solution. Such a system can be written as

$$
\begin{aligned}
a_{11}x_1 + a_{12}x_2 + \cdots + a_{1n}x_n &= b_1 \\
a_{21}x_1 + a_{22}x_2 + \cdots + a_{2n}x_n &= b_2 \\
&\vdots \\
a_{n1}x_1 + a_{n2}x_2 + \cdots + a_{nn}x_n &= b_n.
\end{aligned}
\tag{1}
$$

The system (1) is completely determined by its **coefficient matrix** $A = (a_{ij})$ and the column vector $\mathbf{b}$ with ith entry b_i. The **partitioned matrix** or **augmented matrix**

$$
\left(
\begin{array}{cccc|c}
a_{11} & a_{12} & \cdots & a_{1n} & b_1 \\
a_{21} & a_{22} & \cdots & a_{2n} & b_2 \\
 & & \vdots & & \vdots \\
a_{n1} & a_{n2} & \cdots & a_{nn} & b_n
\end{array}
\right)
\tag{2}
$$

summarizes the system (1). The coefficient matrix has been *augmented* by the column vector of constants. We denote the matrix (2) by $(A \mid \mathbf{b})$. A column vector

$$\mathbf{x} = \begin{pmatrix} x_1 \\ x_2 \\ \cdot \\ \cdot \\ \cdot \\ x_n \end{pmatrix}$$

whose components satisfy the system (1) is a **solution** of the system.

It is easy to see (Exercises 19 and 20) that the solutions of the system (1) are the same as the solutions of any system obtained from (1) by the following procedures:

R1 Interchange two equations.

R2 Multiply an equation in (1) by a nonzero constant.

R3 Replace an equation by the sum of itself and a multiple of a different equation of the system.

A MATRIX-REDUCTION METHOD of solving a system of linear equations occurs in the ancient Chinese work, *Nine Chapters of the Mathematical Art*. The author presents the following solution to the system

$$3x + 2y + z = 39$$
$$2x + 3y + z = 34$$
$$x + 2y + 3x = 26.$$

The diagram of the coefficients is to be set up on a "counting board":

$$\begin{array}{ccc} 1 & 2 & 3 \\ 2 & 3 & 2 \\ 3 & 1 & 1 \\ 26 & 34 & 39 \end{array}$$

He then instructs one to multiply the middle column by 3 and then subtract the right column "as many times as possible"; the same is to be done to the left column. The new diagrams are then:

$$\begin{array}{ccc} 1 & 0 & 3 \\ 2 & 5 & 2 \\ 3 & 1 & 1 \\ 26 & 24 & 39 \end{array} \quad \text{and} \quad \begin{array}{ccc} 0 & 0 & 3 \\ 4 & 5 & 2 \\ 8 & 1 & 1 \\ 39 & 24 & 39 \end{array}$$

The next instruction is to multiply the left column by 5 and again subtract the middle column as many times as possible. This gives

$$\begin{array}{ccc} 0 & 0 & 3 \\ 0 & 5 & 2 \\ 36 & 1 & 1 \\ 99 & 24 & 39 \end{array}$$

The system has thus been reduced to the system $3x + 2y + z = 39$, $5y + z = 24$, $36z = 99$, from which the complete solution is easily found.

Applied to the system (1), these procedures correspond to the **elementary row operations,** listed in the box, applied to the entire partitioned matrix (2).

Elementary Row Operations

R1 Row interchange. Interchange two row vectors in a matrix.

R2 Row scaling. Multiply any row vector in the matrix by a nonzero constant.

R3 Row addition. Replace any row vector in the matrix by the sum of itself and a multiple of a different row vector.

To solve the system (1) using the Gauss method, we attempt to use elementary row operations to reduce the matrix (2) to a matrix of the form

$$\left(\begin{array}{cccc|c} u_{11} & u_{12} & \cdots & u_{1n} & c_1 \\ & u_{22} & \cdots & u_{2n} & c_2 \\ & & \ddots & \vdots & \vdots \\ & & & u_{nn} & c_n \end{array}\right) \tag{3}$$

where the square part U to the left of the partition has zero entries below the **main diagonal,** which extends from the upper left to the lower right corner. The original coefficient matrix A of the system (1) is transformed into the **upper-triangular matrix** U with zeros below the main diagonal. The system becomes

$$U\mathbf{x} = \mathbf{c}. \tag{4}$$

If matrix B can be obtained from matrix A by elementary row operations, then B is **row equivalent** to A. The matrices A and U described above are row equivalent. In the case of a unique solution, all the entries $u_{11}, u_{22}, \ldots, u_{nn}$ on the main diagonal will be nonzero, as we will show in Section 1.6. It is very easy to obtain the unique solution from the upper-triangular partitioned matrix (3), as we show by the following example.

Example 1   Suppose the augmented matrix (2) corresponding to a system of three equations in three unknowns can be row-reduced to

$$\left(\begin{array}{ccc|c} -5 & -1 & 3 & 3 \\ 0 & 3 & 5 & 8 \\ 0 & 0 & 2 & -4 \end{array}\right). \tag{5}$$

Find the solution of the system.

Solution   The equations corresponding to the augmented matrix (5) are

$$\begin{aligned} -5x_1 - x_2 + 3x_3 &= 3 \\ 3x_2 + 5x_3 &= 8 \\ 2x_3 &= -4. \end{aligned}$$

From the last equation, we obtain $x_3 = -2$. Substituting in the second equation, we have

$$3x_2 + 5(-2) = 8, \qquad 3x_2 = 18, \qquad x_2 = 6.$$

Finally, we substitute these values for x_2 and x_3 into the top equation, obtaining

$$-5x_1 - 6 + 3(-2) = 3, \qquad -5x_1 = 15, \qquad x_1 = -3.$$

Thus the unique solution is

$$\mathbf{x} = \begin{pmatrix} -3 \\ 6 \\ -2 \end{pmatrix}$$

or equivalently, $x_1 = -3$, $x_2 = 6$, $x_3 = -2$. ◁

The procedure for finding the solution of a system from an upper-triangular augmented matrix, illustrated in Example 1, is called **back substitution.** Of course, the numbers in Example 1 were selected to make the arithmetic easy, as is the case with all our examples and exercises. Such nice numbers don't occur in practice. But back substitution is easy to execute on a computer, and the computer doesn't know whether the arithmetic is easy or hard!

The box on page 45 describes how to reduce the partitioned matrix (2) to form (3) where the left portion is upper-triangular. The procedure consists of fixing up one column at a time, working in left-to-right order and using elementary row operations.

Example 2 shows this procedure, doing just one elementary row operation at a time. We use the symbol $\sim$ for matrices that are row equivalent.

Example 2   Solve the linear system

$$\begin{array}{rcrcrcr} & & x_2 & - & 3x_3 & = & -5 \\ 2x_1 & + & 3x_2 & - & x_3 & = & 7 \\ 4x_1 & + & 5x_2 & - & 2x_3 & = & 10 \end{array}$$

using the Gauss method with back substitution.

Solution   We reduce the corresponding augmented matrix using elementary row operations. Pivots are in color.

$$\left(\begin{array}{ccc|c} 0 & 1 & -3 & -5 \\ 2 & 3 & -1 & 7 \\ 4 & 5 & -2 & 10 \end{array} \right) \sim \left(\begin{array}{ccc|c} 2 & 3 & -1 & 7 \\ 0 & 1 & -3 & -5 \\ 4 & 5 & -2 & 10 \end{array} \right) \qquad \begin{array}{l} \text{Interchange row} \\ \text{1 and row 2.} \end{array}$$

$$\left(\begin{array}{ccc|c} 2 & 3 & -1 & 7 \\ 0 & 1 & -3 & -5 \\ 4 & 5 & -2 & 10 \end{array} \right) \sim \left(\begin{array}{ccc|c} 2 & 3 & -1 & 7 \\ 0 & 1 & -3 & -5 \\ 0 & -1 & 0 & -4 \end{array} \right) \qquad \begin{array}{l} \text{Add } -2 \text{ times} \\ \text{row 1 to row 3.} \end{array}$$

$$\left(\begin{array}{ccc|c} 2 & 3 & -1 & 7 \\ 0 & 1 & -3 & -5 \\ 0 & -1 & 0 & -4 \end{array} \right) \sim \left(\begin{array}{ccc|c} 2 & 3 & -1 & 7 \\ 0 & 1 & -3 & -5 \\ 0 & 0 & -3 & -9 \end{array} \right) \qquad \begin{array}{l} \text{Add row 2 to} \\ \text{row 3.} \end{array}$$

From the last partitioned matrix, we obtain by back substitution

$$-3x_3 = -9, \qquad x_3 = 3$$
$$x_2 - 3(3) = -5, \qquad x_2 = 4$$
$$2x_1 + 3(4) - 1(3) = 7, \qquad 2x_1 = -2, \qquad x_1 = -1.$$

We have found the unique solution $x_1 = -1, x_2 = 4, x_3 = 3$. ◁

Gauss Reduction to Upper-Triangular Form; Unique-Solution Case

Step 1 If the top entry in column 1 is zero, then perform a row interchange operation to get a nonzero element at the top of the column. Section 1.6 will show that this is always possible in the unique-solution case. We call the selected *nonzero* element the **pivot.**

Step 2 Perform a sequence of row additions, adding multiples of the top row to rows below in such a way that the resulting rows below will have zero as first entries.

Step 3 After step 2 the matrix has the form

$$\left(\begin{array}{cccc|c} u_{11} & u_{12} & \cdots & u_{1n} & c_1 \\ 0 & X & \cdots & X & X \\ 0 & X & \cdots & X & X \\ & \vdots & & & \vdots \\ 0 & X & \cdots & X & X \end{array} \right) \qquad (6)$$

and our first column is the way we want it. Cross off—mentally, that is—the top row and the first column of the matrix (6), leaving the shaded portion of the matrix. Go back to step 1 with this smaller matrix and repeat the procedure to fix up the next column. Continue until the upper-triangular form is obtained.

THE GAUSS SOLUTION METHOD is so named since Gauss described it in a paper detailing the computations he made to determine the orbit of the asteroid Pallas. The parameters of the orbit had to be determined by observations of the asteroid over the six-year period from 1803 to 1809. These led to six linear equations in six unknowns with quite complicated coefficients. Gauss showed how to solve these equations by systematically replacing them with a new system in which only the first equation had all six unknowns, the second equation included five unknowns, the third equation only four, and so on, until the sixth equation had but one. This last equation could, of course, be easily solved; the remaining unknowns were then found by back substitution.

THE GAUSS–JORDAN METHOD

A square matrix is in **diagonal form** if all entries *not* on the main diagonal are zero. The Gauss–Jordan method carries the Gauss method a step further, and uses elementary row operations to reduce the left part of a partitioned matrix to the diagonal form with all pivots 1, that is, to the identity matrix *I*. A discussion in Section 1.6 will show that this is always possible in the unique-solution case.

Suppose, for example, that a partitioned matrix is reduced to the form

$$\begin{pmatrix} 1 & 0 & 0 & \bigm| & 9 \\ 0 & 1 & 0 & \bigm| & 7 \\ 0 & 0 & 1 & \bigm| & -1 \end{pmatrix}. \tag{7}$$

It is immediately apparent from the augmented matrix (7) that the solution of the original system is given by

$$x_1 = 9, \qquad x_2 = 7, \quad \text{and} \quad x_3 = -1.$$

Gauss–Jordan Reduction to Diagonal Form

Execute the Gauss method for upper-triangular form as described above, but make each pivot 1 by a row scaling operation, and use the pivot to create zeros *above* it as well as *below* it in that column.

We illustrate with an example.

Example 3   Find the solution of the system

$$\begin{aligned} x_1 - 2x_2 + x_3 &= 7 \\ 2x_1 - 5x_2 + 2x_3 &= 6 \\ 3x_1 + 2x_2 - x_3 &= 1 \end{aligned}$$

by Gauss–Jordan reduction.

THE "JORDAN" HALF OF THE GAUSS–JORDAN METHOD is essentially a systematic technique of "back substitution." In this form it was described by Wilhelm Jordan (1842–1899), a German professor of geodesy, in his *Handbook of Geodesy*. This work was first published in German in 1873 and has since gone through ten editions as well as translations into several other languages. On the other hand, Jordan himself credits the method to Friedrich Robert Helmert and Peter Andreas Hansen a few years earlier.

Wilhelm Jordan was prominent in his field in the late nineteenth century, being involved in several geodetic surveys in Germany as well as in the first major survey of the Libyan desert. He was also the founding editor of the German geodesy journal. His interest in finding a systematic method of solving large systems of linear equations stems from their frequent appearance in problems of triangulation.

Solution Again, we label the steps and color the pivots.

$$\begin{pmatrix} 1 & -2 & 1 & | & 7 \\ 2 & -5 & 2 & | & 6 \\ 3 & 2 & -1 & | & 1 \end{pmatrix}$$

Add -2 times row 1 to row 2 and add -3 times row 1 to row 3 to obtain the next matrix.

$$\sim \begin{pmatrix} 1 & -2 & 1 & | & 7 \\ 0 & -1 & 0 & | & -8 \\ 0 & 8 & -4 & | & -20 \end{pmatrix}$$

Multiply row 2 by -1 to obtain the next matrix.

$$\sim \begin{pmatrix} 1 & -2 & 1 & | & 7 \\ 0 & 1 & 0 & | & 8 \\ 0 & 8 & -4 & | & -20 \end{pmatrix}$$

Add 2 times row 2 to row 1 and add -8 times row 2 to row 3 to obtain the next matrix.

$$\sim \begin{pmatrix} 1 & 0 & 1 & | & 23 \\ 0 & 1 & 0 & | & 8 \\ 0 & 0 & -4 & | & -84 \end{pmatrix}$$

Multiply row 3 by $-\frac{1}{4}$ to obtain the next matrix.

$$\sim \begin{pmatrix} 1 & 0 & 1 & | & 23 \\ 0 & 1 & 0 & | & 8 \\ 0 & 0 & 1 & | & 21 \end{pmatrix}$$

Add -1 times row 3 to row 1 to obtain the final matrix.

$$\sim \begin{pmatrix} 1 & 0 & 0 & | & 2 \\ 0 & 1 & 0 & | & 8 \\ 0 & 0 & 1 & | & 21 \end{pmatrix}.$$

From this last partitioned matrix, we see that the solution is $x_1 = 2$, $x_2 = 8$, $x_3 = 21$. ◁

SYSTEMS WITH THE SAME COEFFICIENT MATRIX

In Section 1.5, we will have occasion to solve several linear systems $A\mathbf{x} = \mathbf{b}$ having the same coefficient matrix A but different constant column vectors $\mathbf{b}$. Suppose we want to solve two such systems,

$$A\mathbf{x} = \mathbf{b} \quad \text{and} \quad A\mathbf{y} = \mathbf{b}'.$$

Rather than solve one after the other, it is more efficient to form a single partitioned matrix with the coefficient matrix A at the left of the partition and augmented with *two* column vectors, $\mathbf{b}$ and $\mathbf{b}'$, to the right of the partition. We then reduce this larger partitioned matrix so that the left part has upper-triangular or diagonal form, solving both systems at once. Since the entries in the column vectors $\mathbf{b}$ and $\mathbf{b}'$ play no role in the selection of elementary row operations to reduce the matrix A, we are able to avoid doing the same work twice.

In the following example, we use the Gauss–Jordan method, reducing the coefficient matrix to the identity matrix I.

Example 4   Solve the linear systems

$$
\begin{array}{rcl}
2x_1 - 4x_2 & = & -10 \\
x_1 - 3x_2 \qquad + x_4 & = & -4 \\
x_1 \qquad - x_3 + 2x_4 & = & 4 \\
3x_1 - 4x_2 + 3x_3 - x_4 & = & -11
\end{array}
\quad \text{and} \quad
\begin{array}{rcl}
2y_1 - 4y_2 & = & -8 \\
y_1 - 3y_2 \qquad + y_4 & = & -2 \\
y_1 \qquad - y_3 + 2y_4 & = & 9 \\
3y_1 - 4y_2 + 3y_3 - y_4 & = & -15
\end{array}
$$

using the Gauss–Jordan method.

Solution   We put both columns of constants to the right of the partition, and again color the pivots.

$$
\left(\begin{array}{cccc|cc}
2 & -4 & 0 & 0 & -10 & -8 \\
1 & -3 & 0 & 1 & -4 & -2 \\
1 & 0 & -1 & 2 & 4 & 9 \\
3 & -4 & 3 & -1 & -11 & -15
\end{array}\right)
$$

Multiply row 1 by $\frac{1}{2}$ to obtain the next matrix.

$$
\sim
\left(\begin{array}{cccc|cc}
1 & -2 & 0 & 0 & -5 & -4 \\
1 & -3 & 0 & 1 & -4 & -2 \\
1 & 0 & -1 & 2 & 4 & 9 \\
3 & -4 & 3 & -1 & -11 & -15
\end{array}\right)
$$

Multiply row 1 by -1 and add to rows 2 and 3. Multiply row 1 by -3 and add to row 4.

$$
\sim
\left(\begin{array}{cccc|cc}
1 & -2 & 0 & 0 & -5 & -4 \\
0 & -1 & 0 & 1 & 1 & 2 \\
0 & 2 & -1 & 2 & 9 & 13 \\
0 & 2 & 3 & -1 & 4 & -3
\end{array}\right)
$$

Multiply row 2 by -1 to obtain the next matrix.

$$
\sim
\left(\begin{array}{cccc|cc}
1 & -2 & 0 & 0 & -5 & -4 \\
0 & 1 & 0 & -1 & -1 & -2 \\
0 & 2 & -1 & 2 & 9 & 13 \\
0 & 2 & 3 & -1 & 4 & -3
\end{array}\right)
$$

Multiply row 2 by 2 and add to row 1. Multiply row 2 by -2 and add to rows 3 and 4.

$$
\sim
\left(\begin{array}{cccc|cc}
1 & 0 & 0 & -2 & -7 & -8 \\
0 & 1 & 0 & -1 & -1 & -2 \\
0 & 0 & -1 & 4 & 11 & 17 \\
0 & 0 & 3 & 1 & 6 & 1
\end{array}\right)
$$

Multiply row 3 by -1 to obtain the next matrix.

$$
\sim
\left(\begin{array}{cccc|cc}
1 & 0 & 0 & -2 & -7 & -8 \\
0 & 1 & 0 & -1 & -1 & -2 \\
0 & 0 & 1 & -4 & -11 & -17 \\
0 & 0 & 3 & 1 & 6 & 1
\end{array}\right)
$$

Multiply row 3 by -3 and add to row 4 to obtain the next matrix.

$$
\sim
\left(\begin{array}{cccc|cc}
1 & 0 & 0 & -2 & -7 & -8 \\
0 & 1 & 0 & -1 & -1 & -2 \\
0 & 0 & 1 & -4 & -11 & -17 \\
0 & 0 & 0 & 13 & 39 & 52
\end{array}\right)
$$

Multiply row 4 by $\frac{1}{13}$ to obtain the next matrix.

$$\sim \begin{pmatrix} 1 & 0 & 0 & -2 & -7 & -8 \\ 0 & 1 & 0 & -1 & -1 & -2 \\ 0 & 0 & 1 & -4 & -11 & -17 \\ 0 & 0 & 0 & 1 & 3 & 4 \end{pmatrix}$$

Multiply row 4 by 2 and add to row 1, add row 4 to row 2, and add 4 times row 4 to row 3.

$$\sim \begin{pmatrix} 1 & 0 & 0 & 0 & -1 & 0 \\ 0 & 1 & 0 & 0 & 2 & 2 \\ 0 & 0 & 1 & 0 & 1 & -1 \\ 0 & 0 & 0 & 1 & 3 & 4 \end{pmatrix}$$

From this final matrix, we see that the solutions are $x_1 = -1$, $x_2 = 2$, $x_3 = 1$, $x_4 = 3$ and $y_1 = 0$, $y_2 = 2$, $y_3 = -1$, $y_4 = 4$. ◁

The last example illustrates that if an $n \times n$ matrix A is row equivalent to the $n \times n$ identity matrix I, then the system of equations $A\mathbf{x} = \mathbf{b}$ has a unique solution. This result follows from the work we have done in this section; the converse will be established in Section 1.6. We summarize in the following theorem.

Theorem 1.3   A Condition for a Unique Solution

Let $A\mathbf{x} = \mathbf{b}$ be a system of n linear equations in n unknowns. The system has a unique solution if and only if the coefficient matrix A is row equivalent to the $n \times n$ identity matrix I.

ELEMENTARY MATRICES

The elementary row operations we have performed can actually be carried out by matrix multiplication. If we interchange its second and third rows, the 3×3 identity matrix

$$I = \begin{pmatrix} 1 & 0 & 0 \\ 0 & 1 & 0 \\ 0 & 0 & 1 \end{pmatrix} \quad \text{becomes} \quad E = \begin{pmatrix} 1 & 0 & 0 \\ 0 & 0 & 1 \\ 0 & 1 & 0 \end{pmatrix}.$$

If $A = (a_{ij})$ is a 3×3 matrix, we compute EA and find that

$$EA = \begin{pmatrix} 1 & 0 & 0 \\ 0 & 0 & 1 \\ 0 & 1 & 0 \end{pmatrix} \begin{pmatrix} a_{11} & a_{12} & a_{13} \\ a_{21} & a_{22} & a_{23} \\ a_{31} & a_{32} & a_{33} \end{pmatrix} = \begin{pmatrix} a_{11} & a_{12} & a_{13} \\ a_{31} & a_{32} & a_{33} \\ a_{21} & a_{22} & a_{23} \end{pmatrix}.$$

The second and third rows of A have been interchanged by multiplying on the left by E.

Definition 1.7   Elementary Matrix

Any matrix that can be obtained from an identity matrix by applying one elementary row operation is an **elementary matrix**.

We leave a proof of the following theorem as an exercise for the student.

Theorem 1.4   The Use of Elementary Matrices

Let A be an $m \times n$ matrix and let E be an $m \times m$ elementary matrix. Multiplication of A on the left by E effects the same elementary row operation on A that was performed on the identity matrix to obtain E.

Thus the reduction of a square matrix to upper-triangular or diagonal form can be accomplished by successive multiplication on the left by elementary matrices. In other words, if A can be reduced to U by elementary row operations, then there exist elementary matrices $E_1, E_2, \ldots, E_t$ such that

$$U = (E_t \cdots E_2 E_1)A.$$

This is by no means an efficient way to execute row reduction, but it is sometimes a handy thing to know in proving theorems.

Example 5   Let

$$A = \begin{pmatrix} 0 & 1 & -3 \\ 2 & 3 & -1 \\ 4 & 5 & -2 \end{pmatrix},$$

so that A is the coefficient matrix for the system of equations in Example 2. Find a matrix C such that CA is an upper-triangular matrix that is row equivalent to A.

Solution   The elementary matrices corresponding to each of the elementary row operations performed in sequence to A in Example 2 are

$$E_1 = \begin{pmatrix} 0 & 1 & 0 \\ 1 & 0 & 0 \\ 0 & 0 & 1 \end{pmatrix}, \qquad E_2 = \begin{pmatrix} 1 & 0 & 0 \\ 0 & 1 & 0 \\ -2 & 0 & 1 \end{pmatrix}, \qquad E_3 = \begin{pmatrix} 1 & 0 & 0 \\ 0 & 1 & 0 \\ 0 & 1 & 1 \end{pmatrix}.$$

The matrix product $E_3 E_2 E_1 A$ must be the upper-triangular matrix

$$\begin{pmatrix} 2 & 3 & -1 \\ 0 & 1 & -3 \\ 0 & 0 & -3 \end{pmatrix}$$

found in Example 2. Therefore, the required matrix is

$$C = E_3 E_2 E_1 = \begin{pmatrix} 0 & 1 & 0 \\ 1 & 0 & 0 \\ 1 & -2 & 1 \end{pmatrix}. \quad \triangleleft$$

One can execute analogous *elementary column* operations on a matrix by multiplying the matrix on the *right* by an elementary matrix. Column reduction of a matrix A is not important for us in this chapter because it does not preserve the solution set of $A\mathbf{x} = \mathbf{b}$ when applied to the augmented matrix $(A \mid \mathbf{b})$. We will have occasion to refer to column reduction later. The effect of multiplication of a matrix A on the right by elementary matrices is explored in Exercises 19–21 of Section 1.5.

SUMMARY

1. A linear system has an associated partitioned (or augmented) matrix, having the coefficient matrix of the system on the left of the partition and the column vector of constants on the right of the partition.

2. The elementary row operations are:

   R1 Row interchange. Interchange of two rows.

   R2 Row scaling. Multiplication of a row by a nonzero constant.

   R3 Row addition. Addition of a multiple of a row to a different row.

3. In the Gauss method, we solve a square linear system by using elementary row operations and reducing the partitioned matrix so that the portion on the left of the partition becomes upper-triangular. The solution is then found by back substitution.

4. The Gauss–Jordan method is similar to the Gauss method, except that the coefficient matrix of the system is reduced to the identity matrix I. The solution then appears as the column vector on the right of the partition in the reduced augmented matrix.

5. Several linear systems with the same coefficient matrix can all be solved at once by augmenting the coefficient matrix by the column vectors of constants from all the systems.

6. An elementary matrix E is one obtained by applying a single elementary row operation to an identity matrix I. Multiplication of a matrix A on the left by E effects the same elementary row operation on A.

7. A square system $A\mathbf{x} = \mathbf{b}$ has a unique solution if and only if the coefficient matrix A can be row-reduced to the identity matrix.

EXERCISES

In Exercises 1–8, find the unique solution of the given square linear system using the Gauss method with back substitution.

1. $2x - y = 8$
$6x - 5y = 32$

2. $4x_1 - 3x_2 = 10$
$8x_1 - x_2 = 10$

3. $y + z = 6$
$3x - y + z = -7$
$x + y - 3z = -13$

4. $2x + y - 3z = 0$
$6x + 3y - 8z = 0$
$2x - y + 5z = -4$

5. $x_1 + 2x_2 - x_3 = -3$
   $3x_1 + 7x_2 + 2x_3 = 1$
   $4x_1 - 2x_2 + x_3 = -2$

6. $2x_2 - x_3 + x_4 = 6$
   $x_1 - 4x_2 + x_3 - 2x_4 = -9$
   $2x_1 - 7x_2 - 2x_3 + x_4 = 10$
   $3x_2 \qquad - 4x_4 = -16$

7. $x_3 - 2x_4 = 5$
   $3x_1 - 6x_2 + 2x_3 \qquad = 18$
   $x_1 - 2x_2 + x_3 - x_4 = 8$
   $2x_1 - 3x_2 \qquad + 3x_4 = 4$

8. $x_1 - 3x_2 + x_3 + 2x_4 = 2$
   $x_1 - 2x_2 + 2x_3 + 4x_4 = -1$
   $2x_1 - 8x_2 - x_3 \qquad = 3$
   $3x_1 - 9x_2 + 4x_3 \qquad = 7$

In Exercises 9–14, find the unique solution of the linear system using the Gauss–Jordan method.

9. $3x_1 - 2x_2 = -8$
   $4x_1 + 5x_2 = -3$

10. $2x_1 + 3x_2 = -7$
   $5x_1 - 4x_2 = 17$

11. $x_2 - 3x_3 = 3$
   $2x_1 + 4x_2 - x_3 = 5$
   $4x_1 - 2x_2 + 5x_3 = 3$

12. $x_1 - 4x_2 + x_3 = 8$
   $3x_1 - 12x_2 + 5x_3 = 26$
   $2x_1 - 9x_2 - x_3 = 14$

13. $x_1 + 2x_2 - 3x_3 \qquad = 8$
   $2x_1 + 5x_2 - 6x_3 \qquad = 17$
   $-x_1 - 2x_2 + x_3 - x_4 = -8$
   $4x_1 + 10x_2 - 9x_3 + x_4 = 33$

14. $2x_1 + 3x_2 \qquad - x_4 = 0$
   $x_1 + x_2 + x_3 - 2x_4 = 7$
   $3x_1 - x_2 + 2x_3 + x_4 = -19$
   $2x_1 - 4x_2 + 3x_3 \qquad = -12$

In each of Exercises 15–18, two or more linear systems having the same coefficient matrix but different column vectors of constants are given. Find the solution of each system.

15. $x - 2y = -3,\ 12,\ 3$
   $3x + y = -2,\ 1,\ 9$

16. $x_1 - 3x_2 - x_3 = 3,\ 0,\ -10$
   $2x_1 - 5x_2 + x_3 = 8,\ 0,\ -11$
   $x_1 + 4x_2 - 2x_3 = -5,\ 0,\ 9$

17. $2x_1 - x_2 + x_3 = 1,\ 1,\ 8$
   $-2x_1 + x_2 - 3x_3 = -5,\ -3,\ -14$
   $4x_1 - 3x_2 + x_3 = 5,\ 1,\ 14$

18. $x_1 + 4x_2 - x_3 + x_4 = 2,\ -5$
   $2x_1 + 7x_2 + x_3 - 2x_4 = 16,\ 9$
   $x_1 + 4x_2 - x_3 + 2x_4 = 1,\ -8$
   $3x_1 - 10x_2 - 2x_3 + 5x_4 = -15,\ 3$

19. Explain geometrically why the solution set of a linear system remains the same if:
   a) two equations are interchanged;
   b) an equation is multiplied by a nonzero constant.

20. Consider a linear system of four equations in four unknowns; x, y, z, and w. Let the first two equations of the system be

$$x - 2y + z - w = 10$$
$$-2x + y \qquad + 3w = -5.$$

Suppose the second equation of this system is replaced by itself plus 4 times the first equation, that is, by $2x - 7y + 4z + w = 35$. Explain in detail why the resulting system of four equations has the same solution set as the original system.

In Exercises 21–26, let A be a 4 × 4 matrix. Find a matrix C such that the result of applying the given sequence of elementary row operations to A could also be found by computing the product CA.

21. Interchange rows 1 and 2.

22. Interchange rows 1 and 3; multiply row 3 by 4.

23. Multiply row 1 by 5; interchange rows 2 and 3; and add 2 times row 3 to row 4.

24. Add 4 times row 2 to row 4; multiply row 4 by -3; and add 5 times row 4 to row 1.

25. Interchange rows 1 and 4; multiply row 2 by 6 and add to row 1; multiply row 1 by -3 and add to row 3; and multiply row 4 by -2 and add to row 2.

26. Multiply row 2 by 3 and add to row 4; multiply row 4 by -2 and add to row 3; multiply row 3 by 5 and add to row 1; and multiply row 1 by -4 and add to row 2.

A *problem we meet when reducing a matrix with a computer is to determine when a computed entry should be zero. The computer might give an entry as 0.00000001, due to roundoff error, when it really should be zero. If the computer uses this entry as a pivot in a future step, the result is chaotic! For this reason, it is common practice to program the computer to replace all sufficiently small computed entries on the left of the partition by zero. Here "sufficiently small" must be specified in terms of the size of the nonzero entries in the original matrix. The software program YUREDUCE, available to schools using this text, allows the user to specify the steps in reduction of a matrix. The program computes the smallest nonzero coefficient magnitude m and asks the user to enter a number r (for ratio); all computed entries of magnitude less than rm in reduction of the coefficient matrix will be set equal to zero. In Exercises 27–32, use the program YURE-DUCE, or similar software, specifying $r = 0.0001$, to solve the linear system or systems.*

27. The system in Exercise 6

28. The system in Exercise 7

29. The system in Exercise 8

30. The systems in Exercise 16

31. The systems in Exercise 18

32.
$$
\begin{aligned}
x_1 - 2x_2 + x_3 - x_4 + 2x_5 &= 1, & 1 \\
2x_1 + x_2 - 4x_3 - x_4 + 5x_5 &= 16, & 10 \\
8x_1 - x_2 + 3x_3 - x_4 - x_5 &= 1, & -5 \\
4x_1 - 2x_2 + 3x_3 - 8x_4 + 2x_5 &= -5, & -3 \\
5x_1 + 3x_2 - 4x_3 + 7x_4 - 6x_5 &= 7, & 1
\end{aligned}
$$

The program MATCOMP, available to schools using this text, can also be used to find the solutions of one or more square systems with a common coefficient matrix. MATCOMP will reduce the left portion of the augmented matrix to diagonal form with pivots of 1, and display the result on the screen. The user can then find the solutions easily. Use MATCOMP or similar software to solve the linear systems in Exercises 33–37.

33. The system in Exercise 5

34. The system in Exercise 13

35. The system in Exercise 14

36. The systems in Exercise 17

37. The systems in Exercise 32

1.5

Inverses of Square Matrices

MATRIX EQUATIONS AND INVERSES

A system of n equations in n unknowns $x_1, x_2, \ldots, x_n$ can be expressed in matrix form as

$$A\mathbf{x} = \mathbf{b}, \tag{1}$$

where A is the $n \times n$ coefficient matrix, $\mathbf{x}$ is the $1 \times n$ column vector with ith entry x_i, and $\mathbf{b}$ is a $1 \times n$ column vector with constant entries. The analogous equation using scalars is

$$ax = b \tag{2}$$

for scalars a and b. We usually think of solving Eq. (2) for x by dividing by a if $a \neq 0$, but we can just as well think of multiplying by $1/a$. Breaking the solution down into small steps, we have

$$(1/a)(ax) = (1/a)b \qquad \text{Multiplying by } 1/a$$
$$[(1/a)a]x = (1/a)b \qquad \text{Associativity of multiplication}$$
$$1x = (1/a)b \qquad \text{Property of } 1/a$$
$$x = (1/a)b. \qquad \text{Property of } 1$$

Let us see whether we can solve Eq. (1) in a similar way. Matrix multiplication is associative, and the $n \times n$ identity matrix I plays the role for multiplication of $n \times n$ matrices that the number 1 plays for multiplication of numbers. What is crucial is finding an $n \times n$ matrix C such that $CA = I$, so that C plays for matrices the role that $1/a$ does for numbers. If such a matrix C exists, then from Eq. (1) we obtain

$$C(A\mathbf{x}) = C\mathbf{b} \qquad \text{Multiplying by } C$$
$$(CA)\mathbf{x} = C\mathbf{b} \qquad \text{Associativity of multiplication}$$
$$I\mathbf{x} = C\mathbf{b} \qquad \text{Property of } C$$
$$\mathbf{x} = C\mathbf{b}, \qquad \text{Property of } I$$

which shows that our column vector $\mathbf{x}$ of unknowns must be the column vector $C\mathbf{b}$. We emphasize: The only question is whether there exists an $n \times n$ matrix C such that $CA = I$. The equation

$$\begin{pmatrix} -4 & 9 \\ 1 & -2 \end{pmatrix} \begin{pmatrix} 2 & 9 \\ 1 & 4 \end{pmatrix} = \begin{pmatrix} 1 & 0 \\ 0 & 1 \end{pmatrix}$$

shows that

$$A = \begin{pmatrix} 2 & 9 \\ 1 & 4 \end{pmatrix} \quad \text{and} \quad C = \begin{pmatrix} -4 & 9 \\ 1 & -2 \end{pmatrix}$$

satisfies $CA = I$.

Unfortunately, it is *not* true that for each $n \times n$ matrix A, one can find an $n \times n$ matrix C such that $CA = I$. For example, if the jth column of A has only zero entries, then the jth column of CA also has only zero entries for any matrix C, so $CA \neq I$ for any matrix C. However, for many important $n \times n$ matrices A, there does exist an $n \times n$ matrix C such that $CA = I$. It can be shown then that $AC = I$ also. This is not obvious; recall that matrix multiplication is not commutative. Exercise 19 of Section 6.3 requests proof of this. We highlight this fact with a box and proceed to a formal definition of such an *inverse* of a matrix A.

Commutativity with Inverses

Let A and C be $n \times n$ matrices and let I be the $n \times n$ identity matrix. If $CA = I$, then $AC = I$.

Definition 1.8 Inverse of a Square Matrix

Let A be an $n \times n$ matrix. An $n \times n$ matrix C is an **inverse of** A if $CA = AC = I$, the $n \times n$ identity matrix.

We now show that if a square matrix has an inverse, then this inverse is unique.

Theorem 1.5 Uniqueness of an Inverse

Let A be an $n \times n$ matrix. If C and D are matrices such that $CA = AD = I$, then $C = D$.

Proof Since matrix multiplication is associative, we have

$$C(AD) = (CA)D.$$

But since $AD = I$ and $CA = I$, we find that

$$C(AD) = CI = C \quad \text{and} \quad (CA)D = ID = D.$$

Therefore, $C = D$. ■

Theorem 1.5 shows that we may speak simply of *the* inverse of a matrix A. We will denote the inverse of A, when it exists, by A^{-1}. Although A^{-1} plays the same role arithmetically as $a^{-1} = 1/a$ (as we showed at the beginning of this section), we will never write A^{-1} as $1/A$.

THE NOTION OF THE INVERSE OF A MATRIX first appears in an 1855 note of Arthur Cayley (1821–1895) and is made more explicit in a paper three years later titled "A Memoir on the Theory of Matrices." In that work, Cayley outlines the basic properties of matrices, noting that most of these derive from work with sets of linear equations. In particular, the inverse comes from the idea of solving a system

$$X = ax + by + cz$$
$$Y = a'x + b'y + c'z$$
$$Z = a''x + b''y + c''z$$

for x, y, z in terms of X, Y, Z. Cayley gives an explicit construction for the inverse in terms of the determinants of the original matrix and of the minors.

In 1842 Arthur Cayley graduated from Trinity College, Cambridge, but could not find a suitable teaching post. So, like Sylvester, he studied law and was called to the bar in 1849. During his 14 years as a lawyer, he wrote about 300 mathematical papers; finally, in 1863 he became a professor at Cambridge, where he remained until his death. It was during his stint as a lawyer that he met Sylvester; their discussions over the next 40 years were extremely fruitful for the progress of algebra. Over his lifetime, Cayley produced close to 1000 papers in pure mathematics, theoretical dynamics, and mathematical astronomy.

Definition 1.9 Invertible and Singular Matrices

A square matrix that has an inverse is called **invertible**. A square matrix that has no inverse is called **singular**.

INVERSES OF ELEMENTARY MATRICES

Let E_1 be an elementary row-interchange matrix, obtained from the identity matrix I by interchanging rows i and k. Recall that E_1A affects the interchange of rows i and k of A for any matrix A of the same size as E_1. In particular, taking $A = E_1$, we see that E_1E_1 interchanges rows i and k of E_1, and hence changes E_1 back to I. Thus

$$E_1E_1 = I.$$

Consequently, E_1 is an invertible matrix and is its own inverse.

Now let $E_{2,r}$ be an elementary row-scaling matrix, obtained from the identity matrix by multiplying row i by a nonzero scalar r. Let $E_{2,1/r}$ be the matrix obtained from the identity matrix by multiplying row i by $1/r$. It is clear that

$$E_{2,1/r}E_{2,r} = E_{2,r}E_{2,1/r} = I,$$

so $E_{2,r}$ is invertible with inverse $E_{2,1/r}$.

Finally, let $E_{3,r}$ be an elementary row-addition matrix, obtained from I by adding r times row i to row k. If $E_{3,-r}$ is obtained from I by adding $-r$ times row i to row k, then clearly

$$E_{3,r}E_{3,-r} = E_{3,-r}E_{3,r} = I.$$

We have established the following fact.

> Every elementary matrix is invertible.

Example 1 Find the inverses of the elementary matrices

$$E_1 = \begin{pmatrix} 0 & 1 & 0 \\ 1 & 0 & 0 \\ 0 & 0 & 1 \end{pmatrix}, \qquad E_2 = \begin{pmatrix} 3 & 0 & 0 \\ 0 & 1 & 0 \\ 0 & 0 & 1 \end{pmatrix}, \quad \text{and} \quad E_3 = \begin{pmatrix} 1 & 0 & 4 \\ 0 & 1 & 0 \\ 0 & 0 & 1 \end{pmatrix}.$$

Solution Since E_1 is obtained from

$$I = \begin{pmatrix} 1 & 0 & 0 \\ 0 & 1 & 0 \\ 0 & 0 & 1 \end{pmatrix}$$

by interchanging the first and second rows, we see that $E_1^{-1} = E_1$.

The matrix E_2 is obtained from I by multiplying the first row by 3, so we must multiply the first row of I by $\frac{1}{3}$ to form

$$E_2^{-1} = \begin{pmatrix} \frac{1}{3} & 0 & 0 \\ 0 & 1 & 0 \\ 0 & 0 & 1 \end{pmatrix}.$$

Finally, E_3 is obtained from I by adding 4 times row 3 to row 1. To form E_3^{-1}, we add -4 times row 3 to row 1, so

$$E_3^{-1} = \begin{pmatrix} 1 & 0 & -4 \\ 0 & 1 & 0 \\ 0 & 0 & 1 \end{pmatrix}.$$

We could easily verify by direct computation that

$$E_1 E_1 = I, \qquad E_2 E_2^{-1} = E_2^{-1} E_2 = I, \quad \text{and} \quad E_3 E_3^{-1} = E_3^{-1} E_3 = I. \quad \lhd$$

INVERSES OF PRODUCTS

The following theorem is fundamental when working with inverses.

Theorem 1.6   Inverses of Products

Let A and B be invertible $n \times n$ matrices. Then AB is invertible, and $(AB)^{-1} = B^{-1}A^{-1}$.

Proof By assumption, there exist matrices A^{-1} and B^{-1} such that $AA^{-1} = I$ and $BB^{-1} = I$. We make use of the associative law for matrix multiplication, and find that

$$(AB)(B^{-1}A^{-1}) = [A(BB^{-1})]A^{-1} = (AI)A^{-1} = AA^{-1} = I.$$

A similar computation shows that $(B^{-1}A^{-1})(AB) = I$. Therefore, $(AB)^{-1} = B^{-1}A^{-1}$. ∎

It is instructive to illustrate Theorem 1.6 by considering a product $E_t \cdots E_3 E_2 E_1$ of elementary matrices. When applied to a matrix A, as in

$$(E_t \cdots E_3 E_2 E_1)A,$$

this product corresponds to a sequence of elementary row operations. First E_1 acts on A, then E_2 acts on E_1A, and so on. To *undo* this sequence, we must first undo the last elementary row operation, performed by E_t. This is accomplished by E_t^{-1}. Thus we see that we should perform the sequence of operations given by

$$E_1^{-1}E_2^{-1}E_3^{-1} \cdots E_t^{-1}$$

to effect $(E_t \cdots E_3 E_2 E_1)^{-1}$.

COMPUTATION OF INVERSES

Let $A = (a_{ij})$ be an $n \times n$ matrix. To find A^{-1}, if it exists, we must find an $n \times n$ matrix $X = (x_{ij})$ such that $AX = I$, that is, such that

$$
\begin{pmatrix}
a_{11} & a_{12} & \cdots & a_{1n} \\
a_{21} & a_{22} & \cdots & a_{2n} \\
& & \vdots & \\
a_{n1} & a_{n2} & \cdots & a_{nn}
\end{pmatrix}
\begin{pmatrix}
x_{11} & x_{12} & \cdots & x_{1n} \\
x_{21} & x_{22} & \cdots & x_{2n} \\
& & \vdots & \\
x_{n1} & x_{n2} & \cdots & x_{nn}
\end{pmatrix}
=
\begin{pmatrix}
1 & 0 & \cdots & 0 \\
0 & 1 & \cdots & 0 \\
& & \vdots & \\
0 & 0 & \cdots & 1
\end{pmatrix}. \tag{3}
$$

The matrix equation (3) corresponds to n^2 linear equations in the n^2 unknowns x_{ij}; there is one linear equation for each of the n^2 positions in an $n \times n$ matrix. For example, equating the entries in the second-row and first-column position on each side of Eq. (3), we obtain the linear equation

$$a_{21}x_{11} + a_{22}x_{21} + \cdots + a_{2n}x_{n1} = 0.$$

Of these n^2 linear equations, n of them involve the n unknowns x_{i1} for $i = 1, 2,$ $\ldots, n$, and these equations are given by the column-vector equation

$$
A\begin{pmatrix} x_{11} \\ x_{21} \\ \vdots \\ x_{n1} \end{pmatrix} = \begin{pmatrix} 1 \\ 0 \\ \vdots \\ 0 \end{pmatrix}. \tag{4}
$$

There are also n equations involving the n unknowns x_{i2}, for $i = 1, 2, \ldots, n$, and so on. In addition to Eq. (4), we must solve

$$
A\begin{pmatrix} x_{12} \\ x_{22} \\ \vdots \\ x_{n2} \end{pmatrix} = \begin{pmatrix} 0 \\ 1 \\ \vdots \\ 0 \end{pmatrix}, \cdots, A\begin{pmatrix} x_{1n} \\ x_{2n} \\ \vdots \\ x_{nn} \end{pmatrix} = \begin{pmatrix} 0 \\ 0 \\ \vdots \\ 1 \end{pmatrix},
$$

where each system has the same coefficient matrix A. Our work in Section 1.4 shows that we should form the augmented matrix

$$
\left(
\begin{array}{cccc|cccc}
a_{11} & a_{12} & \cdots & a_{1n} & 1 & 0 & \cdots & 0 \\
a_{21} & a_{22} & \cdots & a_{2n} & 0 & 1 & \cdots & 0 \\
& & \vdots & & & & \vdots & \\
a_{n1} & a_{n2} & \cdots & a_{nn} & 0 & 0 & \cdots & 1
\end{array}
\right), \tag{5}
$$

which we abbreviate by $(A \mid I)$. The matrix A is to the left of the partition and the identity matrix I is to the right. We then perform a Gauss–Jordan reduction to solve the system. Remember from Theorem 1.5 that if the inverse of A exists, it is unique. Thus if the inverse exists we are in the unique-solution case in solving (5), and as stated in Theorem 1.3, we are able to reduce the coefficient matrix A to the identity matrix. This reduces (5) to

$$\begin{pmatrix} 1 & 0 & \cdots & 0 & | & d_{11} & d_{12} & \cdots & d_{1n} \\ 0 & 1 & \cdots & 0 & | & d_{21} & d_{22} & \cdots & d_{2n} \\ & & \vdots & & | & & \vdots & \\ 0 & 0 & \cdots & 1 & | & d_{n1} & d_{n2} & \cdots & d_{nn} \end{pmatrix}$$

and shows that $A^{-1} = (d_{ij})$. This is an efficient way to compute A^{-1}. We summarize the computation in the following box and the theory in Theorem 1.7.

Computation of A^{-1}

To find A^{-1} if it exists, we may proceed as follows:

Step 1 Form the partitioned matrix $(A \mid I)$.

Step 2 Apply the Gauss–Jordan method to reduce $(A \mid I)$ to $(I \mid D)$. If the reduction can be carried out, then $A^{-1} = D$. Otherwise, A^{-1} does not exist.

Theorem 1.7   Conditions for A^{-1} to Exist

The following conditions for an $n \times n$ matrix A are equivalent:

i) A is invertible.

ii) A is row equivalent to the identity matrix I.

iii) The system $A\mathbf{x} = \mathbf{b}$ has a solution for each n-component column vector $\mathbf{b}$.

iv) A can be expressed as a product of elementary matrices.

Proof   It only remains to show the equivalence of (iii) and (iv) with (i).

If A is invertible, then $A\mathbf{x} = \mathbf{b}$ clearly has the solution $\mathbf{x} = A^{-1}\mathbf{b}$, so (i) implies (iii). On the other hand, if $A\mathbf{x} = \mathbf{b}$ has a solution for each $\mathbf{b}$, then $AX = I$ has a solution; just regard the jth column vector of X as the solution of $A\mathbf{x} = \mathbf{e}_j$, where $\mathbf{e}_j$ is the jth column vector of the identity matrix.

Turning to the equivalence of (i) and (iv), we know that the matrix A is row equivalent to I if and only if there is a sequence of elementary matrices $E_1, E_2, \ldots, E_t$ such that $E_t \cdots E_2 E_1 A = I$, and this is the case if and only if A is expressible as a product $A = E_1^{-1} E_2^{-1} \cdots E_t^{-1}$ of elementary matrices. ∎

Example 2   For the matrix

$$A = \begin{pmatrix} 2 & 9 \\ 1 & 4 \end{pmatrix},$$

compute the inverse we presented at the beginning of this section.

Solution   Reducing the partitioned matrix, we have

$$\left(\begin{matrix} 2 & 9 \\ 1 & 4 \end{matrix}\ \middle|\ \begin{matrix} 1 & 0 \\ 0 & 1 \end{matrix}\right) \sim \left(\begin{matrix} 1 & 4 \\ 2 & 9 \end{matrix}\ \middle|\ \begin{matrix} 0 & 1 \\ 1 & 0 \end{matrix}\right)$$

$$\underset{R_2 = R_2 - 2R_1}{} \sim \left(\begin{matrix} 1 & 4 \\ 0 & 1 \end{matrix}\ \middle|\ \begin{matrix} 0 & 1 \\ 1 & -2 \end{matrix}\right)$$

$$\sim \left(\begin{matrix} 1 & 0 \\ 0 & 1 \end{matrix}\ \middle|\ \begin{matrix} -4 & 9 \\ 1 & -2 \end{matrix}\right).$$

Therefore,

$$A^{-1} = \begin{pmatrix} -4 & 9 \\ 1 & -2 \end{pmatrix}. \quad \triangleleft$$

Example 3   Using Example 2, express

$$\begin{pmatrix} 2 & 9 \\ 1 & 4 \end{pmatrix}$$

as a product of elementary matrices.

Solution   The steps we performed in Example 2 can be applied in sequence to the 2×2 identity matrix to generate elementary matrices:

$$E_1 = \begin{pmatrix} 0 & 1 \\ 1 & 0 \end{pmatrix} \qquad \textbf{Interchanges rows 1 and 2,}$$

$$E_2 = \begin{pmatrix} 1 & 0 \\ -2 & 1 \end{pmatrix} \qquad \textbf{adds } -2 \textbf{ times row 1 to row 2,}$$

$$E_3 = \begin{pmatrix} 1 & -4 \\ 0 & 1 \end{pmatrix} \qquad \textbf{adds } -4 \textbf{ times row 2 to row 1.}$$

Thus we see that $E_3 E_2 E_1 A = I$ so

$$A = E_1^{-1} E_2^{-1} E_3^{-1} = \begin{pmatrix} 0 & 1 \\ 1 & 0 \end{pmatrix}\begin{pmatrix} 1 & 0 \\ 2 & 1 \end{pmatrix}\begin{pmatrix} 1 & 4 \\ 0 & 1 \end{pmatrix}. \quad \triangleleft$$

Example 3 indicates that we can use the boxed rule below to express an invertible matrix A as a product of elementary matrices.

Expressing an Invertible Matrix _A_ as a Product of Elementary Matrices

Write in left-to-right order the inverses of the elementary matrices corresponding to successive row operations that reduce A to I.

Example 4 Determine whether the matrix

$$A = \begin{pmatrix} 1 & 3 & -2 \\ 2 & 5 & -3 \\ -3 & 2 & -4 \end{pmatrix}$$

is invertible, and find its inverse if it is.

Solution We have

$$\left(\begin{array}{ccc|ccc} 1 & 3 & -2 & 1 & 0 & 0 \\ 2 & 5 & -3 & 0 & 1 & 0 \\ -3 & 2 & -4 & 0 & 0 & 1 \end{array} \right) \sim \left(\begin{array}{ccc|ccc} 1 & 3 & -2 & 1 & 0 & 0 \\ 0 & -1 & 1 & -2 & 1 & 0 \\ 0 & 11 & -10 & 3 & 0 & 1 \end{array} \right)$$

$$\sim \left(\begin{array}{ccc|ccc} 1 & 0 & 1 & -5 & 3 & 0 \\ 0 & 1 & -1 & 2 & -1 & 0 \\ 0 & 0 & 1 & -19 & 11 & 1 \end{array} \right)$$

$$\sim \left(\begin{array}{ccc|ccc} 1 & 0 & 0 & 14 & -8 & -1 \\ 0 & 1 & 0 & -17 & 10 & 1 \\ 0 & 0 & 1 & -19 & 11 & 1 \end{array} \right).$$

Therefore, A is an invertible matrix, and

$$A^{-1} = \begin{pmatrix} 14 & -8 & -1 \\ -17 & 10 & 1 \\ -19 & 11 & 1 \end{pmatrix}. \quad \triangleleft$$

Example 5 Express the matrix A of Example 4 as a product of elementary matrices.

Solution From the box after Example 3 we see that we should write in left-to-right order the successive inverses of the elementary matrices corresponding to the row reduction of A in Example 4. We obtain

$$A = \begin{pmatrix} 1 & 0 & 0 \\ 2 & 1 & 0 \\ 0 & 0 & 1 \end{pmatrix} \begin{pmatrix} 1 & 0 & 0 \\ 0 & 1 & 0 \\ -3 & 0 & 1 \end{pmatrix} \begin{pmatrix} 1 & 0 & 0 \\ 0 & -1 & 0 \\ 0 & 0 & 1 \end{pmatrix} \begin{pmatrix} 1 & 3 & 0 \\ 0 & 1 & 0 \\ 0 & 0 & 1 \end{pmatrix}$$

$$\times \begin{pmatrix} 1 & 0 & 0 \\ 0 & 1 & 0 \\ 0 & 11 & 1 \end{pmatrix} \begin{pmatrix} 1 & 0 & 1 \\ 0 & 1 & 0 \\ 0 & 0 & 1 \end{pmatrix} \begin{pmatrix} 1 & 0 & 0 \\ 0 & 1 & -1 \\ 0 & 0 & 1 \end{pmatrix}. \quad \triangleleft$$

Example 6 Find the inverse of

$$A = \begin{pmatrix} 1 & 3 & 4 \\ -2 & -5 & -3 \\ 1 & 4 & 9 \end{pmatrix},$$

if the inverse exists.

Solution   We have

$$\begin{pmatrix} 1 & 3 & 4 & | & 1 & 0 & 0 \\ -2 & -5 & -3 & | & 0 & 1 & 0 \\ 1 & 4 & 9 & | & 0 & 0 & 1 \end{pmatrix} \sim \begin{pmatrix} 1 & 3 & 4 & | & 1 & 0 & 0 \\ 0 & 1 & 5 & | & 2 & 1 & 0 \\ 0 & 1 & 5 & | & -1 & 0 & 1 \end{pmatrix}$$

$$\sim \begin{pmatrix} 1 & 0 & -11 & | & -5 & -3 & 0 \\ 0 & 1 & 5 & | & 2 & 1 & 0 \\ 0 & 0 & 0 & | & -3 & -1 & 1 \end{pmatrix}.$$

We do not have a nonzero pivot in the 3rd row, 3rd column position, so we are not able to complete the reduction. By Theorem 1.7, the matrix is not invertible. ◁

Our consideration of the inverse of a matrix A arose from the problem of solving a system of equations $A\mathbf{x} = \mathbf{b}$ for a column vector $\mathbf{x}$ of n unknowns. The work involved in finding A^{-1} is essentially the use of the methods described in Section 1.4 for finding the solution of $A\mathbf{x} = \mathbf{b}$, so it is clear that no efficiency results in solving this problem by finding A^{-1}. If we have r systems of equations,

$$A\mathbf{x}_1 = \mathbf{b}_1, \, A\mathbf{x}_2 = \mathbf{b}_2, \ldots, A\mathbf{x}_r = \mathbf{b}_r,$$

all with the *same* invertible $n \times n$ coefficient matrix A, it might seem that for large r it would be efficient to solve all the systems by finding A^{-1} and computing the column vectors

$$\mathbf{x}_1 = A^{-1}\mathbf{b}_1, \, \mathbf{x}_2 = A^{-1}\mathbf{b}_2, \ldots, \mathbf{x}_r = A^{-1}\mathbf{b}_r.$$

Chapter 2 will show that using the Gauss method on the augmented matrix $(A \mid \mathbf{b}_1\mathbf{b}_2 \cdots \mathbf{b}_r)$ with back substitution remains more efficient. Thus inversion of a coefficient matrix is not a good way to solve a linear system. However, inverses of matrices are handy when manipulating matrix equations and proving theorems.

SUMMARY

1. Let A be a square matrix. A square matrix C such that $CA = AC = I$ is the inverse of A and is denoted by $C = A^{-1}$. If such an inverse A^{-1} of A exists, then A is said to be invertible. The inverse of an invertible matrix A is unique. A square matrix that has no inverse is called singular.

2. The inverse of a square matrix A exists if and only if A can be reduced to the identity matrix I by elementary row operations, or equivalently, if and only if A is a product of elementary matrices. In this case A is equal to the product, in left-to-right order, of the inverses of the successive elementary matrices corresponding to the sequence of row operations that reduce A to I.

3. To find A^{-1}, if it exists, form the partitioned matrix $(A \mid I)$ and apply the Gauss–Jordan method to reduce this matrix to $(I \mid D)$. If this can be done, then $A^{-1} = D$. Otherwise, A is not invertible.

4. The inverse of a product of invertible matrices is the product of the inverses in the reverse order.

EXERCISES

In Exercises 1–8, (a) find the inverse of the square matrix if it exists, and (b) express each invertible matrix as a product of elementary matrices.

1. $\begin{pmatrix} 1 & 1 \\ 0 & 1 \end{pmatrix}$
  2. $\begin{pmatrix} 3 & 6 \\ 3 & 8 \end{pmatrix}$
  3. $\begin{pmatrix} 3 & 6 \\ 4 & 8 \end{pmatrix}$
  4. $\begin{pmatrix} 6 & 7 \\ 8 & 9 \end{pmatrix}$

5. $\begin{pmatrix} 1 & 0 & 1 \\ 0 & 1 & 1 \\ 0 & 0 & -1 \end{pmatrix}$
  6. $\begin{pmatrix} 1 & 1 & -1 \\ 2 & 0 & 3 \\ -3 & 1 & -7 \end{pmatrix}$
  7. $\begin{pmatrix} 2 & 1 & 4 \\ 3 & 2 & 5 \\ 0 & -1 & 1 \end{pmatrix}$
  8. $\begin{pmatrix} -1 & 2 & 1 \\ 2 & -3 & 5 \\ 1 & 0 & 12 \end{pmatrix}$

In Exercises 9–12, find the inverse of the matrix, if it exists.

9. $\begin{pmatrix} 1 & 0 & 1 & -1 \\ 0 & -1 & -3 & 4 \\ 1 & 0 & -1 & 2 \\ -3 & 0 & 0 & -1 \end{pmatrix}$
  10. $\begin{pmatrix} 1 & -2 & 1 & 0 \\ -3 & 5 & 0 & 2 \\ 0 & 1 & 2 & -4 \\ -1 & 2 & 4 & -2 \end{pmatrix}$

11. $\begin{pmatrix} 1 & 0 & 0 & 0 & 0 & 0 \\ 0 & -1 & 0 & 0 & 0 & 0 \\ 0 & 0 & 2 & 0 & 0 & 0 \\ 0 & 0 & 0 & 3 & 0 & 0 \\ 0 & 0 & 0 & 0 & 4 & 0 \\ 0 & 0 & 0 & 0 & 0 & 5 \end{pmatrix}$
  12. $\begin{pmatrix} 0 & 0 & 0 & 0 & 0 & 6 \\ 0 & 0 & 0 & 0 & 5 & 0 \\ 0 & 0 & 0 & 4 & 0 & 0 \\ 0 & 0 & 3 & 0 & 0 & 0 \\ 0 & 2 & 0 & 0 & 0 & 0 \\ 1 & 0 & 0 & 0 & 0 & 0 \end{pmatrix}$

13. a) Show that the matrix

$$A = \begin{pmatrix} 2 & -3 \\ 5 & -7 \end{pmatrix}$$

is invertible, and find its inverse.

b) Use the result in (a) to find the solution of the system of equations

$$2x_1 - 3x_2 = 4$$
$$5x_1 - 7x_2 = -3.$$

14. Using the inverse of the matrix in Exercise 7, find the solution of the system of equations

$$2x_1 + x_2 + 4x_3 = 5$$
$$3x_1 + 2x_2 + 5x_3 = 3$$
$$- x_2 + x_3 = 8.$$

15. Show that if A is an invertible $n \times n$ matrix, then A^{-1} is invertible. Describe $(A^{-1})^{-1}$.

16. Show that if A is an invertible $n \times n$ matrix, then A^T is invertible. Describe $(A^T)^{-1}$ in terms of A^{-1}.

17. An $n \times n$ matrix A is **nilpotent** if $A^r = O$, the $n \times n$ zero matrix, for some integer r.
   a) Give an example of a nonzero nilpotent 2×2 matrix.
   b) Show that if A is an invertible $n \times n$ matrix, then A is not nilpotent.

18. Consider the 2×2 matrix
$$A = \begin{pmatrix} a & b \\ c & d \end{pmatrix},$$
and let $h = ad - bc$.
   a) Show that if $h \neq 0$, then
$$\begin{pmatrix} d/h & -b/h \\ -c/h & a/h \end{pmatrix}$$
is the inverse of A.
   b) Show that A is invertible if and only if $h \neq 0$.

Exercises 19–21 develop elementary column operations.

19. For each type of elementary matrix E, explain how E can be obtained from the identity matrix by operating on columns.

20. Let A be a square matrix and let E be an elementary matrix of the same size. Find the effect on A of multiplying A on the right by E. [*Hint:* Use Exercise 19.]

21. Let A be an invertible square matrix. Recall that $(BA)^{-1} = A^{-1}B^{-1}$ and use Exercise 20 to answer the following questions.
   a) If two rows of A are interchanged, how does the inverse of the resulting matrix compare with A^{-1}?
   b) Answer the question in part (a) if instead a row of A is multiplied by a nonzero scalar r.
   c) Answer the question in part (a) if instead the ith row of A is multiplied by a scalar r and added to the jth row.

In Exercises 22–25, use the available software program YUREDUCE, or similar software, to find the inverse of the matrix, if it exists. If a printer is available, make a copy of the results. Otherwise, copy down the answers to three significant figures.

22. $\begin{pmatrix} 3 & -1 & 2 \\ 1 & 2 & 1 \\ 0 & 3 & -4 \end{pmatrix}$

23. $\begin{pmatrix} -2 & 1 & 4 \\ 3 & 6 & 7 \\ 13 & 15 & -2 \end{pmatrix}$

24. $\begin{pmatrix} 2 & -1 & 3 & 4 \\ -5 & 2 & 0 & 11 \\ 12 & 13 & -6 & 8 \\ 18 & -10 & 3 & 0 \end{pmatrix}$

25. $\begin{pmatrix} 4 & -10 & 3 & 17 \\ 2 & 0 & -3 & 11 \\ 14 & 2 & 12 & -15 \\ 0 & -10 & 9 & -5 \end{pmatrix}$

In Exercises 26–31, follow the instructions for Exercises 22–25, but use the software program MATCOMP or similar software. Also check that $AA^{-1} = A^{-1}A = I$ for each matrix A whose inverse is found.

26. The matrix in Exercise 9

27. The matrix in Exercise 10

28. The matrix in Exercise 23

29. The matrix in Exercise 22

30. $\begin{pmatrix} 4 & 1 & -3 & 2 & 6 \\ 0 & 1 & 5 & 2 & 1 \\ 3 & 8 & -11 & 4 & 6 \\ 2 & 1 & -8 & 7 & 2 \\ 1 & 3 & -1 & 4 & 8 \end{pmatrix}$

31. $\begin{pmatrix} 2 & -1 & 0 & 1 & 6 \\ 3 & -1 & 2 & 4 & 6 \\ 0 & 1 & 3 & 4 & 8 \\ -1 & 1 & 1 & 1 & 8 \\ 3 & 1 & 4 & -11 & 10 \end{pmatrix}$

1.6
Solving a General Linear System

This section deals with linear systems in which the number of equations is not necessarily equal to the number of unknowns. That is, the coefficient matrix need not be square. We want to describe how to determine whether or not such a linear system has solutions, and how to find all solutions if any exist. The section concludes with a discussion of the structure of the solution set of a linear system.

REDUCTION TO ROW-ECHELON FORM

Consider a general linear system

$$
\begin{aligned}
a_{11}x_1 + a_{12}x_2 + \cdots + a_{1n}x_n &= b_1 \\
a_{21}x_1 + a_{22}x_2 + \cdots + a_{2n}x_n &= b_2 \\
&\vdots \\
a_{m1}x_1 + a_{m2}x_2 + \cdots + a_{mn}x_n &= b_m
\end{aligned}
\tag{1}
$$

consisting of m equations in n unknowns. To attempt to solve the system, we form the augmented matrix

$$
(A \mid \mathbf{b}) =
\begin{pmatrix}
a_{11} & a_{12} & \cdots & a_{1n} & b_1 \\
a_{21} & a_{22} & \cdots & a_{2n} & b_2 \\
& & \vdots & & \vdots \\
a_{m1} & a_{m2} & \cdots & a_{mn} & b_m
\end{pmatrix}
\tag{2}
$$

just as in the case $m = n$. In fact, computations will be very similar to those in Sections 1.4 and 1.5. We now use elementary row operations to transform $(A \mid \mathbf{b})$ so that the coefficient matrix A is reduced to row-echelon form, which we proceed to define.

Definition 1.10   Row-echelon Form

A matrix is in **row-echelon form** if it satisfies these two conditions:

1. All rows containing only zeros appear below rows with nonzero entries.
2. The first nonzero entry (the **pivot**, if there is one) in any row appears in a column to the right of the first nonzero entry in any preceding row.

Example 1   Determine whether each of the matrices

$$
A = \begin{pmatrix} 1 & 3 & 2 \\ 0 & 0 & 0 \\ 0 & 0 & 1 \end{pmatrix}, \qquad
B = \begin{pmatrix} 2 & 4 & 0 \\ 1 & 3 & 2 \\ 0 & 0 & 0 \end{pmatrix},
$$

$$C = \begin{pmatrix} 0 & -1 & 2 \\ 0 & 0 & 3 \\ 0 & 0 & 0 \\ 0 & 0 & 0 \end{pmatrix}, \qquad D = \begin{pmatrix} 1 & 3 & 2 & 5 \\ 0 & 0 & 1 & 3 \\ 0 & 0 & 0 & 1 \\ 0 & 0 & 0 & 0 \end{pmatrix}$$

is in row-echelon form.

Solution   Matrix A is not in row-echelon form since the second row, consisting of all zero entries, is not below the third row, which has a nonzero entry.

Matrix B is not in row-echelon form since the leading nonzero entry 1 in the second row does not appear in a column to the right of the leading nonzero entry 2 in the first row.

Matrix C is in row-echelon form since both conditions of Definition 1.10 are satisfied.

Matrix D satisfies both conditions, and is in row-echelon form.   ◁

Reduction to Row-Echelon Form

Step 1   If the first column of the matrix contains only zero entries, cross it off—mentally, that is. Continue in this fashion until the left column of the remaining matrix has a nonzero entry, or until the columns are exhausted.

Step 2   Use the Gauss reduction on this left column of the remaining matrix, creating a (nonzero) pivot at the top of the column and zeros below.

Step 3   Cross off this left column and the first row of the matrix, obtaining a smaller matrix. (See the shaded portion of the third matrix in the solution of Example 2.) Go back to Step 1 and continue the process with this smaller matrix until either no rows or no columns are left.

Matrix reduction is a basic technique. It is easy to see that any matrix can be reduced to row-echelon form by a sequence of elementary row operations. The steps of the reduction are shown in the box above. We illustrate reduction to row-echelon form with several examples. We choose in all our examples to make pivots 1, since this tends to reduce arithmetic errors in pencil-and-paper computations. However, row-echelon form is obtained using the same Gaussian elimination we studied for square systems in Section 1.4. Solutions of systems can be determined using back substitution after reduction, and it is not essential to make pivots 1.

Example 2   Reduce the matrix

$$\begin{pmatrix} 2 & -4 & 2 & -2 \\ 2 & -4 & 3 & -4 \\ 4 & -8 & 3 & -2 \\ 0 & 0 & -1 & 2 \end{pmatrix}$$

to row-echelon form.

Solution   We follow the steps in the box on page 66 and color the pivots of 1.

$$\begin{pmatrix} 2 & -4 & 2 & -2 \\ 2 & -4 & 3 & -4 \\ 4 & -8 & 3 & -2 \\ 0 & 0 & -1 & 2 \end{pmatrix} \sim \begin{pmatrix} 1 & -2 & 1 & -1 \\ 2 & -4 & 3 & -4 \\ 4 & -8 & 3 & -2 \\ 0 & 0 & -1 & 2 \end{pmatrix} \sim \begin{pmatrix} 1 & -2 & 1 & -1 \\ 0 & 0 & 1 & -2 \\ 0 & 0 & -1 & 2 \\ 0 & 0 & -1 & 2 \end{pmatrix}$$

$$\sim \begin{pmatrix} 1 & -2 & 1 & -1 \\ 0 & 0 & 1 & -2 \\ 0 & 0 & -1 & 2 \\ 0 & 0 & -1 & 2 \end{pmatrix} \sim \begin{pmatrix} 1 & -2 & 1 & -1 \\ 0 & 0 & 1 & -2 \\ 0 & 0 & 0 & 0 \\ 0 & 0 & 0 & 0 \end{pmatrix}$$

$$\sim \begin{pmatrix} 1 & -2 & 1 & -1 \\ 0 & 0 & 1 & -2 \\ 0 & 0 & 0 & 0 \\ 0 & 0 & 0 & 0 \end{pmatrix}$$

This last matrix is in row-echelon form.   ◁

Sometimes it is convenient to use a Gauss–Jordan reduction on a matrix A and obtain an echelon form that contains zeros *above* as well as below the pivots. This time, we will require that the pivots be 1. Such a form is called a **reduced row-echelon form** for A.

Obtaining Reduced Row-Echelon Form

Proceed as in the instructions for obtaining row-echelon form, but use the Gauss–Jordan method, which makes all pivots 1 and produces zeros above as well as below them.

Example 3   Find a reduced row-echelon form for the matrix

$$\begin{pmatrix} 1 & -3 & 1 & 5 \\ 3 & -8 & 2 & -1 \\ 1 & -1 & -2 & 7 \end{pmatrix}.$$

Solution   We have

$$\begin{pmatrix} 1 & -3 & 1 & 5 \\ 3 & -8 & 2 & -1 \\ 1 & -1 & -2 & 7 \end{pmatrix} \sim \begin{pmatrix} 1 & -3 & 1 & 5 \\ 0 & 1 & -1 & -16 \\ 0 & 2 & -3 & 2 \end{pmatrix}$$

$$\sim \begin{pmatrix} 1 & 0 & -2 & -43 \\ 0 & 1 & -1 & -16 \\ 0 & 0 & -1 & 34 \end{pmatrix}$$

$$\sim \begin{pmatrix} 1 & 0 & 0 & -111 \\ 0 & 1 & 0 & -50 \\ 0 & 0 & 1 & -34 \end{pmatrix}. \quad \triangleleft$$

A matrix A may have several different row-echelon forms; for example, interchanging two rows before the start of the reduction may produce a different row-echelon form than would be obtained if the rows were not interchanged. However, it can be shown that the *reduced* row-echelon form of a matrix, having all pivots 1, is unique. This fact, which we will not prove, will not be used in the rest of the text.

SOLVING A GENERAL LINEAR SYSTEM

Consider the general linear system (1) of m equations in n unknowns, which we write in matrix form as $A\mathbf{x} = \mathbf{b}$. To solve the system, we form the partitioned matrix $(A \mid \mathbf{b})$ and use the Gauss or Gauss–Jordan technique on the entire partitioned matrix to reduce the coefficient matrix A at the left of the partition to a row-echelon form or reduced row-echelon form. Any solutions of the system are then easy to find. This is best illustrated by several examples.

Example 4   Find all solutions of a linear system whose partitioned matrix can be row-reduced to

$$\begin{pmatrix} 1 & -3 & 5 & \mid & 3 \\ 0 & 1 & 2 & \mid & 2 \\ 0 & 0 & 0 & \mid & -1 \end{pmatrix}.$$

Solution   The equation corresponding to the last row of this partitioned matrix is

$$0x_1 + 0x_2 + 0x_3 = -1.$$

This equation has no solution, since the left side is zero for any values of the variables while the right side is -1. Thus the linear system has no solution. $\triangleleft$

Example 4 typifies the no-solution case for linear systems.

No-Solution Case

A linear system has no solution precisely when a row-echelon form reduction of the corresponding partitioned matrix contains a row with only zero entries to the left of the partition and a nonzero entry to the right.

Definition 1.11   Consistency of a Linear System

A linear system having no solution is **inconsistent.** Otherwise, the linear system is said to be **consistent.**

We now illustrate the many-solutions case.

Example 5   Find all solutions of a linear system whose partitioned matrix can be row-reduced to

$$\left(\begin{array}{ccccc|c} 1 & -3 & 0 & 5 & 0 & 4 \\ 0 & 0 & 1 & 2 & 0 & -7 \\ 0 & 0 & 0 & 0 & 1 & 1 \\ 0 & 0 & 0 & 0 & 0 & 0 \end{array}\right).$$

Solution   The reduced linear system corresponding to this partitioned matrix is

$$\begin{aligned} x_1 - 3x_2 \quad\;\; + 5x_4 \quad &= \quad 4 \\ x_3 + 2x_4 \quad &= -7 \\ x_5 &= \quad 1. \end{aligned} \tag{3}$$

We solve each equation for the variable corresponding to the colored pivot in the matrix, and obtain

$$\begin{aligned} x_1 &= 3x_2 - 5x_4 + 4 \\ x_3 &= -2x_4 - 7 \\ x_5 &= 1. \end{aligned} \tag{4}$$

We see that we can assign any values we please to x_2 and x_4, and use the system (4) to determine corresponding values of x_1, x_3, and x_5. Thus the system has an infinite number of solutions. We can describe all solutions by the vector equation

$$\mathbf{x} = \begin{pmatrix} x_1 \\ x_2 \\ x_3 \\ x_4 \\ x_5 \end{pmatrix} = \begin{pmatrix} 3r - 5s + 4 \\ r \\ -2s - 7 \\ s \\ 1 \end{pmatrix} \quad \text{for any scalars } r \text{ and } s.$$

For example, we obtain the **particular** solutions

$$\begin{pmatrix} 2 \\ 1 \\ -9 \\ 1 \\ 1 \end{pmatrix} \text{ for } r = s = 1 \quad \text{and} \quad \begin{pmatrix} 25 \\ 2 \\ -1 \\ -3 \\ 1 \end{pmatrix} \text{ for } r = 2, s = -3. \quad \triangleleft$$

Example 5 is typical of the many-solutions case. The echelon form of the coefficient matrix in Example 5 did not contain (nonzero) pivots in the 2nd column or in the 4th column. We take the variables x_2 and x_4 to be **free variables;** the **bound variables** x_1, x_3, x_5 can be computed when any values are assigned to the free variables.

Many-Solutions Case

A linear system has many solutions if these two conditions hold:

1. The system is consistent.
2. At least one column in an echelon-form reduction does not contain a (nonzero) pivot. The free variables corresponding to such columns may be assigned any values and the resulting equations solved for the bound variables.

The only other situation is the unique-solution case, which is described in the following box.

Unique-Solution Case

A linear system has a unique solution if these two conditions hold:

1. The system is consistent.
2. Every column in an echelon-form reduction contains a (nonzero) pivot.

Note that for a row-echelon reduction of a *square* matrix, every column has a (nonzero) pivot if and only if every row has a (nonzero) pivot. This gives us at once Theorem 1.3 in Section 1.4:

> A square linear system $A\mathbf{x} = \mathbf{b}$ has a unique solution if and only if the coefficient matrix A can be row-reduced to the identity matrix I.

A linear system need not be square to have a unique solution, as the following example shows.

Example 6   Give an example of a partitioned matrix in reduced-echelon form that corresponds to a linear system with a unique solution and a 4×3 coefficient matrix.

Solution   The matrix

$$\begin{pmatrix} 1 & 0 & 0 & | & -3 \\ 0 & 1 & 0 & | & 2 \\ 0 & 0 & 1 & | & 5 \\ 0 & 0 & 0 & | & 0 \end{pmatrix}$$

fills the requirements. ◁

Example 7   Give an example of a reduced partitioned matrix corresponding to a linear system with more equations than unknowns, but having an infinite number of solutions.

Solution   The matrix

$$\begin{pmatrix} 1 & -1 & 2 & | & 5 \\ 0 & 1 & 3 & | & -2 \\ 0 & 0 & 0 & | & 0 \\ 0 & 0 & 0 & | & 0 \\ 0 & 0 & 0 & | & 0 \end{pmatrix}$$

meets the conditions. We easily find that all solutions are described by the equations

$$x_1 = -5r + 3$$
$$x_2 = -3r - 2$$
$$x_3 = r$$

for any scalar r. ◁

Examples 6 and 7 are algebraic illustrations of the geometric considerations at the beginning of Section 1.3. Example 6 corresponds to having four planes through a single point. Example 7 corresponds to having five planes, all containing a single line. However, it is impossible to have just two planes with only one point in common. Our next example generalizes this.

Example 8   Show that one cannot have a unique solution to a linear system with fewer equations than unknowns.

Solution   If a linear system $A\mathbf{x} = \mathbf{b}$ has a unique solution, then a row-echelon reduction of A has a (nonzero) pivot in every column. But in row-echelon form, no two pivots are in the same row, so there must be at least as many rows as columns. That is, the linear system must have at least as many equations as unknowns. ◁

Example 8 has a nice application to a **homogeneous linear system** $A\mathbf{x} = \mathbf{0}$, where the column vector of constants has all components zero. Clearly $\mathbf{x} = \mathbf{0}$, the zero vector, is always a solution of such a homogeneous system. This zero vector is called the **trivial solution.** Example 8 shows that if a homogeneous system has fewer equations than unknowns, there must be an additional **nontrivial** solution vector having some nonzero components.

In the case of a square homogeneous system, we know that there is only one solution (namely, the trivial solution) if and only if the matrix A can be row-reduced to the identity matrix. This happens if and only if A is invertible (Theorem 1.7). Thus a square homogeneous system has a nontrivial solution if and only if its coefficient matrix is *not* invertible, that is, the coefficient matrix must be singular.

We summarize these useful facts in a theorem for easy reference.

Theorem 1.8   Existence of a Nontrivial Solution

1. A homogeneous linear system $A\mathbf{x} = \mathbf{0}$ having fewer equations than unknowns has a nontrivial solution, that is, a solution other than the zero vector.

2. Assume that the number of equations is equal to the number of unknowns. Then the homogeneous system $A\mathbf{x} = \mathbf{0}$ has a nontrivial solution if and only if the coefficient matrix A is singular.

As our work below will show, every *consistent* linear system with fewer equations than unknowns actually has an infinite number of solutions. In particular, such a homogeneous system has an infinite number of nontrivial solutions.

THE STRUCTURE OF SOLUTION SETS IN THE MANY-SOLUTIONS CASE

A linear system is said to be **underdetermined** if there is an infinite number of solutions. Our work in Chapter 3 will show that the number of free variables in the solution set of an underdetermined system $A\mathbf{x} = \mathbf{b}$ depends only on the system, and not on the way in which the matrix A is reduced to row-echelon form. This seems intuitively reasonable. We give a more detailed description of the solution set.

Let $A\mathbf{x} = \mathbf{b}$ be an underdetermined linear system, so there is an infinite number of solutions. Let $\mathbf{0}$ be the column vector with the same length as $\mathbf{b}$ but with all components zero. The system $A\mathbf{x} = \mathbf{0}$ is called the **homogeneous system corresponding to the system** $A\mathbf{x} = \mathbf{b}$. This homogeneous system is also underdetermined with the same number of free variables, for the free variables depend only on the row-echelon form derived from the coefficient matrix A.

Theorem 1.9 **Structure of Solution Sets**

Let $A\mathbf{x} = \mathbf{b}$ be a linear system.

1. If $\mathbf{h}$ and $\mathbf{h}'$ are solutions of the homogeneous system $A\mathbf{x} = \mathbf{0}$, then so is $r\mathbf{h} + s\mathbf{h}'$ for any scalars r and s.

2. If $\mathbf{p}$ is any *particular* solution of $A\mathbf{x} = \mathbf{b}$ and $\mathbf{h}$ is a solution of the corresponding homogeneous system $A\mathbf{x} = \mathbf{0}$, then $\mathbf{p} + \mathbf{h}$ is a solution of $A\mathbf{x} = \mathbf{b}$. Moreover, every solution of $A\mathbf{x} = \mathbf{b}$ has this form.

Proof Let $\mathbf{h}$ and $\mathbf{h}'$ be solutions of $A\mathbf{x} = \mathbf{0}$, so that $A\mathbf{h} = \mathbf{0}$ and $A\mathbf{h}' = \mathbf{0}$. By properties of matrix arithmetic, we have

$$
\begin{aligned}
A(r\mathbf{h} + s\mathbf{h}') &= A(r\mathbf{h}) + A(s\mathbf{h}') \\
&= r(A\mathbf{h}) + s(A\mathbf{h}') \\
&= r\mathbf{0} + s\mathbf{0} = \mathbf{0}
\end{aligned}
$$

for any scalars r and s. This shows that $r\mathbf{h} + s\mathbf{h}'$ is also a solution of $A\mathbf{x} = \mathbf{0}$.

Turning to part (2), let $\mathbf{p}$ be a solution of $A\mathbf{x} = \mathbf{b}$, so $A\mathbf{p} = \mathbf{b}$, and let $\mathbf{h}$ be a solution of $A\mathbf{x} = \mathbf{0}$, so $A\mathbf{h} = \mathbf{0}$. Then

$$A(\mathbf{p} + \mathbf{h}) = A\mathbf{p} + A\mathbf{h} = \mathbf{b} + \mathbf{0} = \mathbf{b}.$$

Finally, if $\mathbf{p}'$ is any other solution of $A\mathbf{x} = \mathbf{b}$, then

$$A(\mathbf{p}' - \mathbf{p}) = A\mathbf{p}' - A\mathbf{p} = \mathbf{b} - \mathbf{b} = \mathbf{0}$$

so $\mathbf{p}' - \mathbf{p}$ is a solution $\mathbf{h}$ of $A\mathbf{x} = \mathbf{0}$. From $\mathbf{p}' - \mathbf{p} = \mathbf{h}$ it follows that $\mathbf{p}' = \mathbf{p} + \mathbf{h}$. This completes the proof of (2). ■

Note how easy it was to write down the proof of Theorem 1.9 using matrix notation. What a chore it would have been to write it all out in the notation (1) for a linear system!

We conclude with a specific illustration of Theorem 1.9.

Example 9 Illustrate Theorem 1.9 for the linear system $A\mathbf{x} = \mathbf{b}$ given by

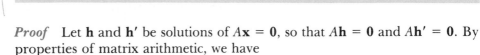

$$
\begin{aligned}
x_1 - 2x_2 + x_3 - x_4 &= 4 \\
2x_1 - 3x_2 + 2x_3 - 3x_4 &= -1 \\
3x_1 - 5x_2 + 3x_3 - 4x_4 &= 3 \\
-x_1 + x_2 - x_3 + 2x_4 &= 5.
\end{aligned}
$$

Solution We must solve both $A\mathbf{x} = \mathbf{0}$ and $A\mathbf{x} = \mathbf{b}$. We reduce the partitioned matrix $(A \mid \mathbf{0}\ \mathbf{b})$ to transform A into reduced row-echelon form, although we will see in a moment that we do not really need to include the column vector $\mathbf{0}$. We have

$$
\begin{pmatrix}
1 & -2 & 1 & -1 & | & 0 & 4 \\
2 & -3 & 2 & -3 & | & 0 & -1 \\
3 & -5 & 3 & -4 & | & 0 & 3 \\
-1 & 1 & -1 & 2 & | & 0 & 5
\end{pmatrix}
\sim
\begin{pmatrix}
1 & -2 & 1 & -1 & | & 0 & 4 \\
0 & 1 & 0 & -1 & | & 0 & -9 \\
0 & 1 & 0 & -1 & | & 0 & -9 \\
0 & -1 & 0 & 1 & | & 0 & 9
\end{pmatrix}
$$

$$
\sim
\begin{pmatrix}
1 & 0 & 1 & -3 & | & 0 & -14 \\
0 & 1 & 0 & -1 & | & 0 & -9 \\
0 & 0 & 0 & 0 & | & 0 & 0 \\
0 & 0 & 0 & 0 & | & 0 & 0
\end{pmatrix}
$$

The reduction is complete. We see why we didn't really need to insert the column vector **0** in the partitioned matrix; it never changes.

From the reduced matrix, we see that the homogeneous system $A\mathbf{x} = \mathbf{0}$ has solution set described by

$$
\mathbf{x} =
\begin{pmatrix} x_1 \\ x_2 \\ x_3 \\ x_4 \end{pmatrix}
=
\begin{pmatrix} -r + 3s \\ s \\ r \\ s \end{pmatrix}
\tag{5}
$$

for any scalars r and s. On the other hand, the solution set of $A\mathbf{x} = \mathbf{b}$ can be described by

$$
\mathbf{x} =
\begin{pmatrix} x_1 \\ x_2 \\ x_3 \\ x_4 \end{pmatrix}
=
\begin{pmatrix} -r + 3s - 14 \\ s - 9 \\ r \\ s \end{pmatrix}
=
\begin{pmatrix} -14 \\ -9 \\ 0 \\ 0 \end{pmatrix}
+
\begin{pmatrix} -r + 3s \\ s \\ r \\ s \end{pmatrix}.
\tag{6}
$$

Equation (6) expresses the **general solution x** of the system $A\mathbf{x} = \mathbf{b}$ as the sum of a particular solution, found by setting $r = s = 0$, and the general solution (5) of $A\mathbf{x} = \mathbf{0}$. This illustrates part (2) of Theorem 1.9. ◁

SUMMARY

1. A matrix is in row-echelon form if:

   a) All rows containing only zero entries are grouped together at the bottom of the matrix.

   b) The leading nonzero element in any row after the first one is in a column to the right of the leading nonzero element in the preceding row.

2. A matrix is in reduced row-echelon form if it is in row-echelon form and if, in addition, each pivot is 1 and is the only nonzero element in its column. Every matrix can be put into reduced row-echelon form using the Gauss–Jordan method, and this form is unique.

3. A linear system $A\mathbf{x} = \mathbf{b}$ has no solution if after $(A \mid \mathbf{b})$ is row-reduced so that A is transformed into row-echelon form, there is a row with only zero entries to the left of the partition, while the entry in that row to the right of the partition is nonzero. The linear system is then *inconsistent*.

4. If $A\mathbf{x} = \mathbf{b}$ is a consistent linear system and if a row-echelon form of A has at least one column containing no (nonzero) pivot, then the system is underdetermined. That is, it has an infinite number of solutions. The free variables corresponding to the columns containing no pivots may be assigned any values, and the reduced linear system may then be solved for the remaining variables.

5. A square system $A\mathbf{x} = \mathbf{b}$ has a unique solution if and only if the coefficient matrix A can be row-reduced to the identity matrix.

6. The solutions of any consistent linear system $A\mathbf{x} = \mathbf{b}$ are precisely the vectors $\mathbf{p} + \mathbf{h}$, where $\mathbf{p}$ is any one particular solution of $A\mathbf{x} = \mathbf{b}$ and $\mathbf{h}$ varies through the solution set of the homogeneous system $A\mathbf{x} = \mathbf{0}$.

7. Sums and scalar multiples of solutions of the homogeneous system $A\mathbf{x} = \mathbf{0}$ are also solutions of $A\mathbf{x} = \mathbf{0}$.

8. A homogeneous system has nontrivial solutions if the number of unknowns exceeds the number of equations.

9. A square homogeneous system has a nontrivial solution if and only if its coefficient matrix is singular.

EXERCISES

In Exercises 1–6, reduce the matrix to (a) row-echelon form and (b) reduced row-echelon form. Answers to (a) are not unique so your answer may differ from the one at the back of the text.

1. $\begin{pmatrix} 2 & 1 & 4 \\ 1 & 3 & 2 \\ 3 & -1 & 6 \end{pmatrix}$

2. $\begin{pmatrix} 2 & 4 & -2 \\ 4 & 8 & 3 \\ -1 & -3 & 0 \end{pmatrix}$

3. $\begin{pmatrix} 0 & 2 & -1 & 3 \\ -1 & 1 & 2 & 0 \\ 1 & 1 & -3 & 3 \\ 1 & 5 & 5 & 9 \end{pmatrix}$

4. $\begin{pmatrix} 0 & 0 & 3 & -2 \\ 0 & 0 & 1 & 2 \\ 1 & 3 & 2 & -4 \end{pmatrix}$

5. $\begin{pmatrix} -1 & 3 & 0 & 1 & 4 \\ 1 & -3 & 0 & 0 & -1 \\ 2 & -6 & 2 & 4 & 0 \\ 0 & 0 & 1 & 3 & -4 \end{pmatrix}$

6. $\begin{pmatrix} 0 & 0 & 1 & 2 & -1 & 4 \\ 0 & 0 & 0 & 1 & -1 & 3 \\ 2 & 4 & -1 & 3 & 2 & -1 \end{pmatrix}$

In Exercises 7–12, describe all solutions of a linear system whose corresponding partitioned matrix can be row-reduced to the given matrix. If requested, also give the indicated particular solution, if it exists.

7. $\begin{pmatrix} 1 & -1 & 2 & \mid & 3 \\ 0 & 1 & 4 & \mid & 2 \end{pmatrix}$,

solution with $x_3 = 2$

8. $\begin{pmatrix} 1 & 2 & 3 & 0 & \mid & 3 \\ 0 & 0 & 1 & 2 & \mid & -1 \\ 0 & 0 & 0 & 0 & \mid & 1 \end{pmatrix}$

9. $\begin{pmatrix} 1 & 0 & 2 & 0 & | & 1 \\ 0 & 1 & 1 & 3 & | & -2 \\ 0 & 0 & 0 & 0 & | & 0 \end{pmatrix}$,

solution with $x_3 = 3$, $x_4 = -2$

10. $\begin{pmatrix} 1 & 1 & 0 & 3 & 0 & | & -4 \\ 0 & 0 & 1 & -1 & 0 & | & 0 \\ 0 & 0 & 0 & 0 & 1 & | & -2 \\ 0 & 0 & 0 & 0 & 0 & | & 0 \end{pmatrix}$,

solution with $x_2 = 2$, $x_4 = 1$

11. $\begin{pmatrix} 1 & 0 & 0 & 0 & | & 0 \\ 0 & 1 & 0 & 0 & | & 2 \\ 0 & 0 & 1 & 0 & | & -5 \\ 0 & 0 & 0 & 1 & | & 2 \\ 0 & 0 & 0 & 0 & | & 0 \end{pmatrix}$

12. $\begin{pmatrix} 1 & -1 & 2 & 0 & 3 & | & 1 \\ 0 & 0 & 0 & 1 & 4 & | & 2 \\ 0 & 0 & 0 & 0 & 0 & | & -1 \\ 0 & 0 & 0 & 0 & 0 & | & 0 \end{pmatrix}$

In Exercises 13–18, describe all solutions of the linear system.

13. $x_1 - 2x_2 = 3$
$3x_1 - x_2 = 14$

14. $x_1 - 3x_2 + x_3 = 2$
$3x_1 - 8x_2 + 2x_3 = 5$
$3x_1 - 7x_2 + x_3 = 1$

15. $x_1 + 4x_2 - 2x_3 = 4$
$2x_1 + 7x_2 - x_3 = -2$
$2x_1 + 9x_2 - 7x_3 = 1$

16. $x_1 - 3x_2 + 2x_3 - x_4 = 8$
$3x_1 - 7x_2 + x_4 = 0$

17. $x_1 - 2x_3 + x_4 = 6$
$2x_1 - x_2 + x_3 - 3x_4 = 0$
$9x_1 - 3x_2 - x_3 - 7x_4 = 4$

18. $x_1 + 2x_2 - 3x_3 + x_4 = 2$
$3x_1 + 6x_2 - 8x_3 - 2x_4 = 1$

In Exercises 19–27, describe all possible values for the unknowns x_i so that the matrix equation is valid.

19. $2(x_1, x_2) - (4, 7) = (-2, 11)$

20. $4(x_1, x_2) + 2(x_1, 3) = (-6, 18)$

21. $(x_1, x_2)\begin{pmatrix} 1 \\ -3 \end{pmatrix} = (2)$

22. $(x_1, x_2)\begin{pmatrix} 3 & -1 \\ 2 & 4 \end{pmatrix} = (0, -14)$

23. $\begin{pmatrix} 1 & -5 \\ 3 & 2 \end{pmatrix}\begin{pmatrix} x_1 \\ x_2 \end{pmatrix} = \begin{pmatrix} 13 \\ 5 \end{pmatrix}$

24. $\begin{pmatrix} x_1 & x_2 \\ x_3 & x_4 \end{pmatrix}\begin{pmatrix} 1 & 1 \\ 1 & 0 \end{pmatrix} = \begin{pmatrix} 0 & 1 \\ 3 & 1 \end{pmatrix}$

25. $(x_1, x_2)\begin{pmatrix} 3 & 0 & 4 \\ 2 & 1 & -1 \end{pmatrix} = (3, 3, -7)$

26. $\begin{pmatrix} x_1 & x_2 \\ x_3 & x_4 \end{pmatrix}\begin{pmatrix} 3 & 5 \\ 2 & 3 \end{pmatrix} = \begin{pmatrix} 1 & 0 \\ 0 & 1 \end{pmatrix}$

27. $(x_1, x_2, x_3)\begin{pmatrix} 2 & -1 \\ 3 & 0 \\ -1 & 4 \end{pmatrix} = (5, -1)$

█ *In Exercises 28 and 29, use either MATCOMP or YUREDUCE, or similar software.*

28. Find all solutions of the linear system

$$3x_1 + 4x_2 - 5x_3 + x_4 + 5x_5 - 2x_6 = 24$$
$$4x_1 + 3x_2 - 23x_3 + x_4 - 4x_6 = -9$$
$$5x_1 + 6x_2 - 13x_3 + 2x_4 + 5x_5 - 8x_6 = 21$$
$$4x_1 + 4x_2 - 16x_3 + x_4 + 5x_5 + 3x_6 = 11.$$

29. Solve these two systems of six equations with a single run of a program. Write down a descr
of all solutions.

$$x_1 + x_2 - x_3 - 2x_4 + 5x_5 + x_6 + 11x_7 = 9, \quad 15$$
$$2x_1 + 3x_2 - x_3 + 2x_4 + 5x_5 - 2x_6 + 14x_7 = -1, \quad 2$$
$$4x_1 - 3x_2 - 5x_3 - 26x_4 + 25x_5 + x_6 - 5x_7 = 12, \quad 17$$
$$-2x_1 + 3x_2 - x_3 + 2x_4 + 5x_5 + 5x_6 + 35x_7 = -4, \quad 1$$
$$- 2x_2 + x_3 - 5x_5 - 4x_6 - 26x_7 = -16, \quad 41$$
$$2x_1 - x_2 + 3x_3 + 10x_4 - 15x_5 - 12x_7 = 57, \quad 53$$

1.7

Application to Population Distribution

Consider situations in which people are split into two or more categories. For example, we might split the citizens of the United States according to income into categories of

> poor,   middle income,   rich.

We might split the inhabitants of North America into categories according to the climate in which they live:

> hot,   temperate,   cold.

As terminology, we will speak of a *population* split into *states*. In the two illustrations above, the populations and states are given by the following.

Population	States
Citizens of the U.S.	poor, middle income, rich
People in North America	hot, temperate, cold

Our *populations* will often consist of people, but this is not essential. For example, at any moment one can classify the population of cars as operational or not operational.

We are interested in how the distribution of a population between states may change over a period of time. Matrices and their multiplication can play an important role in such considerations.

TRANSITION MATRICES

The tendency of a population to move between n states can sometimes be described using an $n \times n$ matrix. Consider a population distributed between $n = 3$ states, which we call state 1, state 2, and state 3. Suppose we know the *proportion* t_{ij} of the population of state j, which moves to state i over a given fixed time period. Note that the direction of movement from state j to state i is the *right-to-left order* of the subscripts in t_{ij}. The matrix $T = (t_{ij})$ is called a **transition matrix**.

Example 1   Let the population of a country be classified according to income as

> **State 1:** poor,
> **State 2:** middle income,
> **State 3:** rich.

Suppose that over each 20-year period (about one generation), we have the following data for people and their offspring:

> of the poor people, 19% become middle income and 1% rich;
> of the middle income people, 15% become poor and 10% rich;
> of the rich people, 5% become poor and 30% middle income.

Give the transition matrix describing these data.

Solution   The entry t_{ij} in the transition matrix T represents the *proportion* of the population moving from state j to state i, not the percentage. Thus the data that 19% of the poor (state 1) will become middle income (state 2) mean that we should take $t_{21} = .19$. Similarly, since 1% of the people in state 1 move to state 3 (rich), we should take $t_{31} = .01$. Now t_{11} represents the proportion of the poor people who remain poor at the end of 20 years. Since this is 80%, we should take $t_{11} = .80$. Continuing in this fashion starting in state 2 and then in state 3, we obtain for the matrix

$$
\begin{array}{ccc}
\text{poor} & \text{mid} & \text{rich}
\end{array}
$$
$$
T = \begin{pmatrix} .80 & .15 & .05 \\ .19 & .75 & .30 \\ .01 & .10 & .65 \end{pmatrix} \begin{array}{l} \text{poor} \\ \text{mid} \\ \text{rich} \end{array} .
$$

We have labeled the columns and rows with the names of the states. Note that an entry of the matrix gives the proportion of the population in the state *above* the entry that moves to the state at the *right* of the entry during one 20-year period. ◁

Note in Example 1 that the sum of the entries in each column of T is 1 since the sum reflects the movement of the *entire* population for the state listed at the top of the column.

In Example 1, suppose that the proportions of the entire population that fall into the various states at the *start* of a time period are given in the column vector

$$
\mathbf{p} = \begin{pmatrix} p_1 \\ p_2 \\ p_3 \end{pmatrix}.
$$

For example, we would have

$$
\mathbf{p} = \begin{pmatrix} \frac{1}{3} \\ \frac{1}{3} \\ \frac{1}{3} \end{pmatrix}
$$

if the whole population were initially equally divided among the states. ⌐
entries in such a **population distribution vector p** must be nonnegative
have sum 1.

Let us find the proportion of the entire population that is in state 1 *after* one
time period of 20 years, knowing that initially the proportion in state 1 is p_1.
The proportion of the state-1 population that remains in state 1 is t_{11}. This
gives a contribution of $t_{11}p_1$ to the proportion of the entire population that will
be found in state 1 after the end of the 20 years. Of course, we also get
contributions to state 1 at the end of 20 years from states 2 and 3. These two
states contribute proportions $t_{12}p_2$ and $t_{13}p_3$ of the entire population to state 1.
Thus after 20 years, the proportion in state 1 is

$$t_{11}p_1 + t_{12}p_2 + t_{13}p_3.$$

Note that this is precisely the first entry in the column vector given by the
product

$$T\mathbf{p} = \begin{pmatrix} t_{11} & t_{12} & t_{13} \\ t_{21} & t_{22} & t_{23} \\ t_{31} & t_{32} & t_{33} \end{pmatrix} \begin{pmatrix} p_1 \\ p_2 \\ p_3 \end{pmatrix}.$$

In a similar fashion, we find that the second and third components of $T\mathbf{p}$ give
the proportions of population in state 2 and state 3 after one time period.

> For an initial population distribution vector **p** and transition
> matrix T, the product vector $T\mathbf{p}$ is the population distribution
> vector after one time period.

MARKOV CHAINS

In Example 1, we found the transition matrix governing the flow of a population between three states over a period of 20 years. Suppose the same transition matrix is valid over the next 20-year period, and for the next 20 years
after that, and so on. That is, suppose there is a sequence or *chain* of 20-year
periods over which the transition matrix is valid. Such a situation is called a
Markov chain. Let us give a formal definition of a transition matrix for a
Markov chain.

Definition 1.12   Transition Matrix

A **transition matrix** for an n-state Markov chain is an $n \times n$ matrix with all
entries nonnegative and having the additional property that the sum of
the entries in each column is 1.

Markov chains arise naturally in biology, psychology, economics, and many
other sciences. Thus they are an important application of linear algebra and of
probability. The entry t_{ij} in a transition matrix T is known as the **probability** of
moving from state j to state i over one time period.

Example 2   Show that the matrix

$$T = \begin{pmatrix} 0 & 0 & 1 \\ 1 & 0 & 0 \\ 0 & 1 & 0 \end{pmatrix}$$

is a transition matrix for a three-state Markov chain, and explain the significance of the zeros and the ones.

Solution   The entries are all nonnegative, and the sum of the entries in each column is 1. Thus the matrix is a transition matrix for a Markov chain.

At least for finite populations, a transition probability $t_{ij} = 0$ means that there is no movement from state j to state i over the time period. That is, transition from state j to state i over the time period is impossible. On the other hand, if $t_{ij} = 1$, then the entire population of state j moves to state i over the time period. That is, transition from state j to state i in the time period is certain.

For the given matrix, we see that over one time period, the entire population of state 1 moves to state 2, the entire population of state 2 moves to state 3, and the entire population of state 3 moves to state 1.   ◁

Let T be the transition matrix over a time period, say 20 years, in a Markov chain. We can form a new Markov chain by looking at the flow of the population over a time period twice as long, that is, over 40 years. Let us see the relationship of the transition matrix for the 40-year time period to the one for the 20-year time period.

We might guess that the transition matrix for 40 years is T^2, for if $\mathbf{p}$ is the initial population distribution vector, then $T\mathbf{p}$ is the population distribution vector after the first time period. Thus $T(T\mathbf{p}) = T^2\mathbf{p}$ is the distribution vector after two periods, which suggests that the two-period transition matrix is T^2. Really, this only shows that $T^2\mathbf{p}$ gives the correct answer for the population distribution vector. We illustrate in the context of an example that T^2 is indeed the correct transition matrix.

MARKOV CHAINS are named for the Russian mathematician Andrei Andreevich Markov (1856–1922), who first defined them in a paper of 1906 dealing with the Law of Large Numbers and subsequently proved many of the standard results about them. His interest in these sequences stemmed from the needs of probability theory; Markov never dealt with their applications to the sciences. The only real examples he used were from literary texts, where the two possible states were vowels and consonants. To illustrate his results, he did a statistical study of the alternation of vowels and consonants in Pushkin's *Eugene Onegin*.

Andrei Markov taught at St. Petersburg University from 1880 to 1905, when he retired to make room for younger mathematicians. Besides his work in probability, he contributed to such fields as number theory, continued fractions, and approximation theory. He was an active participant in the liberal movement in the pre–World War I era in Russia; on many occasions he made public criticisms of the actions of state authorities. In 1913, when as a member of the Academy of Sciences he was asked to participate in the pompous ceremonies celebrating the 300th anniversary of the Romanov dynasty, he instead organized a celebration of the 200th anniversary of Jakob Bernoulli's publication of the Law of Large Numbers.

Example 3 For the Markov chain with time period 20 years described in Example 1, find the transition matrix for a 40-year time period.

Solution Recall that the transition matrix obtained in Example 1 for the 20-year time period is

$$T = \begin{array}{c} \\ \\ \end{array} \begin{matrix} \text{poor} & \text{mid} & \text{rich} \\ \begin{pmatrix} .80 & .15 & .05 \\ .19 & .75 & .30 \\ .01 & .10 & .65 \end{pmatrix} & \begin{matrix} \text{poor} \\ \text{mid} \\ \text{rich} \end{matrix} \end{matrix} \; .$$

Let n_1 be the size of the population in state 1 at the beginning of a time period. We will find the proportion of n_1 in state 2, after *two* time periods. After the first time period, the distribution of the n_1 members of state 1 among all three states is given by

State 1	State 2	State 3
$(.80)n_1$	$(.19)n_1$	$(.01)n_1$.

During the next time period, 19% of the $(.80)n_1$ in state 1 will make a transition to state 2; 75% of the $(.19)n_1$ in state 2 will remain in state 2; and 30% of the $(0.1)n_1$ in state 3 will move to state 2. Thus, following the fortunes of the original n_1 members of state 1, we find that the number of them who end up in state 2 after two time periods is

$$(.19)(.80)n_1 + (.75)(.19)n_1 + (.30)(.01)n_1 = (.2975)n_1 .$$

Thus the entry in the second row and first column of the two-period transition matrix should be

$$(.19)(.80) + (.75)(.19) + (.30)(.01) = .2975.$$

We note that this is precisely the dot product of the second row vector and the first column vector of T, and is consequently the entry in the second row and first column of T^2. If we continue in this fashion, we find that the two-period matrix is indeed T^2, and a calculator shows that

$$T^2 = \begin{array}{c} \\ \\ \end{array} \begin{matrix} \text{poor} & \text{mid} & \text{rich} \\ \begin{pmatrix} .6690 & .2375 & .1175 \\ .2975 & .6210 & .4295 \\ .0335 & .1415 & .4530 \end{pmatrix} & \begin{matrix} \text{poor} \\ \text{mid} \\ \text{rich} \end{matrix} \end{matrix} \; .$$

Note that all column sums in T^2 are 1, so T^2 is indeed a transition matrix. ◁

If we extend the argument in Example 3, we find that the three-period transition matrix for the Markov chain is T^3, and so on. This points out another situation in which matrix multiplication is useful. Of course, raising even a small matrix like T in Example 3 to a power using pencil and paper, or even a calculator, is tedious. However, a computer can do it easily. The software program MATCOMP available with this text can be used to compute a power of a matrix.

---

m-Period Transition Matrix

A Markov chain with transition matrix T has T^m as m-period transition matrix.

Example 4   For the transition matrix

$$T = \begin{pmatrix} 0 & 0 & 1 \\ 1 & 0 & 0 \\ 0 & 1 & 0 \end{pmatrix}$$

in Example 2, show that after three time periods, the distribution of population among the three states is the same as the initial population distribution.

Solution   After three time periods, the population distribution vector is $T^3\mathbf{p}$ and we easily compute that $T^3 = I$, the 3×3 identity matrix. Thus $T^3\mathbf{p} = \mathbf{p}$, as asserted. Alternatively, we could note that the entire population of state 1 moves to state 2 in the first time period, then to state 3 in the next time period, and finally back to state 1 in the third time period. Similarly, the populations of the other two states move around and then back to the beginning state over the three time periods.   ◁

REGULAR MARKOV CHAINS

We turn to those Markov chains where there exists some fixed number m of time periods in which it is possible to get from any state to any other state. This means that the power T^m of the transition matrix has no zero entries.

Definition 1.13   Regular Transition Matrix; Regular Chain

A transition matrix T is **regular** if T^m has no zero entries for some integer m. A Markov chain having a regular transition matrix is called a **regular chain.**

Example 5   Show that the transition matrix

$$T = \begin{pmatrix} 0 & 0 & 1 \\ 1 & 0 & 0 \\ 0 & 1 & 0 \end{pmatrix}$$

of Example 2 is not regular.

Solution   A computation shows that T^2 still has zero entries. We saw in Example 4 that $T^3 = I$, the 3×3 identity matrix, so we must have $T^4 = T$, $T^5 = T^2$, $T^6 = T^3 = I$, and the powers of T repeat in this fashion. We never eliminate all the zeros. Thus T is not a regular transition matrix.   ◁

If T^m has no zero entries, then $T^{m+1} = (T^m)T$ is easily seen to have no zero entries, since the entries in any column vector of T are nonnegative with at least one entry nonzero. In determining whether a transition matrix is regular, it is not really necessary to actually compute the entries in powers of the matrix. We need only determine whether or not they are zero.

Example 6 If **X** denotes a nonzero entry, determine whether a transition matrix T with the zero and nonzero configuration given by

$$T = \begin{pmatrix} X & 0 & X & 0 \\ 0 & 0 & X & 0 \\ X & 0 & 0 & X \\ 0 & X & 0 & 0 \end{pmatrix}$$

is regular.

Solution We compute configurations of high powers of T as rapidly as we can, since once a power has no zero entries, all higher powers must have nonzero entries. We find that

$$T^2 = \begin{pmatrix} X & 0 & X & X \\ X & 0 & 0 & X \\ X & X & X & 0 \\ 0 & 0 & X & 0 \end{pmatrix}, \qquad T^4 = \begin{pmatrix} X & X & X & X \\ X & 0 & X & X \\ X & X & X & X \\ X & X & X & 0 \end{pmatrix},$$

$$T^8 = \begin{pmatrix} X & X & X & X \\ X & X & X & X \\ X & X & X & X \\ X & X & X & X \end{pmatrix}$$

so the matrix T is indeed regular. ◁

It can be shown that if a Markov chain is *regular*, then over many time periods, the distribution of population among the states approaches a fixed *steady-state distribution vector* **s**. That is, the distribution of population among the states no longer changes significantly as time progresses. This is not to say that there is no longer movement of population between states; the transition matrix T continues to effect changes. But the movement of population out of any state over one time period is balanced by the population moving into that state, so the proportion of the total population in that state remains constant. This is a consequence of the following theorem, which we will not prove.

Theorem 1.10 Achievement of Steady State

Let T be a regular transition matrix. There exists a unique column vector **s** with strictly positive entries whose sum is 1 such that the following hold:

1. As m becomes larger and larger, all columns of T^m approach the column vector **s**.

2. $T\mathbf{s} = \mathbf{s}$, and $\mathbf{s}$ is the unique column vector with this property and whose components add up to 1.

From Theorem 1.10, we can show that if $\mathbf{p}$ is *any* initial population distribution vector for a regular Markov chain with transition matrix T, then after many time periods, the population distribution vector approaches the vector $\mathbf{s}$ described in the theorem. Thus $\mathbf{s}$ is called a **steady-state distribution vector.** We indicate the argument using a 3×3 matrix T. We know that the population distribution vector after m time periods is $T^m\mathbf{p}$. If we let

$$\mathbf{p} = \begin{pmatrix} p_1 \\ p_2 \\ p_3 \end{pmatrix} \quad \text{and} \quad \mathbf{s} = \begin{pmatrix} s_1 \\ s_2 \\ s_3 \end{pmatrix},$$

then Theorem 1.10 tells us that $T^m\mathbf{p}$ is approximately

$$\begin{pmatrix} s_1 & s_1 & s_1 \\ s_2 & s_2 & s_2 \\ s_3 & s_3 & s_3 \end{pmatrix} \begin{pmatrix} p_1 \\ p_2 \\ p_3 \end{pmatrix} = \begin{pmatrix} s_1 p_1 + s_1 p_2 + s_1 p_3 \\ s_2 p_1 + s_2 p_2 + s_2 p_3 \\ s_3 p_1 + s_3 p_2 + s_3 p_3 \end{pmatrix}.$$

Since $p_1 + p_2 + p_3 = 1$, this vector becomes

$$\begin{pmatrix} s_1 \\ s_2 \\ s_3 \end{pmatrix}.$$

Thus after many time periods, the population distribution vector is approximately equal to the steady-state vector $\mathbf{s}$ for *any* choice of initial population distribution vector $\mathbf{p}$.

There are two ways we can attempt to compute the steady-state vector $\mathbf{s}$ of a regular transition matrix T. If we have a computer handy, we can simply raise T to a sufficiently high power so that all column vectors are the same, as far as the computer can print them. The program MATCOMP in our software can be used to do this. Alternatively, we can use part (2) of Theorem 1.10 and solve for $\mathbf{s}$ in the equation

$$T\mathbf{s} = \mathbf{s}. \tag{1}$$

In solving Eq. (1), we will be finding our first *eigenvector* in this text. We will have a lot more to say about eigenvectors in Section 4.5.

Using the identity matrix I, we can rewrite Eq. (1) as

$$T\mathbf{s} = I\mathbf{s}$$
$$T\mathbf{s} - I\mathbf{s} = \mathbf{0}$$
$$(T - I)\mathbf{s} = \mathbf{0}.$$

The last equation represents a homogeneous system of linear equations with coefficient matrix $(T - I)$ and column vector $\mathbf{s}$ of unknowns. From all the solutions of this homogeneous system, choose a solution vector with positive entries that add up to 1. Theorem 1.10 assures us that this solution exists and is unique. Of course, the homogeneous system can be solved easily using a computer. Either of the available programs MATCOMP or YUREDUCE will reduce the augmented matrix to a form from which the solutions can be determined easily. We illustrate both methods with examples.

Example 7 Use the program MATCOMP and raise the transition matrix to powers to find the steady-state distribution vector for the Markov chain in Examples 1 and 3 having states labeled

$$\text{poor,} \qquad \text{middle income,} \qquad \text{rich.}$$

Solution Using MATCOMP and experimenting a bit with powers of the matrix T in Examples 1 and 3, we quickly find that

$$
\begin{array}{ccc}
\text{poor} & \text{mid} & \text{rich}
\end{array}
$$
$$
T^{60} = \begin{pmatrix} .3872054 & .3872054 & .3872054 \\ .4680135 & .4680135 & .4680135 \\ .1447811 & .1447811 & .1447811 \end{pmatrix} \begin{array}{l} \text{poor} \\ \text{mid} \\ \text{rich} \end{array}.
$$

Thus eventually about 38.7% of the population is poor, about 46.8% is middle income, and about 14.5% is rich, and these percentages no longer change as time progresses further over 20-year periods. ◁

Example 8 The inhabitants of a vegetarian-prone community agree on the following rules:

1. No one will eat meat two days in a row.

2. A person who eats no meat one day will flip a fair coin and eat meat on the next day if and only if a head appears.

Determine whether this Markov-chain situation is regular, and if so, find the steady-state distribution vector for the proportions of the population eating no meat and eating meat.

Solution The transition matrix T is easily seen to be

$$
\begin{array}{cc}
\text{no meat} & \text{meat}
\end{array}
$$
$$
T = \begin{pmatrix} \frac{1}{2} & 1 \\ \frac{1}{2} & 0 \end{pmatrix} \begin{array}{l} \text{no meat} \\ \text{meat} \end{array}.
$$

Since T^2 has no zero entries, the Markov chain is regular. We solve

$$(T - I)\mathbf{s} = \mathbf{0}, \quad \text{or} \quad \begin{pmatrix} -\frac{1}{2} & 1 \\ \frac{1}{2} & -1 \end{pmatrix} \begin{pmatrix} s_1 \\ s_2 \end{pmatrix} = \begin{pmatrix} 0 \\ 0 \end{pmatrix}.$$

We reduce the augmented matrix:

$$\begin{pmatrix} -\frac{1}{2} & 1 & | & 0 \\ \frac{1}{2} & -1 & | & 0 \end{pmatrix} \sim \begin{pmatrix} 1 & -2 & | & 0 \\ -\frac{1}{2} & 1 & | & 0 \end{pmatrix} \sim \begin{pmatrix} 1 & -2 & | & 0 \\ 0 & 0 & | & 0 \end{pmatrix}.$$

Thus we have

$$\begin{pmatrix} s_1 \\ s_2 \end{pmatrix} = \begin{pmatrix} 2r \\ r \end{pmatrix} \quad \text{for some scalar } r.$$

But we must have $s_1 + s_2 = 1$, so $2r + r = 1$ and $r = \frac{1}{3}$. Consequently, the steady-state population distribution is given by the vector $\begin{pmatrix} \frac{2}{3} \\ \frac{1}{3} \end{pmatrix}$. We see that, eventually, on each day about $\frac{2}{3}$ of the people eat no meat and the other $\frac{1}{3}$ eat meat. This is independent of the initial distribution of population between the states. All might eat meat the first day, or all might eat no meat; the steady-state vector remains $\begin{pmatrix} \frac{2}{3} \\ \frac{1}{3} \end{pmatrix}$ in either case. ◁

If we were to solve Example 8 by reducing an augmented matrix with a computer, we should add a new row corresponding to the condition $s_1 + s_2 = 1$ for the desired steady-state vector. That way, the unique solution can be seen at once from the reduction of the augmented matrix. Of course, this can be done using pencil-and-paper computations just as well. If we insert this as the first condition on **s** and rework Example 8, our work appears as follows:

$$\begin{pmatrix} 1 & 1 & | & 1 \\ -\frac{1}{2} & 1 & | & 0 \\ \frac{1}{2} & -1 & | & 0 \end{pmatrix} \sim \begin{pmatrix} 1 & 1 & | & 1 \\ 0 & \frac{3}{2} & | & \frac{1}{2} \\ 0 & -\frac{3}{2} & | & -\frac{1}{2} \end{pmatrix} \sim \begin{pmatrix} 1 & 0 & | & \frac{2}{3} \\ 0 & 1 & | & \frac{1}{3} \\ 0 & 0 & | & 0 \end{pmatrix}.$$

Again, we obtain the steady-state vector $\begin{pmatrix} \frac{2}{3} \\ \frac{1}{3} \end{pmatrix}$.

SUMMARY

1. A transition matrix for a Markov chain is a square matrix with nonnegative entries such that the sum of the entries in each column is 1.

2. The entry in the ith row and jth column of a transition matrix is the proportion of the population in state j that moves to state i during one time period of the chain.

3. If the column vector **p** is the initial population distribution vector between states in a Markov chain with transition matrix T, then the population distribution vector after one time period of the chain is $T\mathbf{p}$.

4. If T is the transition matrix for one time period of a Markov chain, then T^m is the transition matrix for m time periods.

5. A Markov chain and its associated transition matrix T are called regular if there exists an integer m such that T^m has no zero entries.

6. If T is a regular transition matrix for a Markov chain, then

   a) The columns of T^m all approach the same probability distribution vector **s** as m becomes large;

> **b)** **s** is the unique probability distribution vector satisfying $T\mathbf{s} = \mathbf{s}$; and
>
> **c)** As the number of time periods increases, the population distribution vectors approach **s** regardless of the initial population distribution vector **p**. Thus **s** is the steady-state population distribution vector.

EXERCISES

In Exercises 1–8, determine whether the given matrix is a transition matrix. If it is, determine whether it is regular.

1. $\begin{pmatrix} \frac{1}{2} & \frac{1}{2} \\ \frac{1}{3} & \frac{2}{3} \end{pmatrix}$

2. $\begin{pmatrix} 0 & \frac{1}{4} \\ 1 & \frac{3}{4} \end{pmatrix}$

3. $\begin{pmatrix} .2 & .1 & .3 \\ .4 & .5 & -.1 \\ .4 & .4 & .8 \end{pmatrix}$

4. $\begin{pmatrix} .1 & .2 \\ .3 & .4 \\ .6 & .4 \end{pmatrix}$

5. $\begin{pmatrix} .5 & 0 & 0 & .5 \\ .5 & .5 & 0 & 0 \\ 0 & .5 & .5 & 0 \\ 0 & 0 & .5 & .5 \end{pmatrix}$

6. $\begin{pmatrix} .3 & .2 & 0 & .5 \\ .4 & .2 & 0 & .1 \\ .1 & .2 & 1 & .2 \\ .2 & .4 & 0 & .2 \end{pmatrix}$

7. $\begin{pmatrix} 0 & .5 & .2 & .2 & .1 \\ .3 & 0 & .1 & .8 & .5 \\ 0 & 0 & .4 & 0 & .1 \\ .7 & .5 & .1 & 0 & .1 \\ 0 & 0 & .2 & 0 & .2 \end{pmatrix}$

8. $\begin{pmatrix} 0 & .1 & 0 & 0 & .9 \\ 0 & .2 & 0 & 0 & .1 \\ 0 & .3 & 0 & 1 & 0 \\ 1 & .1 & 0 & 0 & 0 \\ 0 & .3 & 1 & 0 & 0 \end{pmatrix}$

Exercises 9–14 deal with the following Markov chain. We classify the women in a country according to whether they live in an urban (U), suburban (S), or rural (R) area. Suppose each woman has just one daughter, who in turn has just one daughter, and so on. Suppose further that:

for urban women, 10% of their daughters settle in rural areas and 50% in suburban areas;

for suburban women, 20% of their daughters settle in rural areas and 30% in urban areas;

for rural women, 20% settle in the suburbs and 70% in rural areas.

Let this Markov chain have as its period the time to produce the next generation.

9. Give the transition matrix for this Markov chain, taking states in the order U, S, R.

10. Find the proportion of urban women whose granddaughters are suburban women.

11. Find the proportion of rural women whose granddaughters are also rural women.

12. If the initial population distribution vector for all the women is

$$\begin{pmatrix} .4 \\ .5 \\ .1 \end{pmatrix},$$

find the population distribution vector for the next generation.

13. Repeat Exercise 12, but find the population distribution vector for the following (third) generation.

14. Show that this Markov chain is regular, and find the steady-state probability distribution vector.

Exercises 15–20 deal with a simple genetic model involving just two types of genes, G and g. Suppose that a physical trait, like eye color, is controlled by a pair of these genes, one inherited from each parent. A person may be classified in one of three states:

Dominant (type GG),   Hybrid (type Gg),   Recessive (type gg).

We assume that an offspring inherits one gene from each parent, and that the gene that is inherited from a parent is a random choice from the parent's two genes; that is, the gene inherited is as likely to be one of the parent's two genes as the other one. We form a Markov chain by starting with a population and always *crossing with hybrids to produce offspring. We assume the time to produce a subsequent generation is the time period.*

15. Give an intuitive argument that the transition matrix for this Markov chain is

$$T = \begin{matrix} & \begin{matrix} D & H & R \end{matrix} & \\ \begin{pmatrix} \frac{1}{2} & \frac{1}{4} & 0 \\ \frac{1}{2} & \frac{1}{2} & \frac{1}{2} \\ 0 & \frac{1}{4} & \frac{1}{2} \end{pmatrix} & \begin{matrix} D \\ H \\ R \end{matrix} \end{matrix}.$$

16. What proportion of the third-generation offspring (after two time periods) of the recessive population is dominant?

17. What proportion of the third-generation offspring (after two time periods) of the hybrid population is not hybrid?

18. If initially the entire population is hybrid, find the population distribution vector in the next generation.

19. If initially the population is evenly divided between the states, find the population distribution vector in the third generation (after two time periods).

20. Show that this Markov chain is regular, and find the steady-state population distribution vector.

21. A state in a Markov chain is **absorbing** if it is impossible to leave that state over the next time period. What characterizes the transition matrix of a Markov chain with an absorbing state? Can a Markov chain with an absorbing state be regular?

22. Consider the genetic model for Exercises 15–20. Suppose that instead of always crossing with hybrids to produce offspring, we always cross with recessives. Give the transition matrix for this Markov chain, and show that there is an absorbing state. (See Exercise 21.)

23. A Markov chain is **absorbing** if it contains at least one absorbing state (see Exercise 21) and if it is possible to get from any state to an absorbing state in some number of time periods.

a) Give an example of a transition matrix for a three-state absorbing Markov chain.

b) Give an example of a transition matrix for a three-state Markov chain that is not absorbing, but has an absorbing state.

24. With reference to Exercise 23, consider an absorbing Markov chain with transition matrix T and having just one absorbing state. Argue that for any initial distribution vector **p**, the vectors $T^m \mathbf{p}$ for large m approach the vector with 1 in the component corresponding to the absorbing state and zeros elsewhere. [*Suggestion:* Let m be such that it is possible to reach the absorbing state from any state in m time periods, and let q be the smallest entry in the row of T^m corresponding to the absorbing state. Form a new chain with just two states, Absorbing (A) and Free (F), which has as time period m time periods of the original chain, and with probability q of moving from state F to state A in that time period. Argue that starting in any state in the original chain, you are more likely to reach an absorbing state in mr time periods than you are starting from state F in the new chain and going for r time periods. Using the fact that large powers of a positive number less than 1 are almost zero, show that for the two-state chain, the population distribution vector approaches $\begin{pmatrix} 1 \\ 0 \end{pmatrix}$ as the number of time periods increases, regardless of the initial population distribution vector.]

25. Let T be an $n \times n$ transition matrix. Show that if every row and every column has fewer than $n/2$ zero entries, then the matrix is regular.

In Exercises 26–35, find the steady-state population distribution vector for the given transition matrix. Use MATCOMP, YUREDUCE, or similar software for Exercises 31–35. See the comment following Example 8.

26. $\begin{pmatrix} 0 & \frac{1}{4} \\ 1 & \frac{3}{4} \end{pmatrix}$

27. $\begin{pmatrix} \frac{1}{2} & \frac{1}{4} \\ \frac{1}{2} & \frac{3}{4} \end{pmatrix}$

28. $\begin{pmatrix} \frac{1}{5} & 1 \\ \frac{4}{5} & 0 \end{pmatrix}$

29. $\begin{pmatrix} 0 & \frac{1}{4} & \frac{1}{2} \\ \frac{1}{2} & 0 & \frac{1}{2} \\ \frac{1}{2} & \frac{3}{4} & 0 \end{pmatrix}$

30. $\begin{pmatrix} \frac{1}{5} & \frac{3}{4} & \frac{1}{8} \\ \frac{4}{5} & 0 & \frac{3}{8} \\ 0 & \frac{1}{4} & \frac{1}{2} \end{pmatrix}$

31. $\begin{pmatrix} .1 & .3 & .4 \\ .2 & 0 & .2 \\ .7 & .7 & .4 \end{pmatrix}$

32. $\begin{pmatrix} .3 & .3 & .1 \\ .1 & .3 & .5 \\ .6 & .4 & .4 \end{pmatrix}$

33. $\begin{pmatrix} .1 & 0 & .2 & .5 \\ .4 & 0 & .2 & .5 \\ .2 & .5 & .4 & 0 \\ .3 & .5 & .2 & 0 \end{pmatrix}$

34. $\begin{pmatrix} .5 & 0 & 0 & 0 & .9 \\ .5 & .4 & 0 & 0 & 0 \\ 0 & .6 & .3 & 0 & 0 \\ 0 & 0 & .7 & .2 & 0 \\ 0 & 0 & 0 & .8 & .1 \end{pmatrix}$

35. The matrix in Exercise 8

Solving Large Linear Systems*

The Gauss and Gauss–Jordan methods presented in Chapter 1 are fine for solving very small linear systems using pencil and paper. Some applied problems—in particular, those requiring numerical solution of differential equations—can lead to very large linear systems, involving thousands of equations in thousands of unknowns. Of course, such large linear systems must be solved using a computer. That is the primary concern of this chapter. Although a computer can work tremendously faster than we can with pencil and paper, each individual arithmetic operation does take some time, and additional time is used whenever the value of a variable is stored or retrieved. Also, indexing in subscripted arrays requires time. When solving a large linear system with a computer, it becomes important to use as efficient a computational algorithm as we can, so that the number of operations required is as small as practically possible.

We begin this chapter with a discussion of the time required for a computer to execute operations and comparison of the efficiency of the Gauss method including back substitution and the Gauss–Jordan method.

Section 2.2 presents the LU (lower and upper triangular) factorization of the coefficient matrix of a square linear system. This factorization will appear as we develop an efficient algorithm for solving, by repeated computer runs, many systems all having the same coefficient matrix.

Section 2.3 deals with problems of roundoff and discusses ill-conditioned matrices. We will see that there are actually very small linear systems, consisting of only two equations in two unknowns, for which good computer programs may give incorrect solutions.

---

* This chapter can be omitted if desired. Its topics are not used in the remainder of the text.

2.1

Considerations of Time

TIMING DATA FOR ONE PC

The computation involved in solving a linear system is just a lot of arithmetic. Arithmetic takes time to execute even with a computer, although the computer is obviously much faster than an individual with pencil and paper. In reducing a matrix using elementary row operations, we spend most of our time adding a multiple of a row vector to another row vector. A typical step in multiplying row k of a matrix A by r and adding it to row i consists of

$$\text{replacing } a_{ij} \text{ by } a_{ij} + ra_{kj}. \tag{1}$$

In a computer language such as BASIC, the operation (1) might become

$$A(I, J) = A(I, J) + R*A(K, J). \tag{2}$$

The computer instruction (2) involves operations of addition and multiplication, as well as indexing, retrieving values, and storing values. We use the terminology of C. B. Moler and call such an operation a **flop.** These flops require time to execute on any computer, although the times vary widely depending on the computer hardware and the language of the computer program. We experimented with our PC (personal computer), which is not noted for speed, using both interpretive BASIC and compiled BASIC. The computer can execute a compiled instruction faster than an interpretive one. We generated two random numbers, R and S, and then found the time required to add them 10,000 times using the BASIC loop

$$\text{FOR I} = 1 \text{ TO N: C} = R + S: \text{NEXT} \tag{3}$$

where we had earlier set N = 10,000. We obtained the data shown in the top row of Table 2.1. We also had the computer execute the loop (3) with + replaced by − (subtraction), then by ∗ (multiplication), and finally by / (division). We similarly timed the execution of 10,000 flops, using

$$\text{FOR I} = 1 \text{ TO N: A(K, J)} = A(K, J) + R * A(M, J): \text{NEXT} \tag{4}$$

where we had set N = 10,000 and had also assigned values to all other variables except I. All our data is shown in Table 2.1.

Table 2.1 gives us quite a bit of insight into the PC that generated the data. Here are some things we noticed immediately.

Point 1 Multiplication takes a bit, but surprisingly little, more time than addition.

Point 2 Division takes the most time of the four arithmetic operations. Indeed, our PC found double-precision division in interpretive BASIC very time-consuming. We should try to minimize divisions as much as possible. For example, when computing

$$(\tfrac{4}{3})(3, 2, 5, 7, 8) + (-4, 11, 2, 1, 5)$$

Table 2.1   **Time (in seconds) for executing 10,000 operations**

Routine	Interpretive BASIC		Compiled BASIC	
	Single precision	*Double precision*	*Single precision*	*Double precision*
Addition [using (3)]	37	44	8	9
Subtraction [using (3) with −]	37	48	8	9
Multiplication [using (3) with *]	39	53	9	11
Division [using (3) with /]	47	223	9	15
Flops [using (4)]	143	162	15	18

to obtain a row vector with first entry zero, we should *not* compute the remaining entries as

$$(\tfrac{4}{3})2 + 11, \qquad (\tfrac{4}{3})5 + 2, \qquad (\tfrac{4}{3})7 + 1, \quad \text{and} \quad (\tfrac{4}{3})8 + 5,$$

which requires four divisions. We should rather do a single division, finding $r = \tfrac{4}{3}$, and then compute

$$2r + 11, \qquad 5r + 2, \qquad 7r + 1, \quad \text{and} \quad 8r + 5.$$

Point 3 Our program ran much faster in compiled BASIC than in interpretive BASIC. In the compiled version on our PC, the time for double-precision division was not so far out of line with the time for other operations.

Point 4 The indexing in the flops requires significant time in interpretive BASIC on our PC.

The software program TIMING available to schools using this text was used to generate the data in Table 2.1 on our PC. Exercise 21 asks students to obtain analogous data for their PC, using this program or similar software.

COUNTING OPERATIONS

We turn to counting the flops required to solve a square linear system $A\mathbf{x} = \mathbf{b}$ that has a unique solution. We assume that no row interchanges are necessary, which is frequently the case.

Suppose we solve the system $A\mathbf{x} = \mathbf{b}$ using the Gauss method with back substitution. Form the augmented matrix

$$(A \mid \mathbf{b}) = \left(\begin{array}{cccc|c} a_{11} & a_{12} & \cdots & a_{1n} & b_1 \\ a_{21} & a_{22} & \cdots & a_{2n} & b_2 \\ & & \vdots & & \vdots \\ a_{n1} & a_{n2} & \cdots & a_{nn} & b_n \end{array}\right).$$

For the moment let us neglect the flops performed on **b** and count just the flops performed on A. In reducing the $n \times n$ matrix A to upper-triangular form U, we execute $n - 1$ flops in adding a multiple of the first row of $(A \mid \mathbf{b})$ to the second row. (We do not need to compute the zero entry at the beginning of our new second row; we know it will be zero.) We similarly use $n - 1$ flops to obtain the new row 3, and so on. This gives a total of $(n - 1)^2$ flops executed using the pivot in row 1. The pivot in row 2 is then used for the $(n - 1) \times (n - 1)$ matrix obtained by crossing off the first row and column of the modified coefficient matrix. By the count just made for the $n \times n$ matrix, we realize that using this pivot in the second row will result in executing $(n - 2)^2$ flops. Continuing in this fashion, we see that to reduce A to upper-triangular form U will require

$$(n - 1)^2 + (n - 2)^2 + (n - 3)^2 + \cdots + 1 \tag{5}$$

flops, together with some divisions. Let's count the divisions. We expect to use just one division each time a row is multiplied by a constant and added to another row (see point 2 above). There will be $n - 1$ divisions involving the pivot in row 1, $n - 2$ involving the pivot in row 2, and so on, for a total of

$$(n - 1) + (n - 2) + (n - 3) + \cdots + 1 \tag{6}$$

divisions.

There are some handy formulas for finding a sum of consecutive integers and a sum of squares of consecutive integers. It can be shown by induction (see Appendix A) that

$$1 + 2 + 3 + \cdots + n = \frac{n(n + 1)}{2} \tag{7}$$

and

$$1^2 + 2^2 + 3^2 + \cdots + n^2 = \frac{n(n + 1)(2n + 1)}{6}. \tag{8}$$

Replacing n by $n - 1$ in formula (8), we see that the number of flops given by the sum (5) is equal to

$$\frac{(n - 1)n(2n - 1)}{6} = \frac{(n^2 - n)(2n - 1)}{6} = \left(\frac{1}{6}\right)(2n^3 - 3n^2 + n)$$

$$= \frac{n^3}{3} - \frac{n^2}{2} + \frac{n}{6}. \tag{9}$$

We are making the assumption that we are using a computer to solve a *large* linear system, involving hundreds or thousands of equations. With $n = 1000$, the value of (9) becomes

$$\frac{1{,}000{,}000{,}000}{3} - \frac{1{,}000{,}000}{2} + \frac{1{,}000}{6}. \tag{10}$$

The largest term in the expression (10) is the first one, $1{,}000{,}000{,}000/3$, corresponding to the $n^3/3$ term in (9). The lower powers of n in the $n^2/2$ and $n/6$ terms contribute little compared with the $n^3/3$ for large n. It is customary to keep just the $n^3/3$ term as a measure of the **order of magnitude** of (9) for large values of n.

Turning to the count of the divisions in the sum (6), we see from (7) with n replaced by $n - 1$ that we are using

$$\frac{(n - 1)n}{2} = \frac{n^2}{2} - \frac{n}{2} \tag{11}$$

divisions. For large n, this is of order of magnitude $n^2/2$, which is inconsequential compared with the order of magnitude $n^3/3$ for the flops. Exercise 1 shows that the number of flops performed on the column vector $\mathbf{b}$ in reducing $(A \mid \mathbf{b})$ is again given by (11), so it can be neglected for large n compared with the order of magnitude $n^3/3$. The result is shown in the following box.

Flop Count for Reducing $(A \mid$ b$)$ to $(U \mid$ c$)$

If $A\mathbf{x} = \mathbf{b}$ is a square linear system in which A is an $n \times n$ matrix, then the number of flops executed in reducing $(A \mid \mathbf{b})$ to the form $(U \mid \mathbf{c})$ is of order of magnitude $n^3/3$ for large n.

We turn now to finding the number of flops used in back substitution to solve the upper-triangular linear system $U\mathbf{x} = \mathbf{c}.$ This system can be written out as

$$u_{11}x_1 + \cdots + \quad u_{1,n-1}x_{n-1} + \quad u_{1n}x_n = c_1$$

$$\vdots$$

$$u_{n-1,n-1}x_{n-1} + u_{n-1,n}x_n = c_{n-1}$$

$$u_{nn}x_n = c_n.$$

To solve for x_n requires one indexed division, which we consider to be a flop. To solve then for x_{n-1} requires an indexed multiplication and subtraction, followed by an indexed division, which we consider to be two flops. To solve for x_{n-2} requires two flops, each consisting of a multiplication combined with a subtraction, followed by an indexed division, and we consider this to contribute three more flops, and so on. We obtain the count

$$1 + 2 + 3 + \cdots + n = \frac{(n + 1)n}{2} = \frac{n^2}{2} + \frac{n}{2} \tag{12}$$

for the flops in this back substitution. Again, this is of lower order of magnitude than the order of magnitude $n^3/3$ for the flops required to reduce $(A \mid \mathbf{b})$ to $(U \mid \mathbf{c})$.

Flop Count for Back Substitution

If U is an $n \times n$ upper-triangular matrix, then back substitution to solve $U\mathbf{x} = \mathbf{c}$ requires the order of magnitude $n^2/2$ flops for large n.

Combining our results shown in the last two boxes, we arrive at the following flop count.

Flop Count for Solving $A\mathbf{x} = \mathbf{b}$ Using the Gauss Method with Back Substitution

If A is an $n \times n$ matrix, then the number of flops required to solve $A\mathbf{x} = \mathbf{b}$ using the Gauss method with back substitution is of order of magnitude $n^3/3$ for large n.

Example 1   For the computer that produced the execution times shown in Table 2.1, find the approximate time required to solve a system of 100 equations in 100 unknowns using single-precision arithmetic and (a) interpretive BASIC and (b) compiled BASIC.

Solution   From the flop count for the Gauss method, we see that solving such a system with $n = 100$ requires about $100^3/3 = 1,000,000/3$ flops. In interpretive BASIC the time required for 10,000 flops in single precision was about 143 seconds. Thus the 1,000,000/3 flops require about $(1,000,000/30,000)143 = 14,300/3$ seconds, or about 1 hour and 20 minutes. In compiled BASIC where 10,000 flops took about 15 seconds, we find that the time is about $(1,000,000/30,000)15 = 1500/3 = 500$ seconds, or about 8 minutes. The PC that generated the times in Table 2.1 is regarded as slow.   ◁

It is interesting to compare the efficiency of the Gauss method with back substitution to that of the Gauss–Jordan method. Recall that in the Gauss–Jordan method, $(A \mid \mathbf{b})$ is reduced to a form $(I \mid \mathbf{c})$, where I is the identity matrix. Exercise 2 shows that the Gauss–Jordan flop count is of order of magnitude $n^3/2$ for large n. Thus the Gauss–Jordan method is not as efficient as the Gauss method with back substitution if n is large. One expects a Gauss–Jordan program to take about one and a half times as long to execute. The program TIMING can also be used to illustrate this. See Exercises 22–25.

COUNTING FLOPS FOR MATRIX OPERATIONS

The exercises ask you to count the flops involved in matrix addition, multiplication, and exponentiation. Recall that a matrix product $C = AB$ is obtained by taking dot products of row vectors of an $m \times n$ matrix A with column vectors of an $n \times s$ matrix B. We indicate the usual way that a computer finds the dot product c_{ij} in C. First the computer sets $c_{ij} = 0$. Then it replaces c_{ij} by $c_{ij} + a_{i1}b_{1j}$, which gives c_{ij} the value $a_{i1}b_{1j}$. Then the computer replaces c_{ij} by $c_{ij} + a_{i2}b_{2j}$, giving c_{ij} the value $a_{i1}b_{1j} + a_{i2}b_{2j}$. This process continues in the obvious way until finally we have accumulated the desired value

$$c_{ij} = a_{i1}b_{1j} + a_{i2}b_{2j} + \cdots + a_{in}b_{nj}.$$

Each of these replacements of c_{ij} is accomplished by a flop, typically expressed by

$$C(I, J) = C(I, J) + A(I, K) * B(K, J) \tag{13}$$

in BASIC.

Example 2 Find the number of flops required in taking the dot product of two vectors, each with n components.

Solution The discussion above shows that the dot product uses n flops of the form (13), corresponding to the values 1, 2, 3, $\cdots$, n for K. ◁

SUMMARY

1. A flop is a rather vaguely defined execution by a computer, consisting typically of a bit of indexing, retrieving and storing a couple of values, and one or two arithmetic operations. Typical flops might appear in a computer program in instruction lines such as

$$C(I, J) = A(I, J) + B(I, J)$$

or

$$A(I, J) = A(I, J) + R*A(K, J).$$

2. Time is required for a computer to execute a flop, and it is desirable to use as few flops as possible when performing an extensive computation.

3. If the number of flops used to solve a problem is given by a polynomial expression in n, it is customary to keep only the term of highest degree in the polynomial as a measure of the *order of magnitude* for the number of flops when n is large.

4. The order of magnitude for the number of flops used in solving a system $A\mathbf{x} = \mathbf{b}$ for an $n \times n$ matrix A is

$$\frac{n^3}{3} \quad \text{for the Gauss method with back substitution}$$

and

$$\frac{n^3}{2} \quad \text{for the Gauss–Jordan method.}$$

5. The formulas

$$1 + 2 + 3 + \cdots + n = \frac{n(n + 1)}{2}$$

and

$$1^2 + 2^2 + 3^2 + \cdots + n^2 = \frac{n(n + 1)(2n + 1)}{6}$$

are handy in determining the number of flops performed by a computer in matrix computations.

EXERCISES

In all these exercises, assume that no row vectors are interchanged in a matrix.

1. Let A be an $n \times n$ matrix. Show that in reducing $(A \mid \mathbf{b})$ to $(U \mid \mathbf{c})$ using the Gauss method, the number of flops performed on $\mathbf{b}$ is of order of magnitude $n^2/2$ for large n.

2. Let A be an $n \times n$ matrix. Show that in solving $A\mathbf{x} = \mathbf{b}$ using the Gauss–Jordan method, the number of flops has order of magnitude $n^3/2$ for large n.

In Exercises 3–11, let A be an $n \times n$ matrix, and let B and C be $m \times n$ matrices. For each matrix, find the number of flops required for efficient computation of the matrix.

3. $B + C$ **4.** A^2 **5.** BA **6.** A^3 **7.** A^4
8. A^5 **9.** A^6 **10.** A^{63} **11.** A^{64}

12. Which of the following is more efficient with a computer?
  a) Solving a square system $A\mathbf{x} = \mathbf{b}$ by the Gauss–Jordan method which makes each pivot 1 before creating the zeros in the column containing it.
  b) Using a similar technique to reduce the system to a diagonal system $D\mathbf{x} = \mathbf{c}$, where the entries in D are not necessarily all 1, and then dividing by these entries to obtain the solution.

13. Let A be an $n \times n$ matrix where n is large. Find the order of magnitude for the number of flops if A^{-1} is computed using the Gauss–Jordan method on the augmented matrix $(A \mid I)$ without trying to reduce the number of flops used on I due to the zeros that appear in it.

14. Repeat Exercise 13 using the Gauss method with back substitution rather than the Gauss–Jordan method.

15. Repeat Exercise 14, but this time cut down the number of flops performed on I during the Gauss reduction by taking into account the zeros in I.

Exercises 16–20 concern **band matrices.** *In a number of situations, square linear systems $A\mathbf{x} = \mathbf{b}$ occur in which the nonzero entries in the $n \times n$ matrix A are all concentrated near the main diagonal, running from the upper left corner to the lower right corner of A. Such a matrix is called a* **band matrix.** *For example, the matrix*

$$
\overbrace{
\begin{pmatrix}
2 & 1 & 0 & 0 & 0 & 0 \\
1 & 3 & 4 & 0 & 0 & 0 \\
0 & 4 & 1 & 2 & 0 & 0 \\
0 & 0 & 2 & 2 & 7 & 0 \\
0 & 0 & 0 & 7 & 1 & 3 \\
0 & 0 & 0 & 0 & \underbrace{3 & 5}_{w = 2}
\end{pmatrix}}^{w = 2}
\tag{12}
$$

is a symmetric 6×6 band matrix. We say that the **band width** *of a band matrix (a_{ij}) is w if w is the smallest integer such that $a_{ij} = 0$ for $|i - j| \ge w$. Thus the matrix (12) has band width $w = 2$, as indicated. Such a matrix of band width 2 is also called* **tridiagonal.** *We usually assume that the band width is small compared with the size of n. As the band width approaches n the matrix becomes* **full.**

In Exercises 16–20, assume that A is an n × n band matrix with band width w that is small compared with n.

16. What can be said concerning the band width of A^2? of A^3? of A^m?

17. Let A be tridiagonal, so $w = 2$. Find the order of magnitude of the number of flops required to reduce the partitioned matrix $(A \mid \mathbf{b})$ to a form $(U \mid \mathbf{c})$, where U is an upper-triangular matrix, taking into account the banded character of A. (For example, if we count a division as one flop, only two flops are needed to fix up the entire first column!)

18. Repeat Exercise 17 for band width w, expressing the result in terms of w.

19. Repeat Exercise 18, but include also the flops used in back substitution to solve $A\mathbf{x} = \mathbf{b}$.

20. Explain why the Gauss method with back substitution is much more efficient than the Gauss–Jordan method for a banded matrix where w is very small compared with n.

■ *The software available to schools using our text includes a program TIMING, which can be used to time algebraic operations and flops. The program can also be used to time the solution of square systems $A\mathbf{x} = \mathbf{b}$ using both the Gauss method with back substitution and the Gauss–Jordan method. For a user-specified integer $n \leq 60$, the program generates the $n \times n$ matrix A and column vector $\mathbf{b}$ where all entries are in the interval $[-20, 20]$. Use TIMING or similar software for Exercises 21–25.*

21. Run the program TIMING and obtain data for your PC analogous to those in Table 2.1.

22. Run the program TIMING in interpretive BASIC to time the Gauss method and the Gauss–Jordan method, starting with small values of n and increasing them until a few seconds' difference in times for the two methods is obtained. Does the time for the Gauss–Jordan method seem to be about $\frac{3}{2}$ the time for the Gauss method with back substitution? If not, why not? Experiment with both single-precision and double-precision timing.

23. Continuing Exercise 22, increase the size of n until the solutions require two or three minutes. Now does the Gauss–Jordan method seem to take about $\frac{3}{2}$ times as long? If this ratio is significantly different from that obtained in Exercise 22, explain why. (The two ratios may or may not appear approximately the same, depending on the speed of the computer used and whether time in fractions of seconds is displayed.)

24. Repeat Exercise 22 using the compiled option in TIMING. Compare the ratios of times with those obtained in Exercise 22 and explain any difference.

25. Repeat Exercise 23 with the compiled option in TIMING. Compare the ratio of times with those in Exercise 23 and explain any difference.

2.2
The *LU*-Factorization

KEEPING A RECORD OF
ROW OPERATIONS

We continue to work with a square linear system $A\mathbf{x} = \mathbf{b}$ having a unique solution that can be found using Gauss elimination with back substitution without having to interchange any rows. That is, the matrix A can be row-reduced to an upper-triangular matrix U without any row interchanges. This case arises often in applications.

Situations occur in which it is necessary to solve many such systems, all having the same coefficient matrix A but different column vectors $\mathbf{b}$. In Section 1.4, we discussed solving such multiple systems by row-reducing a single augmented matrix

$$(A \mid \mathbf{b}_1, \mathbf{b}_2, \ldots, \mathbf{b}_s)$$

in which we line up all the different column vectors $\mathbf{b}_1, \mathbf{b}_2, \ldots, \mathbf{b}_s$ on the right of the partition. In practice, it may be impossible to solve all these systems using this single augmentation in a single computer run. Here are some possibilities in which a sequence of computer runs may be needed to solve all the systems.

1. Remember that we are concerned with *large* systems. If the number s of vectors $\mathbf{b}_j$ is large, there may not be room in the computer memory to accommodate all these data at one time. Indeed, it might even be necessary to reduce the $n \times n$ matrix A in segments, if n is very large. If we can handle the s vectors $\mathbf{b}_j$ only in groups of m at a time, then we must have at least s/m computer runs to solve all the systems.

2. Perhaps the vectors $\mathbf{b}_j$ are generated over a period of time, and we need to solve systems involving groups of the vectors $\mathbf{b}_j$ as they are generated. For example, we may want to solve $A\mathbf{x} = \mathbf{b}_j$ with r different vectors $\mathbf{b}_j$ each day.

3. Perhaps the vector $\mathbf{b}_{j+1}$ depends on the solution of $A\mathbf{x} = \mathbf{b}_j$. Clearly, we would then have to solve $A\mathbf{x} = \mathbf{b}_1$, determine $\mathbf{b}_2$, then solve $A\mathbf{x} = \mathbf{b}_2$, determine $\mathbf{b}_3$, and so on, until we finally solve $A\mathbf{x} = \mathbf{b}_s$.

From Section 2.1, we know that the magnitude of the number of flops required to reduce A to an upper-triangular matrix U is $n^3/3$ for large n. We want to avoid having to repeat all this work done in reducing A after the first computer run.

We are assuming that A can be reduced to U without interchanging rows, that is, that (nonzero) pivots always appear where we want them as we reduce A to U. This means that only elementary row-addition operations are used: those that add a multiple of a row vector to another row vector. Recall that a row-addition operation can be accomplished by multiplication on the left by an $n \times n$ elementary matrix E, where E is obtained by applying the same row-addition operation to the identity matrix. There is a sequence $E_1, E_2, \ldots, E_h$ of such elementary matrices such that

$$E_h E_{h-1} \cdots E_2 E_1 A = U. \tag{1}$$

Once the matrix U has been found by the computer, the data in it can be stored on a disk or on tape, and simply read in for future computer runs. However, when we are solving $A\mathbf{x} = \mathbf{b}$ by reducing the augmented matrix $(A \mid \mathbf{b})$, the sequence of elementary row operations described by $E_h E_{h-1} \cdots E_2 E_1$ in Eq. (1) must be applied to the *entire* augmented matrix, not merely to A. Thus we need to keep a *record* of this sequence of row operations to perform on column vectors $\mathbf{b}_j$ when they are used in subsequent computer runs.

Here is a way of recording the row-addition operations that is both efficient and algebraically interesting. As we make the reduction of A to U, we create a *lower-triangular matrix L*, which is a record of the row-addition operations performed. We start with the $n \times n$ identity matrix I and as each row-addition operation on A is performed, we change one of the zero entries below the diagonal in I to give a record of that operation. For example, if during the reduction we multiply row *two* by 4 and add the result to row *six*, then we place -4 in the *second* column position in the *sixth* row of the matrix L we are creating as a record. The general formulation is shown in the following box.

Creation of the Matrix L

Start with the $n \times n$ identity matrix I. If during the reduction of A to U, row i is multiplied by r and the result is added to row k, replace the zero in row k and column i of the identity matrix by $-r$. The final result obtained from the identity matrix is L.

Example 1   Reduce the matrix

$$A = \begin{pmatrix} 1 & 3 & -1 \\ 2 & 8 & 4 \\ -1 & 3 & 4 \end{pmatrix}$$

to upper-triangular form U, and create the matrix L described above.

Solution   We proceed in two columns as follows.

Reduction of A to U Creation of L from I

$$A = \begin{pmatrix} 1 & 3 & -1 \\ 2 & 8 & 4 \\ -1 & 3 & 4 \end{pmatrix} \qquad I = \begin{pmatrix} 1 & 0 & 0 \\ 0 & 1 & 0 \\ 0 & 0 & 1 \end{pmatrix}$$

Add -2 times
row 1 to row 2.

$$\sim \begin{pmatrix} 1 & 3 & -1 \\ 0 & 2 & 6 \\ -1 & 3 & 4 \end{pmatrix} \qquad \begin{pmatrix} 1 & 0 & 0 \\ 2 & 1 & 0 \\ 0 & 0 & 1 \end{pmatrix}$$

Add 1 times
row 1 to row 3.

$$\sim \begin{pmatrix} 1 & 3 & -1 \\ 0 & 2 & 6 \\ 0 & 6 & 3 \end{pmatrix} \qquad \begin{pmatrix} 1 & 0 & 0 \\ 2 & 1 & 0 \\ -1 & 0 & 1 \end{pmatrix}$$

Add -3 times
row 2 to row 3.

$$\sim \begin{pmatrix} 1 & 3 & -1 \\ 0 & 2 & 6 \\ 0 & 0 & -15 \end{pmatrix} = U \qquad L = \begin{pmatrix} 1 & 0 & 0 \\ 2 & 1 & 0 \\ -1 & 3 & 1 \end{pmatrix}. \quad \triangleleft$$

We now illustrate how the record kept in L in Example 1 can be used to solve a linear system $A\mathbf{x} = \mathbf{b}$ having the matrix A of Example 1 as coefficient matrix.

Example 2 Use the record in L in Example 1 to solve the linear system $A\mathbf{x} = \mathbf{b}$ given by

$$x_1 + 3x_2 - x_3 = -4$$
$$2x_1 + 8x_2 + 4x_3 = 2$$
$$-x_1 + 3x_2 + 4x_3 = 4.$$

Solution We use the record in L to find the column vector $\mathbf{c}$ that would occur if we were to reduce $(A \mid \mathbf{b})$ to $(U \mid \mathbf{c})$ using these same row operations.

Entry ℓ_{ij} in L	Meaning of the entry	Reduction of $\mathbf{b}$ to $\mathbf{c}$
		$\mathbf{b} = \begin{pmatrix} -4 \\ 2 \\ 4 \end{pmatrix}$
$\ell_{21} = 2$	Multiply row 1 by -2 and add to row 2.	$\sim \begin{pmatrix} -4 \\ 10 \\ 4 \end{pmatrix}$
$\ell_{31} = -1$	Multiply row 1 by $-(-1) = 1$ and add to row 3.	$\sim \begin{pmatrix} -4 \\ 10 \\ 0 \end{pmatrix}$
$\ell_{32} = 3$	Multiply row 2 by -3 and add to row 3.	$\sim \begin{pmatrix} -4 \\ 10 \\ -30 \end{pmatrix} = \mathbf{c}.$

If we put this result together with the matrix U obtained in Example 1, the reduced partitioned matrix for the linear system becomes

$$(U \mid \mathbf{c}) = \begin{pmatrix} 1 & 3 & -1 & \mid & -4 \\ 0 & 2 & 6 & \mid & 10 \\ 0 & 0 & -15 & \mid & -30 \end{pmatrix}.$$

Back substitution then yields

$$x_3 = \frac{-30}{-15} = 2$$

$$x_2 = \frac{10 - 6(2)}{2} = \frac{-2}{2} = -1$$

$$x_1 = -4 + 3 + 2 = 1. \quad \triangleleft$$

We give another example.

Example 3   Let

$$A = \begin{pmatrix} 1 & -2 & 0 & 3 \\ -2 & 3 & 1 & -6 \\ -1 & 4 & -4 & 3 \\ 5 & -8 & 4 & 0 \end{pmatrix} \quad \text{and} \quad \mathbf{b} = \begin{pmatrix} 11 \\ -21 \\ -1 \\ 23 \end{pmatrix}.$$

Generate the matrix L while reducing the matrix A to U. Then use U and the record in L to solve $A\mathbf{x} = \mathbf{b}$.

Solution   We work in two columns again. This time we fix up a whole column of A in each step.

Reduction of A | Generation of L

$$A = \begin{pmatrix} 1 & -2 & 0 & 3 \\ -2 & 3 & 1 & -6 \\ -1 & 4 & -4 & 3 \\ 5 & -8 & 4 & 0 \end{pmatrix} \qquad I = \begin{pmatrix} 1 & 0 & 0 & 0 \\ 0 & 1 & 0 & 0 \\ 0 & 0 & 1 & 0 \\ 0 & 0 & 0 & 1 \end{pmatrix}$$

$$\sim \begin{pmatrix} 1 & -2 & 0 & 3 \\ 0 & -1 & 1 & 0 \\ 0 & 2 & -4 & 6 \\ 0 & 2 & 4 & -15 \end{pmatrix} \qquad \begin{pmatrix} 1 & 0 & 0 & 0 \\ -2 & 1 & 0 & 0 \\ -1 & 0 & 1 & 0 \\ 5 & 0 & 0 & 1 \end{pmatrix}$$

$$\sim \begin{pmatrix} 1 & -2 & 0 & 3 \\ 0 & -1 & 1 & 0 \\ 0 & 0 & -2 & 6 \\ 0 & 0 & 6 & -15 \end{pmatrix} \qquad \begin{pmatrix} 1 & 0 & 0 & 0 \\ -2 & 1 & 0 & 0 \\ -1 & -2 & 1 & 0 \\ 5 & -2 & 0 & 1 \end{pmatrix}$$

$$\sim \begin{pmatrix} 1 & -2 & 0 & 3 \\ 0 & -1 & 1 & 0 \\ 0 & 0 & -2 & 6 \\ 0 & 0 & 0 & 3 \end{pmatrix} = U \quad L = \begin{pmatrix} 1 & 0 & 0 & 0 \\ -2 & 1 & 0 & 0 \\ -1 & -2 & 1 & 0 \\ 5 & -2 & -3 & 1 \end{pmatrix}$$

We now apply the record below the diagonal in L to the vector $\mathbf{b}$, working under the main diagonal down each column of the record in L in turn.

First column of L: $\mathbf{b} = \begin{pmatrix} 11 \\ -21 \\ -1 \\ 23 \end{pmatrix} \sim \begin{pmatrix} 11 \\ 1 \\ -1 \\ 23 \end{pmatrix} \sim \begin{pmatrix} 11 \\ 1 \\ 10 \\ 23 \end{pmatrix} \sim \begin{pmatrix} 11 \\ 1 \\ 10 \\ -32 \end{pmatrix}$

Second column of L: $\begin{pmatrix} 11 \\ 1 \\ 10 \\ -32 \end{pmatrix} \sim \begin{pmatrix} 11 \\ 1 \\ 12 \\ -32 \end{pmatrix} \sim \begin{pmatrix} 11 \\ 1 \\ 12 \\ -30 \end{pmatrix}$

Third column of L:
$$\begin{pmatrix} 11 \\ 1 \\ 12 \\ -30 \end{pmatrix} \sim \begin{pmatrix} 11 \\ 1 \\ 12 \\ 6 \end{pmatrix}$$

The partitioned matrix

$$\left(\begin{array}{cccc|c} 1 & -2 & 0 & 3 & 11 \\ 0 & -1 & 1 & 0 & 1 \\ 0 & 0 & -2 & 6 & 12 \\ 0 & 0 & 0 & 3 & 6 \end{array} \right)$$

yields upon back substitution

$$x_4 = \frac{6}{3} = 2$$
$$x_3 = (12 - 12)/(-2) = 0$$
$$x_2 = (1 - 0)/(-1) = -1$$
$$x_1 = 11 - 2 - 6 = 3. \quad \triangleleft$$

Two obvious questions come to mind.

1. Why bother to put the entries 1 down the main diagonal in L, since they are not used?

2. If we multiply a row by r and add to another row while reducing A, why do we put $-r$ rather than r in the record in L, and then change back to r again when performing the operations on the column vector **b**?

We do these two things only because the matrix L formed in this way has an interesting algebraic property, which we will discuss in a moment. In fact, when solving a large system using a computer, we certainly would not fuss with the entries 1 down the diagonal. Indeed, we can save memory space by not even generating a matrix L separate from the one being reduced. When creating a zero below the main diagonal, place the record $-r$ or r desired, as described in Question 2 above, directly in the matrix being reduced at the position where a zero is being created! The computer already has space reserved for an entry there. Just remember that the final matrix contains the desired entries of U on and above the diagonal and the record for L or $-L$ below the diagonal. *We will always use $-r$ rather than r as record entry in this text.* Thus for the 4×4 matrix in Example 3, we obtain

$$\left(\begin{array}{cccc} 1 & -2 & 0 & 3 \\ -2 & -1 & 1 & 0 \\ -1 & -2 & -2 & 6 \\ 5 & -2 & -3 & 3 \end{array} \right) \qquad \textbf{Combined } L\backslash U \textbf{ display} \qquad (2)$$

where the black entries on or above the main diagonal give the essential data for U and the color entries are the essential data for L.

Let us examine the efficiency of solving a system $A\mathbf{x} = \mathbf{b}$ if U and L are already known. Each entry in the record in L requires one flop to execute when applying this record to reduce a column vector $\mathbf{b}$. The number of entries is

$$1 + 2 + \cdots + n - 1 = \frac{(n-1)n}{2} = \frac{n^2}{2} - \frac{n}{2},$$

which is of order of magnitude $n^2/2$ for large n. We saw in Section 2.1 that back substitution requires about $n^2/2$ flops also, giving a total of n^2 flops for large n to solve $A\mathbf{x} = \mathbf{b}$, once U and L are known. If instead we computed A^{-1} and found $\mathbf{x} = A^{-1}\mathbf{b}$, the product $A^{-1}\mathbf{b}$ also requires n^2 flops. But there are at least two advantages in using the *LU*-technique. First of all, finding U requires about $n^3/3$ flops for large n, whereas finding A^{-1} requires n^3 flops. (See Exercise 15 of Section 2.1.) If $n = 1000$, the difference in computer time is considerable. Also, more computer memory would be used in reducing $(A \mid I)$ to $(I \mid A^{-1})$ than in the efficient way we record L as we find U, illustrated in the combined $L\backslash U$ display (2).

We give a specific illustration in which keeping the record L is useful.

Example 4   Let

$$A = \begin{pmatrix} 1 & 2 & -1 \\ -2 & -5 & 3 \\ -1 & -3 & 0 \end{pmatrix} \quad \text{and} \quad \mathbf{b} = \begin{pmatrix} 9 \\ -17 \\ -44 \end{pmatrix}.$$

Solve the linear system $A^3\mathbf{x} = \mathbf{b}$.

Solution   We view $A^3\mathbf{x} = \mathbf{b}$ as $A(A^2\mathbf{x}) = \mathbf{b}$, and substitute $\mathbf{y} = A^2\mathbf{x}$ to obtain $A\mathbf{y} = \mathbf{b}$. We can solve this equation for $\mathbf{y}$. Then we write $A^2\mathbf{x} = \mathbf{y}$ as $A(A\mathbf{x}) = \mathbf{y}$ or $A\mathbf{z} = \mathbf{y}$, where $\mathbf{z} = A\mathbf{x}$. We then solve $A\mathbf{z} = \mathbf{y}$ for $\mathbf{z}$. Finally, we solve $A\mathbf{x} = \mathbf{z}$ for the desired $\mathbf{x}$. Since we are using the same coefficient matrix A each time, it is efficient to find the matrices L and U and proceed as in Example 3 to find $\mathbf{y}, \mathbf{z},$ and $\mathbf{x}$ in turn.

We easily find that the matrices U and L are given by

$$U = \begin{pmatrix} 1 & 2 & -1 \\ 0 & -1 & 1 \\ 0 & 0 & -2 \end{pmatrix} \quad \text{and} \quad L = \begin{pmatrix} 1 & 0 & 0 \\ -2 & 1 & 0 \\ -1 & 1 & 1 \end{pmatrix}.$$

Applying the record in L to $\mathbf{b}$, we obtain

$$\mathbf{b} = \begin{pmatrix} 9 \\ -17 \\ -44 \end{pmatrix} \sim \begin{pmatrix} 9 \\ 1 \\ -35 \end{pmatrix} \sim \begin{pmatrix} 9 \\ 1 \\ -36 \end{pmatrix}.$$

From the matrix U, we find that

$$\mathbf{y} = \begin{pmatrix} -7 \\ 17 \\ 18 \end{pmatrix}.$$

To solve $A\mathbf{z} = \mathbf{y}$, we apply the record in L to $\mathbf{y}$:

$$\mathbf{y} = \begin{pmatrix} -7 \\ 17 \\ 18 \end{pmatrix} \sim \begin{pmatrix} -7 \\ 3 \\ 11 \end{pmatrix} \sim \begin{pmatrix} -7 \\ 3 \\ 8 \end{pmatrix}.$$

Using U, we obtain

$$\mathbf{z} = \begin{pmatrix} 3 \\ -7 \\ -4 \end{pmatrix}.$$

Finally, to solve $A\mathbf{x} = \mathbf{z}$, we apply the record in L to $\mathbf{z}$:

$$\mathbf{z} = \begin{pmatrix} 3 \\ -7 \\ -4 \end{pmatrix} \sim \begin{pmatrix} 3 \\ -1 \\ -1 \end{pmatrix} \sim \begin{pmatrix} 3 \\ -1 \\ 0 \end{pmatrix}.$$

Using U, we find that

$$\mathbf{x} = \begin{pmatrix} 1 \\ 1 \\ 0 \end{pmatrix}. \quad \triangleleft$$

The program LUFACTOR available with this text can be used to find the matrices L and U, and has an option for iteration to solve a system $A^m\mathbf{x} = \mathbf{b}$ as we did in Example 4.

THE FACTORIZATION $A = LU$

This heading shows why the matrix L we described is algebraically interesting: We have $A = LU$.

Example 5   Illustrate $A = LU$ for the matrices obtained in Example 3.

Solution   From Example 3, we have

$$LU = \begin{pmatrix} 1 & 0 & 0 & 0 \\ -2 & 1 & 0 & 0 \\ -1 & -2 & 1 & 0 \\ 5 & -2 & -3 & 1 \end{pmatrix} \begin{pmatrix} 1 & -2 & 0 & 3 \\ 0 & -1 & 1 & 0 \\ 0 & 0 & -2 & 6 \\ 0 & 0 & 0 & 3 \end{pmatrix} = \begin{pmatrix} 1 & -2 & 0 & 3 \\ -2 & 3 & 1 & -6 \\ -1 & 4 & -4 & 3 \\ 5 & -8 & 4 & 0 \end{pmatrix} = A. \quad \triangleleft$$

It is not difficult to establish that we always have $A = LU$. Recall from Eq. (1) that

$$E_h E_{h-1} \cdot \cdot \cdot E_2 E_1 A = U, \tag{3}$$

for elementary matrices E_i corresponding to elementary row operations consisting of multiplying a row by a constant and adding the result to a lower row. From Eq. (3), we obtain

$$A = E_1^{-1} E_2^{-1} \cdot \cdot \cdot E_{h-1}^{-1} E_h^{-1} U. \tag{4}$$

We proceed to show that the matrix L is equal to the product $E_1^{-1} E_2^{-1} \cdot \cdot \cdot$ $E_{h-1}^{-1} E_h^{-1}$ appearing in (4). Now if E_i is obtained from the identity matrix by multiplying a row by r (possibly $r = 0$) and adding the result to another row, then E_i^{-1} is obtained from the identity matrix by multiplying the same row by $-r$ and adding the result to the other row. Since E_h corresponds to the last row operation performed, $E_h^{-1} I$ places the negative of the last multiplier to the nth row and $(n - 1)$st column of this product, precisely where it should appear in L. Similarly,

$$E_{h-1}^{-1}(E_h^{-1} I)$$

has the appropriate entry desired in L in the nth row and $(n - 2)$nd column;

$$E_{h-2}^{-1}(E_{h-1}^{-1} E_h^{-1} I)$$

has the appropriate entry desired in L in the $(n - 1)$st row and $(n - 2)$nd column; and so on. That is, the record entries of L may be created from the expression

$$E_1^{-1} E_2^{-1} \cdot \cdot \cdot E_{h-1}^{-1} E_h^{-1} I$$

starting at the right with I and working to the left. This is the order shown by the color numbers below the main diagonal in the matrix

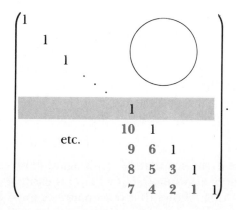

As elementary row operations are performed corresponding to multiplying the shaded row by constants and adding the results to lower rows, none of the entries already created below the main diagonal will be changed, for the zeros

to the right of the main diagonal lie over those entries. Thus

$$E_1^{-1}E_2^{-1} \cdots E_{h-1}^{-1}E_h^{-1}I = L \tag{5}$$

and $A = LU$ follows at once from Eq. (4).

Factorization of an invertible square matrix A into LU with L lower-triangular and U upper-triangular is not unique. For example, if r is a nonzero scalar, then rL is lower-triangular and $(1/r)U$ is upper-triangular, and if $A = LU$, then we also have $A = (rL)[(1/r)U]$. But let D be the $n \times n$ *diagonal* matrix having the same main diagonal as U, that is,

$$D = \begin{pmatrix} u_{11} & & & \\ & u_{22} & & \\ & & \ddots & \\ & & & u_{nn} \end{pmatrix}.$$

Let U^* be the upper-triangular matrix obtained from U by multiplying the ith row by $1/u_{ii}$ for $i = 1, 2, \ldots, n$. Then $U = DU^*$, and we have

$$A = LDU^*.$$

Now both L and U^* have all entries 1 on their main diagonals. This type of factorization is unique, as we now show.

Theorem 2.1 Unique Factorization

Let A be an $n \times n$ matrix. A factorization of the form $A = LDU$ where

L is lower-triangular with all main diagonal entries 1,
U is upper-triangle with all main diagonal entries 1,
D is diagonal with all main diagonal entries nonzero,

is unique.

Proof Suppose $A = L_1D_1U_1$ and $A = L_2D_2U_2$ are two such factorizations. Observe that both L_1^{-1} and L_2^{-1} are also lower-triangular, D_1^{-1} and D_2^{-1} are both diagonal, and U_1^{-1} and U_2^{-1} are both still upper-triangular. Just think how the matrix reductions of $(L_1 \mid I)$ or $(D_1 \mid I)$ or $(U_1 \mid I)$ to find the inverses look.

Now from $L_1D_1U_1 = L_2D_2U_2$, we obtain

$$L_2^{-1}L_1 = D_2U_2U_1^{-1}D_1^{-1}. \tag{6}$$

We easily see that $L_2^{-1}L_1$ is again lower-triangular with entries 1 on its main diagonal, while $D_2U_2U_1^{-1}D_1^{-1}$ is easily seen to be upper-triangular. Equation (6) then shows that both matrices must be I, so $L_1L_2^{-1} = I$ and $L_1 = L_2$. A similar argument starting over with $L_1D_1U_1 = L_2D_2U_2$ rewritten as

$$U_1U_2^{-1} = D_1^{-1}L_1^{-1}L_2D_2 \tag{7}$$

shows that $U_1 = U_2$. We then have $L_1D_1U_1 = L_1D_2U_1$, and multiplication on the left by L_1^{-1} and on the right by U_1^{-1} yields $D_1 = D_2$. ∎

SYSTEMS REQUIRING ROW INTERCHANGES

Let A be an invertible square matrix whose row reduction to an upper-triangular matrix U requires at least one row interchange. Then not all elementary matrices corresponding to the necessary row operations add multiples of row vectors to other row vectors. We can still write

$$E_h E_{h-1} \cdot \cdot \cdot E_2 E_1 A = U$$

so

$$A = E_1^{-1} E_2^{-1} \cdot \cdot \cdot E_{h-1}^{-1} E_h^{-1} U,$$

but now that some of these E_i^{-1} interchange rows, their product may not be lower-triangular. However, after we discover which row interchanges are necessary, we could start over, and make these necessary row interchanges in the matrix A before we start creating zeros below the diagonal. It is easy to see that the upper-triangular matrix U obtained would still be the same. Suppose, for example, that to obtain a nonzero element in pivot position in the ith row we interchange this ith row with a kth row further down in the matrix. The new ith row will be the same as though it had been put in the ith row position before the start of the reduction, for in either case, it has been modified during the reduction only by the addition of multiples of rows *above* the ith row position. As multiples of it are now added to rows below the ith row position, the same rows, except possibly for order, below the ith row position are created whether row interchange is performed during reduction or was completed before reduction started.

Interchanging some rows before the start of the reduction amounts to multiplication of A on the left by a sequence of elementary row-interchange matrices. Any product of elementary row-interchange matrices is called a **permutation matrix.** Thus we can form PA for a permutation matrix P, and PA will then admit a factorization $PA = LU$. We state this as a theorem.

Theorem 2.2 *LU*-Factorization

> Let A be an invertible square matrix. Then there exists a permutation matrix P, a lower-triangular matrix L, and an upper-triangular matrix U such that $PA = LU$.

Example 6   Illustrate Theorem 2.2 for the matrix

$$A = \begin{pmatrix} 1 & 3 & 2 \\ -2 & -6 & 1 \\ 2 & 5 & 7 \end{pmatrix}.$$

Solution   Starting to reduce A to upper-triangular form, we have

$$\begin{pmatrix} 1 & 3 & 2 \\ -2 & -6 & 1 \\ 2 & 5 & 7 \end{pmatrix} \sim \begin{pmatrix} 1 & 3 & 2 \\ 0 & 0 & 5 \\ 0 & -1 & 3 \end{pmatrix}$$

and we now find it necessary to interchange rows 2 and 3, which will then produce the desired U. Thus we may take

$$P = \begin{pmatrix} 1 & 0 & 0 \\ 0 & 0 & 1 \\ 0 & 1 & 0 \end{pmatrix}$$

and we have

$$PA = \begin{pmatrix} 1 & 3 & 2 \\ 2 & 5 & 7 \\ -2 & -6 & 1 \end{pmatrix} \sim \begin{pmatrix} 1 & 3 & 2 \\ 0 & -1 & 3 \\ 0 & 0 & 5 \end{pmatrix} = U.$$

The record matrix L for reduction of PA becomes

$$L = \begin{pmatrix} 1 & 0 & 0 \\ 2 & 1 & 0 \\ -2 & 0 & 1 \end{pmatrix},$$

and we easily check that

$$PA = \begin{pmatrix} 1 & 3 & 2 \\ 2 & 5 & 7 \\ -2 & -6 & 1 \end{pmatrix} = \begin{pmatrix} 1 & 0 & 0 \\ 2 & 1 & 0 \\ -2 & 0 & 1 \end{pmatrix}\begin{pmatrix} 1 & 3 & 2 \\ 0 & -1 & 3 \\ 0 & 0 & 5 \end{pmatrix} = LU. \quad \triangleleft$$

Suppose that when solving a large square linear system $A\mathbf{x} = \mathbf{b}$ with a computer, keeping the record L, we find that row interchanges become necessary. (Fortunately, this does not often happen.) One thing we could do is keep going and make the row interchanges to find an upper-triangular matrix. We could then start over, make those row interchanges first, just as in Example 6, obtaining a matrix PA, and then obtain U and L for the matrix PA instead. Of course, we would first have to compute $P\mathbf{b}$ when solving $A\mathbf{x} = \mathbf{b}$, since the system now becomes $PA\mathbf{x} = P\mathbf{b}$, before we could proceed with the record L and back substitution. This is an undesirable procedure, for it requires reduction of both A and PA, taking a total of about $2n^3/3$ flops, assuming that A is $n \times n$ and n is large. Surely it would be better to devise a method of record keeping that will also keep track of any row interchanges as they occur, and then apply this improved record to $\mathbf{b}$ and use back substitution to find the solution. We toss this suggestion out for enthusiastic programmers to consider on their own.

SUMMARY

1. If A is an $n \times n$ invertible matrix that can be row-reduced to an upper-triangular matrix U without row interchanges, then there exists a lower-triangular $n \times n$ matrix L such that $A = LU$.

2. The matrix L in part (1) can be found as follows: Start with the $n \times n$ identity matrix I. If during the reduction of A to U row i is multiplied

by r and the result is added to row k, replace the zero in row k and column i of the identity matrix by $-r$. The final result obtained from the identity matrix is the matrix L.

3. Once A has been reduced to U and L has been found, the solution of $A\mathbf{x} = \mathbf{b}$ for a new column vector $\mathbf{b}$ can be found by a computer using about n^2 flops for large n.

4. If A is as described in part (1), then A has a unique factorization of the form $A = LDU$, where

   L is lower-triangular with all diagonal entries 1,

   U is upper-triangular with all diagonal entries 1, and

   D is a diagonal matrix with all diagonal entries nonzero.

5. For any invertible matrix A, there exists a permutation matrix P such that PA can be row-reduced to an upper-triangular matrix U, and has the properties described for A in parts (1), (2), (3), and (4).

EXERCISES

1. Discuss briefly the need to worry about the time required to create the record matrix L when solving

$$A\mathbf{x} = \mathbf{b}_1, \mathbf{b}_2, \ldots, \mathbf{b}_s$$

for a large $n \times n$ matrix A.

2. Is there any practical value in creating the record matrix L when one needs to solve only a *single* square linear system $A\mathbf{x} = \mathbf{b}$, as we did in Example 3?

In Exercises 3–7, find the solution of $A\mathbf{x} = \mathbf{b}$ from the given combined $L\backslash U$ display of the matrix A and the given vector $\mathbf{b}$.

3. $L\backslash U = \begin{pmatrix} 1 & 4 \\ 3 & -2 \end{pmatrix}$, $\mathbf{b} = \begin{pmatrix} 2 \\ -4 \end{pmatrix}$

4. $L\backslash U = \begin{pmatrix} -2 & 5 \\ -2 & 1 \end{pmatrix}$, $\mathbf{b} = \begin{pmatrix} 3 \\ -7 \end{pmatrix}$

5. $L\backslash U = \begin{pmatrix} 1 & -3 & 4 \\ -2 & -1 & 9 \\ 0 & 1 & -6 \end{pmatrix}$, $\mathbf{b} = \begin{pmatrix} 2 \\ -2 \\ 8 \end{pmatrix}$

6. $L\backslash U = \begin{pmatrix} 1 & -4 & 2 \\ 0 & 2 & -1 \\ 3 & 2 & -2 \end{pmatrix}$, $\mathbf{b} = \begin{pmatrix} -3 \\ 2 \\ 3 \end{pmatrix}$

7. $L\backslash U = \begin{pmatrix} 1 & 0 & 0 & 1 \\ 0 & -1 & 2 & 1 \\ -1 & -2 & 1 & 3 \\ 2 & 1 & -1 & 3 \end{pmatrix}$, $\mathbf{b} = \begin{pmatrix} 4 \\ 7 \\ -8 \\ 14 \end{pmatrix}$

In Exercises 8–13, let A be the given matrix. Find a permutation matrix P, if necessary, and matrices L and U such that $PA = LU$. Check the answer using matrix multiplication. Then solve the system $A\mathbf{x} = \mathbf{b}$ using P, L, and U.

8. $A = \begin{pmatrix} 2 & -1 \\ 6 & -5 \end{pmatrix}$, $\mathbf{b} = \begin{pmatrix} 8 \\ 32 \end{pmatrix}$

9. $A = \begin{pmatrix} 2 & 1 & -3 \\ 6 & 3 & -8 \\ 2 & -1 & 5 \end{pmatrix}$, $\mathbf{b} = \begin{pmatrix} 6 \\ 17 \\ 0 \end{pmatrix}$

10. $A = \begin{pmatrix} 1 & 1 & -3 \\ 0 & 1 & 1 \\ 3 & -1 & 1 \end{pmatrix}$, $\mathbf{b} = \begin{pmatrix} -13 \\ 6 \\ -7 \end{pmatrix}$

11. $A = \begin{pmatrix} 1 & 2 & -1 \\ 3 & 7 & 2 \\ 4 & -2 & 1 \end{pmatrix}$, $\mathbf{b} = \begin{pmatrix} -3 \\ 1 \\ -2 \end{pmatrix}$

12. $A = \begin{pmatrix} 1 & -4 & 1 & -2 \\ 0 & 2 & -1 & 1 \\ 2 & -7 & -2 & 1 \\ 0 & 3 & 0 & -4 \end{pmatrix}$, $\mathbf{b} = \begin{pmatrix} -9 \\ 6 \\ 10 \\ -16 \end{pmatrix}$

13. $A = \begin{pmatrix} 1 & 2 & -3 & 0 \\ 2 & 5 & -6 & 0 \\ -1 & -2 & 1 & -1 \\ 4 & 10 & -9 & 1 \end{pmatrix}$, $\mathbf{b} = \begin{pmatrix} 8 \\ 17 \\ -8 \\ 33 \end{pmatrix}$

In Exercises 14–17, proceed as in Example 4 to solve the given system.

14. Solve $A^2\mathbf{x} = \mathbf{b}$ for $A = \begin{pmatrix} 1 & 1 \\ 3 & 0 \end{pmatrix}$, $\mathbf{b} = \begin{pmatrix} -11 \\ -6 \end{pmatrix}$.

15. Solve $A^4\mathbf{x} = \mathbf{b}$ for $A = \begin{pmatrix} -1 & 2 \\ 2 & -3 \end{pmatrix}$, $\mathbf{b} = \begin{pmatrix} 144 \\ -233 \end{pmatrix}$.

16. Solve $A^2\mathbf{x} = \mathbf{b}$ for $A = \begin{pmatrix} 2 & -1 & 0 \\ 4 & -1 & 2 \\ -6 & 2 & 0 \end{pmatrix}$, $\mathbf{b} = \begin{pmatrix} -2 \\ 14 \\ 12 \end{pmatrix}$.

17. Solve $A^3\mathbf{x} = \mathbf{b}$ if $A = \begin{pmatrix} -1 & 0 & 1 \\ 3 & 1 & 0 \\ -2 & 0 & 4 \end{pmatrix}$, $\mathbf{b} = \begin{pmatrix} 27 \\ 29 \\ 122 \end{pmatrix}$.

In Exercises 18–21, find the unique factorization LDU for the given matrix A.

18. The matrix in Exercise 8

19. The matrix in Exercise 10

20. The matrix in Exercise 11

21. The matrix in Exercise 13

In Exercises 22–24, use the available program LUFACTOR, or similar software, to find the combined L\U display (2) for the matrix A. Then solve $A\mathbf{x} = \mathbf{b}_i$ for each of the column vectors $\mathbf{b}_i$.

22. $A = \begin{pmatrix} 2 & 1 & -3 & 4 & 0 \\ 1 & 4 & 6 & -1 & 5 \\ -2 & 0 & 3 & 1 & 4 \\ 6 & 1 & 2 & 8 & -3 \\ 4 & 1 & 3 & 2 & 1 \end{pmatrix}$, $\mathbf{b}_i = \begin{pmatrix} 13 \\ 7 \\ 4 \\ 27 \\ 17 \end{pmatrix}$, $\begin{pmatrix} 10 \\ 0 \\ 10 \\ 60 \\ 20 \end{pmatrix}$, and $\begin{pmatrix} -20 \\ 96 \\ 53 \\ 5 \\ 26 \end{pmatrix}$

for $i = 1, 2,$ and 3, respectively.

23. $A = \begin{pmatrix} 1 & 3 & -5 & 2 & 1 \\ 4 & -6 & 10 & 8 & 3 \\ 3 & 6 & -1 & 4 & 7 \\ 2 & 1 & 11 & -3 & 13 \\ -6 & 3 & 1 & 4 & 21 \end{pmatrix}$, $\mathbf{b}_i = \begin{pmatrix} 0 \\ 20 \\ -9 \\ -31 \\ -43 \end{pmatrix}$, $\begin{pmatrix} -48 \\ 218 \\ 0 \\ 100 \\ 143 \end{pmatrix}$, and $\begin{pmatrix} 87 \\ -151 \\ 102 \\ -46 \\ 223 \end{pmatrix}$

for $i = 1, 2,$ and 3, respectively.

24. $A = \begin{pmatrix} -1 & 3 & 2 & 3 & 4 & 6 \\ 3 & 1 & 0 & -2 & 1 & 7 \\ 6 & 8 & -2 & 1 & 4 & 6 \\ 4 & -2 & 3 & 1 & 7 & 6 \\ 3 & -2 & 1 & 10 & 8 & 0 \\ 4 & -5 & 8 & -3 & 2 & 10 \end{pmatrix}$, $\mathbf{b}_i = \begin{pmatrix} 19 \\ 1 \\ -8 \\ 82 \\ 76 \\ 103 \end{pmatrix}$, $\begin{pmatrix} 82 \\ 2 \\ 82 \\ 31 \\ 92 \\ -61 \end{pmatrix}$, and $\begin{pmatrix} 25 \\ -3 \\ -24 \\ -73 \\ -115 \\ 20 \end{pmatrix}$

for $i = 1$, 2, and 3, respectively.

In Exercises 25–29, use the program LUFACTOR to solve the indicated system.

25. $A^5\mathbf{x} = \mathbf{b}$ for A and $\mathbf{b}$ in Exercise 14

26. $A^7\mathbf{x} = \mathbf{b}$ for A and $\mathbf{b}$ in Exercise 15

27. $A^6\mathbf{x} = \mathbf{b}$ for A and $\mathbf{b}$ in Exercise 16

28. $A^8\mathbf{x} = \mathbf{b}$ for A and $\mathbf{b}$ in Exercise 17

29. $A^3\mathbf{x} = \mathbf{b}_1$ for A and $\mathbf{b}_1$ in Exercise 24

The available program TIMING has an option to time the formation of the combined L\U display from a matrix A. The user specifies the magnitude n for an $n \times n$ linear system $A\mathbf{x} = \mathbf{b}$. The program then generates an $n \times n$ matrix A with entries in the interval $[-20, 20]$ and creates the L\U display shown in the matrix (2) in the text. The time required for the reduction is indicated. The user may then specify a number s for solution of $A\mathbf{x} = \mathbf{b}_1, \mathbf{b}_2, \ldots, \mathbf{b}_s$. The computer generates a column vector and solves s systems using the record in L and back substitution; the time to find these s solutions is also indicated. Recall that the reduction of A should require about $n^3/3$ flops for large n, and the solution for each column vector should require about n^2 flops. In Exercises 30–34, run the program TIMING or similar software, and see if the ratios of times obtained seems to conform to the ratios of the numbers of flops required, at least as n increases in size. Compare the time required to solve one system, including the computation of L\U, with the time required to solve a system of the same size using the Gauss method with back substitution. The times should be about the same.

Choose single-precision or double-precision timing and interpretive BASIC or compiled BASIC.

30. $n = 4$; $s = 1, 3, 12$

31. $n = 10$; $s = 1, 5, 10$

32. $n = 16$; $s = 1, 8, 16$

33. $n = 24$; $s = 1, 12, 24$

34. $n = 30$; $s = 1, 15, 30$

2.3
Pivoting, Scaling, and Ill-conditioned Matrices

SOME PROBLEMS ENCOUNTERED WITH COMPUTERS

A computer can't do absolutely precise arithmetic with real numbers. For example, any computer can only compute using an approximation of the number $\sqrt{2}$, never with the number $\sqrt{2}$ itself. Computers work using base-2 arithmetic, representing numbers as strings of zeros and ones. But they encounter the same problems that we encounter if we pretend that we are a computer using base-10 arithmetic, and capable of handling only some fixed finite number of significant digits. We will illustrate computer problems in base-10 notation since it is more familiar to all of us.

Suppose our base-10 computer is asked to computer the quotient $\frac{2}{3}$, and can represent a number in floating-point arithmetic using 8 significant figures. It represents $\frac{2}{3}$ as 0.66666666 without rounding off the last place to a 7, for it does not know that the next digit is also 6. Of course, software is usually designed so that what we see is a few figures less than the computer is actually working with, and the computer does **round off** the answer it prints for us. Our 8-figure computer might print 0.666667 as its answer to our problem of computing $\frac{2}{3}$. However, it works with its **truncated** answer for $\frac{2}{3}$, not with a rounded-off one. We will refer to all errors generated by such truncation as **roundoff errors.**

Most people realize that in an extended arithmetic computation by a computer, the roundoff error can accumulate to such an extent that the final result of the computation becomes meaningless. We will see that there are at least two situations in which a computer cannot meaningfully execute even a single arithmetic operation. In order to avoid working with long strings of digits, we assume that our computer can compute only to three significant figures in its work. Thus our 3-figure, base-10 computer computes the quotient $\frac{2}{3}$ as 0.666. We box the first computer problem that concerns us, and give an illustration.

> Addition of two numbers of very different magnitude may result in the loss of some or even all the significant figures of the smaller number.

Example 1   Evaluate our 3-figure computer's computation of

$$45.1 + .0725.$$

Solution   Since it can handle only three significant figures, our computer represents the actual sum 45.1725 as 45.1. In other words, the second summand .0725 might as well be zero, so far as our computer is concerned. The datum .0725 is completely lost. ◁

The difficulty illustrated in Example 1 can cause serious problems in attempting to solve a linear system $A\mathbf{x} = \mathbf{b}$ with a computer, as we will illustrate in a moment. First we box another problem a computer may have.

> Subtraction of nearly equal numbers can result in a loss of significant figures.

Example 2   Evaluate our 3-figure computer's computation of

$$\frac{2}{3} - \frac{665}{1000}.$$

Solution   The actual difference is

$$.666666 \cdot \cdot \cdot - .665 = .00166666 \cdot \cdot \cdot .$$

However, our 3-figure computer obtains

$$.666 - .665 = .001.$$

Two of the three significant figures with which the computer can work have been lost.   ◁

The difficulty illustrated in Example 2 is encountered if one tries to use a computer to do differential calculus.

The "cure" for both difficulties is to have the computer work with more significant figures. But using more significant figures requires more time to execute a program, and of course the same errors can occur "farther out." For example, the typical microcomputer of this decade, with software designed for routine computations, will compute $10^{10} + 10^{-10}$ as 10^{10}, losing the second summand 10^{-10} entirely.

PARTIAL PIVOTING

In row reduction of a matrix to echelon form a technique called *partial pivoting* is often used. In **partial pivoting,** one interchanges the row in which the pivot is to occur with a row farther down, if necessary, so that the pivot becomes as large in absolute value as possible. To illustrate, suppose that row reduction of a matrix to echelon form leads to an intermediate matrix

$$\begin{pmatrix} 2 & 8 & -1 & 3 & 4 \\ 0 & -2 & 3 & -5 & 6 \\ 0 & 4 & 1 & 2 & 0 \\ 0 & -7 & 3 & 1 & 4 \end{pmatrix}.$$

Using partial pivoting, we would then interchange rows 2 and 4 to use the entry -7 of maximum magnitude among the possibilities -2, 4, and -7 for pivots in the second column. That is, we form the matrix

$$\begin{pmatrix} 2 & 8 & -1 & 3 & 4 \\ 0 & -7 & 3 & 1 & 4 \\ 0 & 4 & 1 & 2 & 0 \\ 0 & -2 & 3 & -5 & 6 \end{pmatrix}$$

and continue with the reduction of this matrix.

We show by example the advantage of partial pivoting. Let us consider the linear system

$$.01x_1 + 100x_2 = 100$$
$$-100x_1 + 200x_2 = 100. \tag{1}$$

Example 3   Find the actual solution of the linear system (1). Then compare the result with that obtained by a 3-figure computer using the Gauss method with back substitution, but without partial pivoting.

Solution   First we find the actual solution:

$$\begin{pmatrix} .01 & 100 & | & 100 \\ -100 & 200 & | & 100 \end{pmatrix} \sim \begin{pmatrix} .01 & 100 & | & 100 \\ 0 & 1{,}000{,}200 & | & 1{,}000{,}100 \end{pmatrix}$$

and back substitution yields

$$x_2 = \frac{1{,}000{,}100}{1{,}000{,}200}$$
$$x_1 = \left(100 - \frac{100{,}010{,}000}{1{,}000{,}200}\right)100 = \frac{1{,}000{,}000}{1{,}000{,}200}.$$

Thus $x_1 \approx .9998$ and $x_2 \approx .9999$. On the other hand, our 3-figure computer obtains

$$\begin{pmatrix} .01 & 100 & | & 100 \\ -100 & 200 & | & 100 \end{pmatrix} \sim \begin{pmatrix} .01 & 100 & | & 100 \\ 0 & 1{,}000{,}000 & | & 1{,}000{,}000 \end{pmatrix}$$

which leads to

$$x_2 = 1$$
$$x_1 = (100 - 100)100 = 0.$$

The x_1-parts of the solution are very different. Our 3-figure computer completely lost the second-row data entries 200 and 100 in the matrix when it added 10,000 times the first row to the second row.   ◁

Example 4   Find the solution to the system (1) that our 3-figure computer would obtain using partial pivoting.

Solution   Using partial pivoting, the 3-figure computer obtains

$$\begin{pmatrix} .01 & 100 & | & 100 \\ -100 & 200 & | & 100 \end{pmatrix} \sim \begin{pmatrix} -100 & 200 & | & 100 \\ .01 & 100 & | & 100 \end{pmatrix}$$
$$\sim \begin{pmatrix} -100 & 200 & | & 100 \\ 0 & 100 & | & 100 \end{pmatrix}$$

which yields

$$x_2 = 1$$
$$x_1 = (100 - 200)/(-100) = 1.$$

This is close to the solution $x_1 \approx .9998$ and $x_2 \approx .9999$ obtained in Example 3, and much better than the erroneous $x_1 = 0$, $x_2 = 1$ obtained in Example 1 without pivoting. The data entries 200 and 100 in the second row of the initial matrix were not lost in this computation. ◁

To understand completely the reason for the difference between the solutions in Example 3 and in Example 4, consider again the linear system

$$.01x_1 + 100x_2 = 100$$
$$-100x_1 + 200x_2 = 100,$$

which was "solved" in Example 3. Multiplication of the first row by 10,000 produced a coefficient of x_2 of such magnitude compared with the second equation coefficient 200 that the coefficient 200 was totally destroyed in the ensuing addition. (Remember that we are using our 3-figure computer. In a less dramatic example, some significant figures of the smaller coefficient might still contribute to the sum.) If we use partial pivoting, the linear system becomes

$$-100x_1 + 200x_2 = 100$$
$$.01x_1 + 100x_2 = 100$$

and multiplication of the first equation by a *number less than 1* does not threaten the significant digits of the numbers in the second equation. This explains the success of partial pivoting in this situation.

Suppose now we multiply the first equation of the system (1) by 10,000, which of course does not alter the solution of the system. We then obtain the linear system

$$100x_1 + 1,000,000x_2 = 1,000,000$$
$$-100x_1 + \qquad 200x_2 = \qquad 100. \tag{2}$$

If we solved the system (2) using *partial pivoting,* we would not interchange the rows, since -100 is of no greater magnitude than 100. Addition of the first row to the second by our 3-figure computer again totally destroys the coefficient 200 in the second row. Exercise 1 asks for a check that the erroneous "solution" $x_1 = 0$, $x_2 = 1$ is again obtained. We could avoid this problem by **full pivoting.** In full pivoting, columns are also interchanged, if necessary, to make pivots as large as possible. That is, if a pivot is to be found in the row i and column j position of an intermediate matrix G, then not only rows but also columns are interchanged as needed to put the entry g_{rs} of greatest magnitude, where $r \geq i$ and $s \geq j$, in the pivot position. Thus full pivoting for the system (2) will lead to the matrix

$$\begin{matrix} x_2 & x_1 & \\ \begin{pmatrix} 1,000,000 & 100 & 1,000,000 \\ 200 & -100 & 100 \end{pmatrix}. \end{matrix} \tag{3}$$

Exercise 2 illustrates that row reduction by our 3-figure computer of the matrix (3) gives a reasonable solution of the system (2). Note, however, that we now have to do some bookkeeping and remember that the entries in column 1 of the matrix (3) are really the coefficients of x_2, not of x_1. For a matrix of any size, the search for elements of maximum magnitude and the bookkeeping required in full pivoting take a lot of computer time. Partial pivoting is frequently used, representing a compromise between time and accuracy. The program MATCOMP that we make available uses partial pivoting in its Gauss–Jordan reduction to reduced echelon form. Thus if the system (1) is modified so that the Gauss–Jordan method without pivoting fails to give a reasonable solution for a 20-figure computer, MATCOMP could probably handle it. However, one can no doubt create a similar modification of the 2×2 system (2) for which MATCOMP would give an erroneous "solution."

SCALING

We display again the system (2):

$$100x_1 + 1,000,000x_2 = 1,000,000$$
$$-100x_1 + 200x_2 = 100.$$

We might recognize that the number 1,000,000 dangerously dominates the data entries in the second row, at least so far as our 3-figure computer is concerned. We might multiply the first equation in the system (2) by .0001 to cut those two numbers down to size, essentially coming back to the system (1). Partial pivoting handles the system (1) well. Multiplication of an equation by a nonzero constant for such a purpose is known as **scaling.** Of course, one could equivalently scale by multiplying the second equation in the system (2) by 10,000 to bring its coefficients into line with the large numbers in the first equation.

We box one other computer problem that can sometimes be helped by scaling. In reducing a matrix to echelon form, we need to know whether the entry that appears in a pivot position as we start work on a column is truly nonzero, and indeed, whether there is any truly nonzero entry from that column to serve as pivot.

Due to roundoff error, a computed number that really should be zero is quite likely to be of small, nonzero magnitude.

Taking this into account, one usually programs row-echelon reduction so that entries of unexpectedly small size are changed to zero. MATCOMP finds the smallest nonzero magnitude m among 1 and all the coefficient data supplied for the linear system, and sets $E = rm$, where r is specified by the user. (Default r is .0001.) In reduction of the coefficient matrix, a computed entry of magnitude less than E is replaced by zero. The same procedure is followed in YUREDUCE. Whatever computed number we program a computer to choose for E in a program like MATCOMP, we will be able to devise some linear system for which E is either too large or too small to give the correct result.

An equivalent procedure to the one in MATCOMP that we just outlined is to *scale* the original data for the linear system in such a way that the smallest nonzero entry is of magnitude roughly 1, and then always use the same value, perhaps 10^{-4}, for E.

When one of the authors first started experimenting with a computer, he was horrified to discover that a matrix inversion routine built into the mainframe BASIC program of a major computer company would refuse to invert a matrix if all the entries were small enough. An error message like

"Nearly singular matrix. Inversion impossible."

would appear. He considers a 2×2 matrix

$$\begin{pmatrix} a & b \\ c & d \end{pmatrix}$$

to be nearly singular if and only if lines in the plane having equations of the form

$$ax + by = r$$
$$cx + dy = s$$

are nearly parallel. Now the lines

$$10^{-9}x + \quad 0y = 1$$
$$0x + 10^{-9}y = 1$$

are actually perpendicular, the first one being vertical and the second horizontal. That is as far away from parallel as one can get! It annoyed the author greatly to have the matrix

$$\begin{pmatrix} 10^{-8} & 0 \\ 0 & 10^{-9} \end{pmatrix}$$

called "nearly singular." The inverse is obviously

$$\begin{pmatrix} 10^{8} & 0 \\ 0 & 10^{9} \end{pmatrix}.$$

A scaling routine was promptly written to be executed before calling the inversion routine. The matrix was multiplied by a constant that would bring the smallest nonzero magnitude to at least one, then the inversion subroutine was used, and the result rescaled to provide the inverse of the original matrix. For example, applied to the matrix just discussed, this procedure becomes

$$\begin{pmatrix} 10^{-8} & 0 \\ 0 & 10^{-9} \end{pmatrix} \text{ multiplied by } 10^{9} \text{ becomes } \begin{pmatrix} 10 & 0 \\ 0 & 1 \end{pmatrix},$$

$$\text{inverted becomes } \begin{pmatrix} \frac{1}{10} & 0 \\ 0 & 1 \end{pmatrix},$$

$$\text{multiplied by } 10^{9} \text{ becomes } \begin{pmatrix} 10^{8} & 0 \\ 0 & 10^{9} \end{pmatrix}.$$

Having more programming experience now, this author is much more charitable and understanding. The user may also find unsatisfactory things about MATCOMP. We hope that this little anecdote has helped explain the notion of scaling.

ILL-CONDITIONED MATRICES

The line $x + y = 100$ in the plane has x-intercept 100 and y-intercept 100, as shown in Fig. 2.1. The line $x + .9y = 100$ also has x-intercept 100, but y-intercept larger than 100. The two lines are almost parallel. The common x-intercept shows that the solution of the linear system

$$\begin{aligned} x + y &= 100 \\ x + .9y &= 100 \end{aligned} \tag{4}$$

is $x = 100$, $y = 0$, as illustrated in Fig. 2.2. Now the line $.9x + y = 100$ has y-intercept 100 but x-intercept larger than 100, so the linear system

$$\begin{aligned} x + y &= 100 \\ .9x + y &= 100 \end{aligned} \tag{5}$$

has the very different solution $x = 0$, $y = 100$, as shown in Fig. 2.3. The systems (4) and (5) are examples of **ill-conditioned** or **unstable** systems; small changes in the coefficients or in the constants on the right-hand sides can produce very great changes in the solutions. We say that a matrix A is **ill-conditioned** if a linear system $A\mathbf{x} = \mathbf{b}$ having A as coefficient matrix is ill-conditioned. For two equations in two unknowns, solving an ill-conditioned system corresponds to finding the intersection of two nearly parallel lines, as shown in Fig. 2.4. Changing a coefficient of x or y slightly in one equation changes the slope of that line slightly, but may generate a big change in the location of the point of intersection of the two lines.

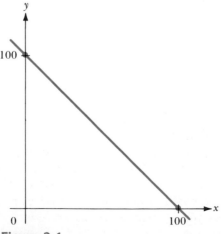

Figure 2.1
The line $x + y = 100$.

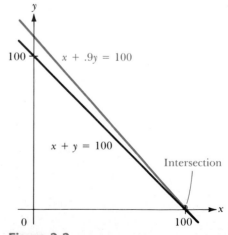

Figure 2.2
The lines are almost parallel.

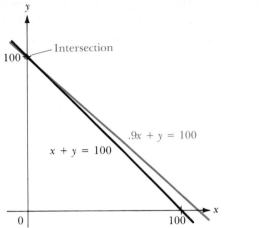

Figure 2.3
A very different solution than in Fig. 2.2.

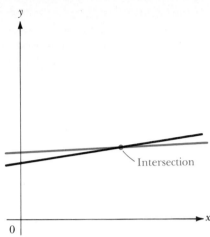

Figure 2.4
The intersection of two nearly parallel lines.

Computers have a lot of trouble finding accurate solutions of ill-conditioned systems like (4) and (5), because the small roundoff errors created by the computer can produce large changes in the solution. Pivoting and scaling usually don't help the situation; the systems are basically unstable. Note that the coefficients of x and y in the systems (4) and (5) are of comparable magnitude.

Among the most famous ill-conditioned matrices are the Hilbert matrices. These are very bad matrices named after a very good mathematician, David Hilbert (1862–1943)! The entry in the ith row and jth column of a Hilbert matrix is $1/(i + j - 1)$. Thus if we let H_n be the $n \times n$ Hilbert matrix, we have

$$H_2 = \begin{pmatrix} 1 & \frac{1}{2} \\ \frac{1}{2} & \frac{1}{3} \end{pmatrix}, \qquad H_3 = \begin{pmatrix} 1 & \frac{1}{2} & \frac{1}{3} \\ \frac{1}{2} & \frac{1}{3} & \frac{1}{4} \\ \frac{1}{3} & \frac{1}{4} & \frac{1}{5} \end{pmatrix}, \quad \text{and so on.}$$

It can be shown that H_n is invertible for all n, so a square linear system $H_n \mathbf{x} = \mathbf{b}$ has a unique solution, but the solution may be very hard to find. When the matrix is reduced to echelon form, entries of surprisingly small magnitude appear. Scaling a row in which all entries are close to zero may help a bit.

Bad as the Hilbert matrices are, powers of them are even worse. The software we make available includes a program called HILBERT, which is modeled on YUREDUCE. The computer generates a Hilbert matrix of the size we specify up to 10 by 10. It will then raise the matrix to the power 2, 4, 8, or 16 if we so request. We may then proceed roughly as in YUREDUCE. Programs like YUREDUCE and HILBERT should help us understand this section, for we can watch and see just what is happening as we reduce the matrix. MAT-COMP, which simply spits out answers, may produce an absurd result, but we have no way of really knowing exactly where things went wrong.

SUMMARY

1. Addition of numbers of very different magnitude by a computer can cause loss of some or all of the significant figures in the number of smaller magnitude.

2. Subtraction of nearly equal numbers by a computer can cause loss of significant figures.

3. Due to roundoff error, a computer may obtain a nonzero value for a number that should be zero. To attempt to handle this problem, a computer program might assign certain computed numbers the value zero whenever the numbers have magnitude less than some predetermined small positive number.

4. In partial pivoting, the pivot in each column is created by changing rows, if necessary, so that the pivot has at least the maximum magnitude of any entry below it in that column. Partial pivoting may be helpful in avoiding the problem stated in part (1) above.

5. In full pivoting, columns are interchanged as well as rows, if necessary, to create pivots of maximum possible magnitude. Using full pivoting requires much more computer time than partial pivoting, and bookkeeping is necessary to keep track of the relationship between the columns and the variables. Partial pivoting is more commonly used.

6. Scaling, multiplication of a row by a nonzero constant, can be used to reduce the size of entries that threaten to dangerously dominate entries in lower rows, or to increase the size of entries in a row where some entries are very small.

7. A linear system $A\mathbf{x} = \mathbf{b}$ is ill-conditioned or unstable if small changes in the numbers can produce a large change in the solution. The matrix A is then also called ill-conditioned.

8. Hilbert matrices, which are square matrices with entry $1/(i + j - 1)$ in the ith row and jth column, are examples of ill-conditioned matrices.

9. With the present technology, it appears hopeless to write a computer program that will successfully handle every linear system involving even very small coefficient matrices, say of size at most 10×10.

EXERCISES

1. Find the "solution" by a 3-figure computer of the system

$$100x_1 + 1,000,000x_2 = 1,000,000$$
$$-100x_1 + \quad\quad 200x_2 = \quad\quad 100$$

using just partial pivoting.

2. Repeat Exercise 1, but use full pivoting.

3. Find the "solution" without pivoting by a 5-figure computer of the linear system in Exercise 1. Is the solution reasonably accurate? If so, modify the system a bit to obtain one for which a 5-figure computer does not find a reasonable solution without pivoting.

4. Modify the linear system in Exercise 1 so that an 8-figure computer (roughly the usual single-precision computer) will not obtain a reasonable solution without partial pivoting.

5. Repeat Exercise 4 for an 18-figure computer (roughly the usual double-precision computer).

6. Find a linear system with two equations in x and y such that a change of .001 in the coefficient of x in the second equation produces a change of at least 1,000,000 in both of the values x and y in a solution.

7. Let A be an invertible square matrix. Show that the following scaling routine for finding A^{-1} using a computer is *mathematically* correct.
   a) Find the minimum absolute value m of all nonzero entries in A.
   b) Multiply A by an integer $n > 1/m$.
   c) Find the inverse of the resulting matrix.
   d) Multiply the matrix obtained in (c) by n to get A^{-1}.

Use the programs MATCOMP, YUREDUCE, and HILBERT, or similar software, for the remaining exercises.

8. Use YUREDUCE in single-precision computational mode to "solve" the system

$$.00001x_1 + 10,000x_2 = 1$$
$$10,000x_1 + 20,000x_2 = 1$$

without using any pivoting. Check the answer mentally to see if it is approximately correct. If it is, put more zeros in the large numbers and more zeros after the decimal point in the small one until a system is obtained in which the "solution" without pivoting is erroneous.

9. Use YUREDUCE to solve the system in Exercise 8, which did not give a reasonable solution without pivoting, but this time use partial pivoting.

10. Repeat Exercise 9, but this time use full pivoting. Compare the answer with that in Exercise 9.

11. Repeat Exercise 8 with the system that did not yield a reasonable solution, but use double-precision mode in YUREDUCE this time. Again, do not use any pivoting.

12. Modify the system in Exercise 8 to obtain one for which YUREDUCE in double-precision mode without pivoting yields an erroneous "solution."

13. Take the system formed in Exercise 12 and solve it using YUREDUCE in double precision, and using (a) partial pivoting and (b) full pivoting. Compare answers. Is the solution reasonable?

14. See how MATCOMP handles the linear system formed in Exercise 12.

15. Experiment with MATCOMP and see if it gives a "nearly singular matrix" message when finding the inverse of a 2×2 diagonal matrix

$$rI = r\begin{pmatrix} 1 & 0 \\ 0 & 1 \end{pmatrix},$$

where r is a sufficiently small nonzero number. Use the default for roundoff control in MATCOMP.

16. Using MATCOMP, use the scaling routine suggested in Exercise 7 to find the inverse of the matrix

$$
\begin{pmatrix}
.000001 & -.000003 & .000011 & .000006 \\
-.000002 & .000013 & .000007 & .000010 \\
.000009 & -.000011 & 0 & -.000005 \\
.000014 & -.000008 & -.000002 & .000003
\end{pmatrix}.
$$

Then see if the same answer is obtained without using the scaling routine. Check the inverse in each case by multiplying by the original matrix.

Use the program HILBERT or similar software in the remaining exercises. Since the Hilbert matrices are nonsingular, diagonal entries computed during the elimination should never be zero. Except in Exercise 24, enter 0 for r when r is requested during a run of HILBERT. Experiment with both single-precision and double-precision computing in the exercises that follow.

17. Solve $H_2\mathbf{x} = \mathbf{b}, \mathbf{c}$, where

$$
\mathbf{b} = \begin{pmatrix} 2 \\ 9 \end{pmatrix} \quad \text{and} \quad \mathbf{c} = \begin{pmatrix} 3 \\ 8 \end{pmatrix}.
$$

Use just one run of HILBERT, that is, solve both systems at once. Note that the components of $\mathbf{b}$ and $\mathbf{c}$ differ by just 1. Find the difference in the components of the two solution vectors.

18. Repeat Exercise 17, changing the coefficient matrix to $H_2{}^4$.

19. Repeat Exercise 17, using as coefficient matrix H_4 with

$$
\mathbf{b} = \begin{pmatrix} 2 \\ 1 \\ 4 \\ 6 \end{pmatrix} \quad \text{and} \quad \mathbf{c} = \begin{pmatrix} 3 \\ 0 \\ 5 \\ 7 \end{pmatrix}.
$$

20. Repeat Exercise 19, using as coefficient matrix $H_4{}^4$.

21. Find the inverse of H_2, and use the I menu option to test whether the computed inverse is correct.

22. Continue Exercise 21 by trying to find the inverses of H_2 raised to the powers 2, 4, 8, and 16. Was it always possible to reduce the power of the Hilbert matrix to diagonal form in both single-precision and double-precision computation? If not, what happened? Why did it happen? Was scaling (1 an entry) of any help? For those cases in which reduction to diagonal form was possible, did the computed inverse seem to be reasonably accurate when tested?

23. Repeat Exercises 21 and 22 using H_4 and the various powers of it. Are problems encountered for lower powers this time?

24. We have seen that in reducing a matrix, we may wish to instruct the computer to assign the value zero to computed entries of sufficiently small magnitude, since entries that should be zero might otherwise be left nonzero due to roundoff error. Use HILBERT to try to solve the linear system

$$
H_3{}^2\mathbf{x} = \mathbf{b}, \quad \text{where} \quad \mathbf{b} = \begin{pmatrix} -3 \\ 1 \\ 4 \end{pmatrix},
$$

entering as r the number .0001 when requested. Will routine Gauss–Jordan elimination using HILBERT solve the system? Will HILBERT solve it using scaling (1 an entry)?

Vector Spaces

For an efficient study of mathematics, we should try to group together objects having the same mathematical structure. We can then study these objects all at once by considering properties and theorems that can be derived just from their common mathematical structure. Organizing mathematics in this way avoids repeating the same arguments each time we make use of an object having this structure. Viewed from this perspective, linear algebra becomes the study of all objects that have a *vector-space structure*.

Let $\mathbb{R}^n$ be the set of all vectors with n components. These Euclidean spaces $\mathbb{R}^n$ for each positive integer n are vector spaces. We all have an intuitive idea that the space $\mathbb{R}$ of real numbers is one-dimensional, the space $\mathbb{R}^2$ is two-dimensional, the space $\mathbb{R}^3$ is three-dimensional, and so on. In advanced work, it can be shown that there is an analogous concept of *dimension* for every vector space. Furthermore, it can be shown that every real vector space V of finite dimension n is algebraically indistinguishable from $\mathbb{R}^n$; the vectors in V can be *coordinatized* using the vectors in $\mathbb{R}^n$. This means that a study of finite-dimensional real vector spaces amounts to a study of the vector spaces $\mathbb{R}^n$.

This text is primarily concerned with the space $\mathbb{R}^n$. We begin this chapter with the study of geometry in $\mathbb{R}^n$, especially the geometric meaning of vector addition and of multiplication of a vector by a scalar. It is always helpful to be able to draw pictures when studying mathematics. Section 3.2 summarizes the algebraic properties of $\mathbb{R}^n$, and uses them as motivation to define the general concept of a vector space. Examples of vector spaces other than $\mathbb{R}^n$ are given. The remainder of the chapter deals with properties of $\mathbb{R}^n$ and other vector spaces. Whenever a proof of a property of $\mathbb{R}^n$ really amounts to a proof of the property for a general vector space, we will phrase the property and proof in terms of general vector spaces. However, the bulk of our work is done in the context of $\mathbb{R}^n$. We believe that this is appropriate for an introductory course.

3.1
The Geometry of $\mathbb{R}^n$

The *vector space* $\mathbb{R}^n$ may be regarded as either a space of row vectors or a space of column vectors. Unless we indicate otherwise, we will give preference to row-vector notation to save space.

GEOMETRIC REPRESENTATION OF VECTORS

We are already familiar with the Euclidean two-space $\mathbb{R}^2$ and the usual manner in which it is represented as the x, y-plane using a rectangular coordinate system. We will generalize our work in $\mathbb{R}^2$ to $\mathbb{R}^n$; consequently, we find it convenient to use (x_1, x_2) as well as (x, y). Thus each vector $\mathbf{v} = (v_1, v_2)$ in $\mathbb{R}^2$ has associated with it a unique point in the plane, as shown in Fig. 3.1. The *vector* $\mathbf{v}$ can be represented geometrically by the arrow that begins at the origin $\mathbf{0} = (0, 0)$ and terminates at the *point* (v_1, v_2). Analytically, there is no distinction between the *vector* $\mathbf{v}$ and the *point* (v_1, v_2), for each consists of the ordered pair (v_1, v_2). In physical applications, we think of a vector as having length and direction, which are concepts that can be represented geometrically by an arrow. Thus it is customary to regard the *vector* $\mathbf{v} = (v_1, v_2)$ geometrically as an arrow, and the *point* (v_1, v_2) as the tip of the arrow.

The **magnitude** $\|\mathbf{v}\|$ of $\mathbf{v} = (v_1, v_2)$ is defined to be the length of the arrow in Fig. 3.1. Using the Pythagorean theorem, we have

$$\|\mathbf{v}\| = \sqrt{v_1{}^2 + v_2{}^2}. \tag{1}$$

THE IDEA OF AN N-DIMENSIONAL SPACE FOR $N > 3$ reached acceptance gradually during the nineteenth century; it is thus difficult to pinpoint a first "invention" of this concept. Among the various early uses of this notion are its appearances in a work on the divergence theorem by the Russian mathematician Mikhail Ostrogradskii (1801–1862) in 1836, in the geometrical tracts of Hermann Grassmann (1809–1877) in the early 1840s, and in a brief paper of Arthur Cayley in 1846. Unfortunately, the first two authors were virtually ignored in their lifetimes. In particular, the work of Grassman was quite philosophical and extremely difficult to read. Cayley's note merely stated that one can generalize certain results to dimensions greater than three "without recourse to any metaphysical notion with regard to the possibility of a space of four dimensions." Sir William Rowan Hamilton (1805–1865), in an 1841 letter, also noted that "it *must* be possible, in *some* way or other, to introduce not only triplets but *polyplets*, so as in some sense to satisfy the symbolical equation

$$a = (a_1, a_2, \ldots a_n);$$

a being here one symbol, as indicative of one (complex) thought; and $a_1, a_2, \ldots a_n$ denoting n real numbers, positive or negative."

Hamilton, whose work on quaternions will be mentioned later, and who spent much of his professional life as the Royal Astronomer of Ireland, is most famous for his work in dynamics. As Erwin Schrödinger wrote, "the Hamiltonian principle has become the cornerstone of modern physics, the thing with which a physicist expects *every* physical phenomenon to be in conformity."

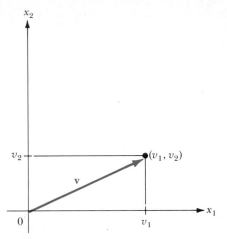

Figure 3.1
Geometric representation of a vector in $\mathbb{R}^2$.

Example 1   Represent the vector $\mathbf{v} = (3, -4)$ geometrically and find its **magnitude**.

Solution   The vector $\mathbf{v} = (3, -4)$ has magnitude

$$\|\mathbf{v}\| = \sqrt{3^2 + (-4)^2} = \sqrt{25} = 5$$

and is shown in Fig. 3.2.   ◁

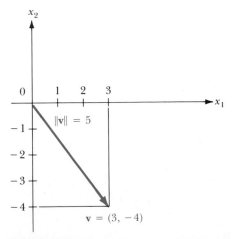

Figure 3.2
The magnitude of v.

To represent $\mathbb{R}^3$ geometrically, we choose three perpendicular lines through the origin $\mathbf{0} = (0, 0, 0)$ as coordinate axes, as shown in Fig. 3.3. The coordinate system in this figure is called a *right-hand system* because when the fingers of the right hand are curved in the direction required to rotate the positive x_1-axis toward the positive x_2-axis, the right thumb points up the x_3-axis.

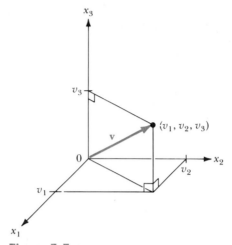

Figure 3.3
Geometric representation of a vector in $\mathbb{R}^3$.

Each vector $\mathbf{v} = (v_1, v_2, v_3)$ in $\mathbb{R}^3$ corresponds to a unique point in space, as shown in Fig. 3.3. The *vector* $\mathbf{v}$ is again represented by an arrow from the origin $\mathbf{0} = (0, 0, 0)$ to the *point* (v_1, v_2, v_3). The **magnitude** $\|\mathbf{v}\|$ of $\mathbf{v}$ is the length of this arrow. This magnitude appears in Fig. 3.3 as the hypotenuse of a right triangle whose altitude is $|v_3|$ and whose base in the x_1, x_2-plane has length $\sqrt{v_1{}^2 + v_2{}^2}$. Using the Pythagorean theorem, we obtain

$$\|\mathbf{v}\| = \sqrt{v_1{}^2 + v_2{}^2 + v_3{}^2}. \tag{2}$$

Example 2   Represent the vector $\mathbf{v} = (2, 3, 4)$ geometrically and find its magnitude.

Solution   The vector $\mathbf{v} = (2, 3, 4)$ has magnitude $\|\mathbf{v}\| = \sqrt{2^2 + 3^2 + 4^2} = \sqrt{29}$ and is represented in Fig. 3.4.   ◁

By analogy, we consider any vector $\mathbf{v} = (v_1, v_2, \cdots v_n)$ in $\mathbb{R}^n$ to be represented by an arrow from the origin $\mathbf{0} = (0, 0, \cdots, 0)$ to the point $\mathbf{v}$, as shown in Fig. 3.5. Note that this picture is less cluttered than Figs. 3.3 and 3.4 because we do not attempt to represent n mutually perpendicular coordinate axes for the variables $x_1, x_2 \cdots, x_n$. Indeed, we will leave out the coordinate axes whenever it is convenient so that our diagrams will be at once cleaner and more general.

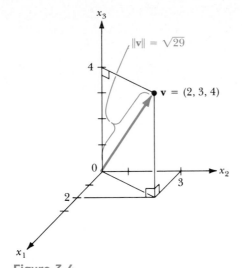

Figure 3.4
The magnitude of v.

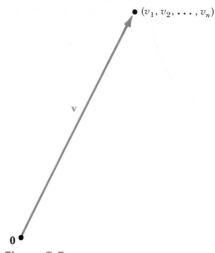

Figure 3.5
Geometric representation of an *n*-component vector.

As suggested by Eqs. (1) and (2), we *define* the magnitude $\|\mathbf{v}\|$ of the vector $\mathbf{v}$ as follows.

Definition 3.1   Magnitude of a Vector in $\mathbb{R}^n$

Let $\mathbf{v} = (v_1, v_2, \cdots, v_n)$ be a vector in $\mathbb{R}^n$. The **magnitude** of $\mathbf{v}$ is

$$\|\mathbf{v}\| = \sqrt{v_1^2 + v_2^2 + \cdots + v_n^2}. \tag{3}$$

Example 3   Find the magnitude of the vector

$$\mathbf{v} = (-2, 1, 3, -1, 4, 2, 1).$$

Solution   We have

$$\|\mathbf{v}\| = \sqrt{(-2)^2 + 1^2 + 3^2 + (-1)^2 + 4^2 + 2^2 + 1^2} = \sqrt{36} = 6. \quad \triangleleft$$

ADDITION, SUBTRACTION, AND SCALAR MULTIPLICATION

We already know how to add and subtract matrices of the same size and how to multiply a matrix by a scalar. Since the vectors in $\mathbb{R}^n$ can be regarded as $1 \times n$ matrices, the following definition is a review.

Definition 3.2   Vector-space Operations in $\mathbb{R}^n$

Let $\mathbf{v} = (v_1, v_2, \cdots, v_n)$ and $\mathbf{w} = (w_1, w_2, \cdots, w_n)$ be vectors in $\mathbb{R}^n$. The vectors are added and subtracted as follows.

> **Vector addition:** $\mathbf{v} + \mathbf{w} = (v_1 + w_1, v_2 + w_2, \cdots, v_n + w_n)$
>
> **Vector subtraction:** $\mathbf{v} - \mathbf{w} = (v_1 - w_1, v_2 - w_2, \cdots, v_n - w_n)$
>
> If r is any scalar, the vector $\mathbf{v}$ is multiplied by r as follows.
>
> **Scalar multiplication:** $r\mathbf{v} = (rv_1, rv_2, \cdots, rv_n)$

The sum of the vectors $\mathbf{v}$ and $\mathbf{w}$ can be represented geometrically as shown in Fig. 3.6. The tip of the vector $\mathbf{v} + \mathbf{w}$ can be located by drawing the vector $\mathbf{w}$ as an arrow emanating from the tip of $\mathbf{v}$. This is nothing more than a translation of axes, regarding the point at the tip of $\mathbf{v}$ as a local origin. Equivalently, one can translate the axes to the point at the tip of $\mathbf{w}$, and draw the vector $\mathbf{v}$ as an arrow emanating from this point. When both of these translated vectors are drawn, one has *completed the parallelogram* shown in Fig. 3.6 and $\mathbf{v} + \mathbf{w}$ is the indicated diagonal.

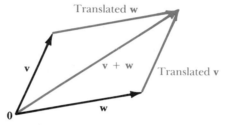

Figure 3.6
Geometric representation of v + w in $\mathbb{R}^n$.

APPLICATION TO FORCE VECTORS

A practical physical model of vector addition is given by force vectors. We illustrate for force vectors in the plane. Suppose we are moving some object in the plane by pushing it. The motion of the object is influenced by the *direction* in which we are pushing and how *hard* we are pushing. Thus it is natural to represent the force with which we are pushing by an arrow pointing in the direction in which we are pushing, and having a length that represents how hard we are pushing. Such an arrow represents the physicist's notion of a *force vector*. If we choose coordinate axes in the plane at the point where the force is applied to the object, then the force can be represented by an arrow emanating from the origin.

Suppose two forces represented by vectors $\mathbf{v} = (v_1, v_2)$ and $\mathbf{w} = (w_1, w_2)$ are acting on a body at $\mathbf{0} = (0, 0)$ in the plane as shown in Fig. 3.7. It can be shown that the effect the two forces have on the body is the same as the effect of a single force represented by the diagonal of the parallelogram determined by the given vectors, as shown in the figure. We see from the figure that the vector along the diagonal is $(v_1 + w_1, v_2 + w_2)$; this vector is the sum of the two given vectors.

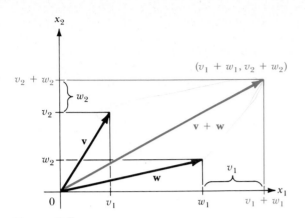

Figure 3.7
Geometric representation of v + w in $\mathbb{R}^2$.

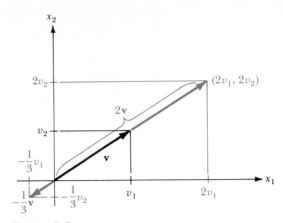

Figure 3.8
Geometric representation of scalar multiplication in $\mathbb{R}^2$.

If $\mathbf{v} = (v_1, v_2)$ is a force vector, then the vector corresponding to double the force in the same direction is $2\mathbf{v} = (2v_1, 2v_2)$, while the vector corresponding to one-third the force in the opposite direction is $(-1/3)\mathbf{v} = (-v_1/3, -v_2/3)$, as shown in Fig. 3.8. Thus this physical model also illustrates scalar multiplication.

GEOMETRIC REPRESENTATION OF v − w

The difference $\mathbf{v} - \mathbf{w}$ of two vectors in $\mathbb{R}^n$ can be represented geometrically as the arrow from the tip of $\mathbf{w}$ to the tip of $\mathbf{v}$, as shown in Fig. 3.9. This can be seen by noting that $\mathbf{v} - \mathbf{w}$ is the vector which when added to $\mathbf{w}$ yields $\mathbf{v}$. The dashed arrow in Fig. 3.9 shows $\mathbf{v} - \mathbf{w}$ in *standard position*, that is, as emanating from the origin.

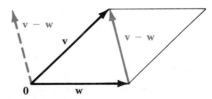

Figure 3.9
Geometric representation of v − w in $\mathbb{R}^n$.

Example 4   Find the distance from $\mathbf{v} = (1, 2, 5, 1)$ to $\mathbf{w} = (2, 1, 0, 3)$ in $\mathbb{R}^4$.

Solution   Glancing at Fig. 3.9, we see that the required distance is

$$\|\mathbf{v} - \mathbf{w}\| = \|(-1, 1, 5, -2)\| = \sqrt{(-1)^2 + 1^2 + 5^2 + (-2)^2} = \sqrt{31}. \quad \triangleleft$$

APPLICATION TO VELOCITY VECTORS AND NAVIGATION

The following two examples are concerned with another physical model for vector arithmetic. A vector is the **velocity vector** of a moving object at an instant if it points in the direction of the motion and if its magnitude is the speed of the object at that instant. Common navigation problems can be solved easily using velocity vectors.

Example 5   Suppose a sloop is sailing at 8 knots following a course of 010°, that is, 10° east of north, on a bay that has a 2-knot current setting in the direction 070°, that is, 70° east of north. Find the course and speed made good. (The expression *made good* is standard navigation terminology for the actual course and speed of a vessel over the bottom.)

Solution   The velocity vectors **s** for the sloop and **c** for the current are shown in Fig. 3.10, in which the vertical axis points due north. We find **s** and **c** by using a calculator and computing

$$\mathbf{s} = (8 \cos 80°, 8 \sin 80°) \approx (1.39, 7.88)$$

and

$$\mathbf{c} = (2 \cos 20°, 2 \sin 20°) \approx (1.88, 0.684).$$

By adding **s** and **c**, we find the vector **v** representing the course and speed of the sloop over the bottom, that is, the course and speed made good. Thus **v** = **s** + **c** = (3.27, 8.56). Therefore, the speed of the sloop is

$$\|\mathbf{v}\| = \sqrt{(3.27)^2 + (8.56)^2} = 9.16$$

and the course made good is given by

$$90° - \arctan(8.56/3.27) \approx 90° - 69° = 21°.$$

That is, the course is 021°.   ◁

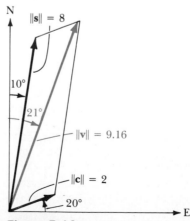

Figure 3.10
The vector v = s + c.

Example 6 Suppose the captain of our sloop realizes the importance of keeping track of the current. He wishes to sail in 5 hours to a harbor that bears 120° and is 35 nautical miles away. That is, he wishes to make good the course 120° and the speed 7 knots. He knows from a tide and current table that the current is setting due south at 2 knots. What should be his course and speed through the water?

Solution Again we represent in a vector diagram (see Fig. 3.11) the course and speed to be made good by a vector **v** and the velocity of the current by **c**. The correct course and speed to follow is represented by the vector **s** and is obtained by computing

$$\mathbf{s} = \mathbf{v} - \mathbf{c}$$
$$= (7 \cos 30°, -7 \sin 30°) - (0, -2)$$
$$= (6.06, -3.5) - (0, -2) = (6.06, -1.5).$$

Thus the captain should steer course 90° − arctan(−1.5/6.06) ≈ 90° + 13.9° = 103.9° and should proceed at

$$\|\mathbf{s}\| = \sqrt{(6.06)^2 + (-1.5)^2} \approx 6.24 \text{ knots.} \quad \triangleleft$$

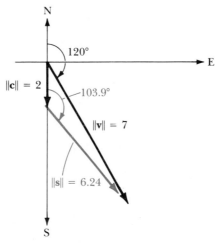

Figure 3.11
The vector s = v − c.

PARALLEL VECTORS

We turn to the geometric representation of scalar multiplication. Our work with force vectors suggests that for a vector **v** in $\mathbb{R}^n$ and scalar r, the vector $r\mathbf{v}$ has the *same direction* as **v** if $r > 0$ and has the *opposite direction* if $r < 0$. The zero vector is considered to have all directions. The magnitude of $r\mathbf{v}$ is

$$\|r\mathbf{v}\| = |r|\,\|\mathbf{v}\|. \tag{4}$$

This was illustrated in Fig. 3.8. We can verify Eq. (4) by computing

$$\|r\mathbf{v}\| = \|(rv_1, rv_2, \cdots, rv_n)\| = \sqrt{(rv_1)^2 + (rv_2)^2 + \cdots + (rv_n)^2}$$
$$= |r|\sqrt{v_1^2 + v_2^2 + \cdots + v_n^2} = |r|\,\|\mathbf{v}\|.$$

With this discussion as motivation, we define the general notion of parallel vectors.

Definition 3.3   Parallel Vectors

Two nonzero vectors $\mathbf{v}$ and $\mathbf{w}$ in $\mathbb{R}^n$ are **parallel** and we write $\mathbf{v} \parallel \mathbf{w}$ if one is scalar multiple of the other. If $\mathbf{v} = r\mathbf{w}$ with $r > 0$, then $\mathbf{v}$ and $\mathbf{w}$ have the **same direction;** if $r < 0$, then $\mathbf{v}$ and $\mathbf{w}$ have **opposite direction.**

Example 7   Determine whether the vectors $\mathbf{v} = (2, 1, 3, -4)$ and $\mathbf{w} = (6, 3, 9, -12)$ are parallel.

Solution   We put $\mathbf{v} = r\mathbf{w}$ and try to solve for r. This gives rise to four component equations:

$$2 = 6r, \qquad 1 = 3r, \qquad 3 = 9r, \qquad -4 = -12r.$$

Since $r = \frac{1}{3} > 0$ is a common solution to the four equations, we conclude that $\mathbf{v}$ and $\mathbf{w}$ are parallel with the same direction. $\lhd$

UNIT VECTORS

A vector in $\mathbb{R}^n$ is a **unit vector** if it has magnitude 1. Given any nonzero vector $\mathbf{v}$ in $\mathbb{R}^n$, a unit vector having the same direction as $\mathbf{v}$ is given by $(1/\|\mathbf{v}\|)\mathbf{v}$.

Example 8   Find a unit vector having the same direction as $\mathbf{v} = (2, 1, -3)$, and find a vector of magnitude 3 having direction opposite to $\mathbf{v}$.

Solution   Since $\|\mathbf{v}\| = \sqrt{2^2 + 1^2 + (-3)^2} = \sqrt{14}$, we see that $\mathbf{u} = (1/\sqrt{14})(2, 1, -3)$ is the unit vector having the same direction as $\mathbf{v}$, and $\mathbf{w} = -3\mathbf{u} = (-3/\sqrt{14})(2, 1, -3)$ is the other required vector. $\lhd$

The two-component unit vectors are precisely the vectors that extend from the origin to the unit circle $x^2 + y^2 = 1$ with center $(0, 0)$ and radius 1 in $\mathbb{R}^2$. See Fig. 3.12(a). The three-component unit vectors extend from $(0, 0, 0)$ to the unit sphere in $\mathbb{R}^3$, as illustrated in Fig. 3.12(b).

The vectors $\mathbf{e}_r = (0, 0, \ldots, 1, 0, \ldots, 0)$ in $\mathbb{R}^n$ with rth component 1 and zeros elsewhere are known as the **unit coordinate vectors** in $\mathbb{R}^n$. See Fig. 3.13. One often writes the unit coordinate vectors in $\mathbb{R}^2$ and $\mathbb{R}^3$ as

$$\mathbf{i} = \mathbf{e}_1, \qquad \mathbf{j} = \mathbf{e}_2, \quad \text{and} \quad \mathbf{k} = \mathbf{e}_3.$$

Each vector $\mathbf{v}$ in $\mathbb{R}^n$ can be expressed in a unique way as a sum of multiples of the unit coordinate vectors. Namely, for $\mathbf{v} = (v_1, v_2, \ldots, v_n)$, we have

$$\mathbf{v} = v_1\mathbf{e}_1 + v_2\mathbf{e}_2 + \cdots v_n\mathbf{e}_n. \tag{5}$$

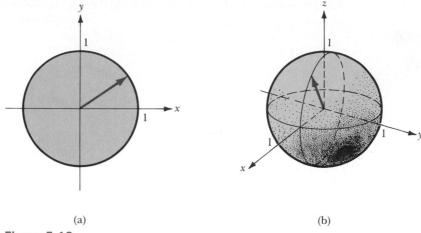

(a)   (b)

Figure 3.12
(a) A typical unit vector in $\mathbb{R}^2$; (b) a typical unit vector in $\mathbb{R}^3$.

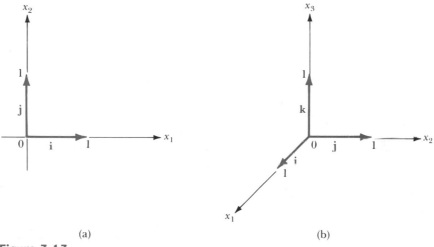

(a)   (b)

Figure 3.13
(a) Unit coordinate vectors in $\mathbb{R}^2$; (b) unit coordinate vectors in $\mathbb{R}^3$.

Each summand of (5) concentrates our attention on a single component of **v**, which is an important aspect of these unit coordinate vectors.

ANGLE BETWEEN VECTORS; THE DOT PRODUCT

We often describe a direction to someone by pointing. Our extended arm then represents a vector whose direction is clear. We describe directions in $\mathbb{R}^n$ by using vectors. The direction of a vector **v** in $\mathbb{R}^n$ is determined by its coordinates. We can specify this direction further by indicating the angle that the vector **v** makes with each unit coordinate vector. More generally, let us find the angle θ that a vector $\mathbf{v} = (v_1, v_2, \ldots, v_n)$ makes with a vector $\mathbf{w} = (w_1, w_2, \ldots, w_n)$ in $\mathbb{R}^n$. The law of cosines is just what we need to do this. If we refer to Fig. 3.14, the law of cosines tells us that

$$\|\mathbf{v}\|^2 + \|\mathbf{w}\|^2 = \|\mathbf{v} - \mathbf{w}\|^2 + 2\|\mathbf{v}\|\|\mathbf{w}\|(\cos \theta)$$

or

$$v_1^2 + \cdots + v_n^2 + w_1^2 + \cdots + w_n^2$$
$$= (v_1 - w_1)^2 + \cdots + (v_n - w_n)^2 + 2\|\mathbf{v}\|\|\mathbf{w}\|(\cos \theta). \quad (6)$$

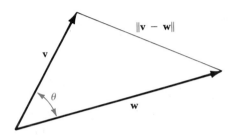

Figure 3.14
The angle between v and w.

After computing the squares on the right-hand side of (6) and simplifying, we obtain

$$\|\mathbf{v}\|\|\mathbf{w}\|(\cos \theta) = v_1 w_1 + \cdots + v_n w_n. \quad (7)$$

We recognize the expression on the right-hand side of (7) as the *dot product* **v** · **w** of **v** and **w**, which was defined in Section 1.3. Thus for nonzero vectors **v** and **w** in $\mathbb{R}^n$, we can attempt to define the angle θ between **a** and **b** by the following.

The **angle** between vectors **v** and **w** is $\arccos\left(\dfrac{\mathbf{v} \cdot \mathbf{w}}{\|\mathbf{v}\|\|\mathbf{w}\|}\right)$. $\quad (8)$

Equation (8) makes sense provided that

$$-1 \leq \frac{\mathbf{v} \cdot \mathbf{w}}{\|\mathbf{v}\|\|\mathbf{w}\|} \leq 1, \quad (9)$$

which allows us to compute the arccos of this quantity. The condition (9) can be rewritten as

$$|\mathbf{v} \cdot \mathbf{w}| \leq \|\mathbf{v}\|\|\mathbf{w}\|. \quad (10)$$

The relation (10) does hold for all **v** and **w** in $\mathbb{R}^n$. It is known as the *Schwarz inequality* and we will prove it in Section 3.2.

Example 9   Find the angle θ between the vectors $(1, 2, 0, 2)$ and $(-3, 1, 1, 5)$ in $\mathbb{R}^4$.

Solution   We have

$$\cos \theta = \frac{(1, 2, 0, 2) \cdot (-3, 1, 1, 5)}{\sqrt{1^2 + 2^2 + 0^2 + 2^2} \sqrt{(-3)^2 + 1^2 + 1^2 + 5^2}} = \frac{9}{(3)(6)} = \frac{1}{2}.$$

Thus $\theta = 60°$. ◁

We see from Eq. (7) that the dot product $\mathbf{v} \cdot \mathbf{w}$ of two vectors in $\mathbb{R}^n$ is described geometrically by

$$\mathbf{v} \cdot \mathbf{w} = \|\mathbf{v}\|\|\mathbf{w}\|(\cos \theta), \tag{11}$$

where θ is the angle between $\mathbf{v}$ and $\mathbf{w}$.

PERPENDICULAR VECTORS

Two vectors $\mathbf{v}$ and $\mathbf{w}$ in $\mathbb{R}^n$ are perpendicular if the angle between them is $90°$. This is the case precisely if the numerator in Eq. (8) is zero.

THE SCHWARZ INEQUALITY is due independently to Augustin-Louis Cauchy (see note on page 200), Hermann Amandus Schwarz (1843–1921), and Viktor Yakovlevich Bunyakovsky (1804–1889).

It was first stated as a theorem about coordinates in an appendix to Cauchy's 1821 text for his course on analysis at the Ecole Polytechnique as follows:

$$|a\alpha + a'\alpha' + a''\alpha'' + \cdots| \leq \sqrt{a^2 + a'^2 + a''^2 + \cdots} \sqrt{\alpha^2 + \alpha'^2 + \alpha''^2 + \cdots}.$$

Cauchy's proof follows from the algebraic identity

$$(a\alpha + a'\alpha' + a''\alpha'' + \cdots)^2 + (a\alpha' - a'\alpha)^2 + (a\alpha'' - a''\alpha)^2 + \cdots + (a'\alpha'' - a''\alpha')^2 + \cdots$$
$$= (a^2 + a'^2 + a''^2 + \cdots)(\alpha^2 + \alpha'^2 + \alpha''^2 + \cdots).$$

Bunyakovsky proved the inequality for functions in 1859; that is, he stated the result

$$\left[\int_a^b f(x)g(x)\, dx \right]^2 \leq \int_a^b f^2(x)\, dx \cdot \int_a^b g^2(x)\, dx,$$

where we can consider $\int_a^b f(x)g(x)\, dx$ to be the inner product of the functions $f(x), g(x)$ in the vector space of continuous functions on $[a, b]$ (see Example 5 on page 142). Bunyakovsky served as vice-president of the St. Petersburg Academy of Sciences from 1864 until his death. In 1875, the Academy established a mathematics prize in his name in recognition of his 50 years of teaching and research.

Schwarz stated the inequality in 1884. In his case, the vectors were functions ϕ, χ of two variables in a region T of the plane, and the inner product of these functions was given by $\iint_T \phi\chi\, dx\, dy$, where this integral is assumed to exist. The inequality then states that

$$\left| \iint_T \phi\chi\, dx\, dy \right| \leq \sqrt{\iint_T \phi^2\, dx\, dy} \cdot \sqrt{\iint_T \chi^2\, dx\, dy}.$$

Schwarz' proof is similar to the one given in the text (page 145). Schwarz was the leading mathematician in Berlin around the turn of the century; the work in which the inequality appears is devoted to a question about minimal surfaces.

Definition 3.4   Perpendicular or Orthogonal Vectors

Two vectors $\mathbf{v}$ and $\mathbf{w}$ in $\mathbb{R}^n$ are **perpendicular** or **orthogonal** and we write $\mathbf{v} \perp \mathbf{w}$ if their dot product is zero, that is, if $\mathbf{v} \cdot \mathbf{w} = 0$.

Example 10   Determine whether the vectors $\mathbf{v} = (4, 1, -2, 1)$ and $\mathbf{w} = (3, -4, 2, -4)$ are perpendicular.

Solution   We have

$$\mathbf{v} \cdot \mathbf{w} = (4)(3) + (1)(-4) + (-2)(2) + (1)(-4) = 0.$$

Thus $\mathbf{v} \perp \mathbf{w}$.   ◁

SUMMARY

Let $\mathbf{v} = (v_1, v_2, \cdots, v_n)$ and $\mathbf{w} = (w_1, w_2, \cdots, w_n)$ be vectors in $\mathbb{R}^n$.

1. The vectors $\mathbf{v}$ and $\mathbf{w}$ can be added and subtracted, and each can be multiplied by any scalar r. The result in each case is a vector in $\mathbb{R}^n$. These operations are special cases of the matrix operations defined in Section 1.3.

2. The angle θ between the vectors $\mathbf{v}$ and $\mathbf{w}$ can be found by using the relation $\mathbf{v} \cdot \mathbf{w} = \|\mathbf{v}\|\|\mathbf{w}\|(\cos \theta)$, where $\mathbf{v} \cdot \mathbf{w}$ is the dot product of $\mathbf{v}$ and $\mathbf{w}$.

3. Nonzero vectors $\mathbf{v}$ and $\mathbf{w}$ are parallel if one is a scalar multiple of the other. In that case, the vectors have the same direction if the scalar is positive and opposite directions if the scalar is negative.

4. The vectors $\mathbf{v}$ and $\mathbf{w}$ are perpendicular or orthogonal if their dot product is zero.

EXERCISES

In Exercises 1–4, find the indicated distance by finding the magnitude of an appropriate vector.

1. The distance from $(-1, 4, 2)$ to $(0, 8, 1)$ in $\mathbb{R}^3$

2. The distance from $(2, -1, 3)$ to $(4, 1, -2)$ in $\mathbb{R}^3$

3. The distance from $(3, 1, 2, 4)$ to $(-1, 2, 1, 2)$ in $\mathbb{R}^4$

4. The distance from $(-1, 2, 1, 4, 7, -3)$ to $(2, 1, -3, 5, 4, 5)$ in $\mathbb{R}^6$

In Exercises 5–22, let $\mathbf{u} = (-1, 3, 4)$, $\mathbf{v} = (2, 1, -1)$, and $\mathbf{w} = (-2, -1, 3)$. Find the indicated quantity.

5. $-\mathbf{u}$

6. $\|\mathbf{v}\|$

7. $\mathbf{u} + \mathbf{v}$

8. $\mathbf{v} - 2\mathbf{u}$

9. $3\mathbf{u} - \mathbf{v} + 2\mathbf{w}$

10. $\mathbf{v} - \mathbf{u}$

11. $(\frac{4}{5})\mathbf{w}$

12. The unit vector parallel to $\mathbf{u}$ having the same direction

13. The unit vector parallel to $\mathbf{w}$ having opposite direction

14. $\mathbf{u} \cdot \mathbf{v}$

15. $\mathbf{u} \cdot (\mathbf{v} + \mathbf{w})$

16. $(\mathbf{u} + \mathbf{v}) \cdot \mathbf{w}$

17. The angle between $\mathbf{u}$ and $\mathbf{v}$

18. The angle between $\mathbf{u}$ and $\mathbf{w}$

19. The value of x such that $(x, -3, 5)$ is perpendicular to $\mathbf{u}$

20. The value of y such that $(-3, y, 10)$ is perpendicular to $\mathbf{u}$

21. A nonzero vector perpendicular to both **u** and **v**

22. A nonzero vector perpendicular to both **u** and **w**

23. Find the angle between **v** = (1, −1, 2, 3, 0, 4) and **w** = (7, 0, 1, 3, 2, 4) in $\mathbb{R}^6$.

24. Show that (2, 0, 4), (4, 1, −1), and (6, 7, 7) are vertices of a right triangle in $\mathbb{R}^3$.

In Exercises 25–30, classify the vectors as parallel, perpendicular, or neither. If they are parallel, state whether they have the same direction or opposite directions.

25. (−1, 4) and (8, 2)

26. (−2, −1) and (5, 2)

27. (3, 2, 1) and (−9, −6, −3)

28. (2, 1, 4, −1) and (0, 1, 2, 4)

29. (10, 4, −1, 8) and (−5, −2, 3, −4)

30. (4, 1, 2, 1, 6) and (8, 2, 4, 2, 3)

31. The captain of a barge wishes to get to a point directly across a straight river that runs from north to south. If the current flows directly downstream at 5 knots and if the barge can steam at 13 knots, in what direction should the captain steer his barge?

32. A 100-lb weight is suspended by a rope passed through an eyelet on top of the weight and making angles of 30° with the vertical, as shown in Fig. 3.15. Find the tension (magnitude of the force vector) along the rope. [*Hint:* The sum of the force vectors along the two halves of the rope at the eyelet must be an upward vertical vector of magnitude 100.]

33. a) Answer Exercise 32 if each half of the rope makes an angle of θ with the vertical at the eyelet.

   b) Find the tension in the rope if both sides are vertical ($\theta = 0$).

   c) What happens if an attempt is made to stretch the rope out straight (horizontal) while the 100-lb weight hangs on it?

34. Suppose a weight of 100 lb is suspended by *two* ropes *tied* at an eyelet on top of the weight, as shown in Fig. 3.16. Let the angles the ropes make with the vertical be θ_1 and θ_2 as shown in the figure. Let the tension in the ropes be T_1 for the right-hand rope and T_2 for the left-hand rope.

   a) Show that the force vector $\mathbf{F}_1$ shown in Fig. 3.16 is $T_1(\sin\theta_1)\,\mathbf{i} + T_1(\cos\theta_1)\mathbf{j}$.

   b) Find the corresponding expression for $\mathbf{F}_2$ in terms of T_2 and θ_2.

   c) If the system is in equilibrium, $\mathbf{F}_1 + \mathbf{F}_2 = 100\mathbf{j}$, so $\mathbf{F}_1 + \mathbf{F}_2$ must have **i**-component 0 and **j**-component 100. Write two equations reflecting this fact, using the answers to parts (a) and (b).

   d) Find T_1 and T_2 if $\theta_1 = 45°$ and $\theta_2 = 30°$.

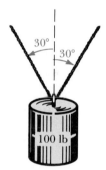

Figure 3.15
Both halves of the rope make an angle of 30° with the vertical.

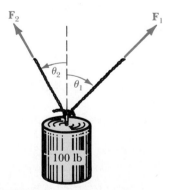

Figure 3.16
Two ropes tied at the eyelet and making angles θ_1 and θ_2 with the vertical.

35. The software available to schools using this text includes a program called VECTGRPH, which is designed to strengthen students' geometric understanding of vector concepts. The program uses graphics for vectors in the plane only. Run the program, and work with the first item involving vector addition and subtraction until a score of 80 percent or better is consistently achieved.

3.2
Vector Spaces and Algebra in $\mathbb{R}^n$

In this section we examine more systematically the operations of vector addition and scalar multiplication that were presented in Section 3.1. The properties given in Theorem 3.1 below describe $\mathbb{R}^n$ as a *vector space*.

Theorem 3.1 Vector-space Properties of $\mathbb{R}^n$

Let **u**, **v**, and **w** be any vectors in $\mathbb{R}^n$ and let r and s be any scalars in $\mathbb{R}$.

PROPERTIES OF VECTOR ADDITION

A1	$(\mathbf{u} + \mathbf{v}) + \mathbf{w} = \mathbf{u} + (\mathbf{v} + \mathbf{w})$	Associative law
A2	$\mathbf{v} + \mathbf{w} = \mathbf{w} + \mathbf{v}$	Commutative law
A3	$\mathbf{0} + \mathbf{v} = \mathbf{v}$	Nature of the zero vector
A4	$\mathbf{v} + (-\mathbf{v}) = \mathbf{0}$	$-\mathbf{v}$ as additive inverse of $\mathbf{v}$

PROPERTIES INVOLVING SCALAR MULTIPLICATION

S1	$r(\mathbf{v} + \mathbf{w}) = r\mathbf{v} + r\mathbf{w}$	A distributive law
S2	$(r + s)\mathbf{v} = r\mathbf{v} + s\mathbf{v}$	A distributive law
S3	$r(s\mathbf{v}) = (rs)\mathbf{v}$	Associative law
S4	$1\mathbf{v} = \mathbf{v}$	Preservation of scale

The eight properties given in Theorem 3.1 are quite easy to prove and we leave most of them to the exercises. The proofs in Examples 1 and 2 below are typical.

Example 1 Prove property A2 of Theorem 3.1.

Solution Writing

$$\mathbf{v} = (v_1, v_2, \ldots, v_n) \quad \text{and} \quad \mathbf{w} = (w_1, w_2, \ldots, w_n),$$

we have

$$\mathbf{v} + \mathbf{w} = (v_1 + w_1, v_2 + w_2, \ldots, v_n + w_n)$$

and

$$\mathbf{w} + \mathbf{v} = (w_1 + v_1, w_2 + v_2, \ldots, w_n + v_n).$$

These two vectors are equal because $v_i + w_i = w_i + v_i$ for each i. In other words, the commutative law of vector addition follows directly from the commutative law of addition of numbers. ◁

Example 2 Prove property S2 of Theorem 3.1.

Solution Writing $\mathbf{v} = (v_1, v_2, \ldots, v_n)$, we have

$$
\begin{aligned}
(r + s)\mathbf{v} &= (r + s)(v_1, v_2, \ldots, v_n) \\
&= ((r + s)v_1, (r + s)v_2, \ldots, (r + s)v_n) \\
&= (rv_1 + sv_1, rv_2 + sv_2, \ldots, rv_n + sv_n) \\
&= (rv_1, rv_2, \ldots, rv_n) + (sv_1, sv_2, \ldots, sv_n) \\
&= r\mathbf{v} + s\mathbf{v}.
\end{aligned}
$$

Thus the property $(r + s)\mathbf{v} = r\mathbf{v} + s\mathbf{v}$ involving vectors follows from the analogous property $(r + s)v_i = rv_i + s_iv_i$ for numbers. ◁

The *vector space* $\mathbb{R}^n$ serves as a model for the general definition of a real vector space.

Definition 3.5 Vector Space

> A **real vector space** is a nonempty set V of objects called **vectors,** together with a rule for adding any two vectors $\mathbf{v}$ and $\mathbf{w}$ to produce a vector $\mathbf{v} + \mathbf{w}$ in V and a rule for multiplying any vector $\mathbf{v}$ in V by any scalar r in $\mathbb{R}$ to produce a vector $r\mathbf{v}$ in V. Moreover, there must exist a vector $\mathbf{0}$ in V and for each $\mathbf{v}$ in V there must exist a vector $-\mathbf{v}$ in V such that properties A1–A4 and S1–S4 of Theorem 3.1 are satisfied for these operations of vector addition and scalar multiplication with all choices for vectors and scalars.

Note that for $\mathbf{v}$ and $\mathbf{w}$ in a real vector space V, we require that $\mathbf{v} + \mathbf{w}$ must again be a *vector in* V. We express this condition by saying that V is **closed under vector addition.** Similarly, $r\mathbf{v}$ is again in V for each $\mathbf{v}$ in V and each scalar r in $\mathbb{R}$, so that V is also **closed under scalar multiplication.** The vector $\mathbf{0}$

THOUGH VECTORS IN $\mathbb{R}^n$ were dealt with by mathematicians and physicists throughout the second half of the nineteenth century (as were other objects that we today consider as vectors), it was not until the appearance of Hermann Weyl's (1885–1955) treatise *Space–Time–Matter* in 1918 that an abstract, axiomatic definition of a vector space appeared in print. Weyl wrote this book as an introduction to Einstein's general theory of relativity; in Chapter 1 he discussed the nature of Euclidean space and as a part of that discussion formulated what are now the standard axioms of a vector space (A1–A4, S1–S4). Since he was dealing only with spaces V of finite dimension, he included the "axiom of dimensionality"—that for some whole number h, there are h linearly independent vectors in V, but every set of $h + 1$ vectors is linearly dependent (see Section 3.4).

The vector-space axioms themselves had been known for years, but they were generally proved as consequences of other definitions of vectors. For example, these appear in Giuseppe Peano's (1858–1932) brief Italian text, *Geometric Calculus* (1888), in which he explains the work of Hermann Grassmann (see note on page 160).

in V is called the **zero vector,** while $-\mathbf{v}$ is called the **additive inverse of v** and is usually read "minus **v**." We write $\mathbf{v} - \mathbf{w}$ for $\mathbf{v} + (-\mathbf{w})$.

One refers to V as a *real* vector space to emphasize that the scalars in Definition 3.5 are in $\mathbb{R}$. We usually drop the term *real* since we will seldom use scalars that are not real numbers.

Example 3   Show that the set M of all $m \times n$ matrices is a vector space, using as vector addition and scalar multiplication the usual addition and multiplication by a scalar for matrices.

Solution   We have seen that addition of $m \times n$ matrices and multiplication of an $m \times n$ matrix by a scalar again yield an $m \times n$ matrix. Thus M is closed under vector addition and scalar multiplication. We take as zero vector in M the usual zero matrix, all of whose entries are zero. For any matrix A in M we consider $-A$ to be the matrix $(-1)A$. The properties of matrix arithmetic on page 36 show that all eight properties A1–A4 and S1–S4 required of a vector space are satisfied. ◁

Example 4   Show that the set P of all polynomials with coefficients in $\mathbb{R}$ is a vector space, using for vector addition and scalar multiplication the usual addition of polynomials and multiplication of a polynomial by a scalar.

Solution   Let p and q be polynomials

$$p = a_0 + a_1x + a_2x^2 + \cdots + a_nx^n$$

and

$$q = b_0 + b_1x + b_2x^2 + \cdots + b_mx^m.$$

If $m \geq n$, we recall that the **sum** of p and q is given by

$$p + q = (a_0 + b_0) + (a_1 + b_1)x + \cdots + (a_n + b_n)x^n \\ + b_{n+1}x^{n+1} + \cdots + b_mx^m.$$

A similar definition is made if $m < n$. The **product** of p by a scalar r is given by

$$rp = ra_0 + ra_1x + ra_2x^2 + \cdots + ra_nx^n.$$

Taking the usual notions of the zero polynomial and of $-p$, we recognize that the eight properties A1–A4 and S1–S4 required of a vector space are familiar properties for these polynomial operations. Thus P is a vector space. ◁

Example 5   Let F be the set of all real-valued functions of a real variable, that is, F is the set of all functions mapping $\mathbb{R}$ into $\mathbb{R}$. The vector **sum** $f + g$ of two functions f and g in F is defined in the usual way to be the function whose value at any x in $\mathbb{R}$ is $f(x) + g(x)$, that is,

$$(f + g)(x) = f(x) + g(x).$$

For any scalar r in $\mathbb{R}$ and function f in F, the **product** rf is the function whose value at x is $rf(x)$, so that

$$(rf)(x) = rf(x).$$

Show that F with these operations is a vector space.

Solution   We observe that for f and g in F, both $f + g$ and rf are functions mapping $\mathbb{R}$ into $\mathbb{R}$, so that $f + g$ and rf are in F. Thus F is closed under vector addition and scalar multiplication. We take as zero vector in F the constant function whose value at each x in $\mathbb{R}$ is 0; we will denote this zero function by $\mathbf{0}$. For each function f in F, we take as $-f$ the function $(-1)f$ in F.

There are four vector-addition properties to verify and they are all easy. We illustrate by verifying condition A4. For f in F, the function $f + (-f) = f + (-1)f$ has as value at x in $\mathbb{R}$ the number $f(x) + (-1)f(x)$, which is 0. Consequently, $f + (-f)$ is the zero function $\mathbf{0}$ and A4 is verified.

The scalar multiplicative properties are just as easy to verify. For example, to verify S4 we must compute $1f$ at any x in $\mathbb{R}$ and compare the result with $f(x)$. We obtain $(1f)(x) = 1f(x) = f(x)$, so that $1f = f$.   ◁

We now indicate that vector addition and scalar multiplication possess still more of the properties we are accustomed to expect.

Theorem 3.2   Elementary Properties of Vector Spaces

Every vector space V has the following properties:

i) The vector $\mathbf{0}$ is the *unique* vector $\mathbf{x}$ satisfying the equation $\mathbf{x} + \mathbf{v} = \mathbf{v}$ for all vectors $\mathbf{v}$ in V.

ii) For each vector $\mathbf{v}$ in V, the vector $-\mathbf{v}$ is the *unique* vector $\mathbf{y}$ satisfying $\mathbf{v} + \mathbf{y} = \mathbf{0}$.

iii) If $\mathbf{u} + \mathbf{v} = \mathbf{u} + \mathbf{w}$ for vectors $\mathbf{u}$, $\mathbf{v}$, and $\mathbf{w}$ in V, then $\mathbf{v} = \mathbf{w}$.

iv) $0\mathbf{v} = \mathbf{0}$ for all vectors $\mathbf{v}$ in V.

v) $r\mathbf{0} = \mathbf{0}$ for all scalars r in $\mathbb{R}$.

vi) $(-r)\mathbf{v} = r(-\mathbf{v}) = -(r\mathbf{v})$ for all scalars r in $\mathbb{R}$ and vectors $\mathbf{v}$ in V.

Proof   We prove only (i), and leave proofs of the remaining properties as Exercises 12–16.

Suppose vectors $\mathbf{0}$ and $\mathbf{0}'$ both satisfy the equation $\mathbf{x} + \mathbf{v} = \mathbf{v}$ for all $\mathbf{v}$ in V. We then obtain

$$\mathbf{0} + \mathbf{0}' = \mathbf{0}' \quad \text{and} \quad \mathbf{0}' + \mathbf{0} = \mathbf{0}.$$

By the commutative property A2, we know that $\mathbf{0} + \mathbf{0}' = \mathbf{0}' + \mathbf{0}$, and we conclude that $\mathbf{0} = \mathbf{0}'$.   ■

PROPERTIES OF THE DOT PRODUCT AND OF MAGNITUDES IN $\mathbb{R}^n$

For the remainder of this section, we restrict our attention to the vector space $\mathbb{R}^n$, and consider properties of the dot product and of magnitude for vectors in $\mathbb{R}^n$.

Theorem 3.3   Properties of the Dot Product and of Magnitude

Let $\mathbf{u}$, $\mathbf{v}$, and $\mathbf{w}$ be vectors in $\mathbb{R}^n$ and let r be a scalar in $\mathbb{R}$.

PROPERTIES OF THE DOT PRODUCT

P1 $\mathbf{v} \cdot \mathbf{w} = \mathbf{w} \cdot \mathbf{v}$ Commutative law

P2 $\mathbf{u} \cdot (\mathbf{v} + \mathbf{w}) = \mathbf{u} \cdot \mathbf{v} + \mathbf{u} \cdot \mathbf{w}$ Distributive law

P3 $r(\mathbf{v} \cdot \mathbf{w}) = (r\mathbf{v}) \cdot \mathbf{w} = \mathbf{v} \cdot (r\mathbf{w})$ Homogeneity

P4 $\mathbf{v} \cdot \mathbf{v} \geq 0$ and $\mathbf{v} \cdot \mathbf{v} = 0$ Positive definite property
     if and only if $\mathbf{v} = \mathbf{0}$

PROPERTIES OF MAGNITUDE

M1 $\|\mathbf{v}\|^2 = \mathbf{v} \cdot \mathbf{v}$ Definition of magnitude

M2 $\|r\mathbf{v}\| = |r|\|\mathbf{v}\|$ Stretch/contract

Again, all the properties in Theorem 3.3 are easy to verify, as illustrated in Example 6 below.

Example 6   Verify P4 of Theorem 3.3.

Solution   We let $\mathbf{v} = (v_1, v_2, \ldots, v_n)$ and we find that

$$\mathbf{v} \cdot \mathbf{v} = v_1{}^2 + v_2{}^2 + \cdots + v_n{}^2.$$

Now a sum of squares is nonnegative and can be zero if and only if each summand is zero. But a summand $v_i{}^2$ is itself a square, and will be zero if and only if $v_i = 0$. This completes the demonstration.   ◁

Property M2 in Theorem 3.3 was established in Section 3.1. Properties M1 and M2 can be useful in computing the magnitude of a vector and in establishing relations between magnitudes, as illustrated in the following two examples.

Example 7   Find the magnitude of the vector

$$\mathbf{v} = (-6, -12, 6, 18, -6).$$

Solution   Since $\mathbf{v} = -6(1, 2, -1, -3, 1)$, property M2 of Theorem 3.3 tells us that

$$\|\mathbf{v}\| = |-6|\|(1, 2, -1, -3, 1)\| = 6\sqrt{16} = 24.   ◁$$

Example 8   Show that the sum of the squares of the lengths of the diagonals of a parallelogram is equal to the sum of the squares of the lengths of the sides. (This is the *parallelogram relation*.)

Solution   We take our parallelogram with vertex at the origin and vectors $\mathbf{v}$ and $\mathbf{w}$ emanating from the origin forming two sides, as shown in Fig. 3.17. The lengths of the diagonals are then $\|\mathbf{v} + \mathbf{w}\|$ and $\|\mathbf{v} - \mathbf{w}\|$. Using the properties in

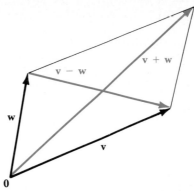

Figure 3.17
The parallelogram has v + w and
v − w as vector diagonals.

Theorem 3.3, we have

$$\|\mathbf{v} + \mathbf{w}\|^2 + \|\mathbf{v} - \mathbf{w}\|^2 = (\mathbf{v} + \mathbf{w}) \cdot (\mathbf{v} + \mathbf{w}) + (\mathbf{v} - \mathbf{w}) \cdot (\mathbf{v} - \mathbf{w})$$
$$= \mathbf{v} \cdot \mathbf{v} + 2\mathbf{v} \cdot \mathbf{w} + \mathbf{w} \cdot \mathbf{w} + \mathbf{v} \cdot \mathbf{v} - 2\mathbf{v} \cdot \mathbf{w} + \mathbf{w} \cdot \mathbf{w}$$
$$= 2(\mathbf{v} \cdot \mathbf{v}) + 2(\mathbf{w} \cdot \mathbf{w})$$
$$= 2\|\mathbf{v}\|^2 + 2\|\mathbf{w}\|^2,$$

which is what we wished to prove. $\triangleleft$

THE SCHWARZ AND TRIANGLE INEQUALITIES

In the preceding section, we defined the angle θ between two nonzero vectors $\mathbf{v}$ and $\mathbf{w}$ in $\mathbb{R}^n$ to be

$$\theta = \arccos \frac{\mathbf{v} \cdot \mathbf{w}}{\|\mathbf{v}\|\|\mathbf{w}\|}$$

provided that

$$-1 \leq \frac{\mathbf{v} \cdot \mathbf{w}}{\|\mathbf{v}\|\|\mathbf{w}\|} \leq 1,$$

that is, provided that $|\mathbf{v} \cdot \mathbf{w}| \leq \|\mathbf{v}\|\|\mathbf{w}\|$. We now prove that this inequality is true for all choices of $\mathbf{v}$ and $\mathbf{w}$ in $\mathbb{R}^n$.

Theorem 3.4   Schwarz Inequality

Let $\mathbf{v}$ and $\mathbf{w}$ be vectors in $\mathbb{R}^n$. Then

$$|\mathbf{v} \cdot \mathbf{w}| \leq \|\mathbf{v}\|\|\mathbf{w}\|.$$

Proof   We will make repeated use of Theorems 3.1 and 3.3. For any scalars r and s, we have

$$\|r\mathbf{v} + s\mathbf{w}\|^2 \geq 0. \tag{1}$$

Now

$$\|r\mathbf{v} + s\mathbf{w}\|^2 = (r\mathbf{v} + s\mathbf{w}) \cdot (r\mathbf{v} + s\mathbf{w})$$
$$= r^2(\mathbf{v} \cdot \mathbf{v}) + 2rs(\mathbf{v} \cdot \mathbf{w}) + s^2(\mathbf{w} \cdot \mathbf{w}). \tag{2}$$

In particular, setting $r = \mathbf{w} \cdot \mathbf{w}$ and $s = -(\mathbf{v} \cdot \mathbf{w})$, we obtain from Eqs. (1) and (2)

$$(\mathbf{w} \cdot \mathbf{w})^2(\mathbf{v} \cdot \mathbf{v}) - 2(\mathbf{w} \cdot \mathbf{w})(\mathbf{v} \cdot \mathbf{w})(\mathbf{v} \cdot \mathbf{w}) + (\mathbf{v} \cdot \mathbf{w})^2(\mathbf{w} \cdot \mathbf{w})$$
$$= (\mathbf{w} \cdot \mathbf{w})^2(\mathbf{v} \cdot \mathbf{v}) - (\mathbf{w} \cdot \mathbf{w})(\mathbf{v} \cdot \mathbf{w})^2 \geq 0 \tag{3}$$

or

$$(\mathbf{w} \cdot \mathbf{w})[(\mathbf{v} \cdot \mathbf{v})(\mathbf{w} \cdot \mathbf{w}) - (\mathbf{v} \cdot \mathbf{w})^2] \geq 0. \tag{4}$$

Now if $\mathbf{w} \cdot \mathbf{w} = 0$, then $\mathbf{w} = \mathbf{0}$ and our result follows. If $\mathbf{w} \cdot \mathbf{w} \neq 0$, then $\mathbf{w} \cdot \mathbf{w} > 0$, so Eq. (4) yields

$$(\mathbf{v} \cdot \mathbf{v})(\mathbf{w} \cdot \mathbf{w}) - (\mathbf{v} \cdot \mathbf{w})^2 \geq 0$$

or

$$(\mathbf{v} \cdot \mathbf{w})^2 \leq (\mathbf{v} \cdot \mathbf{v})(\mathbf{w} \cdot \mathbf{w}) = \|\mathbf{v}\|^2\|\mathbf{w}\|^2.$$

After taking square roots, we obtain the desired result. ◼

There is another important inequality that follows easily from the Schwarz inequality; namely, for vectors $\mathbf{v}$ and $\mathbf{w}$ in $\mathbb{R}^n$, we have

$$\|\mathbf{v} + \mathbf{w}\| \leq \|\mathbf{v}\| + \|\mathbf{w}\|. \qquad \textbf{Triangle inequality} \tag{5}$$

Figure 3.18 indicates the origin of the name "triangle inequality" for Eq. (5). Viewed geometrically, it asserts that the sum of the lengths of two sides of a triangle is at least as great as the length of the third side. We give an algebraic proof.

Theorem 3.5 The Triangle Inequality

For any vectors $\mathbf{v}$ and $\mathbf{w}$ in $\mathbb{R}^n$, we have

$$\|\mathbf{v} + \mathbf{w}\| \leq \|\mathbf{v}\| + \|\mathbf{w}\|.$$

Proof Using the properties of the dot product (Theorem 3.3) as well as the Schwarz inequality, we have

$$\|\mathbf{v} + \mathbf{w}\|^2 = (\mathbf{v} + \mathbf{w}) \cdot (\mathbf{v} + \mathbf{w})$$
$$= \mathbf{v} \cdot \mathbf{v} + 2\mathbf{v} \cdot \mathbf{w} + \mathbf{w} \cdot \mathbf{w}$$
$$\leq \mathbf{v} \cdot \mathbf{v} + 2\|\mathbf{v}\|\|\mathbf{w}\| + \mathbf{w} \cdot \mathbf{w}$$
$$= \|\mathbf{v}\|^2 + 2\|\mathbf{v}\|\|\mathbf{w}\| + \|\mathbf{w}\|^2$$
$$= (\|\mathbf{v}\| + \|\mathbf{w}\|)^2.$$

The desired relation follows at once by taking square roots. ◼

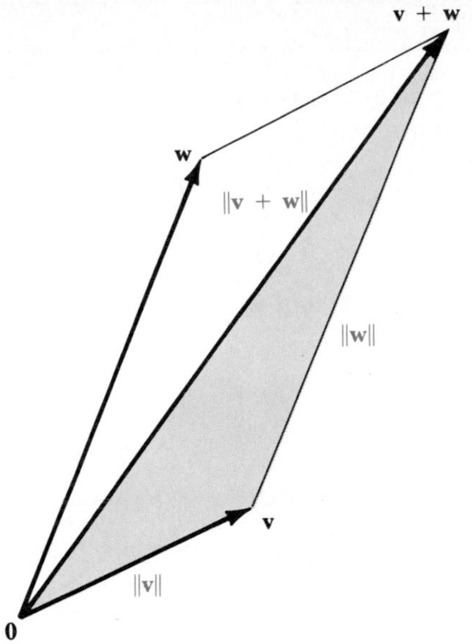

Figure 3.18
The triangle inequality.

SUMMARY

1. A vector space is a nonempty set V of objects called vectors together with rules for adding any two vectors $\mathbf{v}$ and $\mathbf{w}$ in V and for multiplying any vector $\mathbf{v}$ in V by any scalar r in $\mathbb{R}$. It is required that V be closed under this vector addition and scalar multiplication so that $\mathbf{v} + \mathbf{w}$ and $r\mathbf{v}$ are both in V. Moreover, the following axioms must be satisfied for all vectors $\mathbf{u}$, $\mathbf{v}$, and $\mathbf{w}$ in V and all scalars r and s in $\mathbb{R}$:

    A1 $(\mathbf{u} + \mathbf{v}) + \mathbf{w} = \mathbf{u} + (\mathbf{v} + \mathbf{w})$

    A2 $\mathbf{v} + \mathbf{w} = \mathbf{w} + \mathbf{v}$

    A3   There exists a zero vector $\mathbf{0}$ in V such that $\mathbf{0} + \mathbf{v} = \mathbf{v}$ for all vectors $\mathbf{v}$.

    A4   Each vector $\mathbf{v}$ has an additive inverse $-\mathbf{v}$ in V such that $\mathbf{v} + (-\mathbf{v}) = \mathbf{0}$.

    S1 $r(\mathbf{v} + \mathbf{w}) = r\mathbf{v} + r\mathbf{w}$

    S2 $(r + s)\mathbf{v} = r\mathbf{v} + s\mathbf{v}$

    S3 $r(s\mathbf{v}) = (rs)\mathbf{v}$

    S4 $1\mathbf{v} = \mathbf{v}$

2. $\mathbb{R}^n$ is a vector space for each positive integer n.
3. For all vectors $\mathbf{v}$ and $\mathbf{w}$ in $\mathbb{R}^n$, we have:

    Schwarz inequality: $|\mathbf{v} \cdot \mathbf{w}| \leq \|\mathbf{v}\|\|\mathbf{w}\|$;

    Triangle inequality: $\|\mathbf{v} + \mathbf{w}\| \leq \|\mathbf{v}\| + \|\mathbf{w}\|$.

EXERCISES

1. Verify the indicated property of vector addition in $\mathbb{R}^n$, stated in Theorem 3.1.
   a) The property A1
   b) The property A3
   c) The property A4

2. Verify the indicated property of scalar multiplication in $\mathbb{R}^n$, stated in Theorem 3.1.
   a) The property S1
   b) The property S3
   c) The property S4

In Exercises 3–6, let $\mathbf{v} = (2, 1, 4)$ *and* $\mathbf{w} = (1, -3, 2)$. *Compute the indicated quantities in at least two ways.*

3. $4\mathbf{v} - 4\mathbf{w}$

4. $2\mathbf{v} - 5\mathbf{v}$

5. $[3\mathbf{v} - 2(\mathbf{v} + \mathbf{w})] + 3(\mathbf{w} - \mathbf{v})$

6. $\mathbf{v} + \mathbf{w} + \mathbf{v} + \mathbf{w} + \mathbf{v}$

In Exercises 7–11, decide whether or not the given set together with the indicated operations of addition and scalar multiplication is a (real) vector space.

7. The set $\mathbb{R}^2$ with the usual addition, but with scalar multiplication defined by $r(x, y) = (ry, rx)$

8. The set $\mathbb{R}^2$ with the usual scalar multiplication, but with addition defined by $(x, y) + (r, s) = (y + s, x + r)$

9. The set Q of all rational numbers with the usual operations of addition and scalar multiplication

10. The set C of complex numbers, that is,

$$C = \{a + b\sqrt{-1} \mid a, b \text{ in } \mathbb{R}\}$$

with the usual addition of complex numbers and scalar multiplication defined by

$$r(a + b\sqrt{-1}) = ra + rb\sqrt{-1}$$

for any numbers a, b, and r in $\mathbb{R}$

11. The set F of all functions mapping $\mathbb{R}$ into $\mathbb{R}$ with scalar multiplication defined as in Example 5, but with addition defined by $(f + g)(x) = \max\{f(x), g(x)\}$

12. Prove (ii) of Theorem 3.2.

13. Prove (iii) of Theorem 3.2.

14. Prove (iv) of Theorem 3.2.

15. Prove (v) of Theorem 3.2.

16. Prove (vi) of Theorem 3.2.

17. Let V be a vector space. Show that if $\mathbf{v}$ is in V and r is a scalar and if $r\mathbf{v} = \mathbf{0}$, then either $r = 0$ or $\mathbf{v} = \mathbf{0}$.

18. Verify the indicated property of the dot product, stated in Theorem 3.3.
   a) The property P1
   b) The property P2
   c) The property P3

In Exercises 19–22, use Theorem 3.3 to compute the indicated quantities mentally, without pencil or paper.

19. $\|(42, 14)\|$

20. $\|(10, 20, 25, -15)\|$

21. $(14, 21, 28) \cdot (4, 8, 20)$

22. $(12, -36, 24) \cdot (25, 30, 10)$

23. For vectors $\mathbf{v}$ and $\mathbf{w}$ in $\mathbb{R}^n$, show that $\mathbf{v} - \mathbf{w}$ and $\mathbf{v} + \mathbf{w}$ are perpendicular if and only if $\|\mathbf{v}\| = \|\mathbf{w}\|$.

24. For vectors $\mathbf{u}$, $\mathbf{v}$, and $\mathbf{w}$ in $\mathbb{R}^n$ and scalars r and s, show that if $\mathbf{w}$ is perpendicular to both $\mathbf{u}$ and $\mathbf{v}$, then $\mathbf{w}$ is perpendicular to $r\mathbf{u} + s\mathbf{v}$.

25. Use the triangle inequality to prove that

$$\|\mathbf{v} - \mathbf{w}\| \le \|\mathbf{v}\| + \|\mathbf{w}\|$$

for any vectors $\mathbf{v}$ and $\mathbf{w}$ in $\mathbb{R}^n$.

27. Use vector methods to show that the diagonals of a rhombus (a parallelogram with equal sides) are perpendicular. [*Hint:* Use a figure like Fig. 3.17 and an exercise above.]

28. Use vector methods to show that the midpoint of the hypotenuse of a right triangle is equidistant from the three vertices. *Hint:* See Fig. 3.19. Show that

$$\|(\mathbf{v} + \mathbf{w})/2\| = \|(\mathbf{v} - \mathbf{w})/2\|.$$

29. Show that the vectors $\|\mathbf{v}\|\mathbf{w} + \|\mathbf{w}\|\mathbf{v}$ and $\|\mathbf{v}\|\mathbf{w} - \|\mathbf{w}\|\mathbf{v}$ in $\mathbb{R}^n$ are perpendicular.

26. Show that for any vectors $\mathbf{v}$ and $\mathbf{w}$ in $\mathbb{R}^n$, we have

$$\|\mathbf{v} - \mathbf{w}\| \ge \|\mathbf{v}\| - \|\mathbf{w}\|.$$

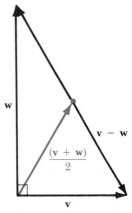

Figure 3.19
The vector $(1/2)(\mathbf{v} + \mathbf{w})$ to the midpoint of the hypotenuse.

---

*An **inner product space** consists of a vector space V together with a rule for multiplying any two vectors $\mathbf{v}$ and $\mathbf{w}$ in V to produce a scalar, written $\mathbf{v} \cdot \mathbf{w}$, that satisfies properties $P_1 - P_4$ of Theorem 3.3. Such a rule is referred to as a (positive definite) **inner product** in V.*

30. Determine if $\mathbb{R}^2$ is a (positive definite) inner product space if we define $(v_1, v_2) \cdot (w_1, w_2) = v_1 w_1 - v_2 w_2$.

32. Determine if the vector space P_3 of all polynomials of degree at most 3 is a (positive definite) inner product space if we define

$$(a_0 + a_1 x + a_2 x^2 + a_3 x^3) \cdot (b_0 + b_1 x + b_2 x^2 + b_3 x^3) = a_0 b_0.$$

34. Two vectors $\mathbf{v}$ and $\mathbf{w}$ in an inner product space V are said to be **orthogonal** if $\mathbf{v} \cdot \mathbf{w} = 0$. For any subset S of V, let $S^\perp$ be the set of all vectors in V that are orthogonal to *every* vector in S.
  a) If $\mathbf{v}$ and $\mathbf{w}$ are in $S^\perp$, verify that $\mathbf{v} + \mathbf{w}$ is $S^\perp$.
  b) If $\mathbf{v}$ is in $S^\perp$ and r is any scalar, verify that $r\mathbf{v}$ is in $S^\perp$.

31. Determine if the vector space M of all 2×2 matrices is a (positive definite) inner product space if we define

$$\begin{pmatrix} a_1 & a_2 \\ a_3 & a_4 \end{pmatrix} \cdot \begin{pmatrix} b_1 & b_2 \\ b_3 & b_4 \end{pmatrix} = a_1 b_1 + a_2 b_2 + a_3 b_3 + a_4 b_4.$$

33. Prove that in an inner product space V, $\mathbf{0} \cdot \mathbf{v} = 0$ for each $\mathbf{v}$ in V.

35. For a vector $\mathbf{v}$ in an inner product space V, we define the **length** $\|\mathbf{v}\|$ of $\mathbf{v}$ to be $\|\mathbf{v}\| = \sqrt{\mathbf{v} \cdot \mathbf{v}}$. Verify the following properties for all $\mathbf{v}$ and $\mathbf{w}$ in V and all scalars r, referring to the proofs relative to the usual dot product in $\mathbb{R}^n$ in the text where necessary.
  a) $\|\mathbf{v}\| \ge 0$ and $\|\mathbf{v}\| = 0$ if and only if $\mathbf{v} = 0$
  b) $\|r\mathbf{v}\| = |r|\|\mathbf{v}\|$
  c) $|\mathbf{v} \cdot \mathbf{w}| \le \|\mathbf{v}\|\|\mathbf{w}\|$
  d) $\|\mathbf{v} + \mathbf{w}\| < \|\mathbf{v}\| + \|\mathbf{w}\|$

3.3
Linear Combinations and Subspaces

We have seen that $\mathbb{R}^n$ is a *vector space*, where vector addition consists of adding corresponding components of row (or column) vectors and scalar multiplication is achieved by multiplying each component by the scalar. Any subset of $\mathbb{R}^n$ with these same operations that is also a vector space is called a *subspace* of $\mathbb{R}^n$. For example, the x_1,x_2-plane in $\mathbb{R}^3$ consisting of all vectors having zero as third entry is a subspace of $\mathbb{R}^3$. However, the subset $\{(m, n, p) \mid m, n, p \text{ any integers}\}$† of $\mathbb{R}^3$ consisting of all vectors in $\mathbb{R}^3$ with integer components is not a subspace, for although this subset is closed under vector addition it is not closed under scalar multiplication. For example, $.5(1, 2, 5) = (.5, 1, 2.5)$ is not in the subset.

Definition 3.6   Subspace

A subset S of a vector space V is a **subspace** of V if S itself fulfills the requirements of a vector space with addition and scalar multiplication producing the same vectors as these operations did in V.

It is crucial to note that in order for a nonempty subset S of a vector space V to be a subspace, the subset together with the operations of vector addition and scalar multiplication must form a self-contained system. That is, any addition or scalar multiplication using vectors in the subset S *always yields a vector that lies again in S*. Then taking any **v** in S, we see that $0\mathbf{v} = \mathbf{0}$ and $(-1)\mathbf{v} = -\mathbf{v}$ are also in S. The eight properties A1–A4 and S1–S4 required of a vector space in Definition 3.5 are sure to be true for the subset, since they hold in all of V. That is, if S is nonempty and closed under addition and scalar multiplication, it is sure to be a vector space in its own right. We have arrived at an efficient test to determine whether a subset is a subspace of a vector space.

Theorem 3.6   Test for a Subspace

A nonempty subset S of a vector space V is a subspace of V if it satisfies the following two conditions:

  i) If **v** and **w** are in S, then $\mathbf{v} + \mathbf{w}$ is in S.

  Closure under vector addition

  ii) If r is any scalar in $\mathbb{R}$ and **v** is in S, then $r\mathbf{v}$ is in S.

  Closure under scalar multiplication

As observed above, condition (ii) of the definition with $r = 0$ shows that the zero vector lies in every subspace. Consequently, a subspace of $\mathbb{R}^n$ always contains the origin.

---

† Recall that the set $\{x \mid P(x)\}$ is read: *the set of all x such that the property P(x) is true.*

The entire vector space V, of course, satisfies the conditions of Theorem 3.6. That is, V is a subspace of itself. Other subspaces of V are called **proper subspaces.** Note that one such subspace is the subset $\{0\}$, consisting of only the zero vector. We call $\{0\}$ the **zero subspace** of V.

Example 1 Verify that the x_1, x_2-plane in $\mathbb{R}^n$ is a subspace of $\mathbb{R}^n$ for $n \geq 2$.

Solution The x_1, x_2-plane in $\mathbb{R}^n$ can be described as the set

$$S = \{\mathbf{x} \in \mathbb{R}^n \mid x_3 = x_4 = \cdots = x_n = 0\}.\dagger$$

Of course S is nonempty. To prove that S is a subspace of $\mathbb{R}^n$, we choose $\mathbf{v}, \mathbf{w}$ in S, writing

$$\mathbf{v} = (v_1, v_2, 0, \ldots, 0) \quad \text{and} \quad \mathbf{w} = (w_1, w_2, 0, \ldots, 0).$$

For any scalar r, both $r\mathbf{v} = (rv_1, rv_2, 0, \ldots, 0)$ and $\mathbf{v} + \mathbf{w} = (v_1 + w_1, v_2 + w_2, 0, \ldots, 0)$ have the required zero entries and are therefore in S. Thus S is indeed a subspace. ◁

Example 2 Verify that the line $y = 2x$ is a subspace of $\mathbb{R}^2$.

Solution The line $y = 2x$ can be described as the set

$$L = \{(x, y) \mid y = 2x \text{ for any scalar } x\}$$

or more simply, $L = \{(x, 2x) \mid x \in \mathbb{R}\}$. Clearly L is nonempty. We choose two vectors $(a, 2a)$ and $(b, 2b)$ in L and compute $(a, 2a) + (b, 2b) = (a + b, 2(a + b))$, which has the form $(x, 2x)$ and consequently is in L. Similarly, $r(a, 2a) = (ra, 2ra)$ is in L for any scalar r. Thus L is a subspace.

The subspace L is shown in Fig. 3.20. Note that it is a line *through the origin*. It is not difficult to see that every line through the origin is a subspace of $\mathbb{R}^2$.

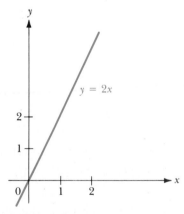

Figure 3.20
The subspace of $\mathbb{R}^2$ where $y = 2x$. ◁

$\dagger$ Recall that $\mathbf{x} \in \mathbb{R}^n$ is read: $\mathbf{x}$ *is a member of* $\mathbb{R}^n$.

Let A be an $m \times n$ matrix and regard $\mathbb{R}^n$ as a space of column vectors. The solution set of the homogeneous system $A\mathbf{x} = \mathbf{0}$ is a subspace of $\mathbb{R}^n$ by Theorem 1.9. It is called the **nullspace** of the matrix A, and is denoted by N_A. The line in Example 2 is the nullspace of the 1×2 matrix $(-2, 1)$, for the equation $y = 2x$ can be expressed as

$$(-2,\ 1) \begin{pmatrix} x \\ y \end{pmatrix} = \mathbf{0}.$$

Example 3   Show that for each integer $n \geq 0$, the set P_n of all polynomials of degree at most n is a subspace of the vector space P of Example 4 in Section 3.2.

Solution   The set P_n is nonempty; for example, every constant polynomial is in P_n. The sum $p + q$ of two polynomials p and q of degrees $\leq n$ is again a polynomial of degree $\leq n$, and the product rp of a polynomial p in P_n by a scalar r is a polynomial of degree at most n. Thus P_n is closed under addition and scalar multiplication so it is a subspace of P. $\lhd$

THE SUBSPACE SPANNED (OR GENERATED) BY VECTORS

Examples 1 and 2 might suggest that any line or plane in $\mathbb{R}^3$ is a subspace. This cannot be the case since a subspace of $\mathbb{R}^3$ must contain the zero vector $\mathbf{0}$. However, we will show that any line or plane containing $\mathbf{0}$ is a subspace of $\mathbb{R}^3$.

Geometrically, we see that a plane in $\mathbb{R}^3$ through the origin is completely determined by any two nonzero and nonparallel vectors $\mathbf{v}_1$ and $\mathbf{v}_2$ that lie in it, as illustrated in Fig. 3.21. The figure shows that any vector in this plane has the form $r_1\mathbf{v}_1 + r_2\mathbf{v}_2$. This expression is a **linear combination** of the vectors $\mathbf{v}_1$ and $\mathbf{v}_2$.

More generally, let $\mathbf{v}_1$ and $\mathbf{v}_2$ be two nonzero and nonparallel vectors in $\mathbb{R}^n$, and consider the set of all linear combinations of $\mathbf{v}_1$ and $\mathbf{v}_2$. We write this set as

$$\mathrm{sp}(\mathbf{v}_1,\ \mathbf{v}_2) = \{r_1\mathbf{v}_1 + r_2\mathbf{v}_2 \mid r_1,\ r_2 \in \mathbb{R}\},$$

called the **span** of $\{\mathbf{v}_1,\ \mathbf{v}_2\}$ or the **span** of $\mathbf{v}_1$ and $\mathbf{v}_2$. Referring to Fig. 3.22, we see that every linear combination of $\mathbf{v}_1$ and $\mathbf{v}_2$ corresponds to some point in the plane determined by the vectors $\mathbf{v}_1$ and $\mathbf{v}_2$.

We claim that $S = \mathrm{sp}(\mathbf{v}_1,\ \mathbf{v}_2)$ is a subspace. Since S is clearly nonempty we simply check the closure conditions of Theorem 3.6:

1. $(r_1\mathbf{v}_1 + r_2\mathbf{v}_2) + (s_1\mathbf{v}_1 + s_2\mathbf{v}_2) = (r_1 + s_1)\mathbf{v}_1 + (r_2 + s_2)\mathbf{v}_2$, which shows that $\mathrm{sp}(\mathbf{v}_1,\ \mathbf{v}_2)$ is closed under addition, and

2. $s(r_1\mathbf{v}_1 + r_2\mathbf{v}_2) = (sr_1)\mathbf{v}_1 + (sr_2)\mathbf{v}_2$, which shows that $\mathrm{sp}(\mathbf{v}_1,\ \mathbf{v}_2)$ is closed under scalar multiplication.

The preceding argument is valid for any vectors $\mathbf{v}_1$ and $\mathbf{v}_2$ in $\mathbb{R}^n$. In particular if $\mathbf{v}_1$ and $\mathbf{v}_2$ are *parallel*, so that $\mathbf{v}_1 = s\mathbf{v}_2$ for some scalar s in $\mathbb{R}$, then $\mathrm{sp}(\mathbf{v}_1,\ \mathbf{v}_2)$ in Fig. 3.22 reduces to a line that contains both vectors $\mathbf{v}_1$ and $\mathbf{v}_2$. In this case, $\mathrm{sp}(\mathbf{v}_1,\ \mathbf{v}_2) = \mathrm{sp}(\mathbf{v}_1)$.

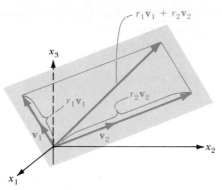

Figure 3.21
The plane sp(v_1, v_2) in $\mathbb{R}^3$.

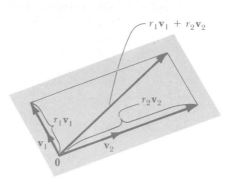

Figure 3.22
The plane sp(v_1, v_2) in $\mathbb{R}^n$.

Since a subspace of $\mathbb{R}^n$ must be closed under addition and scalar multiplication, every subspace containing vectors v_1 and v_2 must contain sp(v_1, v_2). It is clear that sp(v_1, v_2) is the *smallest* subspace of $\mathbb{R}^n$ containing both v_1 and v_2. This subspace is *generated* or *spanned* by these two vectors.

Example 4   Describe geometrically the subspace of $\mathbb{R}^4$ generated by the vector $v = (2, 3, -1, 4)$.

Solution   The subspace sp(v) is a *line* in $\mathbb{R}^4$ containing the origin. Each point on this line is a scalar multiple rv of v, that is, each point has the form (x_1, x_2, x_3, x_4), where $x_1 = 2r$, $x_2 = 3r$, $x_3 = -r$, $x_4 = 4r$ for some scalar r.   ◁

The computations preceding Example 4 that show sp(v_1, v_2) to be a subspace of $\mathbb{R}^n$ can be carried out just as easily with linear combinations of more than two vectors in a general vector space, proving the following theorem.

Theorem 3.7   Subspace Generated or Spanned by Vectors

Let v_1, v_2, . . . , v_k be vectors in a vector space V. The set

$$\text{sp}(v_1, v_2, \ldots, v_k) = \{r_1v_1 + r_2v_2 + \cdots + r_kv_k \mid r_i \in \mathbb{R}\}$$

of all linear combinations of these vectors v_i is a subspace of V.

As the title of Theorem 3.7 indicates, this subspace $S = \text{sp}(v_1, v_2, \ldots, v_k)$ of V is referred to as the subspace of V **generated** or **spanned** by the vectors v_1, v_2, . . . , v_k. The set $\{v_1, v_2, \ldots, v_k\}$ is a **spanning** or **generating** set for S.

Example 5   Describe geometrically the subspace of $\mathbb{R}^3$ spanned by $v_1 = (1, 2, 3)$ and $v_2 = (2, 3, -1)$ in $\mathbb{R}^3$.

Solution   The subspace $sp(\mathbf{v}_1, \mathbf{v}_2)$ is a plane in $\mathbb{R}^3$, part of which is shown in Fig. 3.23. Each point in this plane has the form

$$(x_1, x_2, x_3) = r(1, 2, 3) + s(2, 3, -1),$$

so that $x_1 = r + 2s$, $x_2 = 2r + 3s$, and $x_3 = 3r - s$ for any scalars r and s.   ◁

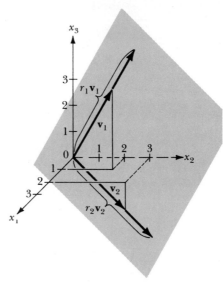

Figure 3.23
The plane sp($\mathbf{v}_1$, $\mathbf{v}_2$).

Example 6   Describe the subspace $sp(1, x, x^2)$ of the vector space P of all polynomials with coefficients in $\mathbb{R}$.

Solution   Since $sp(1, x, x^2) = \{a + bx + cx^2 \mid a, b, c \in \mathbb{R}\}$, we see that $sp(1, x, x^2)$ is the subspace P_2 of all polynomials of degree ≤ 2.   ◁

As illustrated in Example 6, the subspace $sp(1, x, x^2, \ldots, x^n)$ of P is the space P_n of all polynomials of degree $\leq n$.

Example 7   Find a set of vectors in $\mathbb{R}^4$ that spans the solution space of the homogeneous system

$$\begin{aligned} x_1 + x_2 + 3x_3 + x_4 &= 0 \\ 2x_1 + 3x_2 + x_3 + x_4 &= 0 \\ x_1 \qquad + 8x_3 + 2x_4 &= 0. \end{aligned}$$

Solution   We solve the system of equations by reducing its coefficient matrix using the

Gauss-Jordan method, making all pivots 1. We have

$$
\begin{pmatrix} 1 & 1 & 3 & 1 \\ 2 & 3 & 1 & 1 \\ 1 & 0 & 8 & 2 \end{pmatrix} \sim \begin{pmatrix} 1 & 1 & 3 & 1 \\ 0 & 1 & -5 & -1 \\ 0 & -1 & 5 & 1 \end{pmatrix} \sim \begin{pmatrix} 1 & 0 & 8 & 2 \\ 0 & 1 & -5 & -1 \\ 0 & 0 & 0 & 0 \end{pmatrix}.
$$

The system has solution space described by

$$
\mathbf{x} = \begin{pmatrix} x_1 \\ x^2 \\ x_3 \\ x_4 \end{pmatrix} = \begin{pmatrix} -8r - 2s \\ 5r + s \\ r \\ s \end{pmatrix} = r \begin{pmatrix} -8 \\ 5 \\ 1 \\ 0 \end{pmatrix} + s \begin{pmatrix} -2 \\ 1 \\ 0 \\ 1 \end{pmatrix} \tag{1}
$$

for any scalars r and s. The vectors

$$
\mathbf{v}_1 = \begin{pmatrix} -8 \\ 5 \\ 1 \\ 0 \end{pmatrix} \quad \text{and} \quad \mathbf{v}_2 = \begin{pmatrix} -2 \\ 1 \\ 0 \\ 1 \end{pmatrix}
$$

are obtained from the column vector after the second equals sign in (1) if we choose $r = 1, s = 0$ for $\mathbf{v}_1$ and $r = 0, s = 1$ for $\mathbf{v}_2$. Thus the solution space for the given homogeneous system is the subspace $\text{sp}(\mathbf{v}_1, \mathbf{v}_2)$ of $\mathbb{R}^4$ spanned by $\mathbf{v}_1$ and $\mathbf{v}_2$. This is also the nullspace of the coefficient matrix with which we started. ◁

THE COLUMN SPACE AND THE ROW SPACE OF A MATRIX

In addition to the nullspace of an $m \times n$ matrix A, we present two other subspaces related to the matrix. The subspace of $\mathbb{R}^n$ generated by the m rows of A is the **row space** of A. If we regard $\mathbb{R}^m$ as a space of column vectors, the subspace of $\mathbb{R}^m$ generated by the n columns of A is the **column space** of A. The column space of A holds the key to the existence of a solution of a linear system $A\mathbf{x} = \mathbf{b}$. We can write $A\mathbf{x} = \mathbf{b}$ in the form

$$
x_1 \begin{pmatrix} a_{11} \\ a_{21} \\ \vdots \\ a_{m1} \end{pmatrix} + x_2 \begin{pmatrix} a_{12} \\ a_{22} \\ \vdots \\ a_{m2} \end{pmatrix} + \cdots + x_n \begin{pmatrix} a_{1n} \\ a_{2n} \\ \vdots \\ a_{mn} \end{pmatrix} = \begin{pmatrix} b_1 \\ b_2 \\ \vdots \\ b_m \end{pmatrix}. \tag{2}
$$

As $x_1, x_2, \ldots, x_n$ vary over all scalars in $\mathbb{R}$, the left-hand side of (2) runs through the entire column space of A. We obtain at once this criterion for the existence of a solution of $A\mathbf{x} = \mathbf{b}$.

Theorem 3.8 Existence of Solutions

A linear system $A\mathbf{x} = \mathbf{b}$ has a solution if and only if $\mathbf{b}$ lies in the column space of A.

Example 8 Determine if $\mathbf{b} = (7, 6, 1)$ lies in $\mathrm{sp}(\mathbf{v}_1, \mathbf{v}_2, \mathbf{v}_3)$, where $\mathbf{v}_1 = (1, 2, 3)$, $\mathbf{v}_2 = (-2, -5, 2)$, and $\mathbf{v}_3 = (1, 2, -1)$.

Solution We try to solve the system of equations

$$x_1 \begin{pmatrix} 1 \\ 2 \\ 3 \end{pmatrix} + x_2 \begin{pmatrix} -2 \\ -5 \\ 2 \end{pmatrix} + x_3 \begin{pmatrix} 1 \\ 2 \\ -1 \end{pmatrix} = \begin{pmatrix} 7 \\ 6 \\ 1 \end{pmatrix}. \tag{3}$$

Reducing the partitioned matrix for this system, we obtain

$$\left(\begin{array}{ccc|c} 1 & -2 & 1 & 7 \\ 2 & -5 & 2 & 6 \\ 3 & 2 & -1 & 1 \end{array} \right) \sim \left(\begin{array}{ccc|c} 1 & -2 & 1 & 7 \\ 0 & -1 & 0 & -8 \\ 0 & 8 & -4 & -20 \end{array} \right) \sim \left(\begin{array}{ccc|c} 1 & -2 & 1 & 7 \\ 0 & -1 & 0 & -8 \\ 0 & 0 & -4 & -84 \end{array} \right)$$

$$\sim \left(\begin{array}{ccc|c} 1 & -2 & 1 & 7 \\ 0 & 1 & 0 & 8 \\ 0 & 0 & 1 & 21 \end{array} \right).$$

Therefore, $x_3 = 21$, $x_2 = 8$, and $x_1 = 2$, which shows that $\mathbf{b}$ is in $\mathrm{sp}(\mathbf{v}_1, \mathbf{v}_2, \mathbf{v}_3)$. ◁

An important problem is to determine whether n vectors $\mathbf{v}_1, \mathbf{v}_2, \ldots, \mathbf{v}_n$ generate all of $\mathbb{R}^n$. To illustrate, we claim that the vectors $\mathbf{v}_1, \mathbf{v}_2$, and $\mathbf{v}_3$ in Example 8 generate all of $\mathbb{R}^3$. To verify this, we solve the system (3) using a general vector $\mathbf{b} = (b_1, b_2, b_3)$ in $\mathbb{R}^3$ in place of $(7, 6, 1)$. We know that we will obtain a solution again since the coefficient matrix has not changed. We need not find the solution explicitly to conclude that $\mathrm{sp}(\mathbf{v}_1, \mathbf{v}_2, \mathbf{v}_3) = \mathbb{R}^3$.

To test whether n vectors $\mathbf{v}_1, \mathbf{v}_2, \ldots, \mathbf{v}_n$ in $\mathbb{R}^n$ generate $\mathbb{R}^n$, we form the matrix A whose jth column vector is $\mathbf{v}_j$. By Theorem 1.7, we know that $A\mathbf{x} = \mathbf{b}$ has a solution for every choice of $\mathbf{b}$ in $\mathbb{R}^n$ if and only if A is invertible. Thus every $\mathbf{b}$ in $\mathbb{R}^n$ lies in the column space of A if and only if A is invertible. We rephrase this result in the following box.

Criterion for *n* vectors to generate $\mathbb{R}^n$

Vectors $\mathbf{v}_1, \mathbf{v}_2, \ldots, \mathbf{v}_n$ in $\mathbb{R}^n$ generate all of $\mathbb{R}^n$ if and only if the $n \times n$ matrix A having $\mathbf{v}_1, \mathbf{v}_2, \ldots, \mathbf{v}_n$ as column vectors is invertible.

Example 9   Test whether the vectors $(2, 1, 3)$, $(-1, 2, 0)$, and $(1, 8, 6)$ generate $\mathbb{R}^3$.

Solution   Following our criterion above, we form the matrix A having these vectors as column vectors, and row-reduce A to see if it is invertible. We obtain

$$A = \begin{pmatrix} 2 & -1 & 1 \\ 1 & 2 & 8 \\ 3 & 0 & 6 \end{pmatrix} \sim \begin{pmatrix} 1 & 2 & 8 \\ 2 & -1 & 1 \\ 3 & 0 & 6 \end{pmatrix} \sim \begin{pmatrix} 1 & 2 & 8 \\ 0 & -5 & -15 \\ 0 & -6 & -18 \end{pmatrix}$$

$$\sim \begin{pmatrix} 1 & 2 & 8 \\ 0 & 1 & 3 \\ 0 & 0 & 0 \end{pmatrix} = U.$$

Thus A is not invertible, so the given vectors do not generate all of $\mathbb{R}^3$.   ◁

SUMMARY

1. A subset S of a vector space V is a subspace of V if and only if it is nonempty and satisfies the two closure properties:

    $\mathbf{v} + \mathbf{w}$ is contained in S for all vectors $\mathbf{v}$ and $\mathbf{w}$ in S,

    and

    $r\mathbf{v}$ is contained in S for all vectors $\mathbf{v}$ in S and all scalars r.

2. If $\mathbf{v}_1, \mathbf{v}_2, \ldots, \mathbf{v}_k$ are vectors in a vector space V, then the set $\text{sp}(\mathbf{v}_1, \mathbf{v}_2, \ldots, \mathbf{v}_k)$ of all linear combinations of the $\mathbf{v}_i$ is a subspace of V called the span of the $\mathbf{v}_i$ or the subspace generated by the $\mathbf{v}_i$. It is the smallest subspace of V containing all the vectors $\mathbf{v}_i$.

3. The nullspace of an $m \times n$ matrix A is the subspace of $\mathbb{R}^n$ consisting of all solutions of $A\mathbf{x} = \mathbf{0}$.

4. Let A be an $m \times n$ matrix. The row space of A is the subspace of $\mathbb{R}^n$ spanned by the m row vectors of A, while the column space is the subspace of $\mathbb{R}^m$ spanned by the n column vectors of A.

5. The linear system $A\mathbf{x} = \mathbf{b}$ has a solution if and only if $\mathbf{b}$ lies in the column space of A.

6. Vectors $\mathbf{v}_1, \mathbf{v}_2, \ldots, \mathbf{v}_n$ in $\mathbb{R}^n$ span all of $\mathbb{R}^n$ if and only if the matrix A having these vectors as column vectors is invertible.

EXERCISES

In Exercises 1–10, determine whether or not the indicated subset is a subspace of the given vector space.

1. $\{(r, -r) \mid r \in \mathbb{R}\}$ in $\mathbb{R}^2$

2. $\{(x, x + 1) \mid x \in \mathbb{R}\}$ in $\mathbb{R}^2$

3. $\{(n, m) \mid n$ and m are integers$\}$ in $\mathbb{R}^2$

4. $\{(x, y) \mid x, y \in \mathbb{R}$ and $x, y \geq 0\}$ (the first quadrant of $\mathbb{R}^2$)

5. $\{(x, y) \mid x, y \in \mathbb{R}, \text{ and } x,y \geq 0 \text{ or } x,y \leq 0\}$ (the first and third quadrants of $\mathbb{R}^2$)

6. $\{(x, y, z) \mid x, y \in \mathbb{R} \text{ and } z = 3x + 2\}$ in $\mathbb{R}^3$

7. The set of all polynomials of degree greater than 3 in the vector space P of all polynomials with coefficients in $\mathbb{R}$.

8. $\{(x_1, x_2, \ldots, x_n) \mid x_i \in \mathbb{R}, x_2 = 0\}$ in $\mathbb{R}^n$

9. The set of all functions f such that $f(0) = 1$ in the vector space F of all functions mapping $\mathbb{R}$ into $\mathbb{R}$

10. The set of all functions f such that $f(1) = 0$ in the vector space F of all functions mapping $\mathbb{R}$ into $\mathbb{R}$

In Exercises 11–13, let $\mathbf{v}_1 = (1, 3, 4)$, $\mathbf{v}_2 = (2, 7, 2)$, *and* $\mathbf{v}_3 = (-1, 2, 1)$.

11. Determine if $(-3, 1, -2)$, is in $\text{sp}(\mathbf{v}_1, \mathbf{v}_2, \mathbf{v}_3)$.

12. Determine if $(2, 1, 1)$ is in $\text{sp}(\mathbf{v}_1, \mathbf{v}_2, \mathbf{v}_3)$.

13. Determine if $\text{sp}(\mathbf{v}_1, \mathbf{v}_2, \mathbf{v}_3) = \mathbb{R}^3$.

14. Determine if $\text{sp}((1, 2, 1), (2, 1, 0)) = \mathbb{R}^3$.

15. Determine if the vectors $(1, 2, 1)$, $(2, 1, 3)$, $(3, 3, 4)$, and $(-1, 2, 0)$ generate $\mathbb{R}^3$.

16. Determine if the vectors $(1, 2, 1)$, $(2, 1, 3)$, and $(3, 3, 4)$ generate $\mathbb{R}^3$.

17. Let F be the vector space of functions mapping $\mathbb{R}$ into $\mathbb{R}$. Show that:
   a) $\text{sp}(\sin^2 x, \cos^2 x)$ contains all constant functions,
   b) $\text{sp}(\sin^2 x, \cos^2 x)$ contains the function $\cos 2x$,
   c) $\text{sp}(7, \sin^2 2x)$ contains the function $8 \cos 4x$.

In Exercises 18–21, express the given vector $\mathbf{b}$ *as a linear combination of the other vectors* $\mathbf{v}_i$ *in the given space* $\mathbb{R}^n$, *if possible.*

18. $\mathbf{b} = (1, 2)$, $\mathbf{v}_1 = (3, 1)$, $\mathbf{v}_2 = (1, 3)$ in $\mathbb{R}^2$

19. $\mathbf{b} = (1, 2, 1)$, $\mathbf{v}_1 = (1, 1, 1)$, $\mathbf{v}_2 = (1, 1, 2)$, $\mathbf{v}_3 = (0, 1, 1)$ in $\mathbb{R}^3$

20. $\mathbf{b} = (1, 3, 1)$, $\mathbf{v}_1 = (2, 1, 1)$, $\mathbf{v}_2 = (1, 1, 2)$, $\mathbf{v}_3 = (3, 1, 0)$ in $\mathbb{R}^3$

21. $\mathbf{b} = (1, 2, 3)$, $\mathbf{v}_1 = (2, 2, 2)$, $\mathbf{v}_2 = (1, 0, 1)$ in $\mathbb{R}^3$

22. Let P be the vector space of polynomials. Show that $\text{sp}(1, x) = \text{sp}(1 + 2x, x)$. [*Hint:* Show that each of these subspaces is a subset of the other.]

23. Let V be a vector space and let $\mathbf{v}_1$ and $\mathbf{v}_2$ be vectors in V. Follow the hint of Exercise 22 to show that
   a) $\text{sp}(\mathbf{v}_1, \mathbf{v}_2) = \text{sp}(\mathbf{v}_1, 2\mathbf{v}_1 + \mathbf{v}_2)$;
   b) $\text{sp}(\mathbf{v}_1, \mathbf{v}_2) = \text{sp}(\mathbf{v}_1 + \mathbf{v}_2, \mathbf{v}_1 - \mathbf{v}_2)$.

24. Let $\mathbf{v}_1, \mathbf{v}_2, \ldots, \mathbf{v}_k$ and $\mathbf{w}_1, \mathbf{w}_2, \ldots, \mathbf{w}_m$ be vectors in a vector space V. Give a necessary and sufficient condition, involving linear combinations, for

$$\text{sp}(\mathbf{v}_1, \mathbf{v}_2, \ldots, \mathbf{v}_k) = \text{sp}(\mathbf{w}_1, \mathbf{w}_2, \ldots, \mathbf{w}_m).$$

25. Consider the homogeneous system of equations

$$
\begin{aligned}
4x_1 + 9x_2 - x_3 + 9x_4 &= 0 \\
x_1 + 2x_2 - x_3 + 3x_4 &= 0 \\
2x_1 + 5x_2 + x_3 + 3x_4 &= 0 \\
x_1 + x_2 - 4x_3 + 6x_4 &= 0,
\end{aligned}
$$

which has solutions $\mathbf{v} = (7, -3, 1, 0)$ and $\mathbf{w} = (-9, 3, 0, 1)$. Show by direct substitution of $r\mathbf{v} + s\mathbf{w}$ that each linear combination of $\mathbf{v}$ and $\mathbf{w}$ is also a solution to the system.

26. Referring to Exercise 25, verify that the solution space of the given system of equations is $\text{sp}(\mathbf{v}, \mathbf{w})$, using a Gauss–Jordan reduction.

In Exercises 27–34, find as small a set of vectors as you can that generates the solution space of the given homogeneous system of equations.

27. $x - y = 0$
$2x - 2y = 0$

28. $2x + 5y = 0$
$3x + y = 0$

29. $3x_1 - 2x_2 + x_3 = 0$
$2x_1 + x_2 - 5x_3 = 0$
$x_1 + x_2 - 6x_3 = 0$

30. $3x_1 + x_2 + x_3 = 0$
$6x_1 + 2x_2 + 2x_3 = 0$
$-9x_1 - 3x_2 - 3x_3 = 0$

31. $x_1 - x_2 + x_3 - x_4 = 0$
$x_2 + x_3 = 0$
$x_1 + 2x_2 - x_3 + 3x_4 = 0$

32. $2x_1 + x_2 + x_3 + x_4 = 0$
$x_1 - 6x_2 + x_3 = 0$
$3x_1 - 5x_2 + 2x_3 + x_4 = 0$
$5x_1 - 4x_2 + 3x_3 + 2x_4 = 0$

33. $2x_1 + x_2 + x_3 + x_4 = 0$
$3x_1 + x_2 - x_3 + 2x_4 = 0$
$x_1 + x_2 + 3x_3 = 0$
$x_1 - x_2 - 7x_3 + 2x_4 = 0$

34. $x_1 - x_2 + 6x_3 + x_4 - x_5 = 0$
$3x_1 + 2x_2 - 3x_3 + 2x_4 + 5x_5 = 0$
$4x_1 + 2x_2 - x_3 + 3x_4 - x_5 = 0$
$3x_1 - 2x_2 + 14x_3 + x_4 - 8x_5 = 0$
$2x_1 - x_2 + 8x_3 + 2x_4 - 7x_5 = 0$

35. Use the second option with the program VECTGRPH for graphic drill on linear combinations until a score of at least 80 percent can be achieved easily.

In Exercises 36–39, use YUREDUCE or MATCOMP or similar software.

36. Determine if the vectors $(1, 3, 2, 1, 4)$, $(0, 2, 1, 2, 2)$, $(0, 0, 3, 1, 1)$, $(0, 0, 0, 4, 1)$, and $(0, 0, 0, 0, 2)$ generate $\mathbb{R}^5$.

37. Determine if the vectors $(2, 1, 3, -4, 0)$, $(1, 1, 1, -1, 1)$, $(4, -2, -3, 1, 7)$, $(4, -3, -2, -1, 5)$, $(6, -3, 1, 2, 4)$, and $(8, -8, -3, 3, 7)$ generate $\mathbb{R}^5$.

38. Express $(2, -1, 1, 1)$ as a linear combination of $(0, 1, 1, 1)$, $(1, 0, 1, 1)$, $(1, 1, 0, 1)$, and $(1, 1, 1, 0)$ if possible.

39. Express $(1, 3, 2, -1)$ as a linear combination of $(1, 1, 0, 0)$, $(1, 0, 1, 0)$, $(0, 0, 1, 1)$ and $(0, 1, 0, 1)$ if possible.

3.4
Independence and Bases

We now turn to the question of generating a vector space efficiently. Consider a subspace S of a vector space V described by $S = \text{sp}(\mathbf{v}_1, \mathbf{v}_2, \ldots, \mathbf{v}_k)$, where the generating vectors listed are all distinct. The vectors in S are determined completely by these generating vectors $\mathbf{v}_i$. We are interested in seeing whether a shorter list comprised of some of these same vectors $\mathbf{v}_i$ can generate S as well, for the smaller the number of generating vectors, the simpler the description of the subspace S. Assuming $S \neq \{\mathbf{0}\}$, if some $\mathbf{v}_i = \mathbf{0}$, then that $\mathbf{v}_i$ can be omitted. Furthermore, if $\mathbf{v}_2 = r_1\mathbf{v}_1$ for some scalar r_1 in $\mathbb{R}$, then $\mathbf{v}_2$ can be omitted from the list. More generally, if some $\mathbf{v}_j$ can be written in terms of its predecessors as

$$\mathbf{v}_j = r_1\mathbf{v}_1 + r_2\mathbf{v}_2 + \cdots + r_{j-1}\mathbf{v}_{j-1}, \tag{1}$$

then each linear combination

$$\mathbf{v} = s_1\mathbf{v}_1 + s_2\mathbf{v}_2 + \cdots + s_j\mathbf{v}_j + \cdots + s_k\mathbf{v}_k$$

in S can be written without using $\mathbf{v}_j$, namely,

$$\mathbf{v} = (s_1 + s_j r_1)\mathbf{v}_1 + (s_2 + s_j r_2)\mathbf{v}_2 + \cdots + (s_{j-1} + s_j r_{j-1})\mathbf{v}_{j-1}$$
$$+ s_{j+1}\mathbf{v}_{j+1} + s_{j+2}\mathbf{v}_{j+2} + \cdots + s_k \mathbf{v}_k.$$

Therefore, in this case,

$$S = \mathrm{sp}(\mathbf{v}_1, \mathbf{v}_2, \ldots, \mathbf{v}_{j-1}, \mathbf{v}_{j+1}, \ldots, \mathbf{v}_k)$$

and we have indeed succeeded in shortening our list of generating vectors.

The preceding discussion suggests that we study relations between vectors of the form in Eq. (1).

DEPENDENT AND INDEPENDENT SETS OF VECTORS

By moving all terms to the right-hand side of the equation, we can write relation (1) in the form

$$r_1 \mathbf{v}_1 + r_2 \mathbf{v}_2 + \cdots + r_k \mathbf{v}_k = \mathbf{0},$$

where $r_j = -1$ and $r_{j+1} = r_{j+2} = \cdots = r_k = 0$. This is an example of a *dependence relation*.

Definition 3.7 Linearly Dependent Set of Vectors

A set $\{\mathbf{v}_1, \mathbf{v}_2, \ldots, \mathbf{v}_k\}$ of vectors in a vector space V is **linearly dependent** if there exists a **dependence relation**

$$r_1 \mathbf{v}_1 + r_2 \mathbf{v}_2 + \cdots + r_k \mathbf{v}_k = \mathbf{0}, \quad \text{for some } r_j \neq 0. \qquad (2)$$

For convenience, we will often drop the term *linearly* from the description *linearly dependent* and just speak of a *dependent set of vectors*. We will sometimes drop the word *set* and refer to *dependent vectors* $\mathbf{v}_1, \mathbf{v}_2, \ldots, \mathbf{v}_k$.

A COORDINATE-FREE TREATMENT of vector-space concepts appeared in 1862 in the second version of Hermann Grassmann's *Ausdehnungslehre* (*The Calculus of Extension*). In this version he was able to suppress somewhat the philosophical bias that had made his earlier work so unreadable and to concentrate on his new mathematical ideas. These included the basic ideas of the theory of n-dimensional vector spaces, including linear combinations, linear independence, and the notions of a subspace and a basis. He developed the idea of the dimension of a subspace as the maximal number of linearly independent vectors and proved the fundamental relation for two subspaces V and W that $\dim(V + W) = \dim V + \dim W - \dim(V \cap W)$.

Grassmann's notions derived from the attempt to translate geometric ideas about n-dimensional space into the language of algebra without dealing with coordinates, as is done in ordinary analytic geometry. He was the first to produce a complete system in which such concepts as points, line segments, planes, and their analogues in higher dimensions are represented as single elements. Though his ideas were initially difficult to understand, ultimately they entered the mathematical mainstream in such fields as vector analysis and the exterior algebra. Grassmann himself, unfortunately, never attained his goal of becoming a German university professor, spending most of his professional life as a mathematics teacher at a gymnasium (high school) in Stettin. In the final decades of his life, he turned away from mathematics and established himself as an expert in linguistics.

We note at once that any set of vectors that includes the zero vector is a dependent set, for if $\mathbf{v}_j = \mathbf{0}$, then we may take $r_j = 1$ and all other $r_i = 0$ and obtain a dependence relation (2). In practice, we are seldom concerned with a dependent set of vectors that contains the zero vector.

If the set $\{\mathbf{v}_1, \mathbf{v}_2, \ldots, \mathbf{v}_k\}$ does not contain the zero vector and is dependent, then a moment of thought shows that a dependence relation (2) must contain at least two nonzero coefficients r_i. Suppose we let r_j be the nonzero coefficient with maximum subscript. The dependence relation (2) can then be rewritten as

$$\mathbf{v}_j = (-r_1/r_j)\mathbf{v}_1 + (-r_2/r_j)\mathbf{v}_2 + \cdots + (-r_{j-1}/r_j)\mathbf{v}_{j-1},$$

which is a relation of the form (1), expressing $\mathbf{v}_j$ as a linear combination of its predecessors. We immediately obtain the criterion for dependence of a set of nonzero vectors, shown in the following box.

Dependence of a Set of Nonzero Vectors

A finite list of nonzero vectors in a vector space V is linearly dependent if and only if some vector in the list is equal to a linear combination of its predecessors.

For example, two nonzero vectors are dependent if and only if one is a multiple of the other. As another example, note that a set of vectors of the form $\{\mathbf{v}_1, \mathbf{v}_2, 2\mathbf{v}_1 - 3\mathbf{v}_2\}$ is dependent in any vector space V.

A set of vectors that is not dependent is said to be independent. Independent sets of vectors will play an important role for us, and for this reason, we actually write out the negations of Definition 3.7 and the boxed statement above.

Definition 3.8 Independent Set of Vectors

A set $\{\mathbf{v}_1, \mathbf{v}_2, \ldots, \mathbf{v}_k\}$ of vectors in a vector space V is **linearly independent** if no dependence relation of the form (2) exists, so that $r_1\mathbf{v}_1 + r_2\mathbf{v}_2 + \cdots + r_k\mathbf{v}_k = \mathbf{0}$ only if all the coefficients r_i are zero.

The negation of the condition for dependence of a set of nonzero vectors takes this form:

Independence of a Set of Nonzero Vectors

A finite list of nonzero vectors in a vector space V is linearly independent if and only if no vector in the list is equal to a linear combination of its predecessors.

Definition 3.8 may be stated for *column* vectors in $\mathbb{R}^n$ as follows: A set $\{\mathbf{v}_1, \mathbf{v}_2, \ldots, \mathbf{v}_k\}$ of vectors in $\mathbb{R}^n$ is independent if and only if the homogeneous $n \times k$ system of equations

$$x_1\mathbf{v}_1 + x_2\mathbf{v}_2 + \cdots + x_k\mathbf{v}_k = \mathbf{0} \tag{3}$$

has no nontrivial solution, that is, has only the zero solution $\mathbf{x} = \mathbf{0}$. We state this as a theorem.

Theorem 3.9 Application to Linear Systems

Let A be an $n \times k$ matrix. The homogeneous system $A\mathbf{x} = \mathbf{0}$ has no nontrivial solution if and only if the column vectors of A are independent.

We saw in Theorem 1.8 on page 72 that a system such as (3) will have a nontrivial solution if the number of unknowns exceeds the number of equations, that is, if $k > n$ in Eq. (3). In other words, any set $\{\mathbf{v}_1, \mathbf{v}_2, \ldots, \mathbf{v}_k\}$ of vectors in $\mathbb{R}^n$ is dependent if $k > n$.

Any set of more than n vectors in $\mathbb{R}^n$ is linearly dependent.

Example 1 Determine whether the vectors $(1, -2, 1)$, $(3, -5, 2)$, $(2, -3, 6)$, and $(1, 2, 1)$ in $\mathbb{R}^3$ are independent.

Solution Since there are more vectors than the number 3 of components, the vectors are dependent. ◁

In the case $k = n$, the system (3) has no nontrivial solution if and only if the square matrix A containing the vectors $\mathbf{v}_j$ as columns is invertible. Invertibility of A also serves as a test for the $\mathbf{v}_i$ to span $\mathbb{R}^n$; see the box on page 156. We obtain three equivalent conditions, which we state as a theorem.

Theorem 3.10 Equivalent Conditions for *n* Vectors in $\mathbb{R}^n$

Let $\mathbf{v}_1, \mathbf{v}_2, \ldots, \mathbf{v}_n$ be n vectors in $\mathbb{R}^n$. The following conditions are equivalent:

1. The vectors are independent.
2. The vectors generate all of $\mathbb{R}^n$.
3. The matrix A having these vectors as column vectors is invertible.

Example 2 Determine whether the vectors $(1, 0, 0)$, $(1, 1, 0)$, and $(1, 1, 1)$ generate $\mathbb{R}^3$.

Solution It is clear that these vectors are independent, so they must generate $\mathbb{R}^3$ according to Theorem 3.10. Alternatively, the matrix

$$A = \begin{pmatrix} 1 & 1 & 1 \\ 0 & 1 & 1 \\ 0 & 0 & 1 \end{pmatrix}$$

is surely invertible. Again, Theorem 3.10 shows that the vectors must generate $\mathbb{R}^3$. ◁

To establish the dependence or independence of a set of vectors in a vector space V other than $\mathbb{R}^n$, one may attempt to apply Definitions 3.7 and 3.8 directly.

Example 3   Show that $\{x, x^2\}$ is an independent set of functions in the vector space F of all functions mapping $\mathbb{R}$ into $\mathbb{R}$.

Solution   Suppose we have a relation of the form

$$r_1 x + r_2 x^2 = \mathbf{0},$$

where we regard $\mathbf{0}$ as the zero function. Evaluating this equation when $x = 1$ and when $x = -1$, we obtain

$$\begin{aligned} r_1 + r_2 &= 0 & &\text{Taking } x = 1 \\ -r_1 + r_2 &= 0. & &\text{Taking } x = -1 \end{aligned}$$

We easily find that these two equations in the two unknowns r_1 and r_2 have only the trivial solution $r_1 = r_2 = 0$. It follows from Definition 3.8 that x and x^2 are independent functions in the vector space F. ◁

The technique in Example 3 is a standard one for testing the independence of n functions in the vector space F. Consider a dependence relation, and evaluate it at n different values for x. This gives rise to n linear equations with the n coefficients in the dependence relation as unknowns. If the system has only the trivial solution, then the functions are independent. If the system has a nontrivial solution, we have to work harder. Perhaps a different choice of values for x would yield only the trivial solution, or perhaps the functions actually are dependent. Dependence of functions can be difficult to establish unless we are able to spot a dependence relation by inspection.

Example 4   Show that $\{1, \sin^2 x, \cos^2 x\}$ is a dependent set of functions in the vector space F of all functions mapping $\mathbb{R}$ into $\mathbb{R}$.

Solution   From the familiar trigonometric identity $\sin^2 x + \cos^2 x = 1$, we obtain the dependence relation

$$(-1)1 + (1)\sin^2 x + (1)\cos^2 x = \mathbf{0}. ◁$$

BASES

We return to our discussion of a subspace $S = \text{sp}(\mathbf{v}_1, \mathbf{v}_2, \ldots, \mathbf{v}_k)$ of a vector space V. We assume that the vectors $\mathbf{v}_i$ are all distinct and nonzero. We saw that by deleting from this list of generating vectors any vector that is a linear combination of its predecessors, we obtain a shorter list of vectors that still generates S. We can continue this process until no vector remaining in the shortened list is a linear combination of its predecessors. By the boxed statement following Definition 3.8, the vectors then remaining in the list form an independent set of vectors generating S.

We will refer to the process of deleting vectors that can be expressed as a linear combination of predecessors from a list as the *casting-out technique.* It is best to perform this technique systematically, say from left to right in the list.

> **General Casting-Out Technique**
>
> Let $\mathbf{v}_1, \mathbf{v}_2, \ldots, \mathbf{v}_k$ be a list of nonzero vectors generating a subspace S of a vector space V. Start with $\mathbf{v}_2$ in the list and work to the right, casting out any vector that is equal to a linear combination of its remaining predecessors. After all vectors have been tested in this manner, the vectors remaining form an independent generating set for the subspace S.

Independent generating sets of vectors for a vector space are very important in our subsequent work. We make the following definition.

Definition 3.9 Basis for a Subspace

> Let S be a subspace of a vector space V. A set $\{\mathbf{v}_1, \mathbf{v}_2, \ldots, \mathbf{v}_k\}$ of vectors in S is a **basis** for S if:
>
> **1.** the set of vectors generates S, that is, $S = \text{sp}(\mathbf{v}_1, \mathbf{v}_2, \ldots, \mathbf{v}_k)$, and
>
> **2.** the set of vectors is linearly independent.

The set $\{\mathbf{e}_1, \mathbf{e}_2, \ldots, \mathbf{e}_n\}$ of unit coordinate vectors in $\mathbb{R}^n$ is a basis for $\mathbb{R}^n$. This basis is called the **standard basis.** The set $\{1, x, x^2, x^3, \ldots, x^n\}$ is linearly independent in the vector space P of all polynomials with coefficients in $\mathbb{R}$. Thus this set of monomials is a basis for the subspace P_n of all polynomials of degree $\leq n$. We think of a basis for a vector space as a set of reference vectors for the space.

Example 5   Determine if the vectors

$$(1, 2, -1, 0) \quad (0, 1, 0, 1), \quad (-1, -5, 2, 0), \quad \text{and} \quad (2, 3, -2, 7)$$

form a basis for $\mathbb{R}^4$.

Solution   We form the matrix A having these vectors as columns, and obtain

$$\begin{pmatrix} 1 & 0 & -1 & 2 \\ 2 & 1 & -5 & 3 \\ -1 & 0 & 2 & -2 \\ 0 & 1 & 0 & 7 \end{pmatrix} \sim \begin{pmatrix} 1 & 0 & -1 & 2 \\ 0 & 1 & -3 & -1 \\ 0 & 0 & 1 & 0 \\ 0 & 1 & 0 & 7 \end{pmatrix} \sim \begin{pmatrix} 1 & 0 & -1 & 2 \\ 0 & 1 & -3 & -1 \\ 0 & 0 & 1 & 0 \\ 0 & 0 & 3 & 8 \end{pmatrix}$$

$$\sim \begin{pmatrix} 1 & 0 & -1 & 2 \\ 0 & 1 & -3 & -1 \\ 0 & 0 & 1 & 0 \\ 0 & 0 & 0 & 8 \end{pmatrix},$$

which shows A to be invertible. Therefore, the given vectors form a basis for $\mathbb{R}^4$ by Theorem 3.10. ◁

Example 6   Find a basis for the nullspace of the matrix

$$A = \begin{pmatrix} 3 & 1 & 9 \\ 1 & 2 & -2 \\ 2 & 1 & 5 \end{pmatrix}.$$

Solution   The nullspace N_A of the matrix A is the set of solutions of the system $A\mathbf{x} = \mathbf{0}$. Using a Gauss reduction and back substitution, we obtain

$$\left(\begin{array}{ccc|c} 3 & 1 & 9 & 0 \\ 1 & 2 & -2 & 0 \\ 2 & 1 & 5 & 0 \end{array} \right) \sim \left(\begin{array}{ccc|c} 1 & 2 & -2 & 0 \\ 0 & -5 & 15 & 0 \\ 0 & -3 & 9 & 0 \end{array} \right) \sim \left(\begin{array}{ccc|c} 1 & 2 & -2 & 0 \\ 0 & 1 & -3 & 0 \\ 0 & 0 & 0 & 0 \end{array} \right).$$

Therefore,

$$N_A = \left\{ \begin{pmatrix} x_1 \\ x_2 \\ x_3 \end{pmatrix} \middle| \; x_2 = 3x_3, \, x_1 = -4x_3 \text{ for } x_3 \in \mathbb{R}^n \right\}$$

$$= \left\{ x_3 \begin{pmatrix} -4 \\ 3 \\ 1 \end{pmatrix} \middle| \; x_3 \in \mathbb{R} \right\} = \text{sp} \left(\begin{pmatrix} -4 \\ 3 \\ 1 \end{pmatrix} \right).$$

Thus one basis for N_A contains the single vector

$$\begin{pmatrix} -4 \\ 3 \\ 1 \end{pmatrix}. \; ◁$$

Example 7   Find a basis for the subspace

$$S = \{(x_1, x_2, x_3, x_4, x_5) \mid x_2 = 3x_1, x_4 = 2x_1, x_5 = x_3 - x_1\}$$

of $\mathbb{R}^5$.

Solution   We can write

$$
\begin{aligned}
S &= \{(x_1, 3x_1, x_3, 2x_1, x_3 - x_1) \mid x_1, x_3 \in \mathbb{R}\} \\
&= \{x_1(1, 3, 0, 2, -1) + x_3(0, 0, 1, 0, 1) \mid x_1, x_3 \in \mathbb{R}\} \\
&= \mathrm{sp}(\mathbf{v}_1, \mathbf{v}_2),
\end{aligned}
$$

where $\mathbf{v}_1 = (1, 3, 0, 2, -1)$ is obtained from $(x_1, 3x_1, x_3, 2x_1, x_3 - x_1)$ by taking $x_1 = 1$ and $x_3 = 0$, while $\mathbf{v}_2 = (0, 0, 1, 0, 1)$ is obtained by taking $x_1 = 0$ and $x_3 = 1$. Since $\mathbf{v}_2$ is not a multiple of $\mathbf{v}_1$, we conclude that $\{\mathbf{v}_1, \mathbf{v}_2\}$ is a basis for S.   ◁

The casting-out technique described earlier gives us at once this theorem.

Theorem 3.11   Reducing a Generating Set to a Basis

Let S be a subspace of a vector space V generated by a set $\{\mathbf{v}_1, \mathbf{v}_2, \ldots, \mathbf{v}_k\}$ of nonzero vectors. Then S can be generated by a possibly smaller set of these vectors, which is a basis for S. Such a basis can be found by casting out $\{\mathbf{v}_1, \mathbf{v}_2, \ldots, \mathbf{v}_k\}$ those vectors that are linear combinations of predecessors.

Example 8   Consider the subspace S of $\mathbb{R}^3$ given by

$$S = \mathrm{sp}((2, 1, 3), (-1, 2, 0), (1, 8, 6)).$$

Use the casting-out technique to find a basis for S.

Solution   We apply the casting-out technique systematically, starting with the second vector and working to the right. Since $(-1, 2, 0)$ is not a multiple of $(2, 1, 3)$, we keep $(-1, 2, 0)$ in the list. We proceed to the vector $(1, 8, 6)$ and check whether it can be expressed in the form $(1, 8, 6) = r_1(2, 1, 3) + r_2(-1, 2, 0)$ for scalars r_1 and r_2. Rewriting this equation in the order

$$r_1(2, 1, 3) + r_2(-1, 2, 0) = (1, 8, 6),$$

We recognize that it leads to a system of equations in the unknowns r_1 and r_2. We reduce the partitioned matrix of the system:

$$
\left(\begin{array}{rr|r} 2 & -1 & 1 \\ 1 & 2 & 8 \\ 3 & 0 & 6 \end{array}\right) \sim
\left(\begin{array}{rr|r} 1 & 2 & 8 \\ 2 & -1 & 1 \\ 3 & 0 & 6 \end{array}\right) \sim
\left(\begin{array}{rr|r} 1 & 2 & 8 \\ 0 & -5 & -15 \\ 0 & -6 & -18 \end{array}\right) \sim
\left(\begin{array}{rr|r} 1 & 2 & 8 \\ 0 & 1 & 3 \\ 0 & 0 & 0 \end{array}\right).
$$

We obtain the solution $r_2 = 3$ and $r_1 = 2$, and we conclude that $(1, 8, 6)$ is in $\mathrm{sp}((2, 1, 3), (-1, 2, 0))$. Consequently, we cast out $(1, 8, 6)$ and arrive at the basis

$$\{(2, 1, 3), (-1, 2, 0)\}$$

for S.   ◁

Example 9   Use the casting-out technique to find a basis for the subspace

$$S = \text{sp}(x^2 - 1, x^2 + 1, 3, 2x - 1, x)$$

of the vector space P of polynomials.

Solution   Clearly $x^2 + 1 \neq r(x^2 - 1)$ for any scalar r, so we retain $x^2 + 1$ in the casting-out technique. A moment of thought shows that we do have a dependence relation

$$3 = \left(-\frac{3}{2}\right)(x^2 - 1) + \left(\frac{3}{2}\right)(x^2 + 1)$$

so we cast out 3 and work with the list $x^2 - 1, x^2 + 1, 2x - 1, x$. It is impossible to express $2x - 1$ as a linear combination of $x^2 - 1$ and $x^2 + 1$ since no first power of x appears in these two predecessors, so $2x - 1$ is retained in the list. Since an x can be contributed from $(\frac{1}{2})(2x - 1)$ and the undesired $-\frac{1}{2}$ can be eliminated by taking $(\frac{1}{4})[(x^2 + 1) - (x^2 - 1)]$, we realize that x can be expressed as a linear combination of its three predecessors, and should be cast out. Specifically,

$$x = \left(-\frac{1}{4}\right)(x^2 - 1) + \left(\frac{1}{4}\right)(x^2 + 1) + \left(\frac{1}{2}\right)(2x - 1).$$

Thus we arrive at the basis $\{x^2 - 1, x^2 + 1, 2x - 1\}$ for S.   ◁

EXECUTING THE CASTING-OUT TECHNIQUE IN $\mathbb{R}^n$

As we have presented it up to now, the casting-out technique seems potentially laborious. For example, to apply it to a list $\mathbf{v}_1, \mathbf{v}_2, \ldots, \mathbf{v}_6$, it seems that we should test for solutions of five vector equations

$$\mathbf{v}_2 = r_1\mathbf{v}_1, \qquad \mathbf{v}_3 = r_1\mathbf{v}_1 + r_2\mathbf{v}_2, \quad \text{and so on.}$$

In the vector space $\mathbb{R}^n$, each of these equations would lead to a linear system, as illustrated by Example 8. Actually all five linear systems can be solved at once by reducing a single matrix, as stated in the box. We explain the validity of our technique as we work through an example.

The Casting-Out Technique in $\mathbb{R}^n$

Elimination of nonzero vectors that are linear combinations of predecessors in a list $\mathbf{v}_1, \mathbf{v}_2, \ldots, \mathbf{v}_k$ can be accomplished as follows:

1. Form the matrix A with $\mathbf{v}_j$ as the jth column vector.
2. Reduce A to row-echelon form as in the Gauss method.
3. The vectors in the columns of A that give rise to (nonzero) pivots should be retained, and will generate the entire column space of A. That is, cast out from the list those vectors in the columns of A that do not provide (nonzero) pivots.

Example 10   Let S be the subspace of $\mathbb{R}^5$ generated by

$$\mathbf{v}_1 = (1, -1, 0, 2, 1), \qquad \mathbf{v}_2 = (2, 1, -2, 0, 0),$$
$$\mathbf{v}_3 = (0, -3, 2, 4, 2), \qquad \mathbf{v}_4 = (3, 3, -4, -2, -1),$$
$$\mathbf{v}_5 = (2, 4, 1, 0, 1), \qquad \mathbf{v}_6 = (5, 7, -3, -2, 0).$$

Use the casting-out technique to attempt to shorten the list $\mathbf{v}_1, \mathbf{v}_2, \mathbf{v}_3, \mathbf{v}_4, \mathbf{v}_5, \mathbf{v}_6$ of generators of S.

Solution   We reduce the matrix having $\mathbf{v}_j$ as the jth column vector, and obtain an echelon form. Duplication of matrices and the partitions will be explained in a moment.

$$\begin{pmatrix} 1 & 2 & 0 & 3 & 2 & 5 \\ -1 & 1 & -3 & 3 & 4 & 7 \\ 0 & -2 & 2 & -4 & 1 & -3 \\ 2 & 0 & 4 & -2 & 0 & -2 \\ 1 & 0 & 2 & -1 & 1 & 0 \end{pmatrix} \sim \left(\begin{array}{c|ccccc} 1 & 2 & 0 & 3 & 2 & 5 \\ 0 & 3 & -3 & 6 & 6 & 12 \\ 0 & -2 & 2 & -4 & 1 & -3 \\ 0 & -4 & 4 & -8 & -4 & -12 \\ 0 & -2 & 2 & -4 & -1 & -5 \end{array} \right)$$

$$\sim \left(\begin{array}{cc|cccc} 1 & 2 & 0 & 3 & 2 & 5 \\ 0 & 1 & -1 & 2 & 2 & 4 \\ 0 & 0 & 0 & 0 & 5 & 5 \\ 0 & 0 & 0 & 0 & 4 & 4 \\ 0 & 0 & 0 & 0 & 3 & 3 \end{array} \right) \sim \left(\begin{array}{ccc|ccc} 1 & 2 & 0 & 3 & 2 & 5 \\ 0 & 1 & -1 & 2 & 2 & 4 \\ 0 & 0 & 0 & 0 & 5 & 5 \\ 0 & 0 & 0 & 0 & 4 & 4 \\ 0 & 0 & 0 & 0 & 3 & 3 \end{array} \right)$$

$$\sim \left(\begin{array}{cccc|cc} 1 & 2 & 0 & 3 & 2 & 5 \\ 0 & 1 & -1 & 2 & 2 & 4 \\ 0 & 0 & 0 & 0 & 5 & 5 \\ 0 & 0 & 0 & 0 & 4 & 4 \\ 0 & 0 & 0 & 0 & 3 & 3 \end{array} \right) \sim \left(\begin{array}{ccccc|c} 1 & 2 & 0 & 3 & 2 & 5 \\ 0 & 1 & -1 & 2 & 2 & 4 \\ 0 & 0 & 0 & 0 & 1 & 1 \\ 0 & 0 & 0 & 0 & 0 & 0 \\ 0 & 0 & 0 & 0 & 0 & 0 \end{array} \right)$$

$$\sim \begin{pmatrix} \boxed{1} & 2 & 0 & 3 & 2 & 5 \\ 0 & \boxed{1} & -1 & 2 & 2 & 4 \\ 0 & 0 & 0 & 0 & \boxed{1} & 1 \\ 0 & 0 & 0 & 0 & 0 & 0 \\ 0 & 0 & 0 & 0 & 0 & 0 \end{pmatrix}$$

The first column vector in the first matrix is of course retained. When it is determined that a column vector in the first matrix is to be retained, the corresponding column will be shaded in color in the subsequent matrices.

The *second* matrix contains a partition between the first and second columns. If we were to solve the system of equations $\mathbf{v}_2 = r_1\mathbf{v}_1$ using the Gauss method, we would perform exactly the work shown in the first two columns of the first two matrices. The first two columns of the second matrix thus indicate that there is no solution for r_1. Consequently, we will retain the second column vector $\mathbf{v}_2$ so we also shade in color the second column in subsequent matrices. Note that a (nonzero) pivot can be found in the second column.

We draw a partition between the second column and third column in the *third* matrix, and examine the system of equations corresponding to the *first three* columns. Solving the linear system $\mathbf{v}_3 = r_1\mathbf{v}_1 + r_2\mathbf{v}_2$ by the Gauss method amounts precisely to the work in the first three columns of the first three matrices. The first three columns of the third matrix indicate that this time there is a solution, namely, $r_2 = -1$ and $r_1 = 2$. Thus $\mathbf{v}_3$ will be cast out, and the third column will not be color-shaded in subsequent matrices. Note that we did have a solution this time since we were unable to obtain a (nonzero) pivot in the third column.

Looking at the *fourth* matrix, we draw a partition between the third and fourth columns. The work in solving the system $\mathbf{v}_4 = r_1\mathbf{v}_1 + r_2\mathbf{v}_2$ by the Gauss method amounts precisely to the work in columns 1, 2, and 4 of the first four matrices. Columns 1, 2, and 4 of matrix 4 show that this system also has a solution, namely, $r_2 = 2$ and $r_1 = -1$. Thus $\mathbf{v}_4$ will be cast out, and the fourth column will not be color-shaded.

Continuing this argument, we examine the *fifth* matrix, and we see that $\mathbf{v}_5 = r_1\mathbf{v}_1 + r_2\mathbf{v}_2$ has no solution. Therefore, we color-shade the fifth column in the next matrix, and $\mathbf{v}_5$ will be retained.

The *sixth* matrix indicates that $\mathbf{v}_6 = r_1\mathbf{v}_1 + r_2\mathbf{v}_2 + r_5\mathbf{v}_5$ has a solution $r_5 = 1$, $r_2 = 2$, $r_1 = -1$, and so $\mathbf{v}_6$ is cast out.

Our final matrix has no partition and shows three color-shaded columns. They are the columns containing (nonzero) pivots and correspond to $\mathbf{v}_1$, $\mathbf{v}_2$, and $\mathbf{v}_5$ in the original matrix. We conclude that

$$S = \text{sp}(\mathbf{v}_1, \mathbf{v}_2, \mathbf{v}_5).$$

In practice it is not necessary to duplicate matrices as we have done here. One simply obtains an echelon form of the matrix containing as columns the vectors being tested. The columns that contain the (nonzero) pivots correspond to the vectors that are retained; the other vectors are cast out.   ◁

The reduction used in the casting-out technique can be executed easily on a computer. One can use the available programs YUREDUCE or MATCOMP, or similar software.

Example 11   Find a basis for the subspace S of $\mathbb{R}^3$ generated by

$$\{(1, -3, 1), (-2, 6, -2), (2, 1, -4), (-1, 10, -7)\}.$$

Solution   We simply execute the casting-out technique for these four vectors in $\mathbb{R}^3$. We reduce the matrix having them as column vectors, obtaining

$$\begin{pmatrix} 1 & -2 & 2 & -1 \\ -3 & 6 & 1 & 10 \\ 1 & -2 & -4 & -7 \end{pmatrix} \sim \begin{pmatrix} 1 & -2 & 2 & -1 \\ 0 & 0 & 7 & 7 \\ 0 & 0 & -6 & -6 \end{pmatrix} \sim \begin{pmatrix} 1 & -2 & 1 & -1 \\ 0 & 0 & 1 & 1 \\ 0 & 0 & 0 & 0 \end{pmatrix}.$$

Keeping the vectors that lead to columns with (nonzero) pivots, we find that $\{(1, -3, 1), (2, 1, -4)\}$ is a basis for S.   ◁

SUMMARY

> Let V be a vector space.
>
> **1.** A set of vectors $\{\mathbf{v}_1, \mathbf{v}_2, \ldots, \mathbf{v}_k\}$ in V is linearly dependent if there exists a dependence relation
>
> $$r_1\mathbf{v}_1 + r_2\mathbf{v}_2 + \cdots + r_n\mathbf{v}_n = \mathbf{0}, \quad \text{for some } r_i \neq 0.$$
>
> The set is linearly independent if no such dependence relation exists, so that a linear combination of the $\mathbf{v}_i$ is the zero vector only if all coefficients are zero.
>
> **2.** A finite list of nonzero vectors in V forms a linearly independent set if and only if no vector in the list can be expressed as a linear combination of its predecessors.
>
> **3.** A set of vectors $\{\mathbf{v}_1, \mathbf{v}_2, \ldots, \mathbf{v}_k\}$ is a basis for a subspace S of a vector space V if it is both linearly independent and a generating set for S.
>
> **4.** Any finite generating subset of a nonzero subspace S of a vector space V can be reduced, if necessary, to a basis for S by deleting zero vectors and then casting out those that are linear combinations of predecessors.
>
> **5.** The following statements are equivalent for n vectors in $\mathbb{R}^n$:
>   **a)** The vectors form a basis for $\mathbb{R}^n$.
>   **b)** The vectors are linearly independent.
>   **c)** The vectors generate $\mathbb{R}^n$.
>   **d)** A matrix having the vectors as columns is invertible.

EXERCISES

1. Give a geometric criterion for a set of two distinct nonzero vectors in $\mathbb{R}^2$ to be dependent.

2. Argue geometrically that any set of three distinct vectors in $\mathbb{R}^2$ is dependent.

3. Give a geometric criterion for a set of two distinct nonzero vectors in $\mathbb{R}^3$ to be dependent.

4. Give a geometric description of the subspace of $\mathbb{R}^3$ generated by an independent set of two vectors.

5. Give a geometric criterion for a set of three distinct nonzero vectors in $\mathbb{R}^3$ to be dependent.

6. Argue geometrically that every set of four distinct vectors in $\mathbb{R}^3$ is dependent.

In Exercises 7–19, determine whether the given set of vectors is dependent or independent. If it is dependent, find an independent set that generates the same subspace of the indicated vector space as the given set.

7. $\{(1, 3), (-2, -6)\}$ in $\mathbb{R}^2$

8. $\{1, 3), (2, -4)\}$ in $\mathbb{R}^2$

9. $\{(-3, 1), (6, 4)\}$ in $\mathbb{R}^2$

10. $\{(-3, 1), (9, -3)\}$ in $\mathbb{R}^2$

11. $\{(2, 1), (-6, -3), (1, 4)\}$ in $\mathbb{R}^2$

12. $\{(-1, 2, 1), (2, -4, 3)\}$ in $\mathbb{R}^3$

13. $\{(1, -3, 2), (2, -5, 3), (4, 0, 1)\}$ in $\mathbb{R}^3$

14. $\{(1, -4, 3), (3, -11, 2), (1, -3, -4)\}$ in $\mathbb{R}^3$

15. $\{(1, 4, -1, 3), (-1, 5, 6, 2), (1, 13, 4, 7)\}$ in $\mathbb{R}^4$

16. $\{(-2, 3, 1), (3, -1, 2), (1, 2, 3), (-1, 5, 4)\}$ in $\mathbb{R}^3$

17. $\{x^2 - 1, x^2 + 1, 4x, 2x - 3\}$ in P

18. $\{1, 4x + 3, 3x - 4, x^2 + 2, x - x^2\}$ in P

19. $\{1, \sin^2 x, \cos 2x, \cos^2 x\}$ in F

In Exercises 20–23, use the technique discussed after Example 3 to determine whether the given set of functions in the vector space F is independent or dependent.

20. $\{\sin x, \cos x\}$

21. $\{1, x, x^2\}$

22. $\{\sin x, \sin 2x, \sin 3x\}$

23. $\{\sin x, \sin(-x)\}$

24. Show that if $\{\mathbf{v}_1, \mathbf{v}_2, \ldots, \mathbf{v}_k\}$ is an independent subset of a vector space V, then each element of $\mathrm{sp}(\mathbf{v}_1, \mathbf{v}_2, \ldots, \mathbf{v}_k)$ can be expressed *uniquely* as a linear combination of the $\mathbf{v}_i$.

25. Show the converse of Exercise 24; that is, if each vector in $\mathrm{sp}(\mathbf{v}_1, \mathbf{v}_2, \ldots, \mathbf{v}_k)$ can be expressed *uniquely* as a linear combination of the $\mathbf{v}_i$, then $\{\mathbf{v}_1, \mathbf{v}_2, \ldots, \mathbf{v}_k\}$ is an independent set of vectors.

26. Let $\{\mathbf{v}_1, \mathbf{v}_2, \ldots, \mathbf{v}_k\}$ be an independent set of vectors in a subspace S of a vector space V. Show that if S contains a finite generating set, then the given set of vectors can be enlarged, if necessary, to a basis for S. [*Hint:* Let $\{\mathbf{w}_1, \mathbf{w}_2, \ldots, \mathbf{w}_h\}$ be a generating set for S, and apply the casting-out technique to the list of vectors $\mathbf{v}_1, \mathbf{v}_2, \ldots, \mathbf{v}_k, \mathbf{w}_1, \mathbf{w}_2, \ldots, \mathbf{w}_h$.]

In Exercises 27–31, determine whether or not the given set of vectors is a basis for the indicated space $\mathbb{R}^n$.

27. $\{(-1, 1), (1, 2)\}$ for $\mathbb{R}^2$

28. $\{-1, 3, 1), (2, 1, 4)\}$ for $\mathbb{R}^3$

29. $\{(-1, 3, 4), (1, 5, -1), (1, 13, 2)\}$ for $\mathbb{R}^3$

30. $\{(2, 1, -3), (4, 0, 2), (2, -1, 3)\}$ for $\mathbb{R}^3$

31. $\{(2, 1, 0, 2), (2, -3, 1, 0), (3, 2, 0, 0), (5, 0, 0, 0)\}$ for $\mathbb{R}^4$

32. Argue geometrically that if two vectors $\mathbf{v}_1$ and $\mathbf{v}_2$ are chosen *at random* from $\mathbb{R}^2$, it is very likely that $\{\mathbf{v}_1, \mathbf{v}_2\}$ is a basis for $\mathbb{R}^2$. (The probability of obtaining a basis by such a random selection is actually 1.)

33. Repeat Exercise 32 for a *random* selection of vectors $\mathbf{v}_1, \mathbf{v}_2,$ and $\mathbf{v}_3$ in $\mathbb{R}^3$.

In Exercises 34–37, use the programs YUREDUCE or MATCOMP or similar software to cast out vectors in the given list that are linear combinations of predecessors. (See the description of the technique on page 167.) List the vectors retained by number.

34.
$\mathbf{v}_1 = (5, 4, 3),$
$\mathbf{v}_2 = (2, 1, 6),$
$\mathbf{v}_3 = (4, 5, -12),$
$\mathbf{v}_4 = (6, 1, 4),$
$\mathbf{v}_5 = (1, 1, 1)$

35.
$\mathbf{v}_1 = (0, 1, 1, 2),$
$\mathbf{v}_2 = (-3, -2, 4, 5),$
$\mathbf{v}_3 = (1, 2, 0, 1),$
$\mathbf{v}_4 = (-1, 4, 6, 11),$
$\mathbf{v}_5 = (1, 1, 1, 3),$
$\mathbf{v}_6 = (3, 7, 3, 9)$

36.
$\mathbf{v}_1 = (3, 1, 2, 4, 1),$
$\mathbf{v}_2 = (3, -2, 6, 7, -3),$
$\mathbf{v}_3 = (3, 4, -2, 1, 5),$
$\mathbf{v}_4 = (1, 2, 3, 2, 1),$
$\mathbf{v}_5 = (7, 1, 11, 13, -1),$
$\mathbf{v}_6 = (2, -1, 2, 3, 1)$

37.
$\mathbf{v}_1 = (2, -1, 3, 4, 1, 2),$
$\mathbf{v}_2 = (-2, 5, 3, -2, 1, -4),$
$\mathbf{v}_3 = (2, 4, 6, 5, 2, 1),$
$\mathbf{v}_4 = (1, -1, 1, -1, 2, 2),$
$\mathbf{v}_5 = (1, 8, 10, 2, 5, -1),$
$\mathbf{v}_6 = (3, 0, 0, 2, 1, 5),$
$\mathbf{v}_7 = (2, 4, 5, 0, 3, 2)$

3.5

Dimension and Rank

THE DIMENSION OF A SUBSPACE

We think of $\mathbb{R}$ as being one-dimensional, of $\mathbb{R}^2$ as two-dimensional, of $\mathbb{R}^3$ as three-dimensional, and so on. We are finally in a position to give an *algebraic* explanation of this. Indeed, we will establish a notion of dimension for any vector space that can be generated by a finite number of vectors.

Let V be a vector space. A subspace S of V is **finite-dimensional** if there exists a finite set of vectors $B = \{\mathbf{b}_1, \mathbf{b}_2, \ldots, \mathbf{b}_k\}$ that is a basis for S. Although S may have several bases, we will show that they all must have the same number of elements. We demonstrate a key step in the argument.

Theorem 3.12   Relative Size of Generating and Independent Sets

> If S is a subspace of a vector space V, then any finite generating set for S contains at least as many vectors as there are in any finite independent set of vectors in S.

Proof   Let $\{\mathbf{v}_1, \mathbf{v}_2, \ldots, \mathbf{v}_k\}$ be an independent set of vectors in S, and let $\{\mathbf{w}_1, \mathbf{w}_2, \ldots, \mathbf{w}_m\}$ be a generating set for S. We will show that $m \geq k$. We may as well assume $k > 1$, since we certainly have $m \geq 1$.

Consider the set of vectors

$$\{\mathbf{v}_1, \mathbf{w}_1, \mathbf{w}_2, \ldots, \mathbf{w}_m\}.$$

This set generates S because the $\mathbf{w}_j$ generate S. Also, a dependence relation exists among these vectors since $\mathbf{v}_1$ can be expressed as a linear combination of the $\mathbf{w}_j$. Using the casting-out technique on page 164, we obtain, after casting out at least one $\mathbf{w}_j$ and renumbering the rest of the $\mathbf{w}_j$ if necessary, a new set

$$B_1 = \{\mathbf{v}_1, \mathbf{w}_1, \ldots, \mathbf{w}_{m_1}\},$$

which is a basis for S, and where $m_1 \leq m - 1$. Since $k > 1$ and the $\mathbf{v}_i$ are independent, the set $\{\mathbf{v}_1\}$ is not a basis for S, so not all of the $\mathbf{w}_j$ were cast out. We now consider the set of vectors

$$\{\mathbf{v}_1, \mathbf{v}_2, \mathbf{w}_1, \ldots, \mathbf{w}_{m_1}\}$$

and repeat the technique, casting out again at least one $\mathbf{w}_j$ to obtain, after renumbering if necessary, a basis

$$B_2 = \{\mathbf{v}_1, \mathbf{v}_2, \mathbf{w}_1, \ldots, \mathbf{w}_{m_2}\}$$

for S, where $m_2 \leq m - 2$. If $k > 2$, then some $\mathbf{w}_j$ remain in B_2, and we consider

$$\{\mathbf{v}_1, \mathbf{v}_2, \mathbf{v}_3, \mathbf{w}_1, \ldots, \mathbf{w}_{m_2}\}$$

and repeat the process. Note that no $\mathbf{v}_i$ is ever cast out since the set $\{\mathbf{v}_1, \mathbf{v}_2, \ldots, \mathbf{v}_k\}$ is independent, and we must have had at least one $\mathbf{w}_j$ to cast out for each $\mathbf{v}_i$. That is, $m \geq k$, which is precisely what we wanted to show. ■

Theorem 3.12 shows that a finite-dimensional vector space cannot have arbitrarily long lists of independent vectors. Suppose S is a nonzero subspace of a finite-dimensional vector space V. We can find a basis for S in this way: Choose a nonzero vector in S. If possible, choose another vector in S that is not a multiple of the first vector chosen. Continue this process until it is no longer possible to choose another vector in S that is not a linear combination of those already chosen. This process must terminate since V is finite-dimensional. Thus we must eventually obtain a basis for S. In particular, this shows the following:

> Every subspace of a finite-dimensional vector space must itself be finite-dimensional.

We are now ready to show that the number of vectors in a basis for a finite-dimensional subspace S of V is an *invariant* of S. That is, the number of elements in a basis is the same for all choices of bases for S. This invariance will enable us to define the *dimension* of S as the number of elements in any basis for S.

Theorem 3.13 Invariance of Dimension

Let S be a finite-dimensional subspace of a vector space V. Any two bases of S have the same number of elements.

Proof Let

$$B = \{\mathbf{b}_1, \mathbf{b}_2, \ldots, \mathbf{b}_k\} \quad \text{and} \quad B' = \{\mathbf{b}_1', \mathbf{b}_2', \ldots, \mathbf{b}_m'\}$$

be any two bases for S. Viewing B as an independent set in S and B' as a generating set for S, we have $k \leq m$. Now viewing B as a generating set for S and B' as an independent set in S, we obtain $m \leq k$. Therefore, $m = k$. ■

Definition 3.10 Dimension

The **dimension** of a finite-dimensional subspace S of a vector space V is the number of elements in any basis for S, and is denoted by $\dim(S)$. We consider the zero subspace $\{\mathbf{0}\}$ to have dimension zero.

Since the unit coordinate vectors $\mathbf{e}_1, \mathbf{e}_2, \ldots, \mathbf{e}_n$ form a basis for $\mathbb{R}^n$, we see at once that $\dim(\mathbb{R}^n) = n$. Since a basis for the vector space P_n of polynomials of degree $\leq n$ is $\{1, x, x^2, x^3, \ldots, x^n\}$, we see that P_n has dimension $n + 1$.

Example 1   Find the dimension of the subspace

$$S = \text{sp}((1, -3, 1), (-2, 6, -2), (2, 1, -4), (-1, 10, -7))$$

of $\mathbb{R}^3$.

Solution   Example 11 in Section 3.4 showed that

$$\{(1, -3, 1), (2, 1, -4)\}$$

is a basis for S. Consequently, $\dim(S) = 2$. $\triangleleft$

     We now show how any independent set $\{\mathbf{v}_1, \mathbf{v}_2, \ldots, \mathbf{v}_k\}$ of a finite-dimensional subspace S of a vector space V can be expanded, if necessary, to a basis for S. We select a basis $\{\mathbf{b}_1, \mathbf{b}_2, \ldots, \mathbf{b}_m\}$ for S and form the list of vectors

$$\mathbf{v}_1, \mathbf{v}_2, \ldots, \mathbf{v}_k, \mathbf{b}_1, \mathbf{b}_2, \ldots, \mathbf{b}_m.$$

Since these vectors clearly generate S, we use the casting-out technique to arrive at a basis for S. (See Theorem 3.11.) The basis will surely contain the independent set $\{\mathbf{v}_1, \mathbf{v}_2, \ldots, \mathbf{v}_k\}$ because no $\mathbf{v}_i$ can be cast out. We summarize in an "expanding" theorem in contrast to the "reducing" Theorem 3.11.

Theorem 3.14   Expanding an Independent Set to a Basis

Let S be a finite-dimensional subspace of a vector space V. Any independent subset of S can be expanded, if necessary, to a basis for S.

Example 2   Expand the independent set $\{(1, 1, 0), (-1, 2, 0)\}$ to a basis for $\mathbb{R}^3$.

Solution   Since $\{(1, 0, 0), (0, 1, 0), (0, 0, 1)\}$ is a basis for $\mathbb{R}^3$, we apply the casting-out technique to the list of vectors

$$(1, 1, 0), (-1, 2, 0), (1, 0, 0), (0, 1, 0), (0, 0, 1)$$

and it is easily seen that we arrive at the basis

$$\{(1, 1, 0), (-1, 2, 0), (0, 0, 1)\}$$

for $\mathbb{R}^3$. $\triangleleft$

     Let S be a finite-dimensional subspace of a vector space V. For a subset B of S to be a basis for S, two conditions are required:

1. B must generate S, and

2. B must be independent.

If the dimension of S is known and if the number of vectors in B matches the dimension of S, then *either one* of the two conditions for a basis is enough to ensure that the other holds. For if B consists of k independent vectors, then Theorem 3.14 shows that B can be expanded, if necessary, to a basis for S. But if $\dim(S) = k$, then Theorem 3.13 tells us that B cannot be expanded further to become a basis. Consequently, B must already be a basis, and must generate S. Conversely, Theorem 3.11 on page 166 shows that if B generates S, then B can be reduced to a basis for S. But if the number of elements in B matches $\dim(S)$, then again we conclude that B must already be a basis for S, so B is an independent set. We summarize in a theorem that generalizes part of Theorem 3.10 on page 162.

Theorem 3.15   A Test for dim(S) Vectors to Form a Basis for S

Let V be a vector space and let S be a finite-dimensional subspace of V with $\dim(S) = k$. For a set B of k vectors from S, the following statements are equivalent:

1. B is linearly independent, and

2. B generates S.

In either case, B is a basis for S.

Example 3   Use a geometric argument to find a basis for the subspace

$$S = \{(x_1, x_2, x_3) \mid 2x_1 + x_2 - 5x_3 = 0\}$$

of $\mathbb{R}^3$.

Solution   We recognize S as a plane through the origin in $\mathbb{R}^3$ whose equation is $2x_1 + x_2 - 5x_3 = 0$. Since a plane is two-dimensional, we choose two vectors in S as follows:

x_1	x_2	Vector in S
5	0	$\mathbf{v}_1 = (5, 0, 2)$
0	5	$\mathbf{v}_2 = (0, 5, 1)$

Since $\mathbf{v}_1$ and $\mathbf{v}_2$ are independent, they form a basis for S by Theorem 3.15.   ◁

THE RANK OF A MATRIX

There are three important numbers associated with an $m \times n$ matrix A, in addition to m and n:

1. The dimension of the column space of A, which is called the **column rank** of A. We will write it as colrank(A).

2. The dimension of the row space of A, which is called the **row rank** of A and written as rowrank(A).

3. The dimension of the nullspace of A, which is called the **nullity** of A and written null(A).

The relationship among these three numbers can be seen by row-reducing the matrix A to echelon form E. To illustrate, suppose a matrix A has been reduced to the echelon form

$$
E = \begin{pmatrix}
p_1 & X & X & X & X & X & X \\
0 & 0 & p_3 & X & X & X & X \\
0 & 0 & 0 & p_4 & X & X & X \\
0 & 0 & 0 & 0 & 0 & 0 & p_7 \\
0 & 0 & 0 & 0 & 0 & 0 & 0 \\
0 & 0 & 0 & 0 & 0 & 0 & 0
\end{pmatrix}, \tag{1}
$$

where the entries p_i are the (nonzero) pivots. As explained in the casting-out technique on page 167, the column vectors of A that correspond to the columns containing the pivots in (1) form a basis for the column space of A. Consequently, the column rank of A is the number of pivots. On the other hand, we find the dimension of the nullspace of A by counting the columns in (1) that do not contain pivots; they give rise to free variables in the solution of the homogeneous system $A\mathbf{x} = \mathbf{0}$, as explained in Section 1.6. In our illustration, we have free variables x_2, x_5, and x_6 in the system

$$
\begin{pmatrix}
p_1 & X & X & X & X & X & X \\
0 & 0 & p_3 & X & X & X & X \\
0 & 0 & 0 & p_4 & X & X & X \\
0 & 0 & 0 & 0 & 0 & 0 & p_7 \\
0 & 0 & 0 & 0 & 0 & 0 & 0 \\
0 & 0 & 0 & 0 & 0 & 0 & 0
\end{pmatrix}
\begin{pmatrix}
x_1 \\ x_2 \\ x_3 \\ x_4 \\ x_5 \\ x_6 \\ x_7
\end{pmatrix}
=
\begin{pmatrix}
0 \\ 0 \\ 0 \\ 0 \\ 0 \\ 0 \\ 0
\end{pmatrix}. \tag{2}
$$

The bound variables x_1, x_3, x_4, and x_7 can be solved for in terms of the free variables. The technique of back substitution provides numbers a_1, a_2, b_1, b_2,

---

THE RANK OF A MATRIX was defined in 1879 by Georg Frobenius (1849–1917) as follows: If all determinants of the $(r + 1)$st degree vanish, but not all of the rth degree, then r is the rank of the matrix. He used this concept to deal with the questions of canonical forms for certain matrices of integers and with the solutions of certain systems of linear congruences.

On the other hand, the nullity was defined by James Sylvester in 1884 for square matrices as follows: The nullity of an $n \times n$ matrix is i if every minor (determinant) of order $n - i + 1$ (and therefore of every higher order) equals 0 and i is the largest such number for which this is true. Sylvester was interested here, as in much of his mathematical career, in discovering *invariants*, that is, properties of particular mathematical objects that do not change under specified types of transformation. He proceeded to prove what he called one of the cardinal laws in the theory of matrices, that the nullity of the product of two matrices is not less than the nullity of any factor nor greater than the sum of the nullities of the factors.

c_1, c_2, and c_3 such that

$$\begin{pmatrix} x_1 \\ x_2 \\ x_3 \\ x_4 \\ x_5 \\ x_6 \\ x_7 \end{pmatrix} = \begin{pmatrix} c_1 x_2 + c_2 x_5 + c_3 x_6 \\ x_2 \\ b_1 x_5 + b_2 x_6 \\ a_1 x_5 + a_2 x_6 \\ x_5 \\ x_6 \\ 0 \end{pmatrix} = x_2 \underbrace{\begin{pmatrix} c_1 \\ 1 \\ 0 \\ 0 \\ 0 \\ 0 \\ 0 \end{pmatrix}}_{\mathbf{v}_1} + x_5 \underbrace{\begin{pmatrix} c_2 \\ 0 \\ b_1 \\ a_1 \\ 1 \\ 0 \\ 0 \end{pmatrix}}_{\mathbf{v}_2} + x_6 \underbrace{\begin{pmatrix} c_3 \\ 0 \\ b_2 \\ a_2 \\ 0 \\ 1 \\ 0 \end{pmatrix}}_{\mathbf{v}_3}.$$

In this way the nullspace of A is seen to be $sp(\mathbf{v}_1, \mathbf{v}_2, \mathbf{v}_3)$. The vectors $\mathbf{v}_1, \mathbf{v}_2, \mathbf{v}_3$ are independent since in coordinate positions 2, 5, and 6, the color components are the same as those of the independent vectors $\mathbf{e}_1, \mathbf{e}_2, \mathbf{e}_3$ in the standard basis for $\mathbb{R}^3$. Thus $null(A)$ is the number of columns of the echelon matrix E containing no pivots.

Finally, since every column of a row-echelon matrix E either contains a (nonzero) pivot or does not contain a pivot, we see that the sum of the column rank of A and the nullity of A must be the number of columns in A. We summarize our discussion in a theorem.

Theorem 3.16   Rank Equation

Let A be an $m \times n$ matrix with row-echelon form E. Then:

1. $null(A)$ = (Number of free variables in the solution space of $A\mathbf{x} = \mathbf{0}$)
   = (Number of pivot-free columns in E);
2. $colrank(A)$ = Number of (nonzero) pivots in E;
3. *(Rank equation)*: $colrank(A) + null(A)$ = Number of columns of A.

Since $null(A)$ is defined as the number of vectors in a basis, the invariance of dimension shows that the number of free variables obtained in the solution of a linear system $A\mathbf{x} = \mathbf{b}$ is independent of the steps in the row reduction to echelon form, as we asserted in Section 1.6.

Example 4   Find bases for the column space of A and for the null space of A if

$$A = \begin{pmatrix} 1 & 2 & 0 & -1 & 1 \\ 1 & 3 & 1 & 1 & -1 \\ 2 & 5 & 1 & 0 & 0 \\ 3 & 6 & 0 & 0 & -6 \\ 1 & 5 & 3 & 5 & -5 \end{pmatrix}.$$

Solution   We find an echelon form of A:

$$A = \begin{pmatrix} 1 & 2 & 0 & -1 & 1 \\ 1 & 3 & 1 & 1 & -1 \\ 2 & 5 & 1 & 0 & 0 \\ 3 & 6 & 0 & 0 & -6 \\ 1 & 5 & 3 & 5 & -5 \end{pmatrix} \sim \begin{pmatrix} 1 & 2 & 0 & -1 & 1 \\ 0 & 1 & 1 & 2 & -2 \\ 0 & 1 & 1 & 2 & -2 \\ 0 & 0 & 0 & 3 & -9 \\ 0 & 3 & 3 & 6 & -6 \end{pmatrix}$$

$$\sim \begin{pmatrix} 1 & 2 & 0 & -1 & 1 \\ 0 & 1 & 1 & 2 & -2 \\ 0 & 0 & 0 & 0 & 0 \\ 0 & 0 & 0 & 3 & -9 \\ 0 & 0 & 0 & 0 & 0 \end{pmatrix} \sim \begin{pmatrix} 1 & 2 & 0 & -1 & 1 \\ 0 & 1 & 1 & 2 & -2 \\ 0 & 0 & 0 & 3 & -9 \\ 0 & 0 & 0 & 0 & 0 \\ 0 & 0 & 0 & 0 & 0 \end{pmatrix} = E.$$

Since we have pivots in columns 1, 2, and 4 of E, the corresponding column vectors

$$\begin{pmatrix} 1 \\ 1 \\ 2 \\ 3 \\ 1 \end{pmatrix}, \quad \begin{pmatrix} 2 \\ 3 \\ 5 \\ 6 \\ 5 \end{pmatrix}, \quad \text{and} \quad \begin{pmatrix} -1 \\ 1 \\ 0 \\ 0 \\ 5 \end{pmatrix}$$

of A form a basis for the column space of A, and colrank$(A) = 3$. Turning to the nullspace of A, we see that columns 3 and 5 of E contain no pivot. This gives rise to free variables x_3 and x_5 in the solution set of the homogeneous system $A\mathbf{x} = \mathbf{0}$. Looking at the matrix E and using back substitution, we can describe all solutions of this system by the equations

$$\begin{pmatrix} x_1 \\ x_2 \\ x_3 \\ x_4 \\ x_5 \end{pmatrix} = \begin{pmatrix} 2x_3 + 10x_5 \\ -x_3 - 4x_5 \\ x_3 \\ 3x_5 \\ x_5 \end{pmatrix} = x_3 \begin{pmatrix} 2 \\ -1 \\ 1 \\ 0 \\ 0 \end{pmatrix} + x_5 \begin{pmatrix} 10 \\ -4 \\ 0 \\ 3 \\ 1 \end{pmatrix}.$$

The two explicit solutions in this last equation form a basis for the null space of A, and are obtained as follows:

Basis for $\mathbb{R}^2$	x_3	x_5	Basis for the Nullspace
$\mathbf{e}_1$	1	0	$\mathbf{v}_1 = (2,\ -1, 1, 0, 0)$
$\mathbf{e}_2$	0	1	$\mathbf{v}_2 = (10, -4, 0, 3, 1)$

2. The dimensions of the row space, column space, and nullspace of an $m \times n$ matrix A are the row rank, column rank, and nullity of A, respectively.

3. The row rank and column rank of any matrix A are equal, and are referred to as the rank of A. The rank of A is equal to the number of (nonzero) pivots in an echelon reduction of A.

4. The nullity of A plus the rank of A equals the number of columns of A.

5. An $n \times n$ matrix A is invertible if and only if rank$(A) = n$.

EXERCISES

In Exercises 1–4, (a) find a basis for and the dimension of the given subspace of $\mathbb{R}^n$, and (b) expand, if necessary, the basis in part (a) to a basis for $\mathbb{R}^n$.

1. The subspace

$$\{(x_1, x_2, x_3) \mid x_1 + x_2 + x_3 = 0\}$$

in $\mathbb{R}^3$

2. The subspace

$$\text{sp}((1, 2, 0), (2, 1, 1), (0, -3, 1), (3, 0, 2))$$

in $\mathbb{R}^3$

3. The nullspace of the matrix

$$\begin{pmatrix} 2 & 0 & 1 \\ -1 & 1 & 0 \\ 3 & 1 & 2 \end{pmatrix}$$

4. The subspace

$$\{(x_1, x_2, x_3, x_4, x_5) \mid x_1 + x_2 = 0,$$
$$x_2 = x_4, x_5 = x_1 + 3x_3\}$$

of $\mathbb{R}^5$

In Exercises 5–10, find (a) the rank of the given matrix. Also find (b) a basis for its column space and (c) a basis for its nullspace.

5. $\begin{pmatrix} 2 & 0 & -3 & 1 \\ 3 & 4 & 2 & 2 \end{pmatrix}$

6. $\begin{pmatrix} 5 & -1 & 0 & 2 \\ 1 & 2 & 1 & 0 \\ 3 & 1 & -2 & 4 \\ 0 & 4 & -1 & 2 \end{pmatrix}$

7. $\begin{pmatrix} 0 & 6 & 6 & 3 \\ 1 & 2 & 1 & 1 \\ 4 & 1 & -3 & 4 \\ 1 & 3 & 2 & 0 \end{pmatrix}$

8. $\begin{pmatrix} 3 & 1 & 4 & 2 \\ -1 & 0 & -1 & 0 \\ 2 & 1 & 0 & 1 \\ 1 & 0 & -1 & 1 \end{pmatrix}$

9. $\begin{pmatrix} 0 & 1 & 2 & 1 \\ 2 & 1 & 0 & 2 \\ 0 & 2 & 1 & 1 \end{pmatrix}$

10. $\begin{pmatrix} 0 & 2 & 3 & 1 \\ -4 & 4 & 1 & 4 \\ 3 & 3 & 2 & 0 \\ -4 & 0 & 1 & 2 \end{pmatrix}$

In Exercises 11–14, determine whether the given matrix is invertible by finding its rank.

11. $\begin{pmatrix} 0 & -9 & -9 & 2 \\ 1 & 2 & 1 & 1 \\ 4 & 1 & -3 & 4 \\ 1 & 3 & 2 & 0 \end{pmatrix}$

12. $\begin{pmatrix} 2 & 3 & 1 \\ 4 & -1 & 2 \\ 1 & 0 & 1 \end{pmatrix}$

13. $\begin{pmatrix} 2 & 0 & 1 \\ 0 & 0 & 4 \\ 2 & 4 & 0 \end{pmatrix}$

14. $\begin{pmatrix} 3 & 0 & -1 & 2 \\ 4 & 2 & 1 & 8 \\ 1 & 4 & 0 & 1 \\ 2 & 6 & -3 & 1 \end{pmatrix}$

15. Show that the row rank of a matrix is not affected by the three elementary row operations.

16. Show that the column rank of a matrix is not affected by the three elementary column operations.

17. Prove that if S is a subspace of an n-dimensional vector space V and $\dim(S) = n$, then $S = V$.

18. Show that if A is a square matrix, then the nullity of A is the same as the nullity of A^T.

19. Let A be an $m \times n$ matrix and $\mathbf{b}$ be an $n \times 1$ vector. Prove that the system of equations $A\mathbf{x} = \mathbf{b}$ has a solution if and only if $\operatorname{rank}(A) = \operatorname{rank}(A \mid \mathbf{b})$, where $\operatorname{rank}(A \mid \mathbf{b})$ represents the rank of the associated partitioned matrix of the system.

20. Prove that for any $m \times n$ matrix A, both $(A^T)A$ and $A(A^T)$ are symmetric matrices.

21. Let A be an $m \times n$ matrix. Verify that $\operatorname{rank}[A(A^T)] = \operatorname{rank}(A)$. (See Theorem 3.18.)

22. If $\mathbf{a}$ is an $n \times 1$ vector and $\mathbf{b}$ is a $1 \times m$ vector, show that $\mathbf{ab}$ is an $n \times m$ matrix of rank one.

Determinants and Eigenvalues

4.2 — 4.6.

Each square matrix has associated with it a number called the *determinant* of the matrix. We begin this chapter with an introduction to determinants of 2×2 and 3×3 matrices, motivated by computations of area and volume. In Section 4.2 we discuss determinants of order n and their properties. An efficient way to compute determinants is presented in Section 4.3.

In Sections 4.5 and 4.6 we introduce the important topic of eigenvalues and eigenvectors of a square matrix. Determinants enable us to give illustrative computations and applications involving matrices of very small size. Eigenvectors and eigenvalues continue to appear at intervals throughout the rest of the text. In Section 7.4 we discuss some other methods for their computation.

Cramer's rule for solving square linear systems is presented in Section 4.4. An inductive proof of an expansion-by-minors theorem on determinants is given in Section 4.7; many of the properties of determinants in Section 4.2 are derived using this theorem. Both Sections 4.4 and 4.7 are of mostly theoretical interest. Since references and formulas involving Cramer's rule do appear in advanced calculus and other fields, we believe that students should at least read the statement of Cramer's rule and look at an illustration. However, this rule must be one of the most inefficient methods known for solving a square linear system of large size.

4.1
Areas, Volumes, and Cross Products

We introduce determinants using one of their most important applications: finding areas and volumes. We will be finding areas and volumes of very simple box-like regions. In calculus, one finds areas and volumes of regions having more general shapes, using formulas that involve determinants.

THE AREA OF A PARALLELOGRAM

The parallelogram determined by two nonzero and nonparallel vectors $\mathbf{a} = (a_1, a_2)$ and $\mathbf{b} = (b_1, b_2)$ in $\mathbb{R}^2$ is shown in Fig. 4.1. This parallelogram has a vertex at the origin, and we regard the arrows representing $\mathbf{a}$ and $\mathbf{b}$ as forming the two sides of the parallelogram having the origin as a common vertex.

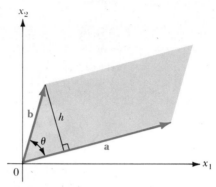

Figure 4.1
The parallelogram determined by a and b.

We can find the area of this parallelogram by multiplying the length $\|\mathbf{a}\|$ of its base by its altitude h, obtaining

$$\text{Area} = \|\mathbf{a}\|\, h = \|\mathbf{a}\|\|\mathbf{b}\|(\sin\theta) = \|\mathbf{a}\|\|\mathbf{b}\|\sqrt{1 - \cos^2\theta}.$$

Recall from page 184 of Section 3.1 that $\mathbf{a} \cdot \mathbf{b} = \|\mathbf{a}\|\|\mathbf{b}\|(\cos\theta)$. Squaring our area equation, we have

$$\begin{aligned}
(\text{Area})^2 &= \|\mathbf{a}\|^2\|\mathbf{b}\|^2 - \|\mathbf{a}\|^2\|\mathbf{b}\|^2\cos^2\theta \\
&= \|\mathbf{a}\|^2\|\mathbf{b}\|^2 - (\mathbf{a}\cdot\mathbf{b})^2 \\
&= (a_1^2 + a_2^2)(b_1^2 + b_2^2) - (a_1b_1 + a_2b_2)^2 \\
&= (a_1b_2 - a_2b_1)^2.
\end{aligned} \tag{1}$$

The last equality should be checked using pencil and paper. On taking square roots, we obtain

$$\text{Area} = |a_1b_2 - a_2b_1|.$$

The number within the absolute value symbol is known as the **determinant** of the matrix

$$\begin{pmatrix} a_1 & a_2 \\ b_1 & b_2 \end{pmatrix}$$

and is denoted by

$$\det(A) = \begin{vmatrix} a_1 & a_2 \\ b_1 & b_2 \end{vmatrix}.$$

That is, if

$$A = \begin{pmatrix} a_1 & a_2 \\ b_1 & b_2 \end{pmatrix},$$

then

$$\det(A) = \begin{vmatrix} a_1 & a_2 \\ b_1 & b_2 \end{vmatrix} = a_1 b_2 - a_2 b_1. \tag{2}$$

One remembers this formula for the determinant by taking the product of the black entries on the main diagonal of the matrix, minus the product of the colored entries on the other diagonal.

Example 1 Find the determinant of the matrix

$$\begin{pmatrix} 2 & 3 \\ 1 & 4 \end{pmatrix}.$$

Solution We have

$$\begin{vmatrix} 2 & 3 \\ 1 & 4 \end{vmatrix} = (2)(4) - (3)(1) = 5. \quad \triangleleft$$

Example 2 Find the area of the parallelogram in $\mathbb{R}^2$ with vertices $(1, 1)$, $(2, 3)$, $(2, 1)$, $(3, 3)$.

Solution The parallelogram is sketched in Fig. 4.2. The sides having $(1, 1)$ as common vertex can be regarded as the vectors

$$\mathbf{a} = (2, 1) - (1, 1) = (1, 0)$$

and

$$\mathbf{b} = (2, 3) - (1, 1) = (1, 2)$$

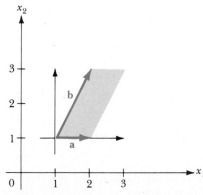

Figure 4.2
The parallelogram determined by
$\mathbf{a} = (1, 0)$ and $\mathbf{b} = (1, 2)$.

as shown in the figure. Therefore, the area of the parallelogram is given by the determinant

$$\begin{vmatrix} 1 & 0 \\ 1 & 2 \end{vmatrix} = (1)(2) - (0)(1)$$

$$= 2. \quad \triangleleft$$

THE CROSS PRODUCT

Equation (2) defines a **second-order determinant,** associated with a 2×2 square matrix. Another application of these second-order determinants appears when we find a vector in $\mathbb{R}^3$ that is perpendicular to each of two given independent vectors $\mathbf{b} = (b_1, b_2, b_3)$ and $\mathbf{c} = (c_1, c_2, c_3)$. Recall that the unit coordinate vectors in $\mathbb{R}^3$ are $\mathbf{i} = (1, 0, 0)$, $\mathbf{j} = (0, 1, 0)$, and $\mathbf{k} = (0, 0, 1)$. We leave as an exercise the verification that

$$\mathbf{p} = \begin{vmatrix} b_2 & b_3 \\ c_2 & c_3 \end{vmatrix} \mathbf{i} - \begin{vmatrix} b_1 & b_3 \\ c_1 & c_3 \end{vmatrix} \mathbf{j} + \begin{vmatrix} b_1 & b_2 \\ c_1 & c_2 \end{vmatrix} \mathbf{k} \tag{3}$$

is a vector perpendicular to both $\mathbf{b}$ and $\mathbf{c}$ (see Exercise 5). This can be seen by computing $\mathbf{p} \cdot \mathbf{b} = \mathbf{p} \cdot \mathbf{c} = 0$. The vector $\mathbf{p}$ in (3) is known as the **cross product** of $\mathbf{b}$ and $\mathbf{c}$, and is denoted $\mathbf{p} = \mathbf{b} \times \mathbf{c}$.

There is a very easy way to remember the formula (3) for the cross product $\mathbf{b} \times \mathbf{c}$. Form the 3×3 *symbolic matrix*

$$\begin{pmatrix} \mathbf{i} & \mathbf{j} & \mathbf{k} \\ b_1 & b_2 & b_3 \\ c_1 & c_2 & c_3 \end{pmatrix}.$$

THE NOTION OF A CROSS PRODUCT grew out of Sir William Rowan Hamilton's attempt to develop a multiplication for "triplets," that is, vectors in R^3. In a paper of 1837, but written four years earlier, he developed the theory of complex numbers as pairs (a, b) of real numbers with the addition $(a, b) + (a', b') = (a + a', b + b')$ and multiplication $(a, b)(a', b') = (aa' - bb', ab' + a'b)$. Over a period of many years, he attempted to generalize this to triples. On October 16, 1843, after many unsuccessful attempts, he finally succeeded in discovering an analogous result, not for triples but for quadruples. As he walked that day in Dublin, he wrote, he could not "resist the impulse . . . to cut with a knife on a stone of Brougham Bridge . . . the fundamental formula with the symbols i, j, k; namely $i^2 = j^2 = k^2 = ijk = -1$." This formula symbolizes his discovery of *quaternions,* elements of the form $w + xi + yj + zk$ with w, x, y, z real numbers, whose multiplication obeys the laws just given. Therefore, in particular, the product of two quaternions

$$\alpha = xi + yj + zk \quad \text{and} \quad \alpha' = x'i + y'j + z'k$$

whose real parts are 0 is given as

$$-xx' - yy' - zz' + (yz' - zy')i + (zx' - xz')j + (xy' - yx')k.$$

Hamilton denoted the imaginary or vector part of this product as $V \cdot \alpha\alpha'$. It is, of course, our modern cross product.

The first appearance of our current symbolism for the cross product is in the brief text *Elements of Vector Analysis* (1881) by the American Josiah Willard Gibbs (1839–1903), professor of mathematical physics at Yale, who wrote it for use in his courses in electricity and magnetism at Yale and in mechanics at Johns Hopkins.

Formula (3) can be obtained from this matrix in a simple way. Multiply the vector $\mathbf{i}$ by the determinant of the 2×2 matrix obtained by crossing out the row and column containing $\mathbf{i}$, as in

$$\begin{pmatrix} \mathbf{i} & \mathbf{j} & \mathbf{k} \\ b_1 & b_2 & b_3 \\ c_1 & c_2 & c_3 \end{pmatrix}.$$

Similarly, multiply $(-\mathbf{j})$ by the determinant of the matrix obtained by crossing out the row and column in which $\mathbf{j}$ appears. Finally, multiply $\mathbf{k}$ by the determinant of the matrix obtained by crossing out the row and column containing $\mathbf{k}$, and add these multiples of $\mathbf{i}$, $\mathbf{j}$, and $\mathbf{k}$ to obtain the formula (3).

Example 3   Find a vector perpendicular to both $(2, 1, 1)$ and $(1, 2, 3)$ in $\mathbb{R}^n$.

Solution   We form the symbolic matrix

$$\begin{pmatrix} \mathbf{i} & \mathbf{j} & \mathbf{k} \\ 2 & 1 & 1 \\ 1 & 2 & 3 \end{pmatrix}$$

and find that

$$(2, 1, 1) \times (1, 2, 3) = \begin{vmatrix} 1 & 1 \\ 2 & 3 \end{vmatrix} \mathbf{i} - \begin{vmatrix} 2 & 1 \\ 1 & 3 \end{vmatrix} \mathbf{j} + \begin{vmatrix} 2 & 1 \\ 1 & 2 \end{vmatrix} \mathbf{k}$$

$$= \mathbf{i} - 5\mathbf{j} + 3\mathbf{k} = (1, -5, 3). \quad \triangleleft$$

The cross product $\mathbf{p} = \mathbf{b} \times \mathbf{c}$ as defined in Eq. (3) not only is perpendicular to both $\mathbf{b}$ and $\mathbf{c}$ but points in the direction determined by the familiar *right-hand rule:* When the fingers of the right hand curve in the direction from $\mathbf{b}$ to $\mathbf{c}$, then the thumb points in the direction of $\mathbf{b} \times \mathbf{c}$ (see Fig. 4.3). We do not attempt to prove this.

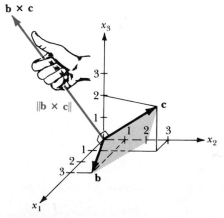

Figure 4.3
The area of the parallelogram determined by b and c is ‖b × c‖.

The magnitude of the vector $\mathbf{p} = \mathbf{b} \times \mathbf{c}$ in (3) is of interest as well: It is the area of the parallelogram with a vertex at the origin in $\mathbb{R}^3$ and edges at that vertex given by the vectors $\mathbf{b}$ and $\mathbf{c}$. To see this, we refer to a diagram like the one in Fig. 4.1, but with $\mathbf{a}$ replaced by $\mathbf{c}$, and we repeat the computation for area. This time, Eq. (1) takes the form

$$(\text{Area})^2 = \|\mathbf{c}\|^2 \|\mathbf{b}\|^2 - (\mathbf{c} \cdot \mathbf{b})^2$$
$$= (c_1^2 + c_2^2 + c_3^2)(b_1^2 + b_2^2 + b_3^2) - (c_1 b_1 + c_2 b_2 + c_3 b_3)^2$$
$$= \begin{vmatrix} b_2 & b_3 \\ c_2 & c_3 \end{vmatrix}^2 + \begin{vmatrix} b_1 & b_3 \\ c_1 & c_3 \end{vmatrix}^2 + \begin{vmatrix} b_1 & b_2 \\ c_1 & c_2 \end{vmatrix}^2.$$

Again pencil and paper are needed to check this last equality. Taking square roots, we obtain

$$\|\mathbf{b} \times \mathbf{c}\| = \text{Area of the parallelogram in } \mathbb{R}^3 \text{ determined by } \mathbf{b} \text{ and } \mathbf{c}.$$

Example 4   Find the area of the parallelogram in $\mathbb{R}^3$ determined by the vectors $\mathbf{b} = (3, 1, 0)$ and $\mathbf{c} = (1, 3, 2)$.

Solution   From the symbolic matrix

$$\begin{pmatrix} \mathbf{i} & \mathbf{j} & \mathbf{k} \\ 3 & 1 & 0 \\ 1 & 3 & 2 \end{pmatrix},$$

we find that

$$\mathbf{b} \times \mathbf{c} = \begin{vmatrix} 1 & 0 \\ 3 & 2 \end{vmatrix} \mathbf{i} - \begin{vmatrix} 3 & 0 \\ 1 & 2 \end{vmatrix} \mathbf{j} + \begin{vmatrix} 3 & 1 \\ 1 & 3 \end{vmatrix} \mathbf{k}$$
$$= (2, -6, 8).$$

Therefore, the area of the parallelogram is

$$\|\mathbf{b} \times \mathbf{c}\| = 2\sqrt{1 + 9 + 16} = 2\sqrt{26},$$

and is shown in Fig. 4.3.   ◁

Example 5   Find the area of the triangle in $\mathbb{R}^3$ with vertices $(-1, 2, 0)$, $(2, 1, 3)$, and $(1, 1, -1)$.

Solution   We think of $(-1, 2, 0)$ as a local origin, and take the vectors corresponding to arrows starting there and reaching to $(2, 1, 3)$ and to $(1, 1, -1)$, namely,

$$\mathbf{a} = (2, 1, 3) - (-1, 2, 0) = (3, -1, 3)$$

and

$$\mathbf{b} = (1, 1, -1) - (-1, 2, 0) = (2, -1, -1).$$

Now $\|\mathbf{a} \times \mathbf{b}\|$ is the area of the parallelogram determined by these vectors, and the area of the triangle is half the area of the parallelogram, as shown in Fig.

4.4. We form the symbolic matrix

$$\begin{pmatrix} \mathbf{i} & \mathbf{j} & \mathbf{k} \\ 3 & -1 & 3 \\ 2 & -1 & -1 \end{pmatrix}$$

and find that $\mathbf{a} \times \mathbf{b} = 4\mathbf{i} + 9\mathbf{j} - \mathbf{k}$. Thus

$$\|\mathbf{a} \times \mathbf{b}\| = \sqrt{16 + 81 + 1} = \sqrt{98} = 7\sqrt{2},$$

so the area of the triangle is $7\sqrt{2}/2$. ◁

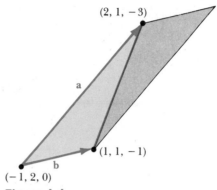

Figure 4.4
The triangle comprises half the parallelogram.

THE VOLUME OF A BOX

The cross product is useful in finding the volume of the box, or parallelepiped, determined by the three nonzero vectors $\mathbf{a} = (a_1, a_2, a_3)$, $\mathbf{b} = (b_1, b_2, b_3)$, and $\mathbf{c} = (c_1, c_2, c_3)$ in $\mathbb{R}^3$, as shown in Fig. 4.5. The volume of the box can be computed by multiplying the area of the base by the altitude h.

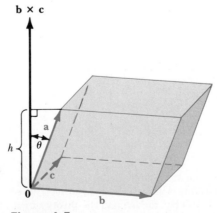

Figure 4.5
The box in $\mathbb{R}^3$ determined by a, b, and c.

We have just seen that the area of the base of the box is equal to $\|\mathbf{b} \times \mathbf{c}\|$ and the altitude can be found by computing

$$h = \|\mathbf{a}\| |\cos \theta| = \frac{\|\mathbf{b} \times \mathbf{c}\| \|\mathbf{a}\| |\cos \theta|}{\|\mathbf{b} \times \mathbf{c}\|} = \frac{|(\mathbf{b} \times \mathbf{c}) \cdot \mathbf{a}|}{\|\mathbf{b} \times \mathbf{c}\|}.$$

The absolute value is used in case $\cos \theta$ is negative. This would be the case if $\mathbf{b} \times \mathbf{c}$ had direction opposite to that shown in Fig. 4.5. Thus

$$\text{Volume} = (\text{Area of base})h = \|\mathbf{b} \times \mathbf{c}\| \frac{|(\mathbf{b} \times \mathbf{c}) \cdot \mathbf{a}|}{\|\mathbf{b} \times \mathbf{c}\|} = |(\mathbf{b} \times \mathbf{c}) \cdot \mathbf{a}|.$$

That is, referring to Eq. (3), which defines $\mathbf{b} \times \mathbf{c}$, we see that

$$\text{Volume} = |a_1(b_2c_3 - b_3c_2) - a_2(b_1c_3 - b_3c_1) + a_3(b_1c_2 - b_2c_1)|. \tag{4}$$

The number within the absolute value symbol is known as a **third-order determinant.** It is the **determinant of the matrix**

$$A = \begin{pmatrix} a_1 & a_2 & a_3 \\ b_1 & b_2 & b_3 \\ c_1 & c_2 & c_3 \end{pmatrix}$$

and is denoted by

$$\det(A) = \begin{vmatrix} a_1 & a_2 & a_3 \\ b_1 & b_2 & b_3 \\ c_1 & c_2 & c_3 \end{vmatrix}.$$

It can be computed as

$$\det(A) = a_1 \begin{vmatrix} b_2 & b_3 \\ c_2 & c_3 \end{vmatrix} - a_2 \begin{vmatrix} b_1 & b_3 \\ c_1 & c_3 \end{vmatrix} + a_3 \begin{vmatrix} b_1 & b_2 \\ c_1 & c_2 \end{vmatrix}. \tag{5}$$

THE VOLUME INTERPRETATION of a determinant first appeared in a 1773 paper on mechanics by Joseph Louis Lagrange (1736–1813). He noted that if the points M, M', M'' have coordinates (x, y, z), (x', y', z'), (x'', y'', z''), respectively, then the tetrahedron with vertices at the origin and those three points will have volume

$$\left(\frac{1}{6}\right)[z(x'y'' - y'x'') + z'(yx'' - xy'') + z''(xy' - yx')],$$

that is,

$$\left(\frac{1}{6}\right) \det \begin{pmatrix} x & y & z \\ x' & y' & z' \\ x'' & y'' & z'' \end{pmatrix}.$$

Lagrange was born in Turin, but spent most of his mathematical career in Berlin and in Paris. He contributed important results to such varied fields as the calculus of variation, celestial mechanics, number theory, and the theory of equations. Among his most famous works are the *Treatise on Analytical Mechanics* (1788), in which he presented the various principles of mechanics from a single point of view, and the *Theory of Analytic Functions* (1797), in which he attempted to base the differential calculus on the theory of power series.

Note the similarity of this formula (5) with our computation of the cross product $\mathbf{b} \times \mathbf{c}$ in Eq. (3). We simply replace $\mathbf{i}$, $\mathbf{j}$, and $\mathbf{k}$ by a_1, a_2, and a_3, respectively.

Example 6 Find the determinant of the matrix

$$A = \begin{pmatrix} 2 & 1 & 3 \\ 4 & 1 & 2 \\ 1 & 2 & -3 \end{pmatrix}.$$

Solution Using (5) we have

$$\begin{vmatrix} 2 & 1 & 3 \\ 4 & 1 & 2 \\ 1 & 2 & -3 \end{vmatrix} = 2 \begin{vmatrix} 1 & 2 \\ 2 & -3 \end{vmatrix} - 1 \begin{vmatrix} 4 & 2 \\ 1 & -3 \end{vmatrix} + 3 \begin{vmatrix} 4 & 1 \\ 1 & 2 \end{vmatrix}$$

$$= 2(-7) - (-14) + 3(7) = 21. \quad \triangleleft$$

Example 7 Find the volume of the box with vertex at the origin determined by the vectors $\mathbf{a} = (4, 1, 1)$, $\mathbf{b} = (2, 1, 0)$, and $\mathbf{c} = (0, 2, 3)$, and sketch the box in a figure.

Solution The box is shown in Fig. 4.6. Its volume is given by the absolute value of the determinant

$$\begin{vmatrix} 4 & 1 & 1 \\ 2 & 1 & 0 \\ 0 & 2 & 3 \end{vmatrix} = 4 \begin{vmatrix} 1 & 0 \\ 2 & 3 \end{vmatrix} - \begin{vmatrix} 2 & 0 \\ 0 & 3 \end{vmatrix} + \begin{vmatrix} 2 & 1 \\ 0 & 2 \end{vmatrix}$$

$$= 10. \quad \triangleleft$$

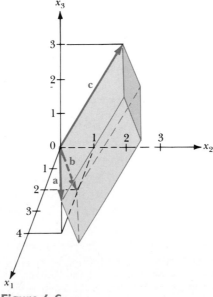

Figure 4.6
The box determined by a, b, and c.

The computation of det(A) in Eq. (5) is referred to as *an expansion of the determinant on the first row*. It is a particular case of a more general procedure for computing det(A), which is described in the next section.

We list the results of our work with cross products and some of their algebraic properties in one place as a theorem. The algebraic properties are easily checked by just computing with the components of the vectors. Example 8 below gives an illustration.

Theorem 4.1   Properties of the Cross Product

Let **a**, **b**, and **c** be vectors in $\mathbb{R}^3$.

1. **b** × **c** = −(**c** × **b**)   Anticommutativity

2. **a** × (**b** × **c**) is generally different   × is not associative
   from (**a** × **b**) × **c**

3. **a** × (**b** + **c**) = (**a** × **b**) + (**a** × **c**)   Distributive properties
   (**a** + **b**) × **c** = (**a** × **c**) + (**b** × **c**)

4. **b** · (**b** × **c**) = (**b** × **c**) · **c** = 0   **b** × **c** is perpendicular
                                                                   to both **b** and **c**

5. ‖**b** × **c**‖ = Area of the parallelogram   Area property
   determined by **b** and **c**

6. **a** · (**b** × **c**) = (**a** × **b**) · **c** = ± Volume of   Volume property
   the box determined by **a**, **b**, and **c**

7. **a** × (**b** × **c**) = (**a** · **c**)**b** − (**a** · **b**)**c**   Formula for computa-
                                                                   tion of **a** × (**b** × **c**)

Example 8   Show that **b** × **c** = −(**c** × **b**) for any vectors **b** and **c** in $\mathbb{R}^3$.

Solution   We compute

$$\mathbf{c} \times \mathbf{b} = \begin{vmatrix} \mathbf{i} & \mathbf{j} & \mathbf{k} \\ c_1 & c_2 & c_3 \\ b_1 & b_2 & b_3 \end{vmatrix}$$

$$= \begin{vmatrix} c_2 & c_3 \\ b_2 & b_3 \end{vmatrix} \mathbf{i} - \begin{vmatrix} c_1 & c_3 \\ b_1 & b_3 \end{vmatrix} \mathbf{j} + \begin{vmatrix} c_1 & c_2 \\ b_1 & b_2 \end{vmatrix} \mathbf{k}.$$

A simple computation shows that interchanging the rows of a 2×2 matrix having a determinant d gives a matrix with determinant $-d$ (see Exercise 11). Comparison of the formula above for **c** × **b** with the formula for **b** × **c** in Eq. (3) then shows that **b** × **c** = −(**c** × **b**).   ◁

SUMMARY

1. A second-order determinant is defined by

$$\begin{vmatrix} a & b \\ c & d \end{vmatrix} = ad - bc.$$

A third-order determinant is defined by Eq. (5).

2. The area of the parallelogram with vertex at the origin determined by nonzero vectors **a** and **b** in $\mathbb{R}^2$ is the absolute value of the determinant of the matrix having row vectors **a** and **b**.

3. The cross product of vectors **b** and **c** in $\mathbb{R}^3$ can be computed using the symbolic determinant

$$\begin{vmatrix} \mathbf{i} & \mathbf{j} & \mathbf{k} \\ b_1 & b_2 & b_3 \\ c_1 & c_2 & c_3 \end{vmatrix}.$$

This cross product **b** × **c** is perpendicular to both **b** and **c**.

4. The area of the parallelogram determined by nonzero vectors **b** and **c** in $\mathbb{R}^3$ is $\|\mathbf{b} \times \mathbf{c}\|$.

5. The volume of the box determined by nonzero vectors **a**, **b**, and **c** in $\mathbb{R}^3$ is the absolute value of the determinant of the matrix having row vectors **a**, **b**, and **c**. This determinant is also equal to $\mathbf{a} \cdot (\mathbf{b} \times \mathbf{c})$.

EXERCISES

In Exercises 1–4, find the indicated determinant.

1. $\begin{vmatrix} -1 & 3 \\ 5 & 0 \end{vmatrix}$

2. $\begin{vmatrix} -1 & 0 \\ 0 & 7 \end{vmatrix}$

3. $\begin{vmatrix} 0 & -3 \\ 5 & 0 \end{vmatrix}$

4. $\begin{vmatrix} 21 & -4 \\ 10 & 7 \end{vmatrix}$

5. Show that the vector $\mathbf{p} = \mathbf{b} \times \mathbf{c}$ given in Eq. (3) is indeed perpendicular to both **b** and **c**.

In Exercises 6–9, find the indicated determinant.

6. $\begin{vmatrix} 1 & 4 & -2 \\ 5 & 13 & 0 \\ 2 & -1 & 3 \end{vmatrix}$

7. $\begin{vmatrix} 2 & -5 & 3 \\ 1 & 3 & 4 \\ -2 & 3 & 7 \end{vmatrix}$

8. $\begin{vmatrix} 1 & -2 & 7 \\ 0 & 1 & 4 \\ 1 & 0 & 3 \end{vmatrix}$

9. $\begin{vmatrix} 2 & -1 & 1 \\ -1 & 0 & 3 \\ 2 & 1 & -4 \end{vmatrix}$

10. Show by direct computation that:

a) $\begin{vmatrix} a_1 & a_2 & a_3 \\ a_1 & a_2 & a_3 \\ c_1 & c_2 & c_3 \end{vmatrix} = 0;$

b) $\begin{vmatrix} a_1 & a_2 & a_3 \\ b_1 & b_2 & b_3 \\ a_1 & a_2 & a_3 \end{vmatrix} = 0.$

11. Show by direct computation that

$$\begin{vmatrix} a_1 & a_2 \\ b_1 & b_2 \end{vmatrix} = - \begin{vmatrix} b_1 & b_2 \\ a_1 & a_2 \end{vmatrix}.$$

12. Show by direct computation that

$$\begin{vmatrix} a_1 & a_2 & a_3 \\ b_1 & b_2 & b_3 \\ c_1 & c_2 & c_3 \end{vmatrix} = - \begin{vmatrix} a_1 & a_2 & a_3 \\ c_1 & c_2 & c_3 \\ b_1 & b_2 & b_3 \end{vmatrix}.$$

In Exercises 13–18, find **a × b**.

13. $\mathbf{a} = 2\mathbf{i} - \mathbf{j} + 3\mathbf{k}, \mathbf{b} = \mathbf{i} + 2\mathbf{j}$

14. $\mathbf{a} = -5\mathbf{i} + \mathbf{j} + 4\mathbf{k}, \mathbf{b} = 2\mathbf{i} + \mathbf{j} - 3\mathbf{k}$

15. $\mathbf{a} = -\mathbf{i} + 2\mathbf{j} + 4\mathbf{k}, \mathbf{b} = 2\mathbf{i} - 4\mathbf{j} - 8\mathbf{k}$

16. $\mathbf{a} = \mathbf{i} - \mathbf{j} + \mathbf{k}, \mathbf{b} = 3\mathbf{i} - 2\mathbf{j} + 7\mathbf{k}$

17. $\mathbf{a} = 2\mathbf{i} - 3\mathbf{j} + 5\mathbf{k}, \mathbf{b} = 4\mathbf{i} - 5\mathbf{j} + \mathbf{k}$

18. $\mathbf{a} = -2\mathbf{i} + 3\mathbf{j} - \mathbf{k}, \mathbf{b} = 4\mathbf{i} - 6\mathbf{j} + \mathbf{k}$

In Exercises 19–23, find the area of the parallelogram with vertex at the origin and having the given vectors as edges.

19. $-\mathbf{i} + 4\mathbf{j}$ and $2\mathbf{i} + 3\mathbf{j}$

20. $-5\mathbf{i} + 3\mathbf{j}$ and $\mathbf{i} + 7\mathbf{j}$

21. $\mathbf{i} + 3\mathbf{j} - 5\mathbf{k}$ and $2\mathbf{i} + 4\mathbf{j} - \mathbf{k}$

22. $2\mathbf{i} - \mathbf{j} + \mathbf{k}$ and $\mathbf{i} + 3\mathbf{j} - \mathbf{k}$

23. $\mathbf{i} - 4\mathbf{j} + \mathbf{k}$ and $2\mathbf{i} + 3\mathbf{j} - 2\mathbf{k}$

In Exercises 24–31, find the area of the given geometric configuration.

24. The triangle with vertices $(-1, 2)$, $(3, -1)$, and $(4, 3)$

25. The triangle with vertices $(3, -4)$, $(1, 1)$, and $(5, 7)$

26. The triangle with vertices $(2, 1, -3)$, $(3, 0, 4)$, and $(1, 0, 5)$

27. The triangle with vertices $(3, 1, -2)$, $(1, 4, 5)$, and $(2, 1, -4)$

28. The triangle in the plane $\mathbb{R}^2$ bounded by the lines $y = x$, $y = -3x + 8$, and $3y + 5x = 0$

29. The parallelogram with vertices $(1, 3)$, $(-2, 6)$, $(1, 11)$, and $(4, 8)$

30. The parallelogram with vertices $(1, 0, 1)$, $(3, 1, 4)$, $(0, 2, 9)$, and $(-2, 1, 6)$

31. The parallelogram in the plane $\mathbb{R}^2$ bounded by the lines $x - 2y = 3$, $x - 2y = 10$, $2x + 3y = -1$, and $2x + 3y = -8$

In Exercises 32–35, find **a · (b × c)** *and* **a × (b × c)**.

32. $\mathbf{a} = \mathbf{i} + 2\mathbf{j} - 3\mathbf{k}, \mathbf{b} = 4\mathbf{i} - \mathbf{j} + 2\mathbf{k}$, and $\mathbf{c} = 3\mathbf{i} + \mathbf{k}$

33. $\mathbf{a} = -\mathbf{i} + \mathbf{j} + 2\mathbf{k}, \mathbf{b} = \mathbf{i} + \mathbf{k}$, and $\mathbf{c} = 3\mathbf{i} - 2\mathbf{j} + 5\mathbf{k}$

34. $\mathbf{a} = \mathbf{i} - 3\mathbf{k}, \mathbf{b} = -\mathbf{i} + 4\mathbf{j}$, and $\mathbf{c} = \mathbf{i} + \mathbf{j} + \mathbf{k}$

35. $\mathbf{a} = 4\mathbf{i} - \mathbf{j} + 2\mathbf{k}, \mathbf{b} = 3\mathbf{i} + 5\mathbf{j} - 2\mathbf{k}$, and $\mathbf{c} = \mathbf{i} - 3\mathbf{j} + \mathbf{k}$

In Exercises 36–39, find the volume of the box having the given vectors as adjacent edges.

36. $-\mathbf{i} + 4\mathbf{j} + 7\mathbf{k}, 3\mathbf{i} - 2\mathbf{j} - \mathbf{k}$, and $4\mathbf{i} + 2\mathbf{k}$

37. $2\mathbf{i} + \mathbf{j} - 4\mathbf{k}, 3\mathbf{i} - \mathbf{j} + 2\mathbf{k}$, and $\mathbf{i} + 3\mathbf{j} - 8\mathbf{k}$

38. $-2\mathbf{i} + \mathbf{j}, 3\mathbf{i} - 4\mathbf{j} + \mathbf{k}$, and $\mathbf{i} - 2\mathbf{k}$

39. $3\mathbf{i} - \mathbf{j} + 4\mathbf{k}, \mathbf{i} - 2\mathbf{j} + 7\mathbf{k}$, and $5\mathbf{i} - 3\mathbf{j} + 10\mathbf{k}$

In Exercises 40–43, find the volume of the tetrahedron having the given vertices. (Consider how the volume of a tetrahedron having three vectors from one point as edges is related to the volume of the box having the same three vectors as adjacent edges.)

40. $(-3, 0, 1)$, $(4, 2, 1)$, $(0, 1, 7)$, and $(1, 1, 1)$

41. $(0, 1, 1)$, $(8, 2, -7)$, $(3, 1, 6)$, and $(-4, -2, 0)$

42. $(-1, 1, 2)$, $(3, 1, 4)$, $(-1, 6, 0)$, and $(2, -1, 5)$

43. $(-1, 2, 4)$, $(2, -3, 0)$, $(-4, 2, -1)$, and $(0, 3, -2)$

In Exercises 44–47, use a determinant to ascertain whether the given points lie on a line in $\mathbb{R}^2$. [Hint: What is the area of a "parallelogram" with collinear vertices?]

44. $(0, 0)$, $(3, 5)$, and $(6, 9)$

45. $(0, 0)$, $(4, 2)$, and $(-6, -3)$

46. $(1, 5)$, $(3, 7)$, and $(-3, 1)$

47. $(2, 3)$, $(1, -4)$, $(6, 2)$

In Exercises 48–51, use a determinant to ascertain whether the given points lie in a plane in $\mathbb{R}^3$. *[Hint: What is the "volume" of a box with coplanar vertices?]*

48. $(0, 0, 0)$, $(1, 4, 3)$, $(2, 5, 8)$, and $(-1, 2, -5)$ **49.** $(0, 0, 0)$, $(2, 1, 1)$, $(3, -2, 1)$, and $(-1, 2, 3)$

50. $(1, -1, 3)$, $(4, 2, 3)$, $(3, 1, -2)$, and $(5, 5, -5)$ **51.** $(1, 2, 1)$, $(3, 3, 4)$, $(2, 2, 2)$, and $(4, 3, 5)$

52. Prove property (2) of Theorem 4.1.

53. Prove property (3) of Theorem 4.1.

54. Prove property (6) of Theorem 4.1.

55. Option 7 of the available program VECTGRPH gives drill on the determinant of a 2×2 matrix as the area of the parallelogram determined by its row vectors, with an associated plus or minus sign. Run this option until a score of 80 percent or better can be achieved with ease.

4.2
The Determinant of a Square Matrix

THE DEFINITION

We defined a third-order determinant in terms of second-order determinants in Eq. (5) on page 190. A second-order determinant can be defined in terms of first-order determinants if we interpret the determinant of a 1×1 matrix to be its sole entry. We define an nth-order determinant in terms of determinants of order $n - 1$. In order to facilitate this, we introduce the **minor matrix** A_{ij} of an $n \times n$ matrix $A = (a_{ij})$; it is the $(n - 1) \times (n - 1)$ matrix obtained by crossing out the ith row and jth column of A. The minor matrix is the portion shown using color shading in

$$
A_{ij} = \begin{pmatrix} a_{11} & \cdots & a_{1j} & \cdots & a_{1n} \\ & \vdots & \vdots & & \vdots \\ a_{i1} & \cdots & a_{ij} & \cdots & a_{in} \\ & \vdots & \vdots & & \vdots \\ a_{n1} & \cdots & a_{nj} & \cdots & a_{nn} \end{pmatrix} \ i\text{th row.} \tag{1}
$$
jth column

Using $|A_{ij}|$ as notation for the determinant of the minor matrix A_{ij}, we can express the determinant of a 3×3 matrix A as

$$
\begin{vmatrix} a_{11} & a_{12} & a_{13} \\ a_{21} & a_{22} & a_{23} \\ a_{31} & a_{32} & a_{33} \end{vmatrix} = a_{11}|A_{11}| - a_{12}|A_{12}| + a_{13}|A_{13}|.
$$

The numbers $a'_{11} = |A_{11}|$, $a'_{12} = -|A_{12}|$, and $a'_{13} = |A_{13}|$ are appropriately called the **cofactors** of a_{11}, a_{12}, and a_{13}. We now proceed to define the determinant of any square matrix by mathematical induction. (See Appendix A for a discussion of mathematical induction.)

Definition 4.1   Cofactors and Determinants

The **determinant** of a 1×1 matrix is its sole entry. Let $n > 1$ and assume that determinants of order less than n have been defined. Let $A = (a_{ij})$ be an $n \times n$ matrix. The **cofactor** of a_{ij} in A is

$$a'_{ij} = (-1)^{i+j} \det(A_{ij}), \tag{2}$$

where A_{ij} is the minor matrix of A given in (1). The **determinant** of A is

$$\det(A) = \begin{vmatrix} a_{11} & a_{12} & \cdots & a_{1n} \\ a_{21} & a_{22} & \cdots & a_{2n} \\ \vdots & \vdots & & \vdots \\ a_{n1} & a_{n2} & \cdots & a_{nn} \end{vmatrix}$$

$$= a_{11}a'_{11} + a_{12}a'_{12} + \cdots + a_{1n}a'_{1n}. \tag{3}$$

In addition to the notation $\det(A)$ for the determinant of A, we will sometimes use $|A|$ when determinants appear in equations to make the equations easier to read.

Example 1   Find the cofactor of the entry 3 in the matrix

$$A = \begin{pmatrix} 2 & 1 & 0 & 1 \\ 3 & 2 & 1 & 2 \\ 4 & 0 & 1 & 4 \\ 1 & 0 & 2 & 1 \end{pmatrix}.$$

THE FIRST APPEARANCE OF THE DETERMINANT OF A SQUARE MATRIX in Western Europe occurs in a 1683 letter from Gottfried von Leibniz to the Marquis de L'Hôpital. Leibniz writes a system of three equations in two unknowns with abstract "numerical" coefficients as follows:

$$10 + 11x + 12y = 0$$
$$20 + 21x + 22y = 0$$
$$30 + 31x + 32y = 0$$

in which he notes that each coefficient number has "two characters, the first marking in which equation it occurs, the second marking which letter it belongs to." He then proceeds to eliminate first y and then x to show that the criterion for the system of equations to have a solution is that $10 \cdot 21 \cdot 32 + 11 \cdot 22 \cdot 30 + 12 \cdot 20 \cdot 31 = 10 \cdot 22 \cdot 31 + 11 \cdot 20 \cdot 32 + 12 \cdot 21 \cdot 30$. This is, of course, equivalent to our condition that the determinant of the matrix of coefficients must vanish. Unfortunately, the letter was not published until 1850 and therefore had no influence on subsequent work.

Determinants also appeared in the contemporaneous work of the Japanese mathematician Takakazu Seki (1642–1708).

Solution Since 3 is in the 2nd row, 1st column position of A, we cross out the 2nd row and 1st column of A and obtain

$$3' = a'_{21} = (-1)^{2+1} \begin{vmatrix} 1 & 0 & 1 \\ 0 & 1 & 4 \\ 0 & 2 & 1 \end{vmatrix} = - \begin{vmatrix} 1 & 4 \\ 2 & 1 \end{vmatrix} - \begin{vmatrix} 0 & 1 \\ 0 & 2 \end{vmatrix}$$

$$= -(-7 - 0) = 7. \quad \triangleleft$$

Example 2 Use (3) in Definition 4.1 to find the determinant of the matrix

$$A = \begin{pmatrix} 5 & -2 & 4 & -1 \\ 0 & 1 & 5 & 2 \\ 1 & 2 & 0 & 1 \\ -3 & 1 & -1 & 1 \end{pmatrix}.$$

Solution We have

$$\det(A) = \begin{vmatrix} 5 & -2 & 4 & -1 \\ 0 & 1 & 5 & 2 \\ 1 & 2 & 0 & 1 \\ -3 & 1 & -1 & 1 \end{vmatrix}$$

$$= 5(-1)^2 \begin{vmatrix} 1 & 5 & 2 \\ 2 & 0 & 1 \\ 1 & -1 & 1 \end{vmatrix} + (-2)(-1)^3 \begin{vmatrix} 0 & 5 & 2 \\ 1 & 0 & 1 \\ -3 & -1 & 1 \end{vmatrix}$$

$$+ 4(-1)^4 \begin{vmatrix} 0 & 1 & 2 \\ 1 & 2 & 1 \\ -3 & 1 & 1 \end{vmatrix} + (-1)(-1)^5 \begin{vmatrix} 0 & 1 & 5 \\ 1 & 2 & 0 \\ -3 & 1 & -1 \end{vmatrix}.$$

Computing the third-order determinants, we have

$$\begin{vmatrix} 1 & 5 & 2 \\ 2 & 0 & 1 \\ 1 & -1 & 1 \end{vmatrix} = 1 \begin{vmatrix} 0 & 1 \\ -1 & 1 \end{vmatrix} - 5 \begin{vmatrix} 2 & 1 \\ 1 & 1 \end{vmatrix} + 2 \begin{vmatrix} 2 & 0 \\ 1 & -1 \end{vmatrix}$$

$$= 1(1) - 5(1) + 2(-2) = -8;$$

$$\begin{vmatrix} 0 & 5 & 2 \\ 1 & 0 & 1 \\ -3 & -1 & 1 \end{vmatrix} = 0 \begin{vmatrix} 0 & 1 \\ -1 & 1 \end{vmatrix} - 5 \begin{vmatrix} 1 & 1 \\ -3 & 1 \end{vmatrix} + 2 \begin{vmatrix} 1 & 0 \\ -3 & -1 \end{vmatrix}$$

$$= 0(1) - 5(4) + 2(-1) = -22;$$

$$\begin{vmatrix} 0 & 1 & 2 \\ 1 & 2 & 1 \\ -3 & 1 & 1 \end{vmatrix} = 0 \begin{vmatrix} 2 & 1 \\ 1 & 1 \end{vmatrix} - 1 \begin{vmatrix} 1 & 1 \\ -3 & 1 \end{vmatrix} + 2 \begin{vmatrix} 1 & 2 \\ -3 & 1 \end{vmatrix}$$

$$= 0(1) - 1(4) + 2(7) = 10;$$

$$\begin{vmatrix} 0 & 1 & 5 \\ 1 & 2 & 0 \\ -3 & 1 & -1 \end{vmatrix} = 0 \begin{vmatrix} 2 & 0 \\ 1 & -1 \end{vmatrix} - 1 \begin{vmatrix} 1 & 0 \\ -3 & -1 \end{vmatrix} + 5 \begin{vmatrix} 1 & 2 \\ -3 & 1 \end{vmatrix}$$

$$= 0(-2) - 1(-1) + 5(7) = 36.$$

Therefore, $\det(A) = 5(-8) + 2(-22) + 4(10) + 1(36) = -8.$ ◁

The preceding example makes one thing plain:

> Computation of determinants of matrices of even moderate size using only Definition 4.1 is a tremendous chore.

According to modern astronomical theory, our solar system would be dead long before a present-day computer could find the determinant of a 50×50 matrix using just the inductive Definition 4.1 (see Exercise 25). Section 4.3 gives an alternative efficient method for computing determinants.

Although it is difficult to explain their importance without using notions from calculus for functions of several variables, determinants are important. Let us just say that in some cases where primary values depend on some secondary values, the single number that best measures the rate that the primary values change as the secondary values change is given by a determinant. This is closely connected with the geometric interpretation of a determinant as a *"volume."* (See Sections 6.6 and 8.3 for a more detailed discussion.) We motivated the determinant of a 2×2 matrix using area and motivated the determinant of a 3×3 matrix using volume in Section 4.1. In Section 5.6 we will show how an nth-order determinant can be interpreted as a "volume."

It is desirable to have an efficient way to compute a determinant. We will spend the remainder of this section developing properties of determinants that will enable us to find a good method for their computation. The computation of $\det(A)$ using Eq. (3) is called **expansion by minors** on the first row. Section 4.7 gives a proof by mathematical induction that $\det(A)$ can be obtained using an expansion by minors *on any row or on any column*. We state this more precisely in a theorem.

Theorem 4.2 General Expansion by Minors

Let A be an $n \times n$ matrix and let r and s be any selections from the list of numbers $1, 2, \ldots , n$. Then

$$\det(A) = a_{r1}a'_{r1} + a_{r2}a'_{r2} + \cdots + a_{rn}a'_{rn} \tag{4}$$

and also

$$\det(A) = a_{1s}a'_{1s} + a_{2s}a'_{2s} + \cdots + a_{ns}a'_{ns}. \tag{5}$$

Equation (4) is the **expansion of** $\det(A)$ **by minors on the rth row** of A, and Eq. (5) is the **expansion of** $\det(A)$ **by minors on the sth column** of A. Theorem

4.2 thus says that $\det(A)$ can be found by expanding by minors on any row or on any column of A.

Example 3   Find the determinant of the matrix

$$A = \begin{pmatrix} 3 & 2 & 0 & 1 & 3 \\ -2 & 4 & 1 & 2 & 1 \\ 0 & -1 & 0 & 1 & -5 \\ -1 & 2 & 0 & -1 & 2 \\ 0 & 0 & 0 & 0 & 2 \end{pmatrix}.$$

Solution   An inductive computation like the one in Definition 4.1 is still the only way we have to compute $\det(A)$ at this point, but we can expedite the computation if we expand by minors at each step on the row or column containing the most zeros. We have

$$\det(A) = 2(-1)^{5+5} \begin{vmatrix} 3 & 2 & 0 & 1 \\ -2 & 4 & 1 & 2 \\ 0 & -1 & 0 & 1 \\ -1 & 2 & 0 & -1 \end{vmatrix} \qquad \textbf{Expanding on the 5th row}$$

$$= (2)(1)(-1)^{2+3} \begin{vmatrix} 3 & 2 & 1 \\ 0 & -1 & 1 \\ -1 & 2 & -1 \end{vmatrix} \qquad \textbf{Expanding on the 3rd column}$$

$$= -2 \left[3(-1)^{1+1} \begin{vmatrix} -1 & 1 \\ 2 & -1 \end{vmatrix} - 1(-1)^{3+1} \begin{vmatrix} 2 & 1 \\ -1 & 1 \end{vmatrix} \right] \qquad \begin{matrix} \textbf{Expanding} \\ \textbf{on the 1st} \\ \textbf{column} \end{matrix}$$

$$= -2[3(1 - 2) - 1(2 + 1)] = -2(-3 - 3) = 12. \quad \triangleleft$$

Example 4   Show that the determinant of an upper- (or lower-) triangular square matrix is the product of its diagonal elements.

Solution   We work with an upper-triangle matrix, the other case being analogous. If

$$U = \begin{pmatrix} u_{11} & u_{12} & \cdots & u_{1n} \\ & u_{22} & \cdots & u_{2n} \\ & & \ddots & \vdots \\ & \bigcirc & & u_{nn} \end{pmatrix}$$

is an upper-triangular matrix, then expanding on first columns each time, we have

$$\det(U) = u_{11} \begin{vmatrix} u_{22} & u_{23} & \cdots & u_{2n} \\ 0 & u_{33} & \cdots & u_{3n} \\ \vdots & \vdots & & \vdots \\ 0 & 0 & \cdots & u_{nn} \end{vmatrix} = u_{11}u_{22} \begin{vmatrix} u_{33} & \cdots & u_{3n} \\ 0 & \cdots & u_{4n} \\ \vdots & & \vdots \\ 0 & \cdots & u_{nn} \end{vmatrix}$$

$$= \cdots = u_{11}u_{22} \cdots u_{nn}. \quad \triangleleft$$

PROPERTIES OF THE DETERMINANT

Using Theorem 4.2, we can establish several properties of determinants that will be of tremendous help in their computation. Since Definition 4.1 was an inductive one, we use mathematical induction as we start to prove properties of determinants. (We remind the reader that mathematical induction is reviewed in Appendix A.) We will always consider A to be an $n \times n$ matrix.

Property 1 · The Transpose Property

For any square matrix A, we have $\det(A) = \det(A^T)$.

Proof Verification of this property is trivial for determinants of orders 1 or 2. Let $n > 2$ and assume the property holds for square matrices of size smaller than $n \times n$. We proceed to prove Property 1 for an $n \times n$ matrix A. We have

$$\det(A) = a_{11}|A_{11}| - a_{12}|A_{12}| + \cdots + (-1)^{n+1}a_{1n}|A_{1n}|.$$

Expanding on the first row of A

Writing $B = A^T$, we have

$$\det(B) = b_{11}|B_{11}| - b_{21}|B_{21}| + \cdots + (-1)^{n+1}b_{n1}|B_{n1}|.$$

Expanding on the first column of B

However, $a_{1j} = b_{j1}$ and $B_{j1} = A_{1j}{}^T$ because $B = A^T$. Applying our induction hypothesis to the $(n-1)$st-order determinant $|A_{1j}|$, we have $|A_{1j}| = |B_{j1}|$. We conclude that $\det(A) = \det(B) = \det(A^T)$. ■

THE THEORY OF DETERMINANTS grew from the efforts of many mathematicians of the late eighteenth and early nineteenth centuries. Besides Gabriel Cramer, whose work we will discuss in the note on page 215, Etienne Bezout (1739–1783) in 1764 and Alexandre-Theophile Vandermonde (1735–1796) in 1771 gave various methods for computing determinants. In a work on integral calculus, Pierre Simon Laplace (1749–1827) had to deal with systems of linear equations. He repeated the work of Cramer, but he also stated and proved the rule that interchanging two adjacent columns of the determinant changes the sign and showed that a determinant with two equal columns will be 0.

The most complete of the early works on determinants is that of Augustin-Louis Cauchy (1789–1857) in 1812. In this work, Cauchy introduced the name "determinant" to replace several older terms, used our current double-subscript notation for a square array of numbers, defined the array of adjoints (or minors) to a given array, and showed that one can calculate the determinant by expanding on any row or column. In addition, Cauchy reproved many of the standard theorems on determinants that had been more or less known for the past 50 years.

Cauchy was the most prolific mathematician of the nineteenth century, contributing to such areas as complex analysis, calculus, differential equations, and mechanics. In particular, he wrote the first calculus text using our modern ε,δ-approach to continuity. Politically he was a conservative; and when the July Revolution of 1830 replaced the Bourbon king Charles X with the Orleans king Louis-Philippe, Cauchy refused to take the oath of allegiance, therefore forfeiting his chairs at the Ecole Polytechnique and the Collège de France and going into exile in Turin and Prague.

This transpose property has a very useful consequence. It guarantees that any property of determinants involving rows of a matrix is equally valid if we replace "rows" by "columns" in the statement of that property. For example, the next property has an analog for columns.

Property 2 **The Row-interchange Property**

If two different rows of a square matrix A are interchanged, the determinant of the resulting matrix is $-\det(A)$.

Proof Again we find that the proof is trivial for the case $n = 2$. Assume that $n > 2$, and that this row-interchange property holds for matrices of size smaller than $n \times n$. Let A be an $n \times n$ matrix, and let B be the matrix obtained from A by interchanging the ith and rth rows, leaving the other rows unchanged. Since $n > 2$, we can choose a kth row for expansion by minors, where k is different from both r and i. Consider the cofactors

$$(-1)^{k+j}|A_{kj}| \quad \text{and} \quad (-1)^{k+j}|B_{kj}|.$$

These numbers must have opposite signs by our induction hypothesis, since the minor matrices A_{kj} and B_{kj} have size $(n - 1) \times (n - 1)$ and B_{kj} can be obtained from A_{kj} by interchanging two rows. That is, $|B_{kj}| = -|A_{kj}|$. Expanding by minors on the kth row to find $\det(A)$ and $\det(B)$, we see at once that $\det(A) = -\det(B)$. ∎

Property 3 **The Equal-rows Property**

If two rows of a matrix A are equal, then $\det(A) = 0$.

Proof Let B be the matrix obtained from A by interchanging the two equal rows of A. By the row-interchange property, we have $\det(B) = -\det(A)$. On the other hand, $B = A$, so that $\det(A) = -\det(A)$. Therefore, $\det(A) = 0$. ∎

Property 4 **The Scalar-multiplication Property**

If a single row of a matrix A is multiplied by a scalar r, then the determinant of the resulting matrix is $r \cdot \det(A)$.

Proof Let r be any scalar and let B be the matrix obtained from A by replacing the kth row $(a_{k1}, a_{k2}, \ldots, a_{kn})$ of A by $(ra_{k1}, ra_{k2}, \ldots, ra_{kn})$. Since the rows of B are equal to those of A except possibly for the kth row, it follows that the minor matrices A_{kj} and B_{kj} are equal for each j. Therefore, $a'_{kj} = b'_{kj}$, and computing $\det(B)$ by expanding by minors on the kth row, we have

$$\det(B) = b_{k1}b'_{k1} + b_{k2}b'_{k2} + \cdots + b_{kn}b'_{kn}$$
$$= r \cdot a_{k1}a'_{k1} + r \cdot a_{k2}a'_{k2} + \cdots + r \cdot a_{kn}a'_{kn}$$
$$= r \cdot \det(A). \quad ∎$$

Example 5   Find the determinant of the matrix

$$A = \begin{pmatrix} 2 & 1 & 3 & 4 & 2 \\ 6 & 2 & 1 & 4 & 1 \\ 6 & 3 & 9 & 12 & 6 \\ 2 & 1 & 3 & 4 & 1 \\ 1 & 4 & 2 & 1 & 1 \end{pmatrix}.$$

Solution   We note that the third row of A is three times the first row. Therefore, we have

$$\det(A) = 3 \begin{vmatrix} 2 & 1 & 3 & 4 & 2 \\ 6 & 2 & 1 & 4 & 1 \\ 2 & 1 & 3 & 4 & 2 \\ 2 & 1 & 3 & 4 & 1 \\ 1 & 4 & 2 & 1 & 1 \end{vmatrix} \qquad \text{Property 4}$$

$$= 3(0) = 0. \quad \triangleleft \qquad \text{Property 3}$$

The equal-row property and the scalar-multiplication property indicate how the determinant of a matrix changes when two of the three elementary row operations are used. The next property deals with the most complicated of the elementary row operations, and lies at the heart of the efficient computation of determinants given in Section 4.3.

Property 5   The Row-addition Property

> If the product of one row of A by a scalar is added to a different row of A, then the determinant of the resulting matrix is the same as $\det(A)$.

Proof   Let $\mathbf{a}_i = (a_{i1}, a_{i2}, \ldots, a_{in})$ be the ith row of A. Suppose that $r\mathbf{a}_i$ is added to the kth row $\mathbf{a}_k$ of A, where r is any scalar and $k \neq i$. We obtain a matrix B whose rows are the same as the rows of A except possibly for the kth row, which is

$$\mathbf{b}_k = (ra_{i1} + a_{k1}, ra_{i2} + a_{k2}, \ldots, ra_{in} + a_{kn}).$$

Clearly the minor matrices A_{kj} and B_{kj} are equal for each j. Therefore, $a'_{kj} = b'_{kj}$, and computing $\det(B)$ by expanding by minors on the kth row, we have

$$\begin{aligned} \det(B) &= b_{k1}b'_{k1} + b_{k2}b'_{k2} + \cdots + b_{kn}b'_{kn} \\ &= (ra_{i1} + a_{k1})a'_{k1} + (ra_{i2} + a_{k2})a'_{k2} + \cdots + (ra_{in} + a_{kn})a'_{kn} \\ &= (ra_{i1}a'_{k1} + ra_{i2}a'_{k2} + \cdots + ra_{in}a'_{kn}) \\ &\quad + (a_{k1}a'_{k1} + a_{k2}a'_{k2} + \cdots + a_{kn}a'_{kn}) \\ &= r \cdot \det(C) + \det(A), \end{aligned}$$

where C is the matrix obtained from A by replacing the kth row of A with the ith row of A. Since C has two equal rows, its determinant is zero, so $\det(B) = \det(A)$. ◼

We conclude our list of properties of determinants with an extremely important one, whose proof we defer until Section 4.3.

Property 6   The Multiplicative Property

If A and B are $n \times n$ matrices, then $\det(AB) = \det(A) \cdot \det(B)$.

Example 6   Find $\det(A)$ if

$$A = \begin{pmatrix} 2 & 0 & 0 \\ 1 & 3 & 0 \\ 4 & 2 & 1 \end{pmatrix} \begin{pmatrix} 1 & 2 & 3 \\ 0 & 1 & 2 \\ 0 & 0 & 2 \end{pmatrix}.$$

Solution   Since the determinant of an upper- or lower-triangular matrix is the product of the diagonal elements (see Example 4), Property 6 shows that

$$\det(A) = \begin{vmatrix} 2 & 0 & 0 \\ 1 & 3 & 0 \\ 4 & 2 & 1 \end{vmatrix} \begin{vmatrix} 1 & 2 & 3 \\ 0 & 1 & 2 \\ 0 & 0 & 2 \end{vmatrix} = (6)(2) = 12. \quad \triangleleft$$

Example 7   If $\det(A) = 3$, find $\det(A^5)$.

Solution   Applying Property 6 several times, we have

$$\det(A^5) = [\det(A)]^5 = 3^5 = 243. \quad \triangleleft$$

SUMMARY

1. The cofactor of an element a_{ij} in a square matrix A is $(-1)^{i+j}|A_{ij}|$, where A_{ij} is the matrix obtained from A by deleting the ith row and the jth column.

2. The determinant of an $n \times n$ matrix may be defined inductively by expansion by minors on the first row. The determinant can be computed by expansion by minors using any row or any column; it is the sum of the products of the entries in that row or column with the cofactors of the entries. Generally, such a computation is hopelessly long.

3. The elementary row operations have the following effect on the determinant of a square matrix A.

   a) If two different rows of A are interchanged, the sign of the determinant is changed.

   b) If a single row of A is multiplied by a scalar, the determinant is multiplied by the scalar.

   c) If a multiple of one row is added to a different row, the determinant is not changed.

4. We have $\det(A) = \det(A^T)$. As a consequence, the properties just listed for elementary row operations are also true for elementary column operations.

5. If two rows or two columns of a matrix are the same, its determinant is zero.

6. The determinant of an upper-triangular matrix or of a lower-triangular matrix is the product of the diagonal entries.

7. If A and B are $n \times n$ matrices, then $\det(AB) = \det(A) \cdot \det(B)$.

EXERCISES

In Exercises 1–10, find the determinant of the given matrix.

1. $\begin{pmatrix} 5 & 2 & 1 \\ 1 & -1 & 4 \\ 3 & 0 & 2 \end{pmatrix}$

2. $\begin{pmatrix} 1 & 0 & 6 \\ 4 & 1 & -1 \\ 5 & 0 & 1 \end{pmatrix}$

3. $\begin{pmatrix} 3 & 2 & 4 \\ 0 & 1 & 2 \\ 1 & 4 & 1 \end{pmatrix}$

4. $\begin{pmatrix} 4 & -1 & 2 \\ 3 & 1 & 0 \\ -1 & 2 & 1 \end{pmatrix}$

5. $\begin{pmatrix} 0 & 1 & 4 \\ 2 & 3 & 1 \\ 1 & 4 & 1 \end{pmatrix}$

6. $\begin{pmatrix} 6 & 2 & 1 \\ 0 & 4 & 1 \\ 0 & 0 & 5 \end{pmatrix}$

7. $\begin{pmatrix} 2 & 3 & 4 & 6 \\ 2 & 0 & -9 & 6 \\ 4 & 1 & 0 & 2 \\ 0 & 1 & -1 & 0 \end{pmatrix}$

8. $\begin{pmatrix} 2 & 0 & -1 & 7 \\ 6 & 1 & 0 & 4 \\ 8 & -2 & 1 & 0 \\ 4 & 1 & 0 & 2 \end{pmatrix}$

9. $\begin{pmatrix} 1 & 2 & 0 & -1 & 2 & 4 \\ 6 & 2 & 8 & 1 & -1 & 1 \\ 4 & 2 & 1 & 2 & 2 & -5 \\ 4 & 5 & 4 & 5 & 1 & 2 \\ 1 & 2 & 0 & -1 & 2 & 4 \\ 1 & 0 & 1 & 8 & 1 & 5 \end{pmatrix}$

10. $\begin{pmatrix} 1 & 0 & 1 & 2 \\ 3 & 4 & 1 & 2 \\ 6 & 1 & 0 & 0 \\ 0 & 1 & 2 & 1 \end{pmatrix}$

11. Find the cofactor of 5 for the matrix in Exercise 2.

12. Find the cofactor of 3 for the matrix in Exercise 4.

13. Find the cofactor of 7 for the matrix in Exercise 8.

14. Find the cofactor of -5 for the matrix in Exercise 9.

In Exercises 15–20, let A be a 3 × 3 matrix with $\det(A) = 2$.

15. Find $\det(A^2)$.

16. Find $\det(A^k)$.

17. Find det(3A).

18. Find det(A + A).

19. Find det(A^{-1}).

20. Find det(A^T).

21. Without using the multiplicative property of determinants (Property 6), prove that det(AB) = det(A) · det(B) for the case where A is a diagonal matrix.

22. Continuing Exercise 21, find two other types of matrix for which it is easy to show that det(AB) = det(A) · det(B).

23. Prove that if three $n \times n$ matrices A, B, and C are identical except for the kth rows $\mathbf{a}_k$, $\mathbf{b}_k$, and $\mathbf{c}_k$, respectively, which are related by $\mathbf{a}_k = \mathbf{b}_k + \mathbf{c}_k$, then

$$\det(A) = \det(B) + \det(C).$$

24. Note that

$$\begin{vmatrix} a_{11} & a_{12} \\ a_{21} & a_{22} \end{vmatrix} = (a_{11}a_{22}) + (-a_{12}\,a_{21})$$

is a sum of signed products where each product contains precisely one factor from each row and one factor from each column of the corresponding matrix. Prove by induction that this is true for an $n \times n$ matrix $A = (a_{ij})$.

25. This exercise is for the reader who is skeptical of our assertion that the solar system would be dead long before a present-day computer could find the determinant of a 50×50 matrix using just Definition 4.1 with expansion by minors.

a) Recall that $n! = n(n - 1)(n - 2)(n - 3) \cdots (3)(2)(1)$. Show by induction that expansion of an $n \times n$ matrix by minors requires at least $n!$ multiplications for $n > 1$.

b) ▪ Run the available program TIMING and find the number of seconds required for 3000 multiplications by your computer. Choose either interpretive or compiled BASIC. Then run the available program EBYMTIME and find the time to perform $n!$ multiplications for $n = 2, 4, 8, 12, 16, 20, 25, 30, 40, 50, 70,$ and 100.

26. ▪ MATCOMP can be used to compute determinants. Check Example 2 and the answers to some of the exercises using MATCOMP.

4.3
Computation of Determinants

We have seen that the computation of determinants of high order is an unreasonable task if it is done directly from Definition 4.1, using just repeated expansion by minors. In the special case where a square matrix is triangular, Example 4 of Section 4.2 shows that the determinant is simply the product of the diagonal entries. We know that a matrix can be reduced to row-echelon form using elementary row operations, and row-echelon form for a square matrix is always triangular. We have seen the effect of each type of elementary row operation on the determinant of a matrix. This suggests that we find the determinant of an $n \times n$ matrix A by reducing it to triangular form. We illustrate this procedure.

Example 1 Find the determinant of the matrix

$$A = \begin{pmatrix} 2 & 2 & 0 & 4 \\ 3 & 3 & 2 & 2 \\ 0 & 1 & 3 & 2 \\ 2 & 0 & 2 & 1 \end{pmatrix}$$

by reducing it to triangular form.

Solution We find that

$$\begin{vmatrix} 2 & 2 & 0 & 4 \\ 3 & 3 & 2 & 2 \\ 0 & 1 & 3 & 2 \\ 2 & 0 & 2 & 1 \end{vmatrix} = 2 \begin{vmatrix} 1 & 1 & 0 & 2 \\ 3 & 3 & 2 & 2 \\ 0 & 1 & 3 & 2 \\ 2 & 0 & 2 & 1 \end{vmatrix}$$ Scalar-multiplication property

$$= 2 \begin{vmatrix} 1 & 1 & 0 & 2 \\ 0 & 0 & 2 & -4 \\ 0 & 1 & 3 & 2 \\ 0 & -2 & 2 & -3 \end{vmatrix}$$ Row-addition property twice

$$= -2 \begin{vmatrix} 1 & 1 & 0 & 2 \\ 0 & 1 & 3 & 2 \\ 0 & 0 & 2 & -4 \\ 0 & -2 & 2 & -3 \end{vmatrix}$$ Row-interchange property

$$= -2 \begin{vmatrix} 1 & 1 & 0 & 2 \\ 0 & 1 & 3 & 2 \\ 0 & 0 & 2 & -4 \\ 0 & 0 & 8 & 1 \end{vmatrix}$$ Row-addition property

$$= (-2)(2) \begin{vmatrix} 1 & 1 & 0 & 2 \\ 0 & 1 & 3 & 2 \\ 0 & 0 & 1 & -2 \\ 0 & 0 & 8 & 1 \end{vmatrix}$$ Scalar-multiplication property

$$= (-2)(2) \begin{vmatrix} 1 & 1 & 0 & 2 \\ 0 & 1 & 3 & 2 \\ 0 & 0 & 1 & -2 \\ 0 & 0 & 0 & 17 \end{vmatrix}$$ Row-addition property

Therefore, $\det(A) = (-2)(2)(17) = -68$. ◁

The preceding example illustrates that if an $n \times n$ matrix A can be reduced to an upper-triangular matrix having n (nonzero) pivots, then $\det(A)$ is equal to $\pm$(product of pivots). On the other hand, if at some point in the row reduction of A a nonzero pivot cannot be found, then the determinant of this matrix as well as det(A) is zero. For example, in a 5×5 case we might have reduced our computation to the determinant

$$
\begin{vmatrix}
p_1 & X & X & X & X \\
0 & p_2 & X & X & X \\
0 & 0 & 0 & X & X \\
0 & 0 & 0 & X & X \\
0 & 0 & 0 & X & X
\end{vmatrix}. \tag{1}
$$

Expanding repeatedly by minors on first columns, we easily see that the determinant (1) is zero.

Theorem 4.3   Computation of the Determinant

Suppose an $n \times n$ matrix A can be row-reduced to an upper-triangular matrix using just row addition and row interchanges. If the reduced matrix has all nonzero diagonal entries, then det(A) = ±(product of these pivots) and is nonzero. If A cannot be so reduced, then det(A) = 0.

Row reduction is an efficient way to program a computer to compute a determinant. If we are using pencil and paper, a slight modification is more practical. We can use elementary row or column operations and the properties of determinants to reduce the computation to the determinant of a matrix having some row or column with a sole nonzero entry. A computer program generally modifies the matrix so the first column has a single nonzero entry, but we can look at the matrix and choose the row or column where this can be achieved most easily. Expanding by minors on that row or column reduces the computation to a determinant of one less order, and we can continue the process until we are left with the computation of a determinant of a 2×2 matrix. Here is an illustration.

Example 2   Find the determinant of the matrix

$$
A = \begin{pmatrix}
2 & -1 & 3 & 5 \\
2 & 0 & 1 & 0 \\
6 & 1 & 3 & 4 \\
-7 & 3 & -2 & 8
\end{pmatrix}.
$$

Solution   Surely it is easiest to create zeros in the second row and then expand by minors on that row. We start by multiplying column 3 by -2 and adding the result to column 1 and continue in a similar way.

$$
\begin{vmatrix}
2 & -1 & 3 & 5 \\
2 & 0 & 1 & 0 \\
6 & 1 & 3 & 4 \\
-7 & 3 & -2 & 8
\end{vmatrix}
=
\begin{vmatrix}
-4 & -1 & 3 & 5 \\
0 & 0 & 1 & 0 \\
0 & 1 & 3 & 4 \\
-3 & 3 & -2 & 8
\end{vmatrix}
= -
\begin{vmatrix}
-4 & -1 & 5 \\
0 & 1 & 4 \\
-3 & 3 & 8
\end{vmatrix}
$$

$$
= -
\begin{vmatrix}
-4 & -1 & 9 \\
0 & 1 & 0 \\
-3 & 3 & -4
\end{vmatrix}
= -
\begin{vmatrix}
-4 & 9 \\
-3 & -4
\end{vmatrix}
$$

$$
= -(16 + 27) = -43. \quad \triangleleft
$$

We recall from Section 1.5 that a square matrix is invertible if and only if it can be reduced to an upper-triangular matrix with nonzero diagonal entries. Recall also that a matrix that is not invertible is said to be singular. According to Theorem 4.3, a square matrix is singular if and only if its determinant is zero. We phrase this in terms of a criterion for invertibility.

Theorem 4.4   Criterion for Invertibility of a Square Matrix

Let A be an $n \times n$ matrix. Then A is invertible if and only if $\det(A) \neq 0$.

In linear algebra, we use matrices in many different ways, and each way provides its own point of view. This results in long lists of equivalent conditions that students should try to keep track of. For example, Theorem 4.4 is only the latest entry in a long list of criteria for a square matrix to be invertible. It is a valuable exercise at this point to write down as many conditions as you can that are equivalent to invertibility for a square matrix (see Exercise 1). We list a few conditions, illustrating what we mean.

A square $n \times n$ matrix A is invertible if and only if any one of the following hold:

1. $\det(A) \neq 0$;

2. the homogeneous linear system $A\mathbf{x} = \mathbf{0}$ has only the trivial solution;

3. the row rank of A is n;

4. the column vectors of A are linearly independent;

and so on.

We conclude with a proof of the multiplicative property of determinants. The proof is a nice application of the computational technique we have developed.

Theorem 4.5   Multiplicative Property of Determinants

If A and B are any two $n \times n$ matrices, then $\det(AB) = \det(A) \cdot \det(B)$.

Proof First we note that if A is a diagonal matrix, then the result follows easily, for the product

$$\begin{pmatrix} a_{11} & & & \\ & a_{22} & & \\ & & \ddots & \\ & & & a_{nn} \end{pmatrix} \begin{pmatrix} b_{11} & b_{12} & \cdots & b_{1n} \\ b_{21} & b_{22} & \cdots & b_{2n} \\ & & \vdots & \\ b_{n1} & b_{n2} & \cdots & b_{nn} \end{pmatrix}$$

has ith row equal to a_{ii} times the ith row of B. Using the scalar-multiplication property in each of these rows, we obtain

$$\det(AB) = (a_{11}a_{22} \cdots a_{nn}) \cdot \det(B) = \det(A) \cdot \det(B).$$

To deal with the nondiagonal case, we can quickly reduce the problem to the case where A is invertible. For if A is singular, then AB is also (see Exercise 20), so both A and AB have a zero determinant by Theorem 4.4, and, of course, $\det(A) \cdot \det(B) = 0$ also.

If we assume A is invertible, it can be row-reduced using just row-interchange and row-addition operations to an upper-triangular matrix with nonzero entries on the diagonal. We continue such row reduction analogous to the Gauss–Jordan method but without making pivots 1, and finally reduce A to a diagonal matrix D with nonzero diagonal entries. We can write $D = EA$, where E is the product of elementary matrices corresponding to the row interchanges and row additions used to reduce A to D. By the properties of determinants, we have $\det(A) = (-1)^r \cdot \det(D)$, where r is the number of row interchanges. The same sequence of steps will reduce the matrix AB to the matrix $E(AB) = (EA)B = DB$ so $\det(AB) = (-1)^r \cdot \det(DB)$. Therefore,

$$\det(AB) = (-1)^r \cdot \det(DB) = (-1)^r \cdot \det(D) \cdot \det(B) = \det(A) \cdot \det(B)$$

and the proof is complete. ■

Example 3 If A is invertible, find $\det(A^{-1})$ in terms of $\det(A)$.

Solution Since $AA^{-1} = I$, we have

$$\det(A) \cdot \det(A^{-1}) = \det(I) = 1$$

by Theorem 4.5. Thus

$$\det(A^{-1}) = 1/[\det(A)]. \quad \triangleleft$$

SUMMARY

> 1. The determinant of an invertible matrix can be found by reducing the matrix to triangular form, using just row-addition and row-interchange operations. The determinant of the original matrix is found by computing $(-1)^r \cdot$ (Product of diagonal elements) for the reduced matrix, where r is the number of row interchanges performed. The determinant of a singular matrix is zero.

2. The determinant of a matrix can be found by row or column reduction of the matrix to one having a sole nonzero entry in some column or row. One then expands by minors on that column or row, and continues this process until the computation is reduced to the determinant of a 2×2 matrix. This is a good way to find a determinant if you have to use pencil and paper.

3. An $n \times n$ matrix A is invertible if and only if $\det(A) \neq 0$.

EXERCISES

1. Make as long a list as you can of conditions equivalent to invertibility for an $n \times n$ matrix A. Don't forget to include the definition of invertibility. The list should surely have more than ten items.

In Exercises 2–11, find the determinant of the given matrix.

2. $\begin{pmatrix} 2 & 3 & -1 \\ 5 & -7 & 1 \\ -3 & 2 & -1 \end{pmatrix}$

3. $\begin{pmatrix} 4 & -3 & 2 \\ -1 & -1 & 1 \\ -5 & 5 & 7 \end{pmatrix}$

4. $\begin{pmatrix} 5 & 2 & 4 & 0 \\ 2 & -3 & -1 & 2 \\ 3 & -4 & 3 & 7 \\ 1 & -1 & 0 & 1 \end{pmatrix}$

5. $\begin{pmatrix} 3 & -5 & -1 & 7 \\ 0 & 3 & 1 & -6 \\ 2 & -5 & -1 & 8 \\ -8 & 8 & 2 & -9 \end{pmatrix}$

6. $\begin{pmatrix} 2 & 1 & 0 & 0 & 0 \\ 3 & -1 & 2 & 0 & 0 \\ 0 & 4 & 1 & -1 & 2 \\ 0 & 0 & -3 & 2 & 4 \\ 0 & 0 & 0 & -1 & 3 \end{pmatrix}$

7. $\begin{pmatrix} 3 & 2 & 0 & 0 & 0 \\ -1 & 4 & 1 & 0 & 0 \\ 0 & -3 & 5 & 2 & 0 \\ 0 & 0 & 0 & 1 & 4 \\ 0 & 0 & 0 & -1 & 2 \end{pmatrix}$

8. $\begin{pmatrix} 0 & 0 & 0 & 3 & 1 \\ 0 & 0 & 2 & 0 & -3 \\ 0 & -2 & 1 & 0 & 0 \\ 5 & -3 & 2 & 0 & 0 \\ -3 & 4 & 0 & 0 & 0 \end{pmatrix}$

9. $\begin{pmatrix} 2 & -1 & 0 & 0 \\ 4 & 5 & 0 & 0 \\ 0 & 0 & 3 & 6 \\ 0 & 0 & -4 & 2 \end{pmatrix}$

10. $\begin{pmatrix} 2 & -1 & 3 & 0 & 0 \\ 0 & 1 & 4 & 0 & 0 \\ -5 & 2 & 6 & 0 & 0 \\ 0 & 0 & 0 & 1 & 4 \\ 0 & 0 & 0 & -2 & 8 \end{pmatrix}$

11. $\begin{pmatrix} 0 & 0 & 0 & 3 & -4 \\ 0 & 0 & 0 & 2 & 1 \\ -1 & 2 & 4 & 0 & 0 \\ 3 & 1 & -2 & 0 & 0 \\ 5 & 1 & 5 & 0 & 0 \end{pmatrix}$

In Exercises 12–15, find the values of λ for which the given matrix is singular.

12. $\begin{pmatrix} 1-\lambda & 2 \\ 3 & 2-\lambda \end{pmatrix}$

13. $\begin{pmatrix} -\lambda & 5 \\ 2 & 3-\lambda \end{pmatrix}$

14. $\begin{pmatrix} 2 - \lambda & 0 & 0 \\ 0 & 1 - \lambda & 4 \\ 0 & 1 & 1 - \lambda \end{pmatrix}$

15. $\begin{pmatrix} 1 - \lambda & 0 & 2 \\ 0 & 2 - \lambda & 3 \\ 0 & 4 & -\lambda \end{pmatrix}$

16. The matrices in Exercises 9 and 10 have zero entries except for entries in an $r \times r$ submatrix R and a separate $s \times s$ submatrix S whose main diagonals lie on the main diagonal of the whole $n \times n$ matrix, and where $r + s = n$. Prove that if A is such a matrix with submatrices R and S, then $\det(A) = \det(R) \cdot \det(S)$.

17. The matrix A in Exercise 11 has structure similar to that discussed in Exercise 16, except that the square submatrices R and S lie along the other diagonal. State a result similar to that in Exercise 16 for such a matrix.

18. State a generalization of the result in Exercise 16 when the matrix A has zero entries except for entries in r submatrices positioned along the diagonal.

19. If A and C are $n \times n$ matrices with C invertible, show that $\det(A) = \det(C^{-1}AC)$.

20. If A and B are $n \times n$ matrices and if A is singular, prove that AB is also singular without using Theorem 4.5. [*Hint:* Assume that AB is invertible and derive a contradiction.]

The available software program YUREDUCE has a menu option D that will compute and display the product of the diagonal elements of a square matrix. The program MATCOMP has a menu option D to compute a determinant. Use YUREDUCE, or similar software, to compute the determinant of the matrices in Exercises 21–23. Write down your results. After this is done, use MATCOMP, or similar software, to compute the determinants of the same matrices again. Compare the answers.

21. $\begin{pmatrix} 11 & -9 & 28 \\ 32 & -24 & 21 \\ 10 & 13 & -19 \end{pmatrix}$

22. $\begin{pmatrix} 13 & -15 & 33 \\ -15 & 25 & 40 \\ 12 & -33 & 27 \end{pmatrix}$

23. $\begin{pmatrix} 7.6 & 2.8 & -3.9 & 19.3 & 25.0 \\ -33.2 & 11.4 & 13.2 & 22.4 & 18.3 \\ 21.4 & -32.1 & 45.7 & -8.9 & 12.5 \\ 17.4 & 11.0 & -6.8 & 20.0 & -35.1 \\ 22.7 & 11.9 & 33.2 & 2.5 & 7.8 \end{pmatrix}$

24. MATCOMP computes determinants in essentially the way described in this section. The matrix

$$A = \begin{pmatrix} 3 & 4 \\ 2 & 3 \end{pmatrix}$$

has determinant 1, so every power of it should have determinant 1. Use MATCOMP with single-precision printing and with the default roundoff control ratio r. Start computing determinants of powers of A. Find the smallest positive integer m such that $\det(A^m) \neq 1$, that is, according to MATCOMP. How bad is the error? What does MATCOMP give for $\det(A^{20})$? At what integer exponent does the break occur between the incorrect value 0 and incorrect values of large magnitude?

Repeat the above, taking zero for roundoff control ratio r. Try to explain why the results are different, and what is happening in each case.

4.4
Cramer's Rule

In this section we describe two classical applications of determinants. One gives a formula for the inverse A^{-1} of an invertible matrix A. The other gives a formula for the components of the solution vector of a square linear system $A\mathbf{x} = \mathbf{b}$ if A is invertible. Both of these formulas are computationally very inefficient compared with the methods we have developed to compute A^{-1} and the solution of $A\mathbf{x} = \mathbf{b}$. However, the structure of the formulas proves useful in theoretical considerations.

THE ADJOINT MATRIX

We begin by finding a formula in terms of determinants for the inverse of an invertible $n \times n$ matrix $A = (a_{ij})$. Recall the definition of the cofactor a'_{ij} from Eq. (2) of Section 4.2. Let $A_{i \to j}$ be the matrix obtained from A by replacing the jth row of A by the ith row. Then

$$\det(A_{i \to j}) = \begin{cases} \det(A) & \text{if } i = j, \\ 0 & \text{if } i \neq j. \end{cases}$$

If we expand $\det(A_{i \to j})$ by minors on the jth row, we have

$$\det(A_{i \to j}) = \sum_{s=1}^{n} a_{is} a'_{js},$$

and we obtain the important relation

$$\sum_{s=1}^{n} a_{is} a'_{js} = \begin{cases} \det(A) & \text{if } i = j, \\ 0 & \text{if } i \neq j. \end{cases} \tag{1}$$

The term on the left-hand side in Eq. (1) is the entry in the ith row and jth column in the product $A(A')^T$, where $A' = (a'_{ij})$ is the matrix whose entries are the cofactors of the entries of A. Thus Eq. (1) can be written in matrix form as

$$A(A')^T = [\det(A)]I,$$

where I is the $n \times n$ identity matrix. Similarly, replacing the jth column of A by the ith column, we have

$$\sum_{r=1}^{n} a'_{ri} a_{rj} = \begin{cases} \det(A) & \text{if } i = j, \\ 0 & \text{if } i \neq j. \end{cases} \tag{2}$$

The relation (2) yields $(A')^T A = [\det(A)]I$.

The matrix $(A')^T$ is called the **adjoint of** A and is denoted by $\mathrm{adj}(A)$. We have established an important relationship between a matrix and its adjoint.

Theorem 4.6 **Property of the Adjoint**

Let A be an $n \times n$ matrix. The adjoint $\operatorname{adj}(A) = (A')^T$ of A satisfies

$$[\operatorname{adj}(A)]A = A[\operatorname{adj}(A)] = [\det(A)]I,$$

where I is the $n \times n$ identity matrix.

Theorem 4.6 provides a formula for the inverse of an invertible matrix, which we present as a corollary.

Corollary *A Formula for the Inverse of an Invertible Matrix*

Let $A = (a_{ij})$ be an $n \times n$ matrix with $\det(A) \neq 0$. Then A is invertible, and

$$A^{-1} = \frac{1}{\det(A)} \operatorname{adj}(A),$$

where $\operatorname{adj}(A) = (a'_{ij})^T$ is the transposed matrix of cofactors.

Example 1 Find the inverse of

$$A = \begin{pmatrix} 4 & 0 & 1 \\ 2 & 2 & 0 \\ 3 & 1 & 1 \end{pmatrix}$$

if the matrix is invertible, using the corollary of Theorem 4.6.

Solution We easily find that $\det(A) = 4$, so that A is invertible. The cofactors a'_{ij} are

$$a'_{11} = (-1)^2 \begin{vmatrix} 2 & 0 \\ 1 & 1 \end{vmatrix} = 2, \qquad a'_{12} = (-1)^3 \begin{vmatrix} 2 & 0 \\ 3 & 1 \end{vmatrix} = -2,$$

$$a'_{13} = (-1)^4 \begin{vmatrix} 2 & 2 \\ 3 & 1 \end{vmatrix} = -4, \qquad a'_{21} = (-1)^3 \begin{vmatrix} 0 & 1 \\ 1 & 1 \end{vmatrix} = 1,$$

$$a'_{22} = (-1)^4 \begin{vmatrix} 4 & 1 \\ 3 & 1 \end{vmatrix} = 1, \qquad a'_{23} = (-1)^5 \begin{vmatrix} 4 & 0 \\ 3 & 1 \end{vmatrix} = -4,$$

$$a'_{31} = (-1)^4 \begin{vmatrix} 0 & 1 \\ 2 & 0 \end{vmatrix} = -2, \qquad a'_{32} = (-1)^5 \begin{vmatrix} 4 & 1 \\ 2 & 0 \end{vmatrix} = 2,$$

$$a'_{33} = (-1)^6 \begin{vmatrix} 4 & 0 \\ 2 & 2 \end{vmatrix} = 8.$$

Hence

$$A' = (a'_{ij}) = \begin{pmatrix} 2 & -2 & -4 \\ 1 & 1 & -4 \\ -2 & 2 & 8 \end{pmatrix}, \quad \text{so} \quad (A')^T = \begin{pmatrix} 2 & 1 & -2 \\ -2 & 1 & 2 \\ -4 & -4 & 8 \end{pmatrix}$$

and

$$A^{-1} = \frac{1}{\det(A)} (A')^T = \left(\frac{1}{4}\right) \begin{pmatrix} 2 & 1 & -2 \\ -2 & 1 & 2 \\ -4 & -4 & 8 \end{pmatrix} = \begin{pmatrix} \frac{1}{2} & \frac{1}{4} & -\frac{1}{2} \\ -\frac{1}{2} & \frac{1}{4} & \frac{1}{2} \\ -1 & -1 & 2 \end{pmatrix}. \quad \triangleleft$$

Obviously the method described in Section 1.5 for finding the inverse of an invertible matrix is more efficient than the method illustrated in the preceding example, especially if the matrix is large. The corollary is often used to find the inverse of a 2×2 matrix. We see at once that if $ad - bc \neq 0$, then

$$\begin{pmatrix} a & b \\ c & d \end{pmatrix}^{-1} = \frac{1}{ad - bc} \begin{pmatrix} d & -b \\ -c & a \end{pmatrix}.$$

CRAMER'S RULE

We turn to the problem of finding formulas in terms of determinants for the components in the solution vector of a square linear system $A\mathbf{x} = \mathbf{b}$, where A is an invertible matrix. We will prove this theorem.

Theorem 4.7   Cramer's Rule

Consider the system of equations $A\mathbf{x} = \mathbf{b}$, where $A = (a_{ij})$ is an $n \times n$ invertible matrix,

$$\mathbf{x} = \begin{pmatrix} x_1 \\ \vdots \\ x_n \end{pmatrix}, \quad \text{and} \quad \mathbf{b} = \begin{pmatrix} b_1 \\ \vdots \\ b_n \end{pmatrix}.$$

The system has a unique solution given by

$$x_k = \frac{\det(B_k)}{\det(A)} \quad \text{for } k = 1, \ldots, n, \tag{3}$$

where B_k is the matrix obtained from A by replacing the kth column of A by the column vector $\mathbf{b}$.

Proof   Since A is invertible, the unique solution of the system $A\mathbf{x} = \mathbf{b}$ can be expressed as

$$\mathbf{x} = A^{-1}\mathbf{b} = \frac{1}{\det(A)} (A')^T\mathbf{b}, \tag{4}$$

where $A' = (a'_{ij})$ is the matrix of cofactors. Comparing Eq. (4) with the desired result (3), we see that we need only show that the kth component of the column vector $(A')^T\mathbf{b}$ is given by the determinant of the matrix

$$B_k = \begin{pmatrix} a_{11} & \cdots & b_1 & \cdots & a_{1n} \\ a_{21} & \cdots & b_2 & \cdots & a_{2n} \\ \vdots & & \vdots & & \vdots \\ a_{n1} & \cdots & b_n & \cdots & a_{nn} \end{pmatrix}$$

obtained from A by replacing the kth column by $\mathbf{b}$, that is, replacing a_{jk} by b_j. Expanding $\det(B_k)$ by minors on the kth column, we have

$$\det(B_k) = \sum_{j=1}^{n} a'_{jk} b_j. \tag{5}$$

But this is indeed just the kth component of the column vector

$$(A')^T\mathbf{b} = \begin{pmatrix} a'_{11} & a'_{21} & \cdots & a'_{n1} \\ \vdots & \vdots & & \vdots \\ a'_{1k} & a'_{2k} & \cdots & a'_{nk} \\ \vdots & \vdots & & \vdots \\ a'_{1n} & a'_{2n} & \cdots & a'_{nn} \end{pmatrix} \begin{pmatrix} b_1 \\ b_2 \\ \vdots \\ b_n \end{pmatrix}.$$

This completes our proof. ∎

Example 2 Solve the linear system

$$5x_1 - 2x_2 + x_3 = 1$$
$$3x_1 + 2x_2 \quad\quad = 3$$
$$x_1 + \quad x_2 - x_3 = 0$$

using Cramer's rule.

CRAMER'S RULE appeared for the first time in full generality in a work of Gabriel Cramer (1704–1752) titled *Introduction to the Analysis of Algebraic Curves* (1750). The problem in which Cramer was interested was that of determining the equation of a plane curve of given degree passing through a certain number of given points. He states the theorem that a curve of nth degree is determined when $(\frac{1}{2})n(n + 3)$ points of the curve are known. For example, a second-degree curve, which he writes as

(*) $A + By + Cx + Dy^2 + Exy + x^2 = 0,$

is determined by five points. The question then is how to determine A, B, C, D, and E given the five points. The obvious method is to substitute into equation (*) the coordinates of each of the five points in turn. This gives us five equations for the five unknown coefficients. Cramer then refers to the appendix of the tract, in which he gives his general rule: "One finds the value of each unknown by forming n fractions of which the common denominator has as many terms as there are permutations of n things." He goes on to explain exactly how one calculates these terms as products of certain coefficients of the n equations, how one determines the appropriate sign for each term, and how one determines the n numerators of the fractions by replacing certain coefficients in this calculation by the constant terms of the system.

Cramer was a Swiss mathematician who taught at the Académie de Calvin in Geneva from 1724 until his death. He was an excellent scholar, but one whose work was overshadowed by such brilliant contemporaries as Euler and the Bernoullis.

Solution   Using the notation in Theorem 4.7, we find that

$$\det(A) = \begin{vmatrix} 5 & -2 & 1 \\ 3 & 2 & 0 \\ 1 & 1 & -1 \end{vmatrix} = -15, \qquad \det(B_1) = \begin{vmatrix} 1 & -2 & 1 \\ 3 & 2 & 0 \\ 0 & 1 & -1 \end{vmatrix} = -5,$$

$$\det(B_2) = \begin{vmatrix} 5 & 1 & 1 \\ 3 & 3 & 0 \\ 1 & 0 & -1 \end{vmatrix} = -15, \qquad \det(B_3) = \begin{vmatrix} 5 & -2 & 1 \\ 3 & 2 & 3 \\ 1 & 1 & 0 \end{vmatrix} = -20,$$

Hence

$$x_1 = (-5)/(-15) = 1/3,$$
$$x_2 = (-15)/(-15) = 1, \quad \text{and}$$
$$x_3 = (-20)/(-15) = 4/3. \quad \triangleleft$$

The most efficient way we have presented to compute a determinant is to row-reduce a matrix to triangular form. This is also the way we solve a square linear system. If A is a 10×10 invertible matrix, solving $A\mathbf{x} = \mathbf{b}$ using Cramer's rule involves row-reducing *eleven* 10×10 matrices $A, B_1, B_2, \ldots,$ B_{10} to triangular form. Solving the linear system by the method of Section 1.4 requires row-reducing just *one* 10×11 matrix so that the first 10 columns are in upper-triangular form. This illustrates the folly of using Cramer's rule to solve linear systems. The *structure* of the components of the solution vector, as given by the Cramer's rule formula $x_k = \det(B_k)/\det(A)$, is of interest in the study of advanced calculus, for example.

SUMMARY

1. Let A be an $n \times n$ matrix and let A' be its matrix of cofactors. The adjoint adj(A) is the matrix $(A')^T$ and satisfies $[\text{adj}(A)]A = A[\text{adj}(A)] = [\det(A)]I$, where I is the $n \times n$ identity matrix.

2. The inverse of an invertible matrix A is given by the explicit formula

$$A^{-1} = \frac{1}{\det(A)} \text{adj}(A).$$

3. If A is invertible, then a system of equations $A\mathbf{x} = \mathbf{b}$ has the unique solution $\mathbf{x}$ whose kth component is given explicitly by the formula

$$x_k = \frac{\det(B_k)}{\det(A)},$$

where the matrix B_k is obtained from A by replacing the kth column of A by $\mathbf{b}$.

4. The methods of Chapter 1 are far more efficient than those described above for actual computation of both the inverse of A and the solution of the system $A\mathbf{x} = \mathbf{b}$.

EXERCISES

In Exercises 1–6, use the corollary to Theorem 4.6 to find A^{-1} if A is invertible.

1. $A = \begin{pmatrix} 2 & 0 \\ 1 & -1 \end{pmatrix}$

2. $A = \begin{pmatrix} 4 & 1 \\ 2 & 1 \end{pmatrix}$

3. $A = \begin{pmatrix} 2 & 1 & 1 \\ 0 & 1 & 1 \\ -2 & 1 & 1 \end{pmatrix}$

4. $A = \begin{pmatrix} 3 & 0 & 4 \\ -2 & 1 & 1 \\ 3 & 1 & 2 \end{pmatrix}$

5. $A = \begin{pmatrix} 3 & 0 & 3 \\ 4 & 1 & -2 \\ -5 & 1 & 4 \end{pmatrix}$

6. $A = \begin{pmatrix} 2 & 1 & 3 \\ 0 & 1 & 4 \\ 1 & 2 & 1 \end{pmatrix}$

In Exercises 7–14, solve the given system of linear equations by Cramer's rule wherever it is possible.

7. $\begin{aligned} x_1 - 2x_2 &= 1 \\ 3x_1 + 4x_2 &= 3 \end{aligned}$

8. $\begin{aligned} 2x_1 - 3x_2 &= 1 \\ -4x_1 + 6x_2 &= -2 \end{aligned}$

9. $\begin{aligned} 3x_1 + x_2 &= 5 \\ 2x_1 + x_2 &= 0 \end{aligned}$

10. $\begin{aligned} x_1 + x_2 &= 1 \\ x_1 + 2x_2 &= 2 \end{aligned}$

11. $\begin{aligned} 5x_1 - 2x_2 + x_3 &= 1 \\ x_2 + x_3 &= 0 \\ x_1 + 6x_2 - x_3 &= 4 \end{aligned}$

12. $\begin{aligned} x_1 + 2x_2 - x_3 &= -2 \\ 2x_1 + x_2 + x_3 &= 0 \\ 3x_1 - x_2 + 5x_3 &= 1 \end{aligned}$

13. $\begin{aligned} x_1 - x_2 + x_3 &= 0 \\ x_1 + 2x_2 - x_3 &= 1 \\ x_1 - x_2 + 2x_3 &= 0 \end{aligned}$

14. $\begin{aligned} 3x_1 + 2x_2 - x_3 &= 1 \\ x_1 - 4x_2 + x_3 &= -2 \\ 5x_1 + 2x_2 &= 1 \end{aligned}$

In Exercises 15 and 16, find the component x_2 of the solution vector for the given linear system.

15. $\begin{aligned} x_1 + x_2 - 3x_3 + x_4 &= 1 \\ 2x_1 + x_2 + 2x_4 &= 0 \\ x_2 - 6x_3 - x_4 &= 5 \\ 3x_1 + x_2 + x_4 - 1 \end{aligned}$

16. $\begin{aligned} 6x_1 + x_2 - x_3 &= 4 \\ x_1 - x_2 + 5x_4 &= -2 \\ -x_1 + 3x_2 + x_3 &= 2 \\ x_1 + x_2 - x_3 + 2x_4 &= 0 \end{aligned}$

![icon] *In Exercises 17–19, use MATCOMP or similar software and the corollary of Theorem 4.6 to find the matrix of cofactors of the given matrix.*

17. $\begin{pmatrix} 1 & 2 & -3 \\ 2 & 3 & 0 \\ 3 & 1 & 4 \end{pmatrix}$

18. $\begin{pmatrix} -52 & 31 & 47 \\ 21 & -11 & 28 \\ 43 & -71 & 87 \end{pmatrix}$

19. $\begin{pmatrix} 6 & -3 & 2 & 14 \\ -3 & 7 & 8 & 1 \\ 4 & 9 & -5 & 3 \\ -8 & -40 & 47 & 29 \end{pmatrix}$

[*Hint:* Entries in the matrix of cofactors are integers. The cofactors of a matrix are continuous functions of its entries, that is, changing an entry a very slight amount will change a cofactor only slightly. Change some entry just a bit to make the determinant nonzero.]

4.5

Eigenvalues and Eigenvectors

ENCOUNTERS WITH $A^k\mathbf{x}$

In Section 1.7 we studied Markov chains dealing with the distribution of a population among states, measured over evenly spaced time intervals. An $n \times n$ transition matrix T describes the movement of the population among the states during one time interval. The matrix T has the property that all entries are nonnegative and the sum of the entries in any column is 1. Suppose $\mathbf{p}$ is the initial population distribution vector, that is, the column vector whose ith component is the proportion of the population in the ith state at the start of the process. Then $T\mathbf{p}$ is the corresponding population distribution vector after one time interval. Similarly, $T^2\mathbf{p}$ is the population distribution vector after two time intervals, and in general, $T^k\mathbf{p}$ gives the distribution of population among the states after k time intervals.

Markov chains provide one example in which we are interested in computing $A^k\mathbf{x}$ for an $n \times n$ matrix A and a column vector $\mathbf{x}$ of n components. We give a famous classical problem that provides another illustration.

Example 1   *(Fibonacci's rabbits).* Suppose that newly born pairs of rabbits produce no offspring during the first month of their lives, but each pair produces one new pair each subsequent month. Starting with $F_1 = 1$ newly born pair in the first month, find the number F_k of pairs in the kth month, assuming that no rabbit dies.

Solution   In the kth month, the number of pairs of rabbits is

$$F_k = \text{(Number of pairs alive the preceding month)}$$
$$+ \text{(Number of newly born pairs for the } k\text{th month).}$$

Since our rabbits do not produce offspring during the first month of their lives, we see that the number of newly born pairs for the kth month is the number F_{k-2} of pairs alive two months before. Thus we can write the equation above as

$$F_k = F_{k-1} + F_{k-2}. \qquad \textbf{Fibonacci's relation} \qquad (1)$$

It is convenient to set $F_0 = 0$, denoting 0 pairs for month 0 before the arrival of the first newly born pair, which is presumably a "gift." Thus the sequence

$$F_0, F_1, F_2, \ldots, F_k, \cdots$$

for the number of pairs of rabbits becomes the **Fibonacci sequence**

$$0, 1, 1, 2, 3, 5, 8, 13, 21, 34, \ldots, \qquad (2)$$

where each term starting with $F_2 = 0 + 1 = 1$ is the sum of the two preceding terms. For any particular k, we could compute F_k by just writing out the sequence far enough.   ◁

Fibonacci published this problem early in the thirteenth century. The Fibonacci sequence (2) occurs naturally in a surprising number of places. For example, leaves appear in a spiral pattern along a branch. Some trees have 5 growths of leaves for every 2 turns, others have 8 growths for every 3 turns, and still others have 13 growths for every 5 turns; note the appearance of these numbers in the sequence (2). A mathematical journal, the *Fibonacci Quarterly*, has published many papers dealing with the Fibonacci sequence.

We said in Example 1 that F_k can be found by simply writing out enough terms of the sequence (2). That can be a tedious task, even if we want to compute only F_{30}. Linear algebra gives us another approach to this problem. The Fibonacci relation (1) can be expressed in matrix form. We easily see that

$$\begin{pmatrix} F_k \\ F_{k-1} \end{pmatrix} = \begin{pmatrix} 1 & 1 \\ 1 & 0 \end{pmatrix} \begin{pmatrix} F_{k-1} \\ F_{k-2} \end{pmatrix}.$$

Thus if we set

$$\mathbf{x}_k = \begin{pmatrix} F_k \\ F_{k-1} \end{pmatrix} \quad \text{and} \quad A = \begin{pmatrix} 1 & 1 \\ 1 & 0 \end{pmatrix},$$

we find that

$$\mathbf{x}_k = A\mathbf{x}_{k-1}. \tag{3}$$

Applying (3) repeatedly, we see that

$$\mathbf{x}_2 = A\mathbf{x}_1, \qquad \mathbf{x}_3 = A\mathbf{x}_2 = A^2\mathbf{x}_1, \qquad \mathbf{x}_4 = A\mathbf{x}_3 = A^3\mathbf{x}_1,$$

and in general

$$\mathbf{x}_k = A^{k-1}\mathbf{x}_1 = \begin{pmatrix} 1 & 1 \\ 1 & 0 \end{pmatrix}^{k-1} \begin{pmatrix} 1 \\ 0 \end{pmatrix}. \tag{4}$$

Thus we can compute the kth Fibonacci number F_k by finding A^{k-1} and multiplying it on the right by the column vector $\mathbf{x}_1$. Raising a matrix to a power is also a bit of a job, but the program MATCOMP can easily find F_{30} for us (see Exercise 22).

Both Markov chains and the Fibonacci sequence lead us to computations of the form $A^k\mathbf{x}$. Other examples leading to $A^k\mathbf{x}$ abound in the physical and social sciences.

> Computations of $A^k\mathbf{x}$ arise in any process in which information given by a column vector gives rise to analogous information at a later time by multiplying the vector by a matrix A.

EIGENVALUES AND EIGENVECTORS

Suppose that A is an $n \times n$ matrix and $\mathbf{v}$ is a column vector with n components such that

$$A\mathbf{v} = \lambda\mathbf{v} \tag{5}$$

for some scalar λ. It is an easy exercise to show that $A^k\mathbf{v} = \lambda^k\mathbf{v}$ (see Exercise 17). Thus $A^k\mathbf{x}$ is easily computed if $\mathbf{x}$ is equal to *this vector* $\mathbf{v}$. For many matrices A, the computation of $A^k\mathbf{x}$ for a *general vector* $\mathbf{x}$ is greatly simplified by finding first all nonzero vectors $\mathbf{v}$ and scalars λ satisfying Eq. (5). In Section 4.6 we will illustrate how this works.

Geometrically, Eq. (5) asserts that $A\mathbf{v}$ is a vector parallel to $\mathbf{v}$. We turn our attention to finding such vectors $\mathbf{v}$ and scalars λ.

Definition 4.2   Eigenvalues and Eigenvectors

Let A be an $n \times n$ matrix. A scalar λ is an **eigenvalue** of A if there is a *nonzero* column vector $\mathbf{v}$ in $\mathbb{R}^n$ such that $A\mathbf{v} = \lambda\mathbf{v}$. The vector $\mathbf{v}$ is then an **eigenvector** of A corresponding to λ. (The terms **characteristic vector** and **characteristic value** or **proper vector** and **proper value** are also used in place of eigenvector and eigenvalue, respectively.)

In this section we show how a determinant can be used to find eigenvalues and eigenvectors; the technique is only practical for relatively small matrices. Some further computational techniques for finding eigenvalues are described in Section 7.4.

We write the matrix equation $A\mathbf{v} = \lambda\mathbf{v}$ as $A\mathbf{v} - \lambda\mathbf{v} = \mathbf{0}$, or as $A\mathbf{v} - \lambda I\mathbf{v} = \mathbf{0}$, where I is the $n \times n$ identity matrix. This last equation can be written as $(A - \lambda I)\mathbf{v} = \mathbf{0}$, so $\mathbf{v}$ must be a solution of the homogeneous linear system

$$(A - \lambda I)\mathbf{x} = \mathbf{0}. \tag{6}$$

An eigenvalue of A is thus a value λ for which the system (6) has a *nontrivial* solution $\mathbf{v}$. (Recall that an eigenvector is *nonzero* by definition.) We know that

---

THE FIRST APPEARANCE OF EIGENVALUES occurs in connection with their use in solving differential equations (see Section 8.4). In 1743 Leonhard Euler first introduced the standard method of solving an nth-order differential equation with constant coefficients

$$y^{(n)} + a_{n-1}y^{(n-1)} + \cdots + a_1y' + a_0y = 0$$

by using functions of the form $y = e^{\lambda t}$, where λ is a root of the characteristic equation

$$z^n + a_{n-1}z^{n-1} + \cdots + a_1z + a_0 = 0.$$

This is the same equation one gets by making the substitutions $y_1 = y, y_2 = y', y_3 = y'', \ldots, y_n = y^{(n-1)}$, replacing the single nth-order equation by a system of n first-order equations

$$y_1' = \quad y_2$$
$$y_2' = \qquad y_3$$
$$\cdots$$
$$y_n' = -a_0y_1 - a_1y_2 - \cdots - a_{n-1}y_n$$

and calculating the characteristic equation of the matrix of coefficients of this system.

About 20 years later, Lagrange gave a more explicit version of this same idea when he found the solution of a system of differential equations by finding the roots of what amounted to the characteristic equation of the matrix of coefficients. The particular system of differential equations came from examining the "infinitesimal movements" of a mechanical system in the neighborhood of its position of equilibrium. Lagrange solved a similar problem in celestial mechanics by using the same technique in 1774.

the system (6) has a nontrivial solution precisely when the determinant of the coefficient matrix is zero, that is, if and only if

$$\det(A - \lambda I) = 0. \tag{7}$$

If $A = (a_{ij})$, then the previous equation can be written

$$\begin{vmatrix} a_{11} - \lambda & a_{12} & \cdots & a_{1n} \\ a_{21} & a_{22} - \lambda & \cdots & a_{2n} \\ \vdots & \vdots & & \vdots \\ a_{n1} & a_{n2} & \cdots & a_{nn} - \lambda \end{vmatrix} = 0. \tag{8}$$

If we expand the determinant in (8), we obtain a polynomial expression $p(\lambda)$ of degree n with coefficients involving the a_{ij}. That is,

$$\det(A - \lambda I) = p(\lambda).$$

The polynomial $p(\lambda)$ is the **characteristic polynomial** of the matrix A. The eigenvalues of A are precisely the solutions of the **characteristic equation** $p(\lambda) = 0$.

Example 2   Find the eigenvalues of the matrix

$$A = \begin{pmatrix} 3 & 2 \\ 2 & 0 \end{pmatrix}.$$

Solution   The characteristic polynomial of A is

$$\det(A - \lambda I) = \begin{vmatrix} 3 - \lambda & 2 \\ 2 & -\lambda \end{vmatrix} = \lambda^2 - 3\lambda - 4.$$

The characteristic equation is

$$\lambda^2 - 3\lambda - 4 = 0$$

and we obtain $(\lambda - 4)(\lambda + 1) = 0$, so $\lambda_1 = -1$ and $\lambda_2 = 4$ are eigenvalues of A.   ◁

Example 3   Show that $\lambda_1 = 1$ is an eigenvalue of every transition matrix for a Markov chain.

Solution   Let T be an $n \times n$ transition matrix for a Markov chain, that is, all entries in T are nonnegative and the sum of the entries in each column of T is 1. We easily see that the sum of the entries in any column of $T - I$ must be zero. Thus the sum of the row vectors in T is the zero vector, so the rows of $T - I$ are linearly dependent. Consequently, the rank of $T - I$ is less than n, and the equation $(T - I)\mathbf{x} = 0$ has a nontrivial solution, so $\lambda_1 = 1$ is an eigenvalue of T.   ◁

The characteristic equation of an $n \times n$ matrix is a polynomial equation of degree n. This equation has n solutions if we allow both real and complex numbers and count the possible multiplicities greater than 1 of some solutions. Linear algebra can be done using complex scalars as well as real scalars. Introducing complex scalars makes the theory simpler and illuminates behavior in the real-scalar case. However, pencil-and-paper computations involving complex numbers can be very laborious, and it is customary not to deal with them in an introductory text such as this one. Perhaps this will change in a few years if programming languages for personal computers also include routines for complex-number arithmetic as a matter of course. We simply state here that complex eigenvalues are important, and are of interest in many applications. Exercises 25 and 30–33 deal with some computations involving complex eigenvalues.

It is possible to have eigenvalues that are multiple roots of the characteristic equation. Our next example illustrates this.

Example 4   Find the eigenvalues of the matrix

$$A = \begin{pmatrix} 2 & 1 & 0 \\ -1 & 0 & 1 \\ 1 & 3 & 1 \end{pmatrix}.$$

Solution   The characteristic polynomial is

$$p(\lambda) = \begin{vmatrix} 2 - \lambda & 1 & 0 \\ -1 & -\lambda & 1 \\ 1 & 3 & 1 - \lambda \end{vmatrix} = (2 - \lambda) \begin{vmatrix} -\lambda & 1 \\ 3 & 1 - \lambda \end{vmatrix} - 1 \begin{vmatrix} -1 & 1 \\ 1 & 1 - \lambda \end{vmatrix}$$

$$= (2 - \lambda)(\lambda^2 - \lambda - 3) - (\lambda - 2) = -(\lambda - 2)(\lambda^2 - \lambda - 2)$$

$$= -(\lambda - 2)(\lambda - 2)(\lambda + 1).$$

Hence the eigenvalues of A are $\lambda_1 = -1$ and $\lambda_2 = \lambda_3 = 2$.   ◁

We turn to the computation of the eigenvectors corresponding to an eigenvalue λ of a matrix A. Having found the eigenvalue, we substitute it in the homogeneous system (6) and solve to find the nontrivial solutions of the system. We will obtain an infinite number of nontrivial solutions, each of which is an eigenvector corresponding to the eigenvalue λ.

Example 5   Find the eigenvectors corresponding to each eigenvalue found in Example 4 for the matrix

$$A = \begin{pmatrix} 2 & 1 & 0 \\ -1 & 0 & 1 \\ 1 & 3 & 1 \end{pmatrix}.$$

Solution   The eigenvalues of A were found to be $\lambda_1 = -1$ and $\lambda_2 = \lambda_3 = 2$. We substitute each of these values in the corresponding homogeneous system (6). The eigenvectors are obtained by reducing the coefficient matrix $A - \lambda I$ in the augmented matrix for the system. For $\lambda_1 = -1$, we obtain

$$(A - \lambda_1 I \mid \mathbf{0}) = (A + I \mid \mathbf{0}) = \begin{pmatrix} 3 & 1 & 0 & \mid & 0 \\ -1 & 1 & 1 & \mid & 0 \\ 1 & 3 & 2 & \mid & 0 \end{pmatrix}$$

$$\sim \begin{pmatrix} 1 & 3 & 2 & \mid & 0 \\ 0 & 4 & 3 & \mid & 0 \\ 0 & -8 & -6 & \mid & 0 \end{pmatrix} \sim \begin{pmatrix} 1 & 3 & 2 & \mid & 0 \\ 0 & 4 & 3 & \mid & 0 \\ 0 & 0 & 0 & \mid & 0 \end{pmatrix}.$$

The solution of the homogeneous system is given by

$$\begin{pmatrix} r/4 \\ -3r/4 \\ r \end{pmatrix} \quad \text{for any scalar } r.$$

Therefore,

$$\mathbf{v}_1 = \begin{pmatrix} r/4 \\ -3r/4 \\ r \end{pmatrix} \quad \text{for any } \textit{nonzero} \text{ scalar } r$$

is an eigenvector corresponding to the eigenvalue $\lambda_1 = -1$. Replacing r by $4r$, we can express this result without fractions as

$$\mathbf{v}_1 = \begin{pmatrix} r \\ -3r \\ 4r \end{pmatrix} \quad \text{for any nonzero scalar } r.$$

For $\lambda_2 = 2$, we obtain

$$(A - \lambda_2 I \mid \mathbf{0}) = (A - 2I \mid \mathbf{0}) = \begin{pmatrix} 0 & 1 & 0 & \mid & 0 \\ -1 & -2 & 1 & \mid & 0 \\ 1 & 3 & -1 & \mid & 0 \end{pmatrix}$$

$$\sim \begin{pmatrix} 1 & 3 & -1 & \mid & 0 \\ 0 & 1 & 0 & \mid & 0 \\ 0 & 0 & 0 & \mid & 0 \end{pmatrix}.$$

This time we find that

$$\mathbf{v}_2 = \begin{pmatrix} r \\ 0 \\ r \end{pmatrix} \quad \text{for any nonzero scalar } r$$

is an eigenvector. As a check, we could compute $A\mathbf{v}_1$ and $A\mathbf{v}_2$. For example, we have

$$A\mathbf{v}_2 = \begin{pmatrix} 2 & 1 & 0 \\ -1 & 0 & 1 \\ 1 & 3 & 1 \end{pmatrix} \begin{pmatrix} r \\ 0 \\ r \end{pmatrix} = \begin{pmatrix} 2r \\ 0 \\ 2r \end{pmatrix} = 2 \begin{pmatrix} r \\ 0 \\ r \end{pmatrix} = 2\mathbf{v}_2 = \lambda_2 \mathbf{v}_2. \quad \triangleleft$$

Example 6 Find the eigenvalues and eigenvectors of the matrix

$$A = \begin{pmatrix} 1 & 0 & 0 \\ -8 & 4 & -6 \\ 8 & 1 & 9 \end{pmatrix}.$$

Solution The characteristic polynomial of A is

$$p(\lambda) = |A - \lambda I| = \begin{vmatrix} 1 - \lambda & 0 & 0 \\ -8 & 4 - \lambda & -6 \\ 8 & 1 & 9 - \lambda \end{vmatrix} = (1 - \lambda) \begin{vmatrix} 4 - \lambda & -6 \\ 1 & 9 - \lambda \end{vmatrix}$$

$$= (1 - \lambda)(\lambda^2 - 13\lambda + 42) = (1 - \lambda)(\lambda - 6)(\lambda - 7).$$

The eigenvalues of A are $\lambda_1 = 1$, $\lambda_2 = 6$, and $\lambda_3 = 7$. We will drop the augmentation of $A - \lambda I$ by the column of zeros as we compute the eigenvectors for each of these eigenvalues. For $\lambda_1 = 1$, we have

$$A - \lambda_1 I = A - I = \begin{pmatrix} 0 & 0 & 0 \\ -8 & 3 & -6 \\ 8 & 1 & 8 \end{pmatrix} \sim \begin{pmatrix} 0 & 0 & 0 \\ 0 & 4 & 2 \\ 8 & 1 & 8 \end{pmatrix} \sim \begin{pmatrix} 8 & 1 & 8 \\ 0 & 2 & 1 \\ 0 & 0 & 0 \end{pmatrix}$$

so

$$\mathbf{v}_1 = \begin{pmatrix} -15r/16 \\ -r/2 \\ r \end{pmatrix} \quad \text{for any nonzero scalar } r$$

is an eigenvector. Replacing r by $-16r$, we can express this result without the fractions as

$$\mathbf{v}_1 = \begin{pmatrix} 15r \\ 8r \\ -16r \end{pmatrix} \quad \text{for any nonzero scalar } r.$$

For $\lambda_2 = 6$, we have

$$A - \lambda_2 I = A - 6I = \begin{pmatrix} -5 & 0 & 0 \\ -8 & -2 & -6 \\ 8 & 1 & 3 \end{pmatrix} \sim \begin{pmatrix} 1 & 0 & 0 \\ 0 & -2 & -6 \\ 0 & 1 & 3 \end{pmatrix} \sim \begin{pmatrix} 1 & 0 & 0 \\ 0 & 1 & 3 \\ 0 & 0 & 0 \end{pmatrix}$$

so

$$\mathbf{v}_2 = \begin{pmatrix} 0 \\ -3r \\ r \end{pmatrix} \quad \text{for any nonzero scalar } r$$

is an eigenvector. Finally, for $\lambda_3 = 7$, we have

$$A - \lambda_3 I = A - 7I = \begin{pmatrix} -6 & 0 & 0 \\ -8 & -3 & -6 \\ 8 & 1 & 2 \end{pmatrix} \sim \begin{pmatrix} 1 & 0 & 0 \\ 0 & -3 & -6 \\ 0 & 1 & 2 \end{pmatrix} \sim \begin{pmatrix} 1 & 0 & 0 \\ 0 & 1 & 2 \\ 0 & 0 & 0 \end{pmatrix}$$

so

$$\mathbf{v}_3 = \begin{pmatrix} 0 \\ -2r \\ r \end{pmatrix} \quad \text{for any nonzero scalar } r$$

is an eigenvector.   ◁

PROPERTIES OF EIGENVALUES AND EIGENVECTORS

We turn now to algebraic properties of eigenvalues and eigenvectors. The properties given in Theorem 4.8 are so easy to prove that we leave the proofs to Exercises 17–19.

Theorem 4.8   Properties of Eigenvalues and Eigenvectors

Let A be an $n \times n$ matrix.

1. If λ is an eigenvalue of A with $\mathbf{v}$ as a corresponding eigenvector, then λ^k is an eigenvalue of A^k, again with $\mathbf{v}$ as a corresponding eigenvector, for any positive integer k.

2. If λ is an eigenvalue of an invertible matrix A with $\mathbf{v}$ as a corresponding eigenvector, then $1/\lambda$ is an eigenvalue of A^{-1}, again with $\mathbf{v}$ as a corresponding eigenvector.

3. If λ is an eigenvalue of A, then the set E_λ consisting of the zero vector together with all eigenvectors of A for this eigenvalue λ is a subspace of $\mathbb{R}^n$, the **eigenspace** of λ.

In the next section, we will make important use of the fact that eigenvectors corresponding to *distinct* eigenvalues are independent. We state this as a theorem, and supply the proof.

Theorem 4.9   Independence of Eigenvectors

Let A be an $n \times n$ matrix. If $\mathbf{v}_1, \mathbf{v}_2, \ldots, \mathbf{v}_m$ are eigenvectors of A corresponding to *distinct* eigenvalues $\lambda_1, \lambda_2, \ldots, \lambda_m$, respectively, then the set $\{\mathbf{v}_1, \mathbf{v}_2, \ldots, \mathbf{v}_m\}$ is linearly independent.

Proof   Suppose the conclusion is false so the eigenvectors $\mathbf{v}_1, \mathbf{v}_2, \ldots, \mathbf{v}_m$ are linearly dependent. Then one of them is a linear combination of its predecessors. Let $\mathbf{v}_k$ be the *first* such vector, so that

$$\mathbf{v}_k = d_1\mathbf{v}_1 + d_2\mathbf{v}_2 + \cdots + d_{k-1}\mathbf{v}_{k-1} \tag{9}$$

and $\{\mathbf{v}_1, \mathbf{v}_2, \ldots, \mathbf{v}_{k-1}\}$ is independent. Multiplying Eq. (9) by λ_k, we obtain

$$\lambda_k \mathbf{v}_k = d_1 \lambda_k \mathbf{v}_1 + d_2 \lambda_k \mathbf{v}_2 + \cdots + d_{k-1} \lambda_k \mathbf{v}_{k-1}. \tag{10}$$

On the other hand, multiplying both sides of Eq. (9) on the left by the matrix A yields

$$\lambda_k \mathbf{v}_k = d_1 \lambda_1 \mathbf{v}_1 + d_2 \lambda_2 \mathbf{v}_2 + \cdots + d_{k-1} \lambda_{k-1} \mathbf{v}_{k-1} \tag{11}$$

since $A\mathbf{v}_i = \lambda_i \mathbf{v}_i$. Subtracting Eq. (11) from Eq. (10), we see that

$$0 = d_1(\lambda_k - \lambda_1)\mathbf{v}_1 + d_2(\lambda_k - \lambda_2)\mathbf{v}_2 + \cdots + d_{k-1}(\lambda_k - \lambda_{k-1})\mathbf{v}_{k-1}.$$

This last equation is a dependence relation since not all the coefficients are zero. (Not all d_i are zero because of Eq. (9) and the λ_i are distinct.) But this contradicts the independence of the set $\{\mathbf{v}_1, \mathbf{v}_2, \ldots, \mathbf{v}_{k-1}\}$. We conclude that $\{\mathbf{v}_1, \mathbf{v}_2, \ldots, \mathbf{v}_m\}$ is independent. ■

Theorem 4.9 tells us that if we take $r = 1$ in Example 6 and form the eigenvectors

$$\mathbf{v}_1 = \begin{pmatrix} 15 \\ 8 \\ -16 \end{pmatrix}, \quad \mathbf{v}_2 = \begin{pmatrix} 0 \\ -3 \\ 1 \end{pmatrix}, \quad \text{and} \quad \mathbf{v}_3 = \begin{pmatrix} 0 \\ -2 \\ 1 \end{pmatrix},$$

which correspond to the distinct eigenvalues 1, 6, and 7, respectively, then these eigenvectors must be independent and consequently must form a basis for $\mathbb{R}^3$. In the following section we will elaborate on cases like this where an $n \times n$ matrix A has eigenvectors that form a basis for $\mathbb{R}^n$.

SUMMARY

Let A be an $n \times n$ matrix.

1. If $A\mathbf{v} = \lambda \mathbf{v}$, where $\mathbf{v}$ is a nonzero column vector and λ is a scalar, then λ is an eigenvalue of A and $\mathbf{v}$ is an eigenvector of A corresponding to λ.

2. The characteristic polynomial $p(\lambda)$ of A is obtained by expanding the determinant $|A - \lambda I|$, where I is the $n \times n$ identity matrix.

3. The eigenvalues λ of A can be found by solving the characteristic equation $p(\lambda) = |A - \lambda I| = 0$. There are at most n real solutions λ of this equation.

4. The eigenvectors of A corresponding to λ are found by solving the homogeneous system $(A - \lambda I)\mathbf{x} = \mathbf{0}$ for the nontrivial solutions, as illustrated in Examples 5 and 6.

5. Let k be a positive integer. If λ is an eigenvalue of A having $\mathbf{v}$ as eigenvector, then λ^k is an eigenvalue of A^k with $\mathbf{v}$ as eigenvector. If A is invertible, this statement is also true for $k = -1$.

6. Let λ be an eigenvalue of an $n \times n$ matrix A. The set E_λ in $\mathbb{R}^n$ consisting of the zero vector and all eigenvectors for λ is a subspace of $\mathbb{R}^n$.

> **7.** Let $\lambda_1, \lambda_2, \ldots, \lambda_m$ be distinct eigenvalues of a matrix A and let $\mathbf{v}_i$ be an eigenvector corresponding to λ_i for $i = 1, 2, \ldots, m$. Then the set $\{\mathbf{v}_1, \mathbf{v}_2, \ldots, \mathbf{v}_m\}$ is a linearly independent set of vectors.

EXERCISES

In Exercises 1–16, find the characteristic polynomial, the real eigenvalues, and the corresponding eigenvectors of the given matrix.

1. $\begin{pmatrix} 1 & 0 \\ 1 & 2 \end{pmatrix}$
2. $\begin{pmatrix} 7 & 5 \\ -10 & -8 \end{pmatrix}$
3. $\begin{pmatrix} -1 & -2 \\ 4 & 5 \end{pmatrix}$
4. $\begin{pmatrix} -7 & -5 \\ 16 & 17 \end{pmatrix}$

5. $\begin{pmatrix} 0 & -1 \\ 1 & 0 \end{pmatrix}$
6. $\begin{pmatrix} 1 & -2 \\ 1 & 2 \end{pmatrix}$
7. $\begin{pmatrix} 2 & 0 & 0 \\ 1 & -1 & -2 \\ -1 & 0 & 1 \end{pmatrix}$
8. $\begin{pmatrix} -1 & 0 & 0 \\ -4 & 2 & -1 \\ 4 & 0 & 3 \end{pmatrix}$

9. $\begin{pmatrix} 8 & 0 & 0 \\ 7 & -1 & -2 \\ -7 & 0 & 1 \end{pmatrix}$
10. $\begin{pmatrix} 1 & 0 & 0 \\ -8 & 4 & -5 \\ 8 & 0 & 9 \end{pmatrix}$
11. $\begin{pmatrix} -2 & 0 & 0 \\ -5 & -2 & -5 \\ 5 & 0 & 3 \end{pmatrix}$
12. $\begin{pmatrix} -4 & 0 & 0 \\ -7 & 2 & -1 \\ 7 & 0 & 3 \end{pmatrix}$

13. $\begin{pmatrix} -1 & 0 & 1 \\ -7 & 2 & 5 \\ 3 & 0 & 1 \end{pmatrix}$
14. $\begin{pmatrix} 4 & 0 & 0 \\ 8 & 4 & 8 \\ 0 & 0 & 4 \end{pmatrix}$
15. $\begin{pmatrix} 0 & 0 & 1 \\ -2 & -2 & 1 \\ 2 & 0 & -1 \end{pmatrix}$
16. $\begin{pmatrix} 2 & 0 & 1 \\ 6 & 4 & -3 \\ 2 & 0 & 3 \end{pmatrix}$

17. Prove part (1) of Theorem 4.8.

18. Prove part (2) of Theorem 4.8.

19. Prove part (3) of Theorem 4.8.

20. Show that a square matrix is invertible if and only if no eigenvalue is zero.

21. Find the eigenvalues and eigenvectors of the matrix

$$A = \begin{pmatrix} 1 & 1 \\ 1 & 0 \end{pmatrix}$$

used to generate the Fibonacci sequence (2).

22. ▣ Use MATCOMP, or similar software, and relation (4) to find the terms given below of the Fibonacci sequence (2) as accurately as possible. (Use double-precision printing if possible.)
a) F_8 (Note that $F_8 = 21$; this part is to check procedure.)
b) F_{30}
c) F_{50}
d) F_{77}
e) F_{150}

23. The first two terms of a sequence are $a_0 = 0$ and $a_1 = 1$. Subsequent terms are generated using the relation

$$a_k = 2a_{k-1} + a_{k-2} \quad \text{for } k \geq 2.$$

a) Write the terms of the sequence through a_8.

b) Find a matrix that can be used to generate the sequence, as the matrix A in Exercise 21 can be used to generate the Fibonacci sequence (2).

c) [icon] Use MATCOMP or similar software to find a_{30}.

24. Repeat Exercise 23 for a sequence where $a_0 = 0$, $a_1 = 1$, $a_2 = 2$, and $a_k = 2a_{k-1} - 3a_{k-2} + a_{k-3}$ for $k \geq 3$.

25. Let

$$A = \begin{pmatrix} 0 & -1 \\ 1 & 0 \end{pmatrix}.$$

It can be shown that if $\mathbf{x}$ is any column vector in $\mathbb{R}^2$, then $A\mathbf{x}$ can be obtained geometrically from $\mathbf{x}$ by rotating $\mathbf{x}$ counterclockwise through an angle of 90°. For example, we find that $A\mathbf{e}_1 = \mathbf{e}_2$ and $A\mathbf{e}_2 = -\mathbf{e}_1$.

a) Argue geometrically that A has no real eigenvalues.
b) Find the complex eigenvalues and eigenvectors of A.
c) Argue geometrically that A^2 should have real eigenvalues of _____ . (Fill in the blank.)
d) Use part (b) and Theorem 4.8 to find the eigenvalues of A^2 and compare with the answer obtained to part (c). What are the real eigenvectors of A^2? the complex eigenvectors of A^2?
e) Find eigenvalues and eigenvectors for A^3.
f) Find eigenvalues and eigenvectors for A^4.

[icon] *In Exercises 26–29, use the program MATCOMP, or similar software, to find the real eigenvalues and corresponding eigenvectors of the given matrix.*

26. $\begin{pmatrix} 7 & 10 & 6 \\ 2 & -1 & -6 \\ -2 & -5 & 0 \end{pmatrix}$

27. $\begin{pmatrix} -1 & 0 & 0 & 0 \\ 3 & 2 & 0 & 0 \\ -3 & 0 & 2 & 0 \\ -3 & 1 & 0 & 3 \end{pmatrix}$

28. $\begin{pmatrix} 0 & 0 & -2 & 2 \\ -3 & 4 & -3 & 3 \\ -4 & 0 & 2 & 4 \\ -2 & 0 & 2 & 4 \end{pmatrix}$

29. $\begin{pmatrix} 4 & 0 & 0 & 0 \\ -6 & 16 & -6 & 6 \\ -16 & 0 & 20 & 16 \\ -16 & 0 & 16 & 20 \end{pmatrix}$

[icon] *The program ALLROOTS can be used to find both real and complex solutions of a polynomial equation. The program uses Newton's method, illustrating that the technique of linear approximation described in Section 1.1 is also useful with complex numbers. The program is designed so that the user can watch approximations approach a solution. Of course, a program designed for research would simply spit out the answers.*

[icon] *In Exercises 30–33, first use MATCOMP to find the characteristic equation of the given matrix. Copy down the equation, and then use ALLROOTS to find both the real and complex eigenvalues of the matrix.*

30. $\begin{pmatrix} -1 & 4 & 6 \\ 2 & 7 & 9 \\ -3 & 11 & 13 \end{pmatrix}$

31. $\begin{pmatrix} 10 & -13 & 8 \\ 3 & -20 & 5 \\ -11 & 7 & -6 \end{pmatrix}$

32. $\begin{pmatrix} -7 & 11 & -7 & 10 \\ 5 & 8 & -13 & 3 \\ -15 & 8 & -9 & 2 \\ 3 & -4 & 20 & -6 \end{pmatrix}$

33. $\begin{pmatrix} 21 & -8 & 0 & 32 \\ -14 & 17 & -6 & 9 \\ 15 & 11 & -13 & 16 \\ -18 & 30 & 43 & 31 \end{pmatrix}$

34. The Cayley–Hamilton theorem states that every square matrix A satisfies its characteristic equation. That is, if the characteristic equation is $p_n\lambda^n + p_{n-1}\lambda^{n-1} + \cdots + p_1\lambda + p_0 = 0$, then $p_nA^n + p_{n-1}A^{n-1} + \cdots + p_1A + p_0I = O$, the zero matrix.

a) Illustrate the Cayley–Hamilton theorem with pencil and paper for the matrix

$$\begin{pmatrix} 2 & -1 \\ 1 & 3 \end{pmatrix}.$$

b) ▣ Use MATCOMP or similar software to illustrate the Cayley–Hamilton theorem for the matrix

$$\begin{pmatrix} -2 & 4 & 6 & -1 \\ 5 & -8 & 3 & 2 \\ 11 & -3 & 7 & 1 \\ 0 & -5 & 9 & 10 \end{pmatrix}.$$

35. Let A be an $n \times n$ matrix and let I be the $n \times n$ identity matrix. Compare the eigenvectors and eigenvalues of A with those of $A + rI$ for a scalar r.

36. Let a square matrix A with real eigenvalues have a unique eigenvalue of greatest magnitude. Accurate numerical computation of this eigenvalue by the *power method* discussed in Section 7.4 can be difficult if there is another eigenvalue of almost equal magnitude.

a) Suppose we know that a 4×4 matrix A has eigenvalues of approximately 20, 2, -3, and -19.5. Using Exercise 35, how might we modify A to find accurately the eigenvalue of maximum magnitude by the power method?

b) Repeat part (a) if the eigenvalues are known to be approximately 19.5, 2, -3, and -20.

4.6
Diagonalization and Applications

DIAGONALIZING A MATRIX

A square matrix is called **diagonal** if all entries not on the main diagonal are zero. In the preceding section we indicated the importance of being able to compute $A^k\mathbf{x}$ for an $n \times n$ matrix A and a column vector $\mathbf{x}$ in $\mathbb{R}^n$. We show that if A has n distinct eigenvalues, then A^k in this computation of $A^k\mathbf{x}$ can be essentially replaced by D^k, where D is a diagonal matrix with the eigenvalues of A as diagonal entries. It is easy to see that D^k is the diagonal matrix obtained from D by raising each diagonal entry to the power k.

Theorem 4.10 **Diagonalization**

Let A be an $n \times n$ matrix having n distinct eigenvalues $\lambda_1, \lambda_2, \ldots, \lambda_n$. Let $\mathbf{v}_1, \mathbf{v}_2, \ldots, \mathbf{v}_n$ be column vectors in $\mathbb{R}^n$ such that $\mathbf{v}_i$ is an eigenvector corresponding to λ_i for $i = 1, 2, \ldots, n$. Let C be the $n \times n$ matrix having $\mathbf{v}_i$ as ith column vector. Then C is invertible and $C^{-1}AC$ is equal to the diagonal matrix

$$
D = \begin{pmatrix} \lambda_1 & & & \\ & \lambda_2 & & \\ & & \ddots & \\ & & & \lambda_n \end{pmatrix}.
$$

Proof   Since the eigenvalues λ_i are distinct, the n vectors $\mathbf{v}_i$ form a linearly independent set by Theorem 4.9. Thus the matrix C has rank n and is invertible.

We show that $CD = AC$, from which $D = C^{-1}AC$ follows at once. We have

$$
CD = \begin{pmatrix} | & | & & | \\ \mathbf{v}_1 & \mathbf{v}_2 & \cdots & \mathbf{v}_n \\ | & | & & | \end{pmatrix} \begin{pmatrix} \lambda_1 & & & \\ & \lambda_2 & & \\ & & \ddots & \\ & & & \lambda_n \end{pmatrix}
$$

$$
= \begin{pmatrix} | & | & & | \\ \lambda_1\mathbf{v}_1 & \lambda_2\mathbf{v}_2 & \cdots & \lambda_n\mathbf{v}_n \\ | & | & & | \end{pmatrix}.
$$

(1)

On the other hand, the ith column vector of the product AC is the product of A and the ith column vector $\mathbf{v}_i$ of C. But $A\mathbf{v}_i = \lambda_i\mathbf{v}_i$ since $\mathbf{v}_i$ is an eigenvector for λ_i. Therefore, AC is also the matrix (1), so $AC = CD$ and $C^{-1}AC = D$. ∎

We now show how the computation of A^k can be essentially reduced to the computation of D^k.

Corollary   *Computation of A^k*

Let A, λ_i, $\mathbf{v}_i$, and C be as in Theorem 4.10. Then $A^k = CD^kC^{-1}$ for each positive integer k.

Proof   From $C^{-1}AC = D$, we obtain $A = CDC^{-1}$. Thus

$$
A^k = \underbrace{(CDC^{-1})(CDC^{-1})(CDC^{-1}) \cdots (CDC^{-1})}_{k \text{ factors}}.
$$

(2)

Clearly the adjacent terms $C^{-1}C$ cancel in Eq. (2) to give $A^k = CD^kC^{-1}$. ∎

Let A be an $n \times n$ matrix. If C is an invertible matrix such that $C^{-1}AC = D$, a diagonal matrix, then we say that A has been **diagonalized by** C. The matrices A and D are an example of *similar matrices*, a term that we now define.

Definition 4.3 Similar Matrices

An $n \times n$ matrix P is **similar** to an $n \times n$ matrix Q if there exists an invertible $n \times n$ matrix C such that $C^{-1}PC = Q$.

The relationship "P is similar to Q" satisfies the properties required for an *equivalence relation* (see Exercise 13). In particular, if P is similar to Q, then Q is also similar to P. This means that similarity need not be stated in a directional way; we can simply say that P and Q are similar matrices.

Example 1   Find a diagonal matrix similar to the matrix

$$A = \begin{pmatrix} 1 & 0 & 0 \\ -8 & 4 & -6 \\ 8 & 1 & 9 \end{pmatrix}$$

of Example 6 in Section 4.5.

Solution   We saw in Example 6 of Section 4.5 that eigenvalues and corresponding eigenvectors of A are given by

$$\lambda_1 = 1, \qquad \lambda_2 = 6, \qquad \lambda_3 = 7,$$

$$\mathbf{v}_1 = \begin{pmatrix} 15 \\ 8 \\ -16 \end{pmatrix}, \qquad \mathbf{v}_2 = \begin{pmatrix} 0 \\ -3 \\ 1 \end{pmatrix}, \qquad \mathbf{v}_3 = \begin{pmatrix} 0 \\ -2 \\ 1 \end{pmatrix}.$$

Theorem 4.10 tells us that if we let

$$C = \begin{pmatrix} 15 & 0 & 0 \\ 8 & -3 & -2 \\ -16 & 1 & 1 \end{pmatrix},$$

then

$$C^{-1}AC = D = \begin{pmatrix} 1 & 0 & 0 \\ 0 & 6 & 0 \\ 0 & 0 & 7 \end{pmatrix}.$$

We are not eager to compute C^{-1} to verify this, although MATCOMP can do so and verify it easily (see Exercise 20). However, $C^{-1}AC = D$ is equivalent to $AC = CD$, and it is easy to check that

$$AC = CD = \begin{pmatrix} 15 & 0 & 0 \\ 8 & -18 & -14 \\ -16 & 6 & 7 \end{pmatrix}. \quad \triangleleft$$

Theorem 4.10 asserts the following.

An $n \times n$ matrix with n distinct eigenvalues is similar to a diagonal matrix.

It is not always essential that the eigenvalues be distinct. Suppose we can find a basis for $\mathbb{R}^n$ comprised of eigenvectors $\mathbf{v}_1, \mathbf{v}_2, \ldots, \mathbf{v}_n$ and we let the matrix C have $\mathbf{v}_i$ as ith column vector. An examination of the proof of Theorem 4.10 shows that $C^{-1}AC = D$, where D is the diagonal matrix having on the diagonal in the ith column the eigenvalue of A associated with the eigenvector $\mathbf{v}_i$.

Example 2   Diagonalize the matrix

$$A = \begin{pmatrix} 1 & -3 & 3 \\ 0 & -5 & 6 \\ 0 & -3 & 4 \end{pmatrix}.$$

Solution   We find that the characteristic equation of A is

$$(1 - \lambda)[(-5 - \lambda)(4 - \lambda) + 18] = (1 - \lambda)(\lambda^2 + \lambda - 2)$$
$$= (1 - \lambda)(\lambda + 2)(\lambda - 1) = 0.$$

Thus the eigenvalues of A are $\lambda_1 = 1$, $\lambda_2 = 1$, and $\lambda_3 = -2$. Note that 1 is a root of the characteristic equation of multiplicity 2; we say that the eigenvalue 1 has *algebraic multiplicity* 2.

Reducing $A - I$, we obtain

$$A - I = \begin{pmatrix} 0 & -3 & 3 \\ 0 & -6 & 6 \\ 0 & -3 & 3 \end{pmatrix} \sim \begin{pmatrix} 0 & 1 & -1 \\ 0 & 0 & 0 \\ 0 & 0 & 0 \end{pmatrix}.$$

We see that the nullspace of $A - I$ has dimension 2 and consists of vectors of the form

$$\begin{pmatrix} s \\ r \\ r \end{pmatrix} \quad \text{for any scalars } r \text{ and } s.$$

Taking $s = 1$ and $r = 0$, and then taking $s = 0$ and $r = 1$, we obtain the independent eigenvectors

$$\mathbf{v}_1 = \begin{pmatrix} 1 \\ 0 \\ 0 \end{pmatrix} \quad \text{and} \quad \mathbf{v}_2 = \begin{pmatrix} 0 \\ 1 \\ 1 \end{pmatrix}$$

corresponding to the eigenvalues $\lambda_1 = \lambda_2 = 1$.

Reducing $A + 2I$, we find that

$$A + 2I = \begin{pmatrix} 3 & -3 & 3 \\ 0 & -3 & 6 \\ 0 & -3 & 6 \end{pmatrix} \sim \begin{pmatrix} 3 & 0 & -3 \\ 0 & 1 & -2 \\ 0 & 0 & 0 \end{pmatrix}.$$

Thus an eigenvector corresponding to $\lambda_3 = -2$ is

$$\mathbf{v}_3 = \begin{pmatrix} 1 \\ 2 \\ 1 \end{pmatrix}.$$

Therefore, if we take

$$C = \begin{pmatrix} 1 & 0 & 1 \\ 0 & 1 & 2 \\ 0 & 1 & 1 \end{pmatrix},$$

we should have

$$C^{-1}AC = D = \begin{pmatrix} 1 & 0 & 0 \\ 0 & 1 & 0 \\ 0 & 0 & -2 \end{pmatrix}.$$

A check shows that indeed

$$AC = CD = \begin{pmatrix} 1 & 0 & -2 \\ 0 & 1 & -4 \\ 0 & 1 & -2 \end{pmatrix}. \quad \triangleleft$$

As we indicated in Example 2, the **algebraic multiplicity** of an eigenvalue λ_i of A is its multiplicity as a root of the characteristic equation of A. Its **geometric multiplicity** is the dimension of the eigenspace E_{λ_i}. Of course, the geometric multiplicity of each eigenvalue must be at least 1, since there always exists a nonzero eigenvector in the eigenspace. However, it is possible for the algebraic multiplicity to be greater than the geometric multiplicity.

Example 3 Referring back to Examples 4 and 5 in Section 4.5, find the algebraic and geometric multiplicities of the eigenvalue 2 of the matrix

$$A = \begin{pmatrix} 2 & 1 & 0 \\ -1 & 0 & 1 \\ 1 & 3 & 1 \end{pmatrix}.$$

Solution Example 4 on page 222 shows that the characteristic equation of A is $-(\lambda - 2)^2(\lambda + 1) = 0$, so 2 is an eigenvalue of algebraic multiplicity 2. Exam-

E_d = nulspace of $A - dI$

ple 5 on page 223 shows that the reduced form of $A - 2I$ is

$$\begin{pmatrix} 1 & 3 & -1 \\ 0 & 1 & 0 \\ 0 & 0 & 0 \end{pmatrix}.$$

Thus the eigenspace E_2, which is the set of vectors of the form

$$\begin{pmatrix} r \\ 0 \\ r \end{pmatrix} \quad \text{for } r \in \mathbb{R},$$

has dimension 1 so the eigenvalue 2 has geometric multiplicity 1. ◁

We state without proof a relationship between the algebraic multiplicity and the geometric multiplicity of a (possibly complex) eigenvalue.

> The geometric multiplicity of an eigenvalue of a matrix A is less than or equal to its algebraic multiplicity.

Let $\lambda_1, \lambda_2, \ldots, \lambda_m$ be the distinct (possibly complex) eigenvalues of an $n \times n$ matrix A. Let B_i be a basis for the eigenspace of λ_i for $i = 1, 2, \ldots, m$. It can be shown by an argument similar to the proof of Theorem 4.9 that the union of these bases B_i is an independent set of vectors in $\mathbb{R}^n$ (see Exercise 14). The remarks preceding Example 2 show that the matrix A is diagonalizable if this union of the B_i is a basis for $\mathbb{R}^n$. Clearly by our boxed statement, this will occur precisely when the geometric multiplicity of each eigenvalue is equal to its algebraic multiplicity. Conversely, it can be shown that if A is diagonalizable, then the algebraic multiplicity of each eigenvalue is the same as its geometric multiplicity. We summarize this in a theorem.

Theorem 4.11   A Criterion for Diagonalization

> An $n \times n$ matrix A is diagonalizable if and only if the algebraic multiplicity of each (possibly complex) eigenvalue is equal to its geometric multiplicity.

Thus the 3×3 matrix in Example 2 is diagonalizable, for its eigenvalue 1 has algebraic and geometric multiplicity 2, and its eigenvalue -2 has algebraic and geometric multiplicity 1. However, the matrix in Example 3 is not diagonalizable, because the eigenvalue 2 has algebraic multiplicity 2 but geometric multiplicity 1.

In more advanced texts in linear algebra, it is shown that every square matrix A is similar to a matrix J, its *Jordan canonical form*. If A is diagonalizable, then J is a diagonal matrix, found precisely as in the examples above. If A is not diagonalizable, then J again has the eigenvalues of A on its main diagonal, but also has entries 1 immediately above some of the diagonal entries. The remaining entries are all zero. For example, the matrix

$$A = \begin{pmatrix} 2 & 1 & 0 \\ -1 & 0 & 1 \\ 1 & 3 & 1 \end{pmatrix}$$

of Example 3 has Jordan canonical form

$$J = \begin{pmatrix} -1 & 0 & 0 \\ 0 & 2 & 1 \\ 0 & 0 & 2 \end{pmatrix}.$$

This Jordan canonical form is as close to diagonalization of A as one can come. The Jordan canonical form has applications to the solution of systems of differential equations.

To conclude this discussion, we state one more result without proof. Symmetric matrices occur often in applications; Chapter 7 gives some illustrations. Theorem 4.12 shows that symmetric matrices are very nice matrices to work with.

Theorem 4.12   Diagonalization of Symmetric Matrices

If a symmetric matrix has real numbers for all its entries, then all its eigenvalues are real numbers. For any symmetric matrix, the algebraic multiplicity of each eigenvalue equals its geometric multiplicity so every symmetric matrix is diagonalizable.

A diagonalizing matrix C for a symmetric matrix A can be chosen to have some very nice properties, as we will show for real symmetric matrices in Chapter 6.

APPLICATIONS: COMPUTING $A^k \mathbf{x}$

Let A be a diagonalizable $n \times n$ matrix and let $\lambda_1, \lambda_2, \ldots, \lambda_n$ be the n, not necessarily distinct, eigenvalues of A. That is, each eigenvalue of A is repeated in this list in accordance with its algebraic multiplicity. Let $\{\mathbf{v}_1, \mathbf{v}_2, \ldots, \mathbf{v}_n\}$

THE JORDAN CANONICAL FORM appears in the *Treatise on Substitutions and Algebraic Equations*, the chief work of the French algebraist Camille Jordan (1838–1921). This text, which appeared in 1870, incorporated the author's group theory work over the preceding decade and became the bible of the field for the remainder of the nineteenth century. The theorem containing the canonical form actually deals not with matrices over the real numbers, but with matrices with entries from the finite field of order p. And as the title of the book indicates, Jordan was not considering matrices as such, but the linear substitutions that they represented.

Camille Jordan, a brilliant student, entered the Ecole Polytechnique in Paris at the age of 17 and practiced engineering from the time of his graduation until 1885. He thus had ample time for mathematical research. From 1873 until 1912 he taught at both the Ecole Polytechnique and the Collège de France. Besides his seminal work on group theory, he is also known for important discoveries in modern analysis and topology.

be a basis for $\mathbb{R}^n$, where $\mathbf{v}_i$ is an eigenvector for λ_i. We have seen that if C is the matrix having $\mathbf{v}_i$ as ith column vector, then

$$C^{-1}AC = D = \begin{pmatrix} \lambda_1 & & & \\ & \lambda_2 & & \\ & & \ddots & \\ & & & \lambda_n \end{pmatrix}.$$

Let $\mathbf{x}$ be any vector in $\mathbb{R}^n$. The corollary of Theorem 4.10 shows that

$$A^k\mathbf{x} = CD^kC^{-1}\mathbf{x} = \begin{pmatrix} | & | & & | \\ \lambda_1{}^k\mathbf{v}_1 & \lambda_2{}^k\mathbf{v}_2 & \cdots & \lambda_n{}^k\mathbf{v}_n \\ | & | & & | \end{pmatrix} C^{-1}\mathbf{x}. \tag{3}$$

Now $C^{-1}\mathbf{x}$ is a column vector; it may be any column vector in $\mathbb{R}^n$ because C^{-1} has rank n so its columns form a basis for $\mathbb{R}^n$. We set

$$C^{-1}\mathbf{x} = \begin{pmatrix} d_1 \\ d_2 \\ \vdots \\ d_n \end{pmatrix}.$$

Then Eq. (3) takes the form

$$A^k\mathbf{x} = d_1\lambda_1{}^k\mathbf{v}_1 + d_2\lambda_2{}^k\mathbf{v}_2 + \cdots + d_n\lambda_n{}^k\mathbf{v}_n. \tag{4}$$

Equation (4) expresses $A^k\mathbf{x}$ as a linear combination of the eigenvectors $\mathbf{v}_i$.

Let us regard $\mathbf{x}$ as an initial *information vector* in a process in which the information vector at the next stage of the process is found by multiplying the present information vector on the left by a matrix A. Illustrations of this situation are provided by Markov chains and the generation of the Fibonacci sequence in Section 4.5. We are interested in the long-term outcome of the process. That is, we wish to study $A^k\mathbf{x}$ for large values of k.

Suppose that all the eigenvalues of A are real numbers with $|\lambda_1| = |\lambda_2| = \cdots = |\lambda_r|$ and the remaining eigenvalues are of smaller magnitude. For large integers k, the kth powers $\lambda_1{}^k, \lambda_2{}^k, \ldots, \lambda_r{}^k$ dominate the kth powers of the remaining eigenvalues in Eq. (4). Thus the first r terms of the sum in Eq. (4) dominate the remaining terms of the sum, unless it should happen that $d_1 = d_2 = \cdots = d_r = 0$. Since the vectors $\mathbf{v}_1, \mathbf{v}_2, \cdots, \mathbf{v}_r$ are independent, a linear combination of them cannot be the zero vector unless all coefficients are zero. In particular, suppose that $|\lambda_1| > |\lambda_2|$ so that λ_1 is the unique eigenvalue of maximum magnitude. If $d_1 \neq 0$ and k is large, the vector $A^k\mathbf{x}$ is approximately $d_1\lambda_1{}^k\mathbf{v}_1$ in the sense that $\|A^k\mathbf{x} - d_1\lambda_1{}^k\mathbf{v}_1\|$ is small compared with $\|A^k\mathbf{x}\|$.

Example 4   Show that a diagonalizable transition matrix T for a Markov chain has no eigenvalues of magnitude > 1.

Solution Example 3 of Section 4.5 shows that 1 is an eigenvalue for every transition matrix of a Markov chain. For every choice of population distribution vector **p**, the vector $T^k\mathbf{p}$ is again a vector with nonnegative entries having sum 1. The discussion above shows that all eigenvalues of T must have magnitude ≤ 1; otherwise, entries in some $T^k\mathbf{p}$ would have very large magnitude as k increases. ◁

Example 5 Find the order of magnitude of the term F_k of the Fibonacci sequence $F_0, F_1,$ $F_2, \ldots$, that is, the sequence

$$0, 1, 1, 2, 3, 5, 8, 13, \ldots$$

for large values of k.

Solution We saw in Section 4.5 that if we let

$$\mathbf{x}_k = \begin{pmatrix} F_k \\ F_{k-1} \end{pmatrix},$$

then

$$\mathbf{x}_k = \begin{pmatrix} 1 & 1 \\ 1 & 0 \end{pmatrix} \mathbf{x}_{k-1}.$$

We compute the relation (4) for

$$A = \begin{pmatrix} 1 & 1 \\ 1 & 0 \end{pmatrix} \quad \text{and} \quad \mathbf{x} = \mathbf{x}_1 = \begin{pmatrix} 1 \\ 0 \end{pmatrix}.$$

The characteristic equation of A is

$$(1 - \lambda)(-\lambda) - 1 = \lambda^2 - \lambda - 1 = 0.$$

Using the quadratic formula, we find the eigenvalues

$$\lambda_1 = \frac{1 + \sqrt{5}}{2} \quad \text{and} \quad \lambda_2 = \frac{1 - \sqrt{5}}{2}.$$

Reducing $A - \lambda_1 I$, we obtain

$$(A - \lambda_1 I) \sim \begin{pmatrix} \dfrac{1 - \sqrt{5}}{2} & 1 \\ 0 & 0 \end{pmatrix} \quad \text{so} \quad \mathbf{v}_1 = \begin{pmatrix} 2 \\ \sqrt{5} - 1 \end{pmatrix}$$

is an eigenvector for λ_1. In an analogous fashion, we find that

$$\mathbf{v}_2 = \begin{pmatrix} -2 \\ \sqrt{5} + 1 \end{pmatrix}$$

is an eigenvalue for λ_2. Thus we take

$$C = \begin{pmatrix} 2 & -2 \\ \sqrt{5} - 1 & \sqrt{5} + 1 \end{pmatrix}.$$

A computation shows that

$$C^{-1} = [1/(4\sqrt{5})]\begin{pmatrix} \sqrt{5} + 1 & 2 \\ 1 - \sqrt{5} & 2 \end{pmatrix}.$$

Thus

$$\begin{pmatrix} d_1 \\ d_2 \end{pmatrix} = C^{-1}\begin{pmatrix} 1 \\ 0 \end{pmatrix} = [1/(4\sqrt{5})]\begin{pmatrix} \sqrt{5} + 1 \\ 1 - \sqrt{5} \end{pmatrix}.$$

Equation (4) takes the form

$$A^k\mathbf{x}_1 = \begin{pmatrix} F_{k+1} \\ F_k \end{pmatrix} = \left(\frac{\sqrt{5} + 1}{4\sqrt{5}}\right)\left(\frac{1 + \sqrt{5}}{2}\right)^k \begin{pmatrix} 2 \\ \sqrt{5} - 1 \end{pmatrix}$$
$$- \left(\frac{\sqrt{5} - 1}{4\sqrt{5}}\right)\left(\frac{1 - \sqrt{5}}{2}\right)^k \begin{pmatrix} -2 \\ \sqrt{5} + 1 \end{pmatrix}.$$

Computing the second component of this vector, we find that

$$F_k = (1/\sqrt{5})\left[\left(\frac{1 + \sqrt{5}}{2}\right)^k - \left(\frac{1 - \sqrt{5}}{2}\right)^k\right]. \tag{5}$$

For large k, the kth power of the eigenvalue $\lambda_1 = (1 + \sqrt{5})/2$ dominates, so

$$F_k \approx (1/\sqrt{5})\left(\frac{1 + \sqrt{5}}{2}\right)^k \quad \text{for large } k. \tag{6}$$

Indeed, since $|\lambda_2| = |(1 - \sqrt{5})/2| < 1$, we see that the contribution from this second eigenvalue to the expression (5) approaches zero as k increases. Since $|\lambda_2{}^k/\sqrt{5}| < \frac{1}{2}$ for $k = 1$ and hence for all k, we see that F_k can be characterized as the closest integer to $(1/\sqrt{5})[(1 + \sqrt{5})/2]^k$ for all k. The approximation (6) verifies that F_k increases exponentially with k as expected for a population of rabbits. $\lhd$

Example 5 is a typical analysis of a process in which an information vector after the kth stage is equal to $A^k\mathbf{x}_1$ for an initial information vector $\mathbf{x}_1$ and a diagonalizable matrix A. Clearly an important consideration is whether any of the eigenvalues of A have magnitude greater than 1. When this is the case, the components of the information vector may grow exponentially in magnitude, as illustrated by the Fibonacci sequence in Example 5, where $|\lambda_1| > 1$. On the other hand, if all the eigenvalues have magnitude less than 1, the components of the information vector must approach zero as k increases.

Stability

The process just described is called **unstable** if A has an eigenvalue of magnitude greater than 1, **stable** if all eigenvalues have magnitude less than 1, and **neutrally stable** if the maximum magnitude of the eigenvalues is 1.

Thus a Markov chain is a neutrally stable process, whereas generation of the Fibonacci sequence is unstable. The eigenvectors are called the **normal modes** of the process.

Chapter 7 contains applications of eigenvalues and eigenvectors to geometry and maximization, and Chapter 8 contains an application to the solution of a system of differential equations. Except in Chapter 8, we will not deal with any applications involving calculus. In the type of process described above, we study information at evenly spaced time intervals. If we study such a process as the number of time intervals increases and their duration approaches zero, we find ourselves in calculus. Eigenvalues and eigenvectors play an important role in applications of calculus, especially in studying any sort of vibration. In these applications of calculus, components of vectors are functions of time.

SUMMARY

Let A be an $n \times n$ matrix.

1. If A has n distinct eigenvalues and C is an $n \times n$ matrix having as jth column vector an eigenvector corresponding to an eigenvalue λ_j, then $C^{-1}AC$ is the diagonal matrix having λ_j on the diagonal in the jth column.

2. If $C^{-1}AC = D$, then $A^k = CD^kC^{-1}$.

3. A is similar to an $n \times n$ matrix B if there exists an invertible $n \times n$ matrix C such that $C^{-1}AC = B$.

4. The algebraic multiplicity of an eigenvalue λ of A is its multiplicity as a root of the characteristic equation; its geometric multiplicity is the dimension of the corresponding eigenspace E_λ.

5. For any eigenvalue, its geometric multiplicity is less than or equal to its algebraic multiplicity.

6. The matrix A is diagonalizable if and only if the geometric multiplicity of each of its eigenvalues is the same as the algebraic multiplicity.

7. Every symmetric matrix is diagonalizable. All eigenvalues of a real symmetric matrix are real numbers.

8. Let A be diagonalizable by a matrix C, let $\mathbf{x}$ be any column vector, and let $\mathbf{d} = C^{-1}\mathbf{x}$. Then

$$A^k\mathbf{x} = d_1\lambda_1{}^k\mathbf{v}_1 + d_2\lambda_2{}^k\mathbf{v}_2 + \cdots + d_n\lambda_n{}^k\mathbf{v}_n,$$

where $\mathbf{v}_j$ is the jth column vector of C.

9. Let multiplication of a column information vector by A give the information vector for the next stage of a process as described on page 236. The process is stable, neutrally stable, or unstable depending on whether the maximum magnitude of the eigenvalues of A is less than 1, equal to 1, or greater than 1, respectively.

EXERCISES

In Exercises 1–8, find the eigenvalues λ_i and corresponding eigenvectors $\mathbf{v}_i$ of the given matrix A, and also find an invertible matrix C and a diagonal matrix D such that $D = C^{-1}AC$.

1. $A = \begin{pmatrix} 1 & 2 \\ 0 & 3 \end{pmatrix}$

2. $A = \begin{pmatrix} 3 & 2 \\ 1 & 4 \end{pmatrix}$

3. $A = \begin{pmatrix} 7 & 8 \\ -4 & -5 \end{pmatrix}$

4. $A = \begin{pmatrix} 6 & 3 & -3 \\ -2 & -1 & 2 \\ 16 & 8 & -7 \end{pmatrix}$

5. $A = \begin{pmatrix} -3 & 10 & -6 \\ 0 & 7 & -6 \\ 0 & 0 & 1 \end{pmatrix}$

6. $A = \begin{pmatrix} -3 & 5 & -20 \\ 2 & 0 & 8 \\ 2 & 1 & 7 \end{pmatrix}$

7. $A = \begin{pmatrix} -2 & 0 & -1 \\ 0 & 2 & 0 \\ 3 & 0 & 2 \end{pmatrix}$

8. $A = \begin{pmatrix} -4 & 6 & -12 \\ 3 & -1 & 6 \\ 3 & -3 & 8 \end{pmatrix}$

In Exercises 9–12, determine whether or not the given matrix is diagonalizable.

9. $\begin{pmatrix} 2 & 4 \\ 0 & 2 \end{pmatrix}$

10. $\begin{pmatrix} 2 & -5 \\ -1 & 2 \end{pmatrix}$

11. $\begin{pmatrix} 1 & 2 & 6 \\ 2 & 0 & -4 \\ 6 & -4 & 3 \end{pmatrix}$

12. $\begin{pmatrix} 3 & 1 & 0 \\ 0 & 3 & 1 \\ 0 & 0 & 3 \end{pmatrix}$

13. Let P, Q, and R be $n \times n$ matrices. Recall that P is similar to Q if there exists an invertible $n \times n$ matrix C such that $C^{-1}PC = Q$. This exercise shows that similarity is an *equivalence relation*.
   a) *Reflexive.* Show that P is similar to itself.
   b) *Symmetric.* Show that if P is similar to Q, then Q is similar to P.
   c) *Transitive.* Show that if P is similar to Q and Q is similar to R, then P is similar to R.

14. Show that if $\lambda_1, \lambda_2, \ldots, \lambda_m$ are distinct real eigenvalues of an $n \times n$ matrix A and if B_i is a basis for the eigenspace E_{λ_i}, then the union of the bases B_i is an independent set of vectors in $\mathbb{R}^n$. Give an argument analogous to the proof of Theorem 4.9.

15. Let the sequence $a_0, a_1, a_2, \ldots$ be given by $a_0 = 0$, $a_1 = 1$, and $a_k = (a_{k-1} + a_{k-2})/2$ for $k \geq 2$.
   a) Find the matrix A that can be used to generate this sequence as we used a matrix to generate the Fibonacci sequence.
   b) Classify this generation process as stable, neutrally stable, or unstable.
   c) Compute the expression (4) for this process. Check computations with the first few terms of the sequence.
   d) Use the answer to (c) to estimate a_k for large k.

16. Repeat Exercise 15 if $a_k = a_{k-1} - (\frac{3}{16})a_{k-2}$ for $k \geq 2$.

17. Repeat Exercise 15 but change the initial data to $a_0 = 1$, $a_1 = 0$.

18. Repeat Exercise 15 if $a_k = a_{k-1}/2 + (\frac{3}{16})a_{k-2}$ for $k \geq 2$.

19. Repeat Exercise 15 if $a_k = a_{k-1} + (\frac{3}{4})a_{k-2}$ for $k \geq 2$.

20. ■ Use MATCOMP, computing C^{-1}, to check that $C^{-1}AC = D$ in Example 1.

In Exercises 21–30, use MATCOMP or similar software to determine whether the given matrix is diagonalizable.

21. $\begin{pmatrix} 18 & 25 & -25 \\ 1 & 6 & -1 \\ 18 & 34 & -25 \end{pmatrix}$

22. $\begin{pmatrix} 8.3 & 8.0 & -6.0 \\ -2.0 & 0.3 & 3.0 \\ 0.0 & 0.0 & 4.3 \end{pmatrix}$

23. $\begin{pmatrix} 13.7 & 34.8 & -11.6 \\ -5.8 & -15.3 & 5.8 \\ 0.0 & 0.0 & 2.1 \end{pmatrix}$

24. $\begin{pmatrix} 24.55 & 46.60 & 46.60 \\ -4.66 & -8.07 & -9.32 \\ -9.32 & -18.64 & -17.39 \end{pmatrix}$

25. $\begin{pmatrix} 7 & -20 & -5 & 5 \\ 5 & -13 & -5 & 0 \\ -5 & 10 & 7 & 5 \\ 5 & -10 & -5 & -3 \end{pmatrix}$

26. $\begin{pmatrix} -22.7 & -26.9 & -6.3 & -46.5 \\ -59.7 & -40.9 & 20.9 & -99.5 \\ 15.9 & 9.6 & -8.4 & 26.5 \\ 43.8 & 36.5 & -7.3 & 78.2 \end{pmatrix}$

27. $\begin{pmatrix} 66.2 & 58.0 & -11.6 & 116.0 \\ 120.6 & 89.6 & -42.6 & 201.0 \\ -21.0 & -15.0 & 7.6 & -35.0 \\ -99.6 & -79.0 & 28.6 & -169.4 \end{pmatrix}$

28. $\begin{pmatrix} -253 & -232 & -96 & 1088 & 280 \\ 213 & 204 & 93 & -879 & -225 \\ -90 & -90 & -47 & 360 & 90 \\ -38 & -36 & -18 & 162 & 40 \\ 62 & 64 & 42 & -251 & -57 \end{pmatrix}$

29. $\begin{pmatrix} 154 & -24 & -36 & -1608 & -336 \\ -126 & 16 & 18 & 1314 & 270 \\ 54 & 0 & 4 & -540 & -108 \\ 24 & 0 & 0 & -236 & -48 \\ -42 & -12 & -18 & 366 & 70 \end{pmatrix}$

30. $\begin{pmatrix} -2513 & 596 & -414 & -2583 & 1937 \\ 127 & -32 & 33 & 132 & -81 \\ -421 & 94 & -83 & -434 & 306 \\ 2610 & -615 & 443 & 2684 & -1994 \\ 90 & -19 & 29 & 94 & -50 \end{pmatrix}$

*4.7
Proof of Theorem 4.2 on Expansion by Minors

The demonstration of the various properties of determinants in Section 4.2 depended on the ability to compute a determinant by expanding it by minors on *any row or column,* as stated in Theorem 4.2. This section is devoted to a proof of Theorem 4.2. We will need to look more closely at the form of the terms that appear in an expanded determinant.

Determinants of orders 2 and 3 can be written as

$$\begin{vmatrix} a_{11} & a_{12} \\ a_{21} & a_{22} \end{vmatrix} = (1)(a_{11}a_{22}) + (-1)(a_{12}a_{21})$$

---

* This section just fills a gap in the theory of determinants, and its omission will cause no difficulty in the succeeding chapters.

and

$$\begin{vmatrix} a_{11} & a_{12} & a_{13} \\ a_{21} & a_{22} & a_{23} \\ a_{31} & a_{32} & a_{33} \end{vmatrix} = (1)(a_{11}a_{22}a_{33}) + (-1)(a_{11}a_{23}a_{32}) \\ + (1)(a_{12}a_{23}a_{31}) + (-1)(a_{12}a_{21}a_{33}) \\ + (1)(a_{13}a_{21}a_{32}) + (-1)(a_{13}a_{22}a_{31}).$$

Note that each determinant appears as a sum of products, each with an associated sign given by (1) or (−1), determined by the formula $(-1)^{1+j}$ as we expand the determinant across the first row. Furthermore, each product contains exactly one factor from each row and exactly one factor from each column of the matrix. That is, the row indices in each product run through all row numbers, and the column indices run through all column numbers. This is an illustration of a general theorem, which we now prove by induction.

Theorem 4.13   Structure of an Expanded Determinant

The determinant of an $n \times n$ matrix A can be expressed as a sum of signed products, where each product contains exactly one factor from each row and exactly one factor from each column. The expansion of $\det(A)$ on any row or column also has this form.

Proof   We consider the expansion of $\det(A)$ on the first row and give a proof by induction. We have just shown that our result is true for determinants of orders 2 and 3. Let $n > 3$ and assume our result holds for all square matrices of size smaller than $n \times n$. Let A be an $n \times n$ matrix. In expanding $\det(A)$ by minors across the first row, the only expression involving a_{1j} is $(-1)^{1+j}a_{1j}|A_{1j}|$. We apply our induction hypothesis to the determinant $|A_{1j}|$ of order $n - 1$: It is a sum of signed products, each of which has one factor from each row and column of A except for row 1 and column j. Clearly, as we multiply this sum term by term by a_{1j}, we obtain a sum of products having a_{1j} as factor from row 1 and column j, and one factor from each other row and from each other column. Thus an expression of the stated form is indeed obtained as we expand $\det(A)$ by minors across the first row.

It is clear that essentially the same argument shows that expansion across any row or down any column yields the same type of sum of signed products. ∎

Our illustration for 2×2 and 3×3 matrices indicates that we might always have the same number of products appearing in $\det(A)$ with a sign given by 1 as by −1. This is indeed the case for determinants of order greater than 1, and the easy induction proof is left as an exercise.

We now prove Theorem 4.2. That is, we show that for an $n \times n$ matrix A, we have

$$|A| = (-1)^{r+1}a_{r1}|A_{r1}| + (-1)^{r+2}a_{r2}|A_{r2}| + \cdots + (-1)^{r+n}a_{rn}|A_{rn}| \tag{1}$$

for any r from 1 to n, and

$$|A| = (-1)^{1+s}a_{1s}|A_{1s}| + (-1)^{2+s}a_{2s}|A_{2s}| + \cdots + (-1)^{n+s}a_{ns}|A_{ns}| \tag{2}$$

for any s from 1 to n.

Proof   We first prove Eq. (1) for any choice of r from 1 to n. Clearly, Eq. (1) holds for $n = 1$ and $n = 2$. Proceeding by induction, let $n > 2$ and assume that determinants of order less than n can be computed using an expansion on *any row*. Let A be an $n \times n$ matrix. We show that expansion of det(A) by minors on row r is the same as expansion on row i for $i < r$. From Theorem 4.13, we know that each of the expansions gives a sum of signed products, where each product contains a single factor from each row and from each column of A. We will compare the products containing as factors both a_{ij} and a_{rs} in each of the expansions. We consider two cases as illustrated in Figs. 4.7 and 4.8.

Figure 4.7
The case $j < s$.

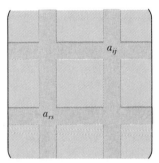

Figure 4.8
The case $s < j$.

If det(A) is expanded on the ith row, the sum of signed products containing $a_{ij}a_{rs}$ is part of $(-1)^{i+j}a_{ij}|A_{ij}|$. In computing $|A_{ij}|$ we may, by our induction assumption, expand on the rth row. For $j < s$, terms of $|A_{ij}|$ involving a_{rs} are then $(-1)^{(r-1)+(s-1)}d$, where d is the determinant of the matrix obtained from A by crossing out rows i and r and columns j and s, as shown in Fig. 4.7. The exponent $(r - 1) + (s - 1)$ occurs because a_{rs} is in row $r - 1$ and column $s - 1$ of A_{ij}. Thus the part of our expansion of det(A) across the ith row that contains $a_{ij}a_{rs}$ is equal to

$$(-1)^{i+j}(-1)^{(r-1)+(s-1)}a_{ij}a_{rs}d \quad \text{for } j < s. \tag{3}$$

For $j > s$, we consult Fig. 4.8 and use similar reasoning to see that the part of our expansion of det(A) across the ith row, which contains $a_{ij}a_{rs}$, is equal to

$$(-1)^{i+j}(-1)^{(r-1)+s}a_{ij}a_{rs}d \quad \text{for } j > s. \tag{4}$$

We now expand det(A) by minors on the rth row, obtaining $(-1)^{r+s}a_{rs}|A_{rs}|$ as the portion involving a_{rs}. Expanding $|A_{rs}|$ on the ith row using our induction assumption, we obtain $(-1)^{i+j}a_{ij}d$ if $j < s$ and $(-1)^{i+(j-1)}a_{ij}d$ if $j > s$. Thus the part of the expansion of $\det(A)$ on the rth row, which contains $a_{ij}a_{rs}$, is equal to

$$(-1)^{r+s}(-1)^{i+j}a_{rs}a_{ij}d \quad \text{for } j < s \tag{5}$$

or

$$(-1)^{r+s}(-1)^{i+(j-1)}a_{rs}a_{ij}d \quad \text{for } j > s. \tag{6}$$

Expressions (3) and (5) are equal since $(-1)^{r+s+i+j} = (-1)^{r+s+i+j-2}$ and expressions (4) and (6) are equal since $(-1)^{r+s+i+j-1}$ is the algebraic sign of each. This concludes the proof that the expansion of the $\det(A)$ by minors across rows i and r are equal.

A similar argument shows that expansion of $\det(A)$ down columns j and s are the same.

Finally, we must show that an expansion of $\det(A)$ on a row is equal to expansion on a column. It is sufficient for us to prove that the expansion of $\det(A)$ on the first row is the same as the expansion on the first column, in view of what we have proved above. Again, we use induction and dispose of the cases $n = 1$ and $n = 2$ as trivial to check. Let $n > 2$ and assume that our result holds for matrices of size smaller than $n \times n$. Let A be an $n \times n$ matrix. Expanding $\det(A)$ on the first row yields

$$a_{11}|A_{11}| + \sum_{j=2}^{n} (-1)^{1+j}a_{1j}|A_{1j}|.$$

For $j > 1$, we expand $|A_{1j}|$ on the first column, using our induction assumption, and obtain $|A_{1j}| = \sum_{i=2}^{n}(-1)^{(i-1)+1}a_{i1}d$, where d is the determinant of the matrix obtained from A by crossing out rows 1 and i and columns 1 and j. Thus the terms in the expansion of $\det(A)$ containing $a_{1j}a_{i1}$ are

$$(-1)^{1+j+i}a_{1j}a_{i1}d. \tag{7}$$

On the other hand, if we expand $\det(A)$ on the first column, we obtain

$$a_{11}|A_{11}| + \sum_{i=2}^{n} (-1)^{i+1}a_{i1}|A_{i1}|.$$

For $i > 1$, expanding on the first row using our induction assumption shows that $|A_{i1}| = \sum_{j=2}^{n}(-1)^{1+(j-1)}a_{1j}d$. This results in

$$(-1)^{i+1+j}a_{i1}a_{1j}d$$

as the part of the expansion of $\det(A)$ containing the sum $a_{i1}a_{1j}$, and this agrees with the expression in formula (7). This concludes our proof. ∎

Applications of Vector Geometry and of Determinants

This chapter is devoted to presenting more applications of linear algebra and extending some applications already presented. In Chapters 1 and 2, we addressed the problem of solving linear systems. In practice, linear systems having no exact solution arise quite often. The study of *projections* in Section 5.1 enables us in Section 5.2 to find *approximate solutions* that are optimal in the *least-squares sense* of such inconsistent linear systems. In Sections 5.3–5.5, we continue to work with projections and to study vector geometry.

We have seen how to use determinants to find the area of a parallelogram in the plane and the volume of a parallelepiped in space. Using the cross product, whose computation involves determinants, we are able to find the area of a parallelogram in space. We will refer to all these configurations as *boxes*. In particular, we consider a parallelogram in space to be a two-dimensional box in $\mathbb{R}^3$. In Section 5.6, we will give a general formula for the *volume* of an m-dimensional box in $\mathbb{R}^n$ for $m \leq n$. This fabulous formula lies at the heart of much of integral calculus. By itself, it justifies the study of determinants.

5.1
Projections

THE PROJECTION OF ONE VECTOR ON ANOTHER

A practical concern in vector applications is to determine what portion of a vector **b** can be considered to act in the direction given by another vector **a**. A force vector acting in a certain direction may be moving a body along a line having a different direction. For example, suppose you are trying to roll your stalled car off the road by pushing on the door jamb at the side, so you can reach in and control the steering wheel when necessary. You are not applying the force in quite the same direction that the car moves, as you would be if you could push from directly behind the car. Such considerations lead to the notion of the *projection of* **b** *on* **a**.

Consider a force vector **b** in $\mathbb{R}^3$ and a line through the origin in the direction of a vector **a**. This line through the origin can be conveniently represented as the subspace sp(**a**), as shown in Fig. 5.1. The length of the side parallel to **a** of the right triangle in Fig. 5.1 represents the component of the force vector **b** that actually moves the body along **a**. The side perpendicular to **a** of the triangle represents the portion of the force acting perpendicular to **a** that is of no help in moving the body along **a**. The **work** done by the force vector **b** in moving the body through a distance *s* along the line sp(**a**) is defined to be the product of *s* and the component of **b** in the direction of **a**, as labeled in Fig. 5.1.

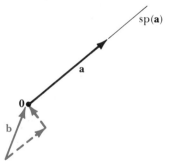

Figure 5.1
A force vector b moving a body along the line sp(a).

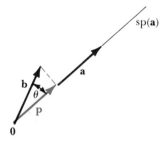

Figure 5.2
The vector projection p of b on a.

For vectors **a** and **b** in $\mathbb{R}^n$ with **a** $\neq$ **0**, the *component of* **b** *along* **a** is the magnitude of the *vector projection* **p** of **b** on **a**, which is shown in Fig. 5.2. Note that in Fig. 5.2, we have drawn vectors of the same magnitude and direction as those in Fig. 5.1, but we have them both starting at the origin **0**.

Let us compute the magnitude of $\|\mathbf{p}\|$ for **p** shown in Fig. 5.2. We see that

$$\|\mathbf{p}\| = \|\mathbf{b}\|(\cos\theta),$$

where θ is the angle between **b** and **a**. From Section 3.1, we know that

$$\mathbf{a} \cdot \mathbf{b} = \|\mathbf{a}\| \|\mathbf{b}\| (\cos \theta).$$

Therefore,

$$\|\mathbf{p}\| = \frac{\mathbf{a} \cdot \mathbf{b}}{\|\mathbf{a}\|}.$$

(Remember that we are assuming $\mathbf{a} \neq \mathbf{0}$.)

We would like to find the vector **p** as well as $\|\mathbf{p}\|$. Since **p** is parallel to **a**, we can find **p** by first forming the *unit* vector $(1/\|\mathbf{a}\|)\mathbf{a}$ having the same direction as **a**, and then multiplying by $\|\mathbf{p}\| = (\mathbf{a} \cdot \mathbf{b})/\|\mathbf{a}\|$. We obtain

$$\mathbf{p} = \frac{\mathbf{a} \cdot \mathbf{b}}{\|\mathbf{a}\|^2} \mathbf{a} = \frac{\mathbf{a} \cdot \mathbf{b}}{\mathbf{a} \cdot \mathbf{a}} \mathbf{a}.$$

We summarize these computations in the following box.

Projection of One Vector on Another

Let **a** and **b** be vectors in $\mathbb{R}^n$ with $\mathbf{a} \neq \mathbf{0}$. The **vector projection** of **b** on **a** is

$$\mathbf{p} = \frac{\mathbf{a} \cdot \mathbf{b}}{\mathbf{a} \cdot \mathbf{a}} \mathbf{a}. \tag{1}$$

The **(scalar) component** of **b** along **a** is

$$\|\mathbf{p}\| = \frac{\mathbf{a} \cdot \mathbf{b}}{\|\mathbf{a}\|}. \tag{2}$$

Example 1   Let $\mathbf{a} = (2, 4, 3)$. Find the vector projection on **a** of a force vector $\mathbf{b} = (1, 2, 3)$ that moves a body along the line sp(**a**) through the origin. Also find the work done by the force in moving the body s units.

Solution   The required vector projection **p** of **b** on **a** is found from Eq. (1) to be

$$\mathbf{p} = \frac{\mathbf{a} \cdot \mathbf{b}}{\mathbf{a} \cdot \mathbf{a}} \mathbf{a} = \frac{19}{29} (2, 4, 3).$$

The scalar component of **b** along **a** can be found by computing $(\mathbf{a} \cdot \mathbf{b})/\|\mathbf{a}\|$, which is the same as

$$\|\mathbf{p}\| = \frac{19}{29} \sqrt{29} = 19/\sqrt{29}.$$

The work done in moving the body s units is therefore $19s/\sqrt{29}$.   ◁

PROJECTION OF A VECTOR ON A SUBSPACE

We wish to find the projection of a vector $\mathbf{b}$ in $\mathbb{R}^n$ on a subspace S of $\mathbb{R}^n$ of dimension higher than one. For example, suppose we move a box across a floor by pushing forward but slightly downward in the natural way on a top edge of the box. The portion of the force vector we apply that actually moves the box across the floor is its projection on the plane of the floor. To see how we might find the projection of a vector $\mathbf{b}$ on a general subspace S, we return to the projection of $\mathbf{b}$ on $\mathrm{sp}(\mathbf{a})$, but analyze it somewhat differently than we did above.

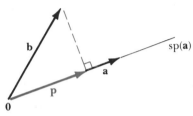

Figure 5.3
The vector projection p of b on sp(a).

As indicated in Fig. 5.3, the vector projection $\mathbf{p}$ of $\mathbf{b}$ on $\mathrm{sp}(\mathbf{a})$, where $\mathbf{a} \neq \mathbf{0}$, has two properties, which we state in quite general terms in the following box.

Properties of the Vector Projection p of Vector b on the Subspace S

1. The vector $\mathbf{p}$ must lie in the subspace S.
2. The vector $\mathbf{b} - \mathbf{p}$ must be perpendicular to *every* vector $\mathbf{a}$ in S.

For the case of the projection $\mathbf{p}$ of $\mathbf{b}$ on the line $\mathrm{sp}(\mathbf{a})$, the first property means that $\mathbf{p} = r\mathbf{a}$ for some scalar r. The second condition requires that the dot product $(\mathbf{b} - r\mathbf{a}) \cdot s\mathbf{a} = 0$ for all scalars s in $\mathbb{R}$; that is, $s[\mathbf{b} \cdot \mathbf{a} - r(\mathbf{a} \cdot \mathbf{a})] = 0$ for all scalars s. Taking $s = 1$, we obtain $r = (\mathbf{a} \cdot \mathbf{b})/(\mathbf{a} \cdot \mathbf{a})$. Thus the projection of $\mathbf{b}$ on the subspace $S = \mathrm{sp}(\mathbf{a})$ is $\mathbf{p} = r\mathbf{a} = [(\mathbf{a} \cdot \mathbf{b})/(\mathbf{a} \cdot \mathbf{a})]\mathbf{a}$ in accord with Eq. (1). We will use the notation $\mathbf{b}_S$ for the projection of $\mathbf{b}$ on a subspace S. Thus the projection of $\mathbf{b}$ on $S = \mathrm{sp}(\mathbf{a})$ is

$$\mathbf{b}_S = \frac{\mathbf{a} \cdot \mathbf{b}}{\mathbf{a} \cdot \mathbf{a}} \mathbf{a}. \tag{3}$$

We now turn to the projection $\mathbf{p}$ of a vector $\mathbf{b}$ on a general subspace $S = \mathrm{sp}(\mathbf{a}_1, \mathbf{a}_2, \ldots, \mathbf{a}_k)$ of $\mathbb{R}^n$, where the $\mathbf{a}_i$ are independent vectors in $\mathbb{R}^n$. Details are worked out for the case $k = 2$ but the computations are equally valid in the general case. We want **p** to satisfy the properties for a projection of **b** on S that we boxed above. Recall that $\text{sp}(\mathbf{a}_1, \mathbf{a}_2)$ corresponds to a plane in $\mathbb{R}^n$ containing the origin, and its members consist of all linear combinations of $\mathbf{a}_1$ and $\mathbf{a}_2$. Figure 5.4 shows the vector projection **p** of **b** on $\text{sp}(\mathbf{a}_1, \mathbf{a}_2)$, which satisfies the two required properties:

1. **p** must lie in the subspace $S = \text{sp}(\mathbf{a}_1, \mathbf{a}_2)$; and
2. $\mathbf{b} - \mathbf{p}$ must be perpendicular to *each* vector in S.

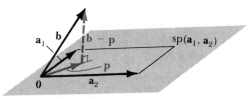

Figure 5.4
The vector projection p of b on $\text{sp}(\mathbf{a}_1, \mathbf{a}_2)$.

If we write our vectors in $\mathbb{R}^n$ as column vectors, the subspace $\text{sp}(\mathbf{a}_1, \mathbf{a}_2)$ is the column space of the $n \times 2$ matrix A whose columns are $\mathbf{a}_1$ and $\mathbf{a}_2$. According to Theorem 3.8 on page 156, all vectors in $\text{sp}(\mathbf{a}_1, \mathbf{a}_2)$ have the form $A\mathbf{x}$, where

$$\mathbf{x} = \begin{pmatrix} x_1 \\ x_2 \end{pmatrix} \quad \text{for any scalars } x_1 \text{ and } x_2.$$

Since **p** lies in the space $S = \text{sp}(\mathbf{a}_1, \mathbf{a}_2)$, we see that

$$\mathbf{p} = A\mathbf{x}_p, \quad \text{where } \mathbf{x}_p = \begin{pmatrix} r_1 \\ r_2 \end{pmatrix} \quad \text{for some scalars } r_1 \text{ and } r_2.$$

Since $\mathbf{b} - A\mathbf{x}_p$ must be perpendicular to each vector in $\text{sp}(\mathbf{a}_1, \mathbf{a}_2)$, the dot product of $(\mathbf{b} - A\mathbf{x}_p)$ and $A\mathbf{x}$ must be zero for all vectors **x**. This dot-product condition can be written as the matrix equation

$$(A\mathbf{x})^T(\mathbf{b} - A\mathbf{x}_p) = \mathbf{x}^T(A^T\mathbf{b} - A^TA\mathbf{x}_p) = \mathbf{0}.$$

In other words, the dot product of the vectors **x** and $A^T\mathbf{b} - A^TA\mathbf{x}_p$ must be zero *for all vectors* **x**. This can happen only if the vector $A^T\mathbf{b} - A^TA\mathbf{x}_p$ is itself the zero vector (see Exercise 13). Therefore, we have

$$A^T\mathbf{b} - A^TA\mathbf{x}_p = \mathbf{0}. \tag{4}$$

Now the 2×2 matrix A^TA appearing in Eq. (4) is invertible because it has the same rank as A (see Theorem 3.18 on page 180), and A has rank 2 because its two columns are independent. Solving Eq. (4) for $\mathbf{x}_p$, we obtain

$$\mathbf{x}_p = (A^TA)^{-1}A^T\mathbf{b}. \tag{5}$$

Denoting the vector projection $\mathbf{p} = A\mathbf{x}_p$ of **b** on $S = \text{sp}(\mathbf{a}_1, \mathbf{a}_2)$ by $\mathbf{p} = \mathbf{b}_S$ and writing **b** as a column vector, we have obtained the case $k = 2$ of the general formula (6) in the box on the following page.

Projection $\mathbf{b}_S$ of $\mathbf{b}$ on the Subspace S

Let $S = \mathrm{sp}(\mathbf{a}_1, \mathbf{a}_2, \ldots, \mathbf{a}_k)$ be a subspace of $\mathbb{R}^n$ and let A have as columns the vectors $\mathbf{a}_1, \mathbf{a}_2, \ldots, \mathbf{a}_k$. The vector projection of $\mathbf{b}$ in $\mathbb{R}^n$ on S is given by

$$\mathbf{b}_S = A(A^T A)^{-1} A^T \mathbf{b}. \tag{6}$$

We leave to Exercise 12 the demonstration that the result in Eq. (3) can be written in the form (6), using matrices of appropriate size.

The analysis that produced the formula (6) is equally valid for the projection $\mathbf{b}_S$ of a vector $\mathbf{b}$ in $\mathbb{R}^n$ on a subspace $S = \mathrm{sp}(\mathbf{a}_1, \mathbf{a}_2, \ldots, \mathbf{a}_k)$, provided that the vectors $\mathbf{a}_1, \mathbf{a}_2, \ldots, \mathbf{a}_k$ are linearly independent. This independence ensures that the $k \times k$ matrix $A^T A$ in formula (6) is invertible.

Example 2   Repeat Example 1 using (6), and find the projection of the vector $\mathbf{b}$ on the subspace $S = \mathrm{sp}(\mathbf{a})$, where

$$\mathbf{b} = \begin{pmatrix} 1 \\ 2 \\ 3 \end{pmatrix} \quad \text{and} \quad \mathbf{a} = \begin{pmatrix} 2 \\ 4 \\ 3 \end{pmatrix}.$$

Solution   Let A be the matrix whose single column is $\mathbf{a}$. We compute

$$A^T A = (2, 4, 3) \begin{pmatrix} 2 \\ 4 \\ 3 \end{pmatrix} = (29)$$

and

$$\mathbf{b}_S = A(A^T A)^{-1} A^T \mathbf{b} = \begin{pmatrix} 2 \\ 4 \\ 3 \end{pmatrix} (1/29)(2, 4, 3) \begin{pmatrix} 1 \\ 2 \\ 3 \end{pmatrix}$$

$$= (1/29) \begin{pmatrix} 4 & 8 & 6 \\ 8 & 16 & 12 \\ 6 & 12 & 9 \end{pmatrix} \begin{pmatrix} 1 \\ 2 \\ 3 \end{pmatrix} = (1/29) \begin{pmatrix} 38 \\ 76 \\ 57 \end{pmatrix}. \quad \triangleleft$$

We refer to the matrix $P_S = A(A^T A)^{-1} A^T$ in Eq. (6) as the **projection matrix for the subspace** S. It takes any vector $\mathbf{b}$ in $\mathbb{R}^n$ and, by left multiplication, projects it onto the vector $\mathbf{b}_S$, which lies in S. We will show shortly that this matrix P_S is uniquely determined by S, which allows us to use the definite article *the* when talking about it. Before demonstrating this uniqueness, we box the formula and give two more examples.

> ### The Projection Matrix P_S for a subspace S
>
> Let $S = \text{sp}(\mathbf{a}_1, \mathbf{a}_2, \ldots, \mathbf{a}_k)$ be a subspace of $\mathbb{R}^n$ and let A have as columns the vectors $\mathbf{a}_1, \mathbf{a}_2, \ldots, \mathbf{a}_k$. The projection matrix for the subspace S is given by
>
> $$P_S = A(A^TA)^{-1}A^T.$$

Example 3 Find the projection matrix for the x_2, x_3-plane in $\mathbb{R}^3$.

Solution The x_2, x_3-plane is the subspace $S = \text{sp}(\mathbf{e}_2, \mathbf{e}_3)$, where $\mathbf{e}_2$ and $\mathbf{e}_3$ are the column vectors of the matrix

$$A = \begin{pmatrix} 0 & 0 \\ 1 & 0 \\ 0 & 1 \end{pmatrix}.$$

We find that $A^TA = I$, the 2×2 identity matrix, and that

$$P_S = A(A^TA)^{-1}A^T = \begin{pmatrix} 0 & 0 \\ 1 & 0 \\ 0 & 1 \end{pmatrix} \begin{pmatrix} 0 & 1 & 0 \\ 0 & 0 & 1 \end{pmatrix} = \begin{pmatrix} 0 & 0 & 0 \\ 0 & 1 & 0 \\ 0 & 0 & 1 \end{pmatrix}.$$

Thus P_S projects each vector

$$\begin{pmatrix} b_1 \\ b_2 \\ b_3 \end{pmatrix} \text{ in } \mathbb{R}^3 \text{ onto } \quad P_S \begin{pmatrix} b_1 \\ b_2 \\ b_3 \end{pmatrix} = \begin{pmatrix} 0 \\ b_2 \\ b_3 \end{pmatrix}$$

in the x_2, x_3-plane. ◁

Example 4 Find the matrix that projects vectors in $\mathbb{R}^3$ onto the plane $2x - y - 3z = 0$.

Solution We observe that the given plane contains the zero vector and can therefore be written as the subspace $S = \text{sp}(\mathbf{a}_1, \mathbf{a}_2)$, where $\mathbf{a}_1$ and $\mathbf{a}_2$ are any two nonzero and nonparallel vectors in the plane. We choose

$$\mathbf{a}_1 = \begin{pmatrix} 0 \\ 3 \\ -1 \end{pmatrix} \text{ and } \mathbf{a}_2 = \begin{pmatrix} 1 \\ 2 \\ 0 \end{pmatrix},$$

so that

$$A = \begin{pmatrix} 0 & 1 \\ 3 & 2 \\ -1 & 0 \end{pmatrix}.$$

Then

$$(A^TA)^{-1} = \begin{pmatrix} 10 & 6 \\ 6 & 5 \end{pmatrix}^{-1} = (1/14) \begin{pmatrix} 5 & -6 \\ -6 & 10 \end{pmatrix}$$

and

$$P_S = (1/14) \begin{pmatrix} 0 & 1 \\ 3 & 2 \\ -1 & 0 \end{pmatrix} \begin{pmatrix} 5 & -6 \\ -6 & 10 \end{pmatrix} \begin{pmatrix} 0 & 3 & -1 \\ 1 & 2 & 0 \end{pmatrix}$$

$$= (1/14) \begin{pmatrix} -6 & 10 \\ 3 & 2 \\ -5 & 6 \end{pmatrix} \begin{pmatrix} 0 & 3 & -1 \\ 1 & 2 & 0 \end{pmatrix} = (1/14) \begin{pmatrix} 10 & 2 & 6 \\ 2 & 13 & -3 \\ 6 & -3 & 5 \end{pmatrix}.$$

Each vector $\mathbf{b}$ in $\mathbb{R}^3$ projects onto the vector

$$\mathbf{b}_S = P_S\mathbf{b} = (1/14) \begin{pmatrix} 10 & 2 & 6 \\ 2 & 13 & -3 \\ 6 & -3 & 5 \end{pmatrix} \begin{pmatrix} b_1 \\ b_2 \\ b_3 \end{pmatrix}$$

$$= (1/14) \begin{pmatrix} 10b_1 + 2b_2 + 6b_3 \\ 2b_1 + 13b_2 - 3b_3 \\ 6b_1 - 3b_2 + 5b_3 \end{pmatrix}. \quad \triangleleft$$

It might appear from the formula $P_S = A(A^TA)^{-1}A^T$ that P_S depends on the particular choice of basis for S to use for the column vectors of A. However, this is not so. If C is any other matrix of the same size *whose column space is the same as the column space S of A*, then two of our exercises will show that $\bar{P}_S = C(C^TC)^{-1}C^T$ is the same matrix as $P_S = A(A^TA)^{-1}A^T$. As a first step, Exercise 16 indicates that the properties of a projection (boxed on page 248) *uniquely* characterize the projection. That is, for any subspace S of $\mathbb{R}^n$ and vector $\mathbf{b}$ in $\mathbb{R}^n$, there is a *unique* vector $\mathbf{p}$ in S such that $\mathbf{b} - \mathbf{p}$ is perpendicular to *every* vector in S. Once this has been shown, we know that $\bar{P}_S\mathbf{x} = P_S\mathbf{x}$ for *every* vector $\mathbf{x}$ in $\mathbb{R}^n$, since multiplication of $\mathbf{x}$ by either $\bar{P}_S$ or P_S must give this unique projection vector. Exercise 17 indicates that this implies that $\bar{P}_S = P_S$.

In view of this uniqueness, we may refer to the matrix $P_S = A(A^TA)^{-1}A^T$ in formula (6) as *the* projection matrix for the subspace S. We summarize our work in a theorem.

Theorem 5.1   Projection Matrix

Let S be a subspace of $\mathbb{R}^n$ and let $\mathbf{a}_1, \mathbf{a}_2, \ldots, \mathbf{a}_k$ in $\mathbb{R}^n$ form a basis for S. The projection matrix for the subspace S is $P_S = A(A^TA)^{-1}A^T$, where A is the $n \times k$ matrix having columns $\mathbf{a}_1, \mathbf{a}_2, \ldots, \mathbf{a}_k$. If $\mathbf{b}$ is any column vector in $\mathbb{R}^n$, then the vector projection of $\mathbf{b}$ on S is given by $\mathbf{b}_S = P_S\mathbf{b}$. The matrix P_S is the unique matrix with the property that for every vector $\mathbf{b}$ in $\mathbb{R}^n$, the vector $P_S\mathbf{b}$ lies in S and the vector $\mathbf{b} - P_S\mathbf{b}$ is perpendicular to every vector in S.

Exercise 15 indicates that the projection matrix P_S given in Theorem 5.1 satisfies these two properties:

Properties of a Projection Matrix *P*

1. $P^2 = P$ (P is **idempotent**).

2. $P^T = P$ (P is symmetric).

We can use Property 2 as a partial check for errors in long computations that lead to P_S, as in Example 4. These two properties completely characterize the projection matrices, as we now show.

Theorem 5.2 **Characterization of Projection Matrices**

The projection matrix P_S of a subspace S of $\mathbb{R}^n$ is both idempotent and symmetric. Conversely, every $n \times n$ matrix that is both idempotent and symmetric is a projection matrix: namely, it is the projection matrix for its column space.

Proof Exercise 15 indicates that a projection matrix is both idempotent and symmetric.

To establish the converse, let P be an $n \times n$ matrix that is both symmetric and idempotent. We show that P is the projection matrix for its own column space S. Let **b** be any vector in $\mathbb{R}^n$. By Theorem 5.1, we need show only that $P\mathbf{b}$ satisfies the characterizing properties of the projection of **b** on S given in the box on page 248. Now $P\mathbf{b}$ surely lies in the column space S of P, since S consists of all vectors $P\mathbf{x}$ for any vector **x** in $\mathbb{R}^n$. The second requirement is that $\mathbf{b} - P\mathbf{b}$ must be perpendicular to each vector $P\mathbf{x}$ in the column space of P. Writing the dot product of $\mathbf{b} - P\mathbf{b}$ and $P\mathbf{x}$ in matrix form, and using the hypotheses $P^2 = P = P^T$, we have

$$(\mathbf{b} - P\mathbf{b})^T P\mathbf{x} = [(I - P)\mathbf{b}]^T P\mathbf{x} = \mathbf{b}^T(I - P)^T P\mathbf{x}$$
$$= \mathbf{b}^T(I - P)^T P^T\mathbf{x} = \mathbf{b}^T(P - P^2)^T\mathbf{x}$$
$$= \mathbf{b}^T(P - P)^T\mathbf{x} = \mathbf{b}^T O\mathbf{x} = \mathbf{0}.$$

Since their dot product is zero, we see that $\mathbf{b} - P\mathbf{b}$ and $P\mathbf{x}$ are indeed perpendicular, and our proof is complete. ■

SUMMARY

---

1. The vector projection of **b** in $\mathbb{R}^n$ on a nonzero vector **a** in $\mathbb{R}^n$ is given by $[(\mathbf{a} \cdot \mathbf{b})/(\mathbf{a} \cdot \mathbf{a})]\mathbf{a}$. The (scalar) component of **b** along **a** is $(\mathbf{a} \cdot \mathbf{b})/\|\mathbf{a}\|$.

Let $\{\mathbf{a}_1, \mathbf{a}_2, \ldots, \mathbf{a}_k\}$ be a basis for a subspace S of $\mathbb{R}^n$ and let A be the $n \times k$ matrix having the $\mathbf{a}_i$ as column vectors so that S is the column space of A.

2. The vector projection of a column vector **b** in $\mathbb{R}^n$ on S is the unique vector **p** in S such that $\mathbf{b} - \mathbf{p}$ is perpendicular to every vector in S.

3. The vector projection of a column vector **b** in $\mathbb{R}^n$ on S is $\mathbf{b}_S = A(A^TA)^{-1}A^T\mathbf{b}$.

4. The matrix $P_S = A(A^TA)^{-1}A^T$ is the projection matrix for the subspace S. It is the unique matrix such that for every vector **b** in $\mathbb{R}^n$, the vector $P_S\mathbf{b}$ lies in S and the vector $\mathbf{b} - P_S\mathbf{b}$ is perpendicular to every vector in S.

5. The projection matrix P_S of a subspace S is idempotent and symmetric. Every symmetric idempotent matrix is the projection matrix for its column space.

EXERCISES

In Exercises 1–4, find each indicated vector projection and also find the corresponding scalar component of the first vector along the second.

1. The vector projection of $(2, 1)$ on $(3, 4)$ in $\mathbb{R}^2$

2. The vector projection of $(3, 4)$ on $(2, 1)$ in $\mathbb{R}^2$

3. The vector projection of $(1, 2, 1)$ on each of the unit coordinate vectors in $\mathbb{R}^3$

4. The vector projection of $(1, 2, 1)$ on $(3, 1, 2)$ in $\mathbb{R}^3$

5. Find the vector projection of $(1, 2, 1)$ on the subspace $\text{sp}((3, 1, 2), (1, 0, 1))$ of $\mathbb{R}^3$.

6. Find the vector projection of $(1, 0, 0)$ on the subspace $\text{sp}((2, 1, 1), (1, 0, 2))$ of $\mathbb{R}^3$.

In Exercises 7–11, find the projection matrix for the indicated subspace.

7. The x_1,x_2-plane in $\mathbb{R}^3$

8. The subspace $\text{sp}((3, 1, 2), (1, 0, 1))$ of $\mathbb{R}^3$ (see Exercise 5)

9. The subspace $\text{sp}((2, 1, 1), (1, 0, 2))$ of $\mathbb{R}^3$ (see Exercise 6)

10. The x_1,x_3-coordinate subspace in $\mathbb{R}^4$

11. The x_1,x_2,x_4-coordinate subspace in $\mathbb{R}^4$

12. Show that Eq. (6) in the text includes Eq. (3) as a special case, that is, write Eq. (3) in the form of Eq. (6).

13. Let **a** be a vector in $\mathbb{R}^n$ and suppose that $\mathbf{x} \cdot \mathbf{a} = 0$ for all vectors **x** in $\mathbb{R}^n$. Show that $\mathbf{a} = \mathbf{0}$.

14. Let **a** be a unit column vector in $\mathbb{R}^n$. Show that $\mathbf{aa}^T$ is the projection matrix for the subspace $\text{sp}(\mathbf{a})$.

15. Show that the projection matrix $P = A(A^TA)^{-1}A^T$ given in Theorem 5.1 satisfies the following two conditions:
    a) $P^2 = P$;
    b) $P^T = P$.

16. Let S be a subspace of $\mathbb{R}^n$ and let **b** be a vector in $\mathbb{R}^n$. Show that there is only one vector **p** in S such that $\mathbf{b} - \mathbf{p}$ is perpendicular to every vector in S. [*Hint:* The standard way to show uniqueness of something is to suppose you have two of the things, and then show that they are equal. Start by assuming that both **p** and **q** are in S, and that both $\mathbf{b} - \mathbf{p}$ and $\mathbf{b} - \mathbf{q}$ are perpendicular to every vector in S. Then show that $\mathbf{p} - \mathbf{q}$ is perpendicular to every vector in S and hence to itself.]

17. Let P and Q be two $n \times n$ matrices and suppose that $P\mathbf{x} = Q\mathbf{x}$ for *every* vector $\mathbf{x}$ in $\mathbb{R}^n$. Show that $P = Q$.

18. What is the projection matrix of the subspace $\mathbb{R}^n$ of $\mathbb{R}^n$?

The formula $A(A^TA)^{-1}A^T$ for a projection matrix can be tedious to compute using pencil and paper, but the software MATCOMP can do it easily. In Exercises 19–23, use MATCOMP, or similar software, to find the indicated vector projections.

19. The projections in $\mathbb{R}^6$ of $(-1, 2, 3, 1, 6, 2)$ and $(2, 0, 3, -1, 4, 5)$ on $\text{sp}((1, -2, 3, 1, 4, 0))$

20. The projections in $\mathbb{R}^3$ of $(1, -1, 4)$, $(3, 3, -1)$, and $(-2, 4, 7)$ on $\text{sp}((1, 3, -4), (2, 0, 3))$

21. The projections in $\mathbb{R}^4$ of $(-1, 3, 2, 0)$ and $(4, -1, 1, 5)$ on $\text{sp}((0, 1, 2, 1), (-1, 2, 1, 4))$

22. The projections in $\mathbb{R}^4$ of $(2, 1, 0, 3)$, $(1, 1, -1, 2)$, and $(4, 3, 1, 3)$ on $\text{sp}((1, 0, -1, 0), (1, 2, -1, 4), (2, 1, 3, -1))$

23. The projections in $\mathbb{R}^5$ of $(2, 1, -3, 2, 4)$ and $(1, -4, 0, 1, 5)$ on $\text{sp}((3, 1, 4, 0, 1), (2, 1, 3, -5, 1))$

24. Work with Topic 3 of the available program VECTGRPH until you are able to get a score of at least 80 percent most of the time.

5.2
The Method of Least Squares

THE NATURE OF THE PROBLEM

In this section we apply our work on projections to problems of data analysis. Suppose data measurements of the form (a_i, b_i) are obtained from observation or experimentation, and are plotted as data points in the x,y-plane. It is desirable to find a mathematical relationship $y = f(x)$ that represents the data reasonably well, so that we can make predictions of data values that were not measured. Geometrically, this means that we would like the graph of $y = f(x)$ in the plane to pass very close to our data points. Depending on the nature of the experiment and the configuration of the plotted data points, we might decide on an appropriate type of function $y = f(x)$ such as a linear function, a quadratic function, an exponential function, etc. We illustrate with three types of problem.

Problem 1   According to Hooke's law, the distance that a spring stretches is proportional to the force applied. Suppose we attach four different weights a_1, a_2, a_3, and a_4 in turn to the bottom of a spring. We measure the four lengths b_1, b_2, b_3, and b_4 of the stretched spring. Suppose the data in Table 5.1 are obtained. Because of Hooke's law, we expect the data points (a_i, b_i) to be close to some line with equation

$$y = f(x) = r_0 + r_1 x,$$

where r_0 is the length of the spring and r_1 is the *spring constant*. That is, if our measurements were exact and the spring ideal, we would have $b_i = r_0 + r_1 a_i$ for specific values r_0 and r_1. ◁

Table 5.1

a_i = weight in ounces	2.0	4.0	5.0	6.0
b_i = length in inches	2.5	4.5	7.0	8.5

In Problem 1 above, we have only the two unknowns r_0 and r_1, and in theory, just two measurements should suffice to find them. In practice, however, we expect to have some error in physical measurements. It is standard procedure to make more measurements than are theoretically necessary in the hope that the errors will roughly cancel each other out in accordance with the laws of probability theory. Substitution of each data point (a_i, b_i) from Problem 1 into the equation $y = r_0 + r_1 x$ gives a single linear equation in the two unknowns r_0 and r_1. The four data points of Problem 1 thus give rise to a linear system of *four* equations in only *two* unknowns. Such a linear system with more equations than unknowns is called **overdetermined,** and one expects to find that the system is *inconsistent,* having no actual solution. It will be our task to find values for the unknowns r_0 and r_1 that will come as close as possible, in some sense, to satisfying all four of the equations.

We have used the illustration presented in Problem 1 to introduce our goal in this section, and we will solve the problem in a moment. We first present two more hypothetical problems, which we will also solve later in the section.

Problem 2 At a recent boat show, the observations listed in Table 5.2 were made relating the prices b_i of sailboats and their weights a_i. Plotting the data points (a_i, b_i), as shown in Fig. 5.5, we might expect a quadratic function of the form

$$y = f(x) = r_0 + r_1 x + r_2 x^2$$

to fit the data fairly well. ◁

Table 5.2

a_i = weight in tons	2	4	5	8
b_i = price in units of \$10,000	1	3	5	12

Problem 3 A population of rabbits on a large island was estimated each year from 1981 to 1984 and produced the data in Table 5.3. Knowing that population growth is

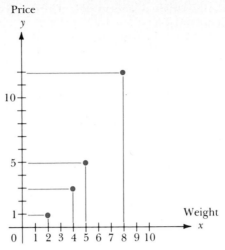

Figure 5.5
Problem 2 data.

exponential in the absence of disease, predators, famine, and so on, we expect an exponential function

$$y = f(x) = re^{sx}$$

to best represent these data. Note that by using logarithms, we can convert this exponential function to the linear form

$$\ln(y) = \ln(r) + sx. \quad \triangleleft$$

Table 5.3

a_i = (Year observed) − 1980	1	2	3	4
b_i = number of rabbits in units of 1000	3	4.5	8	17

THE METHOD OF LEAST SQUARES

Consider now the problem of finding a linear function $f(x) = r_0 + r_1 x$ "best" fitting data points (a_i, b_i) for $i = 1, 2, \ldots, m$, where $m > 2$. Geometrically, this amounts to finding the line in the plane that comes closest, in some sense, to passing through the m data points. If there were no error in our measurements and our data were truly linear, then for some r_0 and r_1 we would have

$$b_i = r_0 + r_1 a_i \quad \text{for } i = 1, 2, \ldots, m.$$

These m linear equations in the two unknowns r_0 and r_1 form an overdetermined system of equations that probably has no solution. Out data points actually satisfy a system of linear approximations, which can be expressed in matrix form as

$$
\begin{pmatrix} b_1 \\ b_2 \\ \vdots \\ b_m \end{pmatrix} \approx \begin{pmatrix} 1 & a_1 \\ 1 & a_2 \\ \vdots & \vdots \\ 1 & a_m \end{pmatrix} \begin{pmatrix} r_0 \\ r_1 \end{pmatrix} \tag{1}
$$
$$
\mathbf{b} \qquad\qquad A \qquad\quad \mathbf{r}
$$

or simply as $\mathbf{b} \approx A\mathbf{r}$. We try to find an optimal solution vector $\bar{\mathbf{r}}$ for the system (1) of approximations. For each vector $\mathbf{r}$, the **error vector** $A\mathbf{r} - \mathbf{b}$ measures how far our system (1) is from being a system of equations with solution vector $\mathbf{r}$. The absolute values of the components of the vector $A\mathbf{r} - \mathbf{b}$ represent the distances $d_i = |r_0 + r_1 a_i - b_i|$, shown in Fig. 5.6.

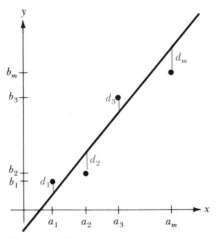

Figure 5.6
The distances d_i.

A TECHNIQUE VERY CLOSE TO THAT OF LEAST SQUARES was developed by Roger Cotes (1682–1716), the gifted mathematician who edited the second edition of Isaac Newton's *Principia*, in a work dealing with errors in astronomical observations, written around 1715.

The complete principle, however, was first formulated by Carl Gauss around the age of 16 while he was adjusting approximations in dealing with the distribution of prime numbers. Gauss later stated that he used the method frequently over the years—for example, when he did calculations concerning the orbits of asteroids. Gauss published the method in 1809 and gave a definitive exposition 14 years later.

On the other hand, it is to Adrien-Marie Legendre (1752–1833), founder of the theory of elliptic functions, that we owe the first publication of the method of least squares in an 1806 work on determining the orbits of comets. After Gauss's 1809 publication, Legendre wrote to him, censuring him for claiming the method as his own. Even as late as 1827, Legendre was still berating Gauss for "appropriating the discoveries of others." In fact, the problem lay in Gauss's failure to publish many of his discoveries promptly; he mentioned them only after they were published by others.

We want to minimize, in some sense, our error vector $A\mathbf{r} - \mathbf{b}$. There are a number of different methods for minimization that are very useful. For example, one might want to minimize the maximum of the distances d_i. We study just one sense of minimization, the one that probably seems most natural at this point. We will minimize the length $\|A\mathbf{r} - \mathbf{b}\|$ of our error vector. Minimizing $\|A\mathbf{r} - \mathbf{b}\|$ is equivalent to minimizing $\|A\mathbf{r} - \mathbf{b}\|^2$, but the latter is computationally easier since it involves no square root. Minimizing $\|A\mathbf{r} - \mathbf{b}\|^2$ means minimizing the sum

$$d_1{}^2 + d_2{}^2 + \cdots + d_m{}^2 \tag{2}$$

of the squares of the distances in Fig. 5.6. Hence the name **"method of least squares"** given to this procedure.

If $\mathbf{a}_1$ and $\mathbf{a}_2$ denote the columns of A in the system (1), then the vector $A\mathbf{r} = r_0\mathbf{a}_1 + r_1\mathbf{a}_2$ lies in the column space $S = \mathrm{sp}(\mathbf{a}_1, \mathbf{a}_2)$ of A. From Fig. 5.7, it is geometrically clear that of all the vectors $A\mathbf{r}$ in S, the one that minimizes $\|A\mathbf{r} - \mathbf{b}\|$ is the *projection* $\mathbf{b}_S = A\bar{\mathbf{r}}$ of $\mathbf{b}$ on S. Equation (5) of Section 5.1 shows that then $A\bar{\mathbf{r}} = A(A^TA)^{-1}A^T\mathbf{b}$ and provides us with this boxed formula for the solution vector $\bar{\mathbf{r}}$, which is optimal in this least-squares sense:

$$
\boxed{
\begin{array}{c}
\textbf{Least-squares Solution} \\[4pt]
\bar{\mathbf{r}} = (A^TA)^{-1}A^T\mathbf{b}
\end{array}
} \tag{3}
$$

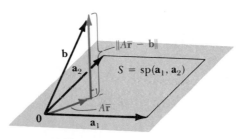

Figure 5.7
The length $\|A\bar{\mathbf{r}} - \mathbf{b}\|$.

The 2×2 matrix A^TA in formula (3) is invertible so long as the columns of A are independent, as shown by Theorem 3.18. For our matrix A shown in system (1), this just means that not all the values a_i are the same. Geometrically, this corresponds to saying that our data points in the plane do not all lie on a vertical line.

Example 1   Find the least-squares fit to the data points in Problem 1 by a straight line, that is, by a linear function $y = r_0 + r_1x$.

Solution   We form the system $\mathbf{b} \approx A\mathbf{r}$ in (1) for the data in Table 5.1:

$$\underset{\mathbf{b}}{\begin{pmatrix} 2.5 \\ 4.5 \\ 7.0 \\ 8.5 \end{pmatrix}} \approx \underset{A}{\begin{pmatrix} 1 & 2 \\ 1 & 4 \\ 1 & 5 \\ 1 & 6 \end{pmatrix}} \underset{\mathbf{r}}{\begin{pmatrix} r_0 \\ r_1 \end{pmatrix}}.$$

We have

$$A^T A = \begin{pmatrix} 1 & 1 & 1 & 1 \\ 2 & 4 & 5 & 6 \end{pmatrix} \begin{pmatrix} 1 & 2 \\ 1 & 4 \\ 1 & 5 \\ 1 & 6 \end{pmatrix} = \begin{pmatrix} 4 & 17 \\ 17 & 81 \end{pmatrix}$$

and

$$(A^T A)^{-1} = (1/35) \begin{pmatrix} 81 & -17 \\ -17 & 4 \end{pmatrix}.$$

From Eq. (3), we obtain

$$\bar{\mathbf{r}} = (A^T A)^{-1} A^T \mathbf{b} = (1/35) \begin{pmatrix} 81 & -17 \\ -17 & 4 \end{pmatrix} \begin{pmatrix} 1 & 1 & 1 & 1 \\ 2 & 4 & 5 & 6 \end{pmatrix} \begin{pmatrix} 2.5 \\ 4.5 \\ 7.0 \\ 8.5 \end{pmatrix}$$

$$= (1/35) \begin{pmatrix} 47 & 13 & -4 & -21 \\ -9 & -1 & 3 & 7 \end{pmatrix} \begin{pmatrix} 2.5 \\ 4.5 \\ 7.0 \\ 8.5 \end{pmatrix} = (1/35) \begin{pmatrix} -30.5 \\ 53.5 \end{pmatrix} \approx \begin{pmatrix} -0.9 \\ 1.5 \end{pmatrix}.$$

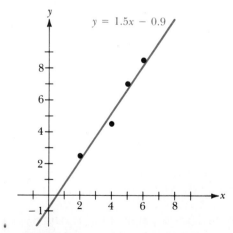

Figure 5.8
The least-squares fit of data points.

Therefore, the equation of the line that best fits the data in the least-squares sense is $y = 1.5x - 0.9$. This line and the data points are shown in Fig. 5.8. ◁

Example 2 Use the method of least squares to fit the data in Problem 3 by an exponential function $y = f(x) = re^{sx}$.

Solution We use logarithms and convert the exponential equation to the linear equation

$$\ln(y) = \ln(r) + sx.$$

Table 5.4

$x = a_i$	1	2	3	4
$y = b_i$	3	4.5	8	17
$z = \ln(b_i)$	1.1	1.5	2.08	2.83

From Table 5.2, we obtain the data in Table 5.4 to use for our logarithmic equation. We obtain

$$A^TA = \begin{pmatrix} 1 & 1 & 1 & 1 \\ 1 & 2 & 3 & 4 \end{pmatrix} \begin{pmatrix} 1 & 1 \\ 1 & 2 \\ 1 & 3 \\ 1 & 4 \end{pmatrix} = \begin{pmatrix} 4 & 10 \\ 10 & 30 \end{pmatrix}$$

$$(A^TA)^{-1} = \begin{pmatrix} \frac{3}{2} & -\frac{1}{2} \\ -\frac{1}{2} & \frac{1}{5} \end{pmatrix}$$

$$(A^TA)^{-1}A^T = \begin{pmatrix} \frac{3}{2} & -\frac{1}{2} \\ -\frac{1}{2} & \frac{1}{5} \end{pmatrix} \begin{pmatrix} 1 & 1 & 1 & 1 \\ 1 & 2 & 3 & 4 \end{pmatrix}$$

$$= \begin{pmatrix} 1 & \frac{1}{2} & 0 & -\frac{1}{2} \\ -\frac{3}{10} & -\frac{1}{10} & \frac{1}{10} & \frac{3}{10} \end{pmatrix}.$$

Multiplying this last matrix on the right by

$$\begin{pmatrix} 1.1 \\ 1.5 \\ 2.08 \\ 2.83 \end{pmatrix},$$

we obtain from Eq. (3),

$$\bar{\mathbf{r}} = \begin{pmatrix} \ln r \\ s \end{pmatrix} = \begin{pmatrix} .435 \\ .577 \end{pmatrix}.$$

Thus $r = e^{.435} \approx 1.54$, and we obtain $y = f(x) = 1.54e^{.577x}$ as a fitting exponential function.

The graph of the function and of the data points in Table 5.5, is shown in Fig. 5.9. On the basis of the function $f(x)$ obtained, we can project the population of rabbits on the island in 1990 to be about

$$f(10) \cdot 1000 = 493{,}628 \text{ rabbits}$$

unless predators, disease, or lack of food interferes. ◁

Table 5.5

a_i	b_i	$f(a_i)$
1	3	2.7
2	4.5	4.9
3	8	8.7
4	17	15.5

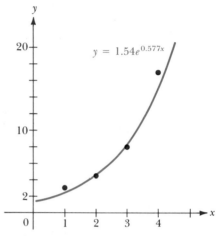

Figure 5.9
Data points and the exponential fit.

In Example 2, it is important to note that we used the least-squares method with the *logarithm* of our original y-coordinate data. Thus it is the equation $\ln y = r + sx$ that is a best least-squares fit rather than the equation $y = f(x) = re^{sx}$. Using the least-squares method to fit the logarithm of the y-coordinate data produces an exponential function that approximates the smaller y-values in the data better than the larger y-values, as illustrated by Table 5.5. It can be shown that the fit $y = f(x)$ amounts roughly to minimizing the *percent of error* in taking $f(a_i)$ for b_i. The least-squares logarithmic fit does not yield the least-squares exponential fit. The supplied program YOUFIT can be used to illustrate this (see Exercises 23 and 24).

LEAST-SQUARES SOLUTIONS FOR LARGER SYSTEMS

It should be clear that data points of more than two components may lead to overdetermined linear systems in more than two unknowns. Suppose that an experiment is repeated m times, and data values

$$b_j, a_{j1}, \ldots, a_{jn}$$

are obtained from measurements on the jth experiment. For example, the data values a_{ij} might be ones that can be controlled and the value b_j measured.

Of course, there may be errors in the controlled as well as in the measured values. Suppose we have reason to believe that the data obtained for each experiment should *theoretically* satisfy the same linear relation

$$y = r_0 + r_1 x_1 + r_2 x_2 + \cdots + r_n x_n \tag{4}$$

with $y = b_j$ and $x_i = a_{ij}$. We then obtain a system of m linear approximations in $n + 1$ unknowns $r_0, r_1, r_2, \ldots, r_n$:

$$
\underbrace{\begin{pmatrix} b_1 \\ b_2 \\ \vdots \\ b_m \end{pmatrix}}_{\mathbf{b}} \approx \underbrace{\begin{pmatrix} 1 & a_{11} & a_{12} & \cdots & a_{1n} \\ 1 & a_{21} & a_{22} & \cdots & a_{2n} \\ \vdots & \vdots & \vdots & & \vdots \\ 1 & a_{m1} & a_{m2} & \cdots & a_{mn} \end{pmatrix}}_{A} \underbrace{\begin{pmatrix} r_0 \\ r_1 \\ r_2 \\ \vdots \\ r_n \end{pmatrix}}_{\mathbf{r}}. \tag{5}
$$

If $m > n + 1$, then the system (5) corresponds to an overdetermined linear system $\mathbf{b} = A\mathbf{r}$, which probably has no exact solution. If the rank of A is $n + 1$, then a repetition of our geometric argument above indicates that the least-squares solution $\bar{\mathbf{r}}$ for the system $\mathbf{b} \approx A\mathbf{r}$ is given by

$$\bar{\mathbf{r}} = (A^T A)^{-1} A^T \mathbf{b}. \tag{6}$$

A linear system of the form (5) arises if m data points (a_i, b_i) are found and a least-squares fit is sought for a polynomial function

$$y = r_0 + r_1 x + \cdots + r_{n-1} x^{n-1} + r_n x^n.$$

The data point (a_i, b_i) leads to the linear approximation

$$b_i \approx r_0 + r_1 a_i + \cdots + r_{n-1} a_i^{n-1} + r_n a_i^n.$$

The m data points thus give rise to a linear system of the form (5), where the matrix A is given by

$$
A = \begin{pmatrix} 1 & a_1 & a_1^2 & \cdots & a_1^n \\ 1 & a_2 & a_2^2 & \cdots & a_2^n \\ \vdots & \vdots & \vdots & & \vdots \\ 1 & a_m & a_m^2 & \cdots & a_m^n \end{pmatrix}. \tag{7}
$$

Example 3   Find the least-squares fit to the data in Problem 2 by a parabola, that is, by a quadratic function

$$y = r_0 + r_1 x + r_2 x^2.$$

Solution   We write the data in the form $\mathbf{b} \approx A\mathbf{r}$, where A has the form (7):

$$
\begin{pmatrix} 1 \\ 3 \\ 5 \\ 12 \end{pmatrix} \approx \begin{pmatrix} 1 & 2 & 4 \\ 1 & 4 & 16 \\ 1 & 5 & 25 \\ 1 & 8 & 64 \end{pmatrix} \begin{pmatrix} r_0 \\ r_1 \\ r_2 \end{pmatrix}.
$$

Then

$$A^T A = \begin{pmatrix} 1 & 1 & 1 & 1 \\ 2 & 4 & 5 & 8 \\ 4 & 16 & 25 & 64 \end{pmatrix} \begin{pmatrix} 1 & 2 & 4 \\ 1 & 4 & 16 \\ 1 & 5 & 25 \\ 1 & 8 & 64 \end{pmatrix} = \begin{pmatrix} 4 & 19 & 109 \\ 19 & 109 & 709 \\ 109 & 709 & 4993 \end{pmatrix}$$

and

$$(A^T A)^{-1} = \frac{1}{5400} \begin{pmatrix} 41556 & -17586 & 1590 \\ -17586 & 8091 & -765 \\ 1590 & -765 & 75 \end{pmatrix}.$$

We compute

$$(A^T A)^{-1} A^T = \frac{1}{5400} \begin{pmatrix} 41556 & -17586 & 1590 \\ -17586 & 8091 & -765 \\ 1590 & -765 & 75 \end{pmatrix} \begin{pmatrix} 1 & 1 & 1 & 1 \\ 2 & 4 & 5 & 8 \\ 4 & 16 & 25 & 64 \end{pmatrix}$$

$$= \frac{1}{5400} \begin{pmatrix} 12744 & -3348 & -6624 & 2628 \\ -4464 & 2538 & 3744 & -1818 \\ 360 & -270 & -360 & 270 \end{pmatrix}$$

$$= \begin{pmatrix} 2.36 & -.62 & -1.22667 & .4866667 \\ -.826667 & .47 & .6933333 & -.3366667 \\ .06666667 & -.05 & -.06666667 & .05 \end{pmatrix}.$$

Then $\bar{\mathbf{r}}$ is obtained by multiplying this last matrix on the right by

$$\begin{pmatrix} 1 \\ 3 \\ 5 \\ 12 \end{pmatrix} \quad \text{resulting in} \quad \bar{\mathbf{r}} = \begin{pmatrix} .207 \\ .01 \\ .183 \end{pmatrix}.$$

Thus the quadratic function that best approximates the data in the least-squares sense is

$$y = f(x) = .207 + .01x + .183x^2.$$

In Fig. 5.10, we show the graph of this quadratic function and plot the data points. Data are given in Table 5.6. On the basis of the least-squares fit, we might project that the price of a 10-ton sailing yacht would be about 18.607 times \$10,000 or \$186,070, and that the price of a 20-ton yacht would be about $f(20)$ times \$10,000 or \$736,070. However, one should be very wary of using a function $f(x)$ that seems to fit data quite well for measured values of x to project data for values of x far from those measured values. Our quadratic function seems to fit our data quite well, but we should not expect the cost we might project for a 100-ton ship to be at all accurate. ◁

Table 5.6

a_i	b_i	$f(a_i)$
2	1	.959
4	3	3.175
5	5	4.832
8	12	11.999

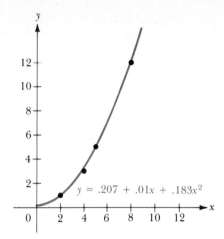

$$y = .207 + .01x + .183x^2$$

Figure 5.10
The graph and data points for Example 5.

Example 4   Show that the least-squares linear fit to data points (a_i, b_i) for $i = 1, 2, \ldots, m$ by a constant function $f(x) = r_0$ is achieved when r_0 is the average y-value

$$(b_1 + b_2 + \cdots + b_m)/m$$

of the data values b_i.

Solution   Such a constant function $y = f(x) = r_0$ of course has a horizontal line as its graph. In this situation, we have $n = 1$ as the number of unknowns and the overdetermined system (5) becomes

$$\underbrace{\begin{pmatrix} b_1 \\ b_2 \\ \vdots \\ b_m \end{pmatrix}}_{\mathbf{b}} \approx \underbrace{\begin{pmatrix} 1 \\ 1 \\ \vdots \\ 1 \end{pmatrix}}_{A} \underbrace{(r_0)}_{\mathbf{r}}.$$

Thus A is the $m \times 1$ matrix with all entries 1. We easily find that $(A^T A) = (m)$, so $(A^T A)^{-1} = (1/m)$. We also find that $A^T \mathbf{b} = b_1 + b_2 + \cdots + b_m$. Thus the least-squares solution (6) is

$$\bar{r} = (b_1 + b_2 + \cdots + b_m)/m$$

as asserted.   ◁

OVERDETERMINED SYSTEMS OF LINEAR EQUATIONS

We have presented examples in which least-squares solutions were found to systems of linear approximations arising from applications. We used $\mathbf{r}$ as the column vector of unknowns in those examples, since we were using x as the independent variable in the formula for a fitting function $f(x)$. We now discuss the mathematics of overdetermined systems, and we use $\mathbf{x}$ as the column vector of unknowns again.

Consider a general overdetermined system of linear equations

$$A\mathbf{x} = \mathbf{b},$$

where A is an $m \times n$ matrix of rank n and $m > n$. We expect such a system to be inconsistent, that is, to have no solution. As we have seen, the system of linear approximations $A\mathbf{x} \approx \mathbf{b}$ has least-squares solution

$$\bar{\mathbf{x}} = (A^T A)^{-1} A^T \mathbf{b}. \tag{8}$$

The vector $\bar{\mathbf{x}}$ is thus the **least-squares solution of the overdetermined system** $A\mathbf{x} = \mathbf{b}$. Equation (8) can be written in the equivalent form

$$A^T A \bar{\mathbf{x}} = A^T \mathbf{b}, \tag{9}$$

which exhibits $\bar{\mathbf{x}}$ as the solution of a consistent linear system with coefficient matrix $A^T A$.

Example 5   Find the least-squares approximate solution of the overdetermined linear system

$$\begin{pmatrix} 1 & 0 & 1 \\ 2 & 1 & 1 \\ 1 & 3 & 0 \\ 0 & 2 & 1 \\ -1 & 2 & -1 \end{pmatrix} \begin{pmatrix} x_1 \\ x_2 \\ x_3 \end{pmatrix} = \begin{pmatrix} 1 \\ 0 \\ 1 \\ 2 \\ 0 \end{pmatrix}$$

by transforming it into the consistent form (9).

Solution   The consistent system (9) is

$$\begin{pmatrix} 1 & 2 & 1 & 0 & -1 \\ 0 & 1 & 3 & 2 & 2 \\ 1 & 1 & 0 & 1 & -1 \end{pmatrix} \begin{pmatrix} 1 & 0 & 1 \\ 2 & 1 & 1 \\ 1 & 3 & 0 \\ 0 & 2 & 1 \\ -1 & 2 & -1 \end{pmatrix} \begin{pmatrix} x_1 \\ x_2 \\ x_3 \end{pmatrix} = \begin{pmatrix} 1 & 2 & 1 & 0 & -1 \\ 0 & 1 & 3 & 2 & 2 \\ 1 & 1 & 0 & 1 & -1 \end{pmatrix} \begin{pmatrix} 1 \\ 0 \\ 1 \\ 2 \\ 0 \end{pmatrix}$$

or

$$\begin{pmatrix} 7 & 3 & 4 \\ 3 & 18 & 1 \\ 4 & 1 & 4 \end{pmatrix} \begin{pmatrix} x_1 \\ x_2 \\ x_3 \end{pmatrix} = \begin{pmatrix} 2 \\ 7 \\ 3 \end{pmatrix}$$

whose solution is found to be

$$\bar{\mathbf{x}} = \begin{pmatrix} x_1 \\ x_2 \\ x_3 \end{pmatrix} = \begin{pmatrix} -.614 \\ .421 \\ 1.259 \end{pmatrix}$$

accurate to three decimal places. This is the least-squares approximate solution to the given overdetermined system. Note that the error vector

$$A\bar{\mathbf{x}} - \mathbf{b} = \begin{pmatrix} .645 \\ .452 \\ .649 \\ 2.101 \\ .198 \end{pmatrix} - \begin{pmatrix} 1 \\ 0 \\ 1 \\ 2 \\ 0 \end{pmatrix} = \begin{pmatrix} -.355 \\ .452 \\ -.351 \\ .101 \\ .198 \end{pmatrix}$$

is perpendicular to each of the columns of A. ◁

SUMMARY

Let $A\mathbf{x} = \mathbf{b}$ be a linear system of m equations in n unknowns, where $m > n$ (an overdetermined system) and where the rank of A is n.

1. The least-squares solution of the corresponding system $A\mathbf{x} \approx \mathbf{b}$ of linear approximations is the vector $\mathbf{x} = \bar{\mathbf{x}}$, which minimizes the magnitude of the error vector $A\mathbf{x} - \mathbf{b}$.

2. The least-squares solution $\bar{\mathbf{x}}$ of $A\mathbf{x} \approx \mathbf{b}$ is given by the formula

$$\bar{\mathbf{x}} = (A^TA)^{-1}A^T\mathbf{b}.$$

Geometrically, $A\bar{\mathbf{x}}$ is the projection of $\mathbf{b}$ on the column space of A.

3. An alternative to using the formula for $\bar{\mathbf{x}}$ in number (2) is to convert the overdetermined system $A\mathbf{x} = \mathbf{b}$ to the consistent system $(A^TA)\mathbf{x} = A^T\mathbf{b}$ by multiplying both sides of $A\mathbf{x} = \mathbf{b}$ by A^T, and then to solve the consistent system.

EXERCISES

1. Let the length b_i of a spring with an attached weight a_i be determined by measurements as shown in Table 5.7.
   a) Find the least-squares linear fit, in accordance with Hooke's law.
   b) Use the answer to (a) to estimate the length of the spring if a weight of 5 ounces is attached.

Table 5.7

a_i = weight in ounces	1	2	4	6
b_i = length in inches	3	4.1	5.9	8.2

2. A company had profits (in units of $10,000) of .5 in 1980, of 1 in 1982, and of 2 in 1985. Let time t be measured in years with $t = 0$ in 1980.
   a) Find the least-squares linear fit of the data.
   b) Using the answer to (a), estimate the profit in 1986.

3. Repeat Exercise 2, but find an exponential fit of the data, working with logarithms as explained in Example 2.

4. A publishing company specializing in college texts starts with a field sales force of 10 persons, and has profits of $100,000. On increasing this sales force to 20, it has profits of $300,000, and increasing its sales force to 30 produces profits of $400,000.
   a) Find the least-squares linear fit for this data. [*Hint:* Express the numbers of salespeople in multiples of 10 and the profit in multiples of $100,000.]
   b) Use the answer to (a) to estimate the profit if the sales force is reduced to 25.
   c) Does the profit obtained using the answer to (a) for a sales force of zero persons seem in any way plausible?

In Exercises 5–7, find the least-squares fit to the given data by a linear function $f(x) = r_0 + r_1 x$, having a straight line as graph in the plane. Graph the line and the data points.

5. (0, 1), (2, 6), (3, 11), (4, 12)   6. (1, 1), (2, 4), (3, 6), (4, 9)   7. (0, 0), (1, 1), (2, 3), (3, 8)

8. Find the least-squares fit to the data in Exercise 6 by a parabola (a quadratic polynomial function).

9. Repeat Exercise 8, but use the data in Exercise 7 instead.

In Exercises 10–14, find the least-squares approximate solution of the given overdetermined system $A\mathbf{x} = \mathbf{b}$ by converting it to a consistent system and then solving, as illustrated in Example 5.

10. $\begin{pmatrix} 1 & 2 \\ 1 & 1 \\ 2 & 3 \end{pmatrix} \begin{pmatrix} x_1 \\ x_2 \end{pmatrix} = \begin{pmatrix} 3 \\ 1 \\ 3 \end{pmatrix}$

11. $\begin{pmatrix} 3 & 1 \\ 1 & 2 \\ 2 & -1 \end{pmatrix} \begin{pmatrix} x_1 \\ x_2 \end{pmatrix} = \begin{pmatrix} 1 \\ 0 \\ -2 \end{pmatrix}$

12. $\begin{pmatrix} 1 & 1 & 1 \\ -1 & 0 & 1 \\ 1 & -1 & 0 \\ 0 & 1 & -1 \end{pmatrix} \begin{pmatrix} x_1 \\ x_2 \\ x_3 \end{pmatrix} = \begin{pmatrix} 0 \\ 1 \\ -1 \\ -2 \end{pmatrix}$

13. $\begin{pmatrix} 1 & 1 & 3 \\ -1 & 0 & 5 \\ 0 & 1 & -2 \\ 1 & -1 & 1 \\ 1 & 0 & 1 \end{pmatrix} \begin{pmatrix} x_1 \\ x_2 \\ x_3 \end{pmatrix} = \begin{pmatrix} 1 \\ -1 \\ 3 \\ -2 \\ 0 \end{pmatrix}$

14. $\begin{pmatrix} 2 & -1 & 1 \\ 1 & 0 & 1 \\ -1 & 0 & 1 \\ 0 & -5 & -1 \\ 1 & 2 & -2 \end{pmatrix} \begin{pmatrix} x_1 \\ x_2 \\ x_3 \end{pmatrix} = \begin{pmatrix} 2 \\ 1 \\ 2 \\ 5 \\ 0 \end{pmatrix}$

The formula $\mathbf{x} = (A^T A)^{-1} A^T \mathbf{b}$ for the least-squares solution of an overdetermined system $A\mathbf{x} = \mathbf{b}$ can easily be evaluated using MATCOMP. In Exercises 15–20, find the least-squares solution requested in the exercise indicated using MATCOMP.

15. Exercise 1   16. Exercise 2   17. Exercise 4
18. Exercise 8   19. Exercise 9   20. Exercise 14

The supplied program YOUFIT can be used to illustrate graphically the fitting of data points by linear, quadratic, or exponential functions. In Exercises 21–25, use YOUFIT to try to visually fit the given data with the indicated type of graph. When this is done, enter the zero data suggested to see the computer's fit. Run twice more with the same data points but without trying to fit the data visually, and determine whether the data is best fitted by a linear, quadratic, or (logarithmically fitted) exponential function by comparing the least-squares sums for the three cases.

21. Fit (1, 2), (4, 6), (7, 10), (10, 14), (14, 19) by a linear graph.

22. Fit (2, 2), (6, 10), (10, 12), (16, 2) by a quadratic graph.

23. Fit (1, 1), (10, 8), (14, 12), (16, 20) by an exponential graph. Try to achieve a lower squares sum than the computer obtains with its least-squares fit that uses logarithms of y-values.

24. Repeat Exercise 23 with data (1, 9), (5, 1), (6, .5), (9, .01).

25. Fit (2, 9), (4, 6), (7, 1), (8, .1) by a linear graph.

5.3
Projections Using Orthonormal Bases

We return to our study of the projection of a vector on a subspace. In Example 3 of Section 5.1, we saw that the projection matrix for the x_2, x_3-plane in $\mathbb{R}^3$ has the very simple description

$$P_S = \begin{pmatrix} 0 & 0 & 0 \\ 0 & 1 & 0 \\ 0 & 0 & 1 \end{pmatrix}.$$

The usually complicated formula $P_S = A(A^TA)^{-1}A^T$ simplified when we used the standard unit coordinate vectors in our basis for S. Namely, we had

$$A = \begin{pmatrix} 0 & 0 \\ 1 & 0 \\ 0 & 1 \end{pmatrix} \quad \text{so} \quad A^TA = \begin{pmatrix} 0 & 1 & 0 \\ 0 & 0 & 1 \end{pmatrix}\begin{pmatrix} 0 & 0 \\ 1 & 0 \\ 0 & 1 \end{pmatrix} = \begin{pmatrix} 1 & 0 \\ 0 & 1 \end{pmatrix} = I.$$

Thus $(A^TA)^{-1} = I$ also, which simplifies what is normally the worst part of the computation in our formula for P_S. We will show that this simplification $(A^TA)^{-1} = I$ can be made in computing P_S for any subspace S of $\mathbb{R}^n$ provided that we can find a basis for S consisting of mutually perpendicular unit vectors to serve as column vectors of A. In Section 5.4 we will see how to find such a basis for any subspace of $\mathbb{R}^n$.

Definition 5.1   Orthonormal Basis

A basis $\{\mathbf{a}_1, \mathbf{a}_2, \ldots, \mathbf{a}_k\}$ for a subspace S of $\mathbb{R}^n$ is **orthonormal** if the vectors $\mathbf{a}_i$ are unit vectors and are mutually perpendicular.

For example, the standard basis $\{\mathbf{e}_1, \mathbf{e}_2, \ldots, \mathbf{e}_n\}$ is an orthonormal basis for $\mathbb{R}^n$.

Expressing Definition 5.1 in terms of dot products, we see that a basis $\{\mathbf{a}_1, \mathbf{a}_2, \ldots, \mathbf{a}_k\}$ for a subspace S is orthonormal if and only if

$$\mathbf{a}_i \cdot \mathbf{a}_j = 0 \quad \text{if } i \neq j \qquad \mathbf{a}_i \perp \mathbf{a}_j \tag{1}$$

and

$$\mathbf{a}_i \cdot \mathbf{a}_i = 1. \qquad \|\mathbf{a}_i\| = 1 \tag{2}$$

Conditions (1) and (2) can be summarized by the single matrix equation

$$A^T A = \begin{pmatrix} \mathbf{a}_1 \cdot \mathbf{a}_1 & \mathbf{a}_1 \cdot \mathbf{a}_2 & \cdots & \mathbf{a}_1 \cdot \mathbf{a}_k \\ \mathbf{a}_2 \cdot \mathbf{a}_1 & \mathbf{a}_2 \cdot \mathbf{a}_2 & \cdots & \mathbf{a}_2 \cdot \mathbf{a}_k \\ \vdots & \vdots & & \vdots \\ \mathbf{a}_k \cdot \mathbf{a}_1 & \mathbf{a}_k \cdot \mathbf{a}_2 & \cdots & \mathbf{a}_k \cdot \mathbf{a}_k \end{pmatrix} = I. \tag{3}$$

The projection matrix for S is then

$$P_S = A(A^T A)^{-1} A^T = A I A^T = A A^T.$$

The projection on S of any vector $\mathbf{b}$ in $\mathbb{R}^n$ is the vector

$$\mathbf{b}_S = P_S \mathbf{b} = A(A^T \mathbf{b}) = A \begin{pmatrix} \mathbf{b} \cdot \mathbf{a}_1 \\ \mathbf{b} \cdot \mathbf{a}_2 \\ \vdots \\ \mathbf{b} \cdot \mathbf{a}_k \end{pmatrix},$$

which can be written as

$$\mathbf{b}_S = (\mathbf{b} \cdot \mathbf{a}_1)\mathbf{a}_1 + (\mathbf{b} \cdot \mathbf{a}_2)\mathbf{a}_2 + \cdots + (\mathbf{b} \cdot \mathbf{a}_k)\mathbf{a}_k. \tag{4}$$

Note that since $\|\mathbf{a}_i\| = 1$, each summand $(\mathbf{b} \cdot \mathbf{a}_i)\mathbf{a}_i$ in Eq. (4) is the projection of $\mathbf{b}$ on the subspace $\text{sp}(\mathbf{a}_i)$, as shown by Eq. (3) of Section 5.1. In other words, Eq. (4) expresses the projection of the vector $\mathbf{b}$ on the subspace S as a *sum of projections of* $\mathbf{b}$ *on each of the vectors in an orthonormal basis for* S. We summarize in a theorem.

Theorem 5.3 Projection in Terms of an Orthonormal Basis

Let $\{\mathbf{a}_1, \mathbf{a}_2, \ldots, \mathbf{a}_k\}$ be an orthonormal basis for a subspace S of $\mathbb{R}^n$. The projection matrix for S is $P_S = A A^T$, where A is the matrix having columns $\mathbf{a}_1, \mathbf{a}_2, \ldots, \mathbf{a}_k$. Let $\mathbf{b}$ be any vector in $\mathbb{R}^n$. The vector projection of $\mathbf{b}$ on S is

$$\mathbf{b}_S = (\mathbf{b} \cdot \mathbf{a}_1)\mathbf{a}_1 + (\mathbf{b} \cdot \mathbf{a}_2)\mathbf{a}_2 + \cdots + (\mathbf{b} \cdot \mathbf{a}_k)\mathbf{a}_k.$$

Example 1 Find the projection matrix for the subspace $S = \text{sp}(\mathbf{a}_1, \mathbf{a}_2)$ of $\mathbb{R}^3$ if

$$\mathbf{a}_1 = \begin{pmatrix} 1/\sqrt{3} \\ -1/\sqrt{3} \\ 1/\sqrt{3} \end{pmatrix} \quad \text{and} \quad \mathbf{a}_2 = \begin{pmatrix} 1/\sqrt{2} \\ 1/\sqrt{2} \\ 0 \end{pmatrix}.$$

Illustrate Theorem 5.3 in this case.

Solution   Let

$$A = \begin{pmatrix} 1/\sqrt{3} & 1/\sqrt{2} \\ -1/\sqrt{3} & 1/\sqrt{2} \\ 1/\sqrt{3} & 0 \end{pmatrix}.$$

Then

$$A^TA = \begin{pmatrix} 1 & 0 \\ 0 & 1 \end{pmatrix}$$

so

$$P_S = AA^T = \begin{pmatrix} \frac{5}{6} & \frac{1}{6} & \frac{1}{3} \\ \frac{1}{6} & \frac{5}{6} & -\frac{1}{3} \\ \frac{1}{3} & -\frac{1}{3} & \frac{1}{3} \end{pmatrix}.$$

Each column vector **b** in $\mathbb{R}^3$ projects onto

$$\mathbf{b}_S = P_S\mathbf{b} = \begin{pmatrix} \frac{5}{6} & \frac{1}{6} & \frac{1}{3} \\ \frac{1}{6} & \frac{5}{6} & -\frac{1}{3} \\ \frac{1}{3} & -\frac{1}{3} & \frac{1}{3} \end{pmatrix}\begin{pmatrix} b_1 \\ b_2 \\ b_3 \end{pmatrix} = (\tfrac{1}{6})\begin{pmatrix} 5b_1 + b_2 + 2b_3 \\ b_1 + 5b_2 - 2b_3 \\ 2b_1 - 2b_2 + 2b_3 \end{pmatrix}.$$

To illustrate Theorem 5.3, we compute the projection of **b** on $\text{sp}(\mathbf{a}_1)$ and $\text{sp}(\mathbf{a}_2)$ in turn, obtaining

$$(\mathbf{b} \cdot \mathbf{a}_1)\mathbf{a}_1 = (\tfrac{1}{3})\begin{pmatrix} b_1 - b_2 + b_3 \\ -b_1 + b_2 - b_3 \\ b_1 - b_2 + b_3 \end{pmatrix}$$

and

$$(\mathbf{b} \cdot \mathbf{a}_2)\mathbf{a}_2 = (\tfrac{1}{2})\begin{pmatrix} b_1 + b_2 \\ b_1 + b_2 \\ 0 \end{pmatrix}.$$

We then see that $\mathbf{b}_S = (\mathbf{b} \cdot \mathbf{a}_1)\mathbf{a}_1 + (\mathbf{b} \cdot \mathbf{a}_2)\mathbf{a}_2$ is the sum of these projections.   ◁

As we will show in Section 5.4, every subspace S of $\mathbb{R}^n$ has an orthonormal basis. If $\{\mathbf{a}_1, \mathbf{a}_2, \ldots , \mathbf{a}_n\}$ is an orthonormal basis for $\mathbb{R}^n$, then the square matrix A having the vector $\mathbf{a}_j$ as jth column is invertible, and $A^{-1} = A^T$ as the equation $A^TA = I$ shows. Since the inverse of A acts both as a right and as a left inverse, we have $AA^T = I$ also, that is, $[(A^T)^T]A^T = I$. This means that the column vectors of A^T, which are the row vectors of A, also form an orthonormal basis for $\mathbb{R}^n$. Conversely, if the row vectors of A form an orthonormal basis for $\mathbb{R}^n$, so do the column vectors.

Definition 5.2   Orthogonal Matrix

An $n \times n$ matrix A is **orthogonal** if it is invertible and $A^{-1} = A^T$.

Our remarks before this definition indicate two alternate defining properties for an orthogonal matrix; we summarize them in the following theorem.

Theorem 5.4   Characterizing Properties of an Orthogonal Matrix

Let A be an $n \times n$ matrix. The following conditions are equivalent:

1. The rows of A form an orthonormal basis for $\mathbb{R}^n$.
2. The columns of A form an orthonormal basis for $\mathbb{R}^n$.
3. The matrix A is orthogonal, that is, invertible with $A^{-1} = A^T$.

The term *orthogonal* applied to a matrix is just a bit misleading. One must remember that not only are the rows (columns) mutually orthogonal, but they also have length one, which is not indicated by the name.

Example 2   Verify that the matrix

$$A = (\tfrac{1}{7}) \begin{pmatrix} 2 & 3 & 6 \\ 3 & -6 & 2 \\ 6 & 2 & -3 \end{pmatrix}$$

is an orthogonal matrix and find A^{-1}.

PROPERTIES OF ORTHOGONAL MATRICES for square systems of coefficients appear in various works of the early nineteenth century. For example, in 1833 Carl Gustav Jacob Jacobi (see the note on page 434) sought to find a linear substitution

$$y_1 = \sum \alpha_{1i} x_i, \, y_2 = \sum \alpha_{2i} x_i, \, \ldots, \, y_n = \sum \alpha_{ni} x_i$$

such that $\sum y_i^2 = \sum x_i^2$. He found that the coefficients of the substitution must satisfy the orthogonality property

$$\sum_i \alpha_{ij} \alpha_{ik} = \begin{cases} 0, & j \neq k, \\ 1, & j = k. \end{cases}$$

One can even trace orthogonal systems of coefficients back to seventeenth- and eighteenth-century works in analytic geometry, when rotations of the plane or of three space are given in order to transform the equations of curves or surfaces. Expressed as matrices, these rotations would give orthogonal ones.

The formal definition of an orthogonal matrix, however, and a comprehensive discussion appeared in an 1878 paper of Georg Ferdinand Frobenius (1849–1917) titled "On Linear Substitutions and Bilinear Forms." In particular, Frobenius dealt with the eigenvalues of such a matrix.

Frobenius, who was a full professor in Zurich and later in Berlin, made his major mathematical contribution in the area of group theory. He was instrumental in developing the concept of an abstract group as well as investigating the theory of finite matrix groups and group characters.

Solution We have

$$AA^T = (\tfrac{1}{49}) \begin{pmatrix} 2 & 3 & 6 \\ 3 & -6 & 2 \\ 6 & 2 & -3 \end{pmatrix} \begin{pmatrix} 2 & 3 & 6 \\ 3 & -6 & 2 \\ 6 & 2 & -3 \end{pmatrix} = \begin{pmatrix} 1 & 0 & 0 \\ 0 & 1 & 0 \\ 0 & 0 & 1 \end{pmatrix}.$$

In this example, A is symmetric so

$$A^{-1} = A^T = A = (\tfrac{1}{7}) \begin{pmatrix} 2 & 3 & 6 \\ 3 & -6 & 2 \\ 6 & 2 & -3 \end{pmatrix}. \quad \triangleleft$$

Chapter 6 is devoted to studying the functional relationship of a column vector $\mathbf{x}$ in $\mathbb{R}^n$ to the column vector $A\mathbf{x}$, where A is a matrix with n columns. While we are talking about orthogonal matrices, we point out important relationships between $\mathbf{x}$ and $A\mathbf{x}$ if A is an orthogonal matrix. Exercises 16–18 request the easy proofs of these properties.

Theorem 5.5 Properties of $A\mathbf{x}$ for an Orthogonal Matrix A

Let A be an orthogonal $n \times n$ matrix and let $\mathbf{x}$ and $\mathbf{y}$ be any column vectors in $\mathbb{R}^n$.

i) $(A\mathbf{x}) \cdot (A\mathbf{y}) = \mathbf{x} \cdot \mathbf{y}$ Preservation of dot product
ii) $\|A\mathbf{x}\| = \|\mathbf{x}\|$ Preservation of length
iii) The angle between $\mathbf{x}$ and $\mathbf{y}$ Preservation of angle
equals the angle between $A\mathbf{x}$
and $A\mathbf{y}$.

We close with an example indicating that a square matrix A whose columns (or rows) are mutually orthogonal is invertible. We describe a technique for finding A^{-1}.

Example 3 Find the inverse of the matrix

$$A = \begin{pmatrix} 4 & 0 & -3 \\ 0 & 1 & 0 \\ 3 & 0 & 4 \end{pmatrix}.$$

by working with an associated orthogonal matrix.

Solution Note that the column vectors of A are mutually orthogonal. Since the column vectors have magnitudes 5, 1, and 5, respectively, the matrix

$$U = AD = \begin{pmatrix} 4 & 0 & -3 \\ 0 & 1 & 0 \\ 3 & 0 & 4 \end{pmatrix} \begin{pmatrix} \tfrac{1}{5} & 0 & 0 \\ 0 & 1 & 0 \\ 0 & 0 & \tfrac{1}{5} \end{pmatrix} = \begin{pmatrix} \tfrac{4}{5} & 0 & -\tfrac{3}{5} \\ 0 & 1 & 0 \\ \tfrac{3}{5} & 0 & \tfrac{4}{5} \end{pmatrix}$$

is an orthogonal matrix with inverse $U^{-1} = U^T$. Of course, D is invertible, and from the equation $U = AD$, we conclude that A is also invertible and $A^{-1} = DU^{-1}$. Thus

$$A^{-1} = DU^{-1} = DU^T = \begin{pmatrix} \frac{1}{5} & 0 & 0 \\ 0 & 1 & 0 \\ 0 & 0 & \frac{1}{5} \end{pmatrix} \begin{pmatrix} \frac{4}{5} & 0 & \frac{3}{5} \\ 0 & 1 & 0 \\ -\frac{3}{5} & 0 & \frac{4}{5} \end{pmatrix}$$

$$= (1/25) \begin{pmatrix} 4 & 0 & 3 \\ 0 & 25 & 0 \\ -3 & 0 & 4 \end{pmatrix}. \quad \triangleleft$$

APPLICATION TO THE METHOD OF LEAST SQUARES

To obtain the least-squares linear fit $y = r_0 + r_1 x$ of k data points (a_i, b_i), we form the matrix

$$A = \begin{pmatrix} 1 & a_1 \\ 1 & a_2 \\ & \vdots \\ 1 & a_k \end{pmatrix} \tag{5}$$

and compute the least-squares solution vector $\bar{\mathbf{r}} = (A^TA)^{-1}A^T\mathbf{b}$. If the set of column vectors in A were orthonormal, then A^TA would be the 2×2 identity matrix I. Of course, this is never the case since the first column vector of A has length $\sqrt{k}$. However, it may happen that the column vectors in A are mutually perpendicular; it is easy to see then that

$$A^TA = \begin{pmatrix} k & 0 \\ 0 & \mathbf{a} \cdot \mathbf{a} \end{pmatrix} \quad \text{so} \quad (A^TA)^{-1} = \begin{pmatrix} 1/k & 0 \\ 0 & 1/(\mathbf{a} \cdot \mathbf{a}) \end{pmatrix}.$$

The matrix A has this property if the x-values $a_1, a_2, \ldots, a_k$ for the data are symmetrically positioned about zero. We illustrate with an example.

Example 4   Find the least-squares linear fit of the data points $(-3, 8)$, $(-1, 5)$, $(1, 3)$, and $(3, 0)$.

Solution   The matrix A is given by

$$A = \begin{pmatrix} 1 & -3 \\ 1 & -1 \\ 1 & 1 \\ 1 & 3 \end{pmatrix}.$$

It is clear why the symmetry of the x-values about zero causes the column vectors of this matrix to be orthogonal. We easily find that

$$A^T A = \begin{pmatrix} 1 & 1 & 1 & 1 \\ -3 & -1 & 1 & 3 \end{pmatrix} \begin{pmatrix} 1 & -3 \\ 1 & -1 \\ 1 & 1 \\ 1 & 3 \end{pmatrix} = \begin{pmatrix} 4 & 0 \\ 0 & 20 \end{pmatrix}.$$

Then

$$\bar{r} = \begin{pmatrix} r_0 \\ r_1 \end{pmatrix} = (A^T A)^{-1} A^T \mathbf{b} = \begin{pmatrix} \frac{1}{4} & 0 \\ 0 & \frac{1}{20} \end{pmatrix} \begin{pmatrix} 1 & 1 & 1 & 1 \\ -3 & -1 & 1 & 3 \end{pmatrix} \begin{pmatrix} 8 \\ 5 \\ 3 \\ 0 \end{pmatrix}$$

$$= \begin{pmatrix} \frac{1}{4} & 0 \\ 0 & \frac{1}{20} \end{pmatrix} \begin{pmatrix} 16 \\ -26 \end{pmatrix} = \begin{pmatrix} 4 \\ -1.3 \end{pmatrix}.$$

Thus the least-squares linear fit is given by $y = 4 - 1.3x$. ◁

Suppose now that the x-values in the data are not symmetrically positioned about zero, but that there is a number $x = c$ about which they are symmetrically located. If we make the variable transformation $t = x - c$, then the t-values for the data are symmetrically positioned about $t = 0$. We can find the t,y-equation for the least-squares linear fit, and then replace t by $x - c$ to obtain the x,y-equation. (Exercises 23 and 24 refine this idea still further.)

Example 5   The number of sales of a particular item by a manufacturer in each of the first four months of 1985 is given in Table 5.8. Find the least-squares linear fit of these data, and use it to project sales in the fifth month.

Table 5.8

a_i = month	1	2	3	4
b_i = thousands sold	2.5	3	3.8	4.5

Solution   Our x-values a_i are symmetrically located about $c = 2.5$. If we take $t = x - 2.5$, the data points (t, y) become

$$(-1.5, 2.5), \quad (-.5, 3), \quad (.5, 3.8), \quad \text{and} \quad (1.5, 4.5).$$

We find that

$$A^T A = \begin{pmatrix} 1 & 1 & 1 & 1 \\ -1.5 & -.5 & .5 & 1.5 \end{pmatrix} \begin{pmatrix} 1 & -1.5 \\ 1 & -.5 \\ 1 & .5 \\ 1 & 1.5 \end{pmatrix} = \begin{pmatrix} 4 & 0 \\ 0 & 5 \end{pmatrix}.$$

Thus

$$(A^TA)^{-1}A^T\mathbf{b} = \begin{pmatrix} \frac{1}{4} & 0 \\ 0 & \frac{1}{5} \end{pmatrix} \begin{pmatrix} 1 & 1 & 1 & 1 \\ -1.5 & -.5 & .5 & 1.5 \end{pmatrix} \begin{pmatrix} 2.5 \\ 3 \\ 3.8 \\ 4.5 \end{pmatrix}$$

$$= \begin{pmatrix} \frac{1}{4} & 0 \\ 0 & \frac{1}{5} \end{pmatrix} \begin{pmatrix} 13.8 \\ 3.4 \end{pmatrix} = \begin{pmatrix} 3.45 \\ .68 \end{pmatrix}.$$

Thus the t,y-equation is $y = 3.45 + .68t$. Replacing t by $x - 2.5$, we obtain

$$y = 3.45 + .68(x - 2.5) = 1.75 + .68x.$$

Setting $x = 5$, we estimate sales in the fifth month to be 5.15 units, or 5150 items.   ◁

SUMMARY

1. A basis for a subspace S is orthonormal if the vectors have length 1 and are mutually perpendicular.

2. Let $\{\mathbf{a}_1, \mathbf{a}_2, \ldots, \mathbf{a}_k\}$ be an orthonormal basis for a subspace S of $\mathbb{R}^n$. Then the projection matrix for S is $P_S = AA^T$, where A is the $n \times k$ matrix having the $\mathbf{a}_i$ as column vectors.

3. A square $n \times n$ matrix A is called orthogonal if it satisfies any one (and hence all) of these three equivalent conditions:

   i) The rows of A form an orthonormal basis for $\mathbb{R}^n$.

   ii) The columns of A form an orthonormal basis for $\mathbb{R}^n$.

   iii) The matrix A is invertible and $A^{-1} = A^T$.

4. Multiplication of column vectors in $\mathbb{R}^n$ on the left by an $n \times n$ orthogonal matrix preserves length, dot product, and the angle between vectors.

5. Suppose the x-values a_i in a set of k data points (a_i, b_i) are symmetrically positioned about zero. Let A be the $k \times 2$ matrix with first column vector having all entries 1 and second column vector $\mathbf{a}$. The columns of A are orthogonal and

$$A^TA = \begin{pmatrix} k & 0 \\ 0 & \mathbf{a} \cdot \mathbf{a} \end{pmatrix}.$$

Computation of the least-squares fit for the data points is then greatly simplified. If the x-values of the data points are symmetrically positioned about $x = c$, then the substitution $t = x - c$ gives data points with t-values symmetrically positioned about zero, and the above simplification applies. See Example 5.

EXERCISES

In Exercises 1–7, use Theorem 5.3 to find the projection matrix for the subspace S with the given orthonormal basis. The vectors are given in row notation to save space.

1. $S = \text{sp}(\mathbf{a}_1, \mathbf{a}_2)$ in $\mathbb{R}^3$, where $\mathbf{a}_1 = (1/\sqrt{2}, 0, -1/\sqrt{2})$ and $\mathbf{a}_2 = (1/\sqrt{3}, -1/\sqrt{3}, 1/\sqrt{3})$

2. $S = \text{sp}(\mathbf{a}_1, \mathbf{a}_2)$ in $\mathbb{R}^3$, where $\mathbf{a}_1 = (3/4, 4/5, 0)$ and $\mathbf{a}_2 = (0, 0, 1)$

3. $S = \text{sp}(\mathbf{a}_1, \mathbf{a}_2)$ in $\mathbb{R}^4$, where $\mathbf{a}_1 = (3/(5\sqrt{2}), 4/(5\sqrt{2}), 1/2, -1/2)$ and
    $\mathbf{a}_2 = (4/(5\sqrt{2}), -3/(5\sqrt{2}), 1/2, 1/2)$

4. $S = \text{sp}(\mathbf{a}_1, \mathbf{a}_2)$ in $\mathbb{R}^4$, where $\mathbf{a}_1 = (2/7, 0, 3/7, -6/7)$ and $\mathbf{a}_2 = (-3/7, 6/7, 2/7, 0)$

5. $S = \text{sp}(\mathbf{a}_1, \mathbf{a}_2)$ in $\mathbb{R}^4$, where $\mathbf{a}_1 = (0, -2/3, 2/3, 1/3)$ and $\mathbf{a}_2 = (2/3, 0, -1/3, 2/3)$

6. $S = \text{sp}(\mathbf{a}_1, \mathbf{a}_2, \mathbf{a}_3)$ in $\mathbb{R}^4$, where $\mathbf{a}_1 = (1/\sqrt{3}, 0, 1/\sqrt{3}, 1/\sqrt{3})$, $\mathbf{a}_2 = (1/\sqrt{3}, 1/\sqrt{3}, -1/\sqrt{3}, 0)$, and
    $\mathbf{a}_3 = (1/\sqrt{3}, -1/\sqrt{3}, 0, -1/\sqrt{3})$

7. $S = \text{sp}(\mathbf{a}_1, \mathbf{a}_2, \mathbf{a}_3)$ in $\mathbb{R}^4$, where $\mathbf{a}_1 = (1/2, 1/2, 1/2, 1/2)$, $\mathbf{a}_2 = (-1/2, 1/2, -1/2, 1/2)$, and $\mathbf{a}_3 = (1/2, 1/2, -1/2, -1/2)$

*In Exercises 8–11, compute the projection of **c** on S in two ways: (a) using the projection matrix found above and (b) using the last sentence of Theorem 5.3. The vectors are given in row notation to save space.*

8. The subspace S in Exercise 1,
    $\mathbf{c} = (6, -12, -6)$

9. The subspace S in Exercise 2,
    $\mathbf{c} = (20, -15, 5)$

10. The subspace S in Exercise 5,
    $\mathbf{c} = (9, 0, -9, 18)$

11. The subspace S in Exercise 7,
    $\mathbf{c} = (4, -12, -4, 0)$

In Exercises 12–15, find the inverse of each matrix by working with an associated orthogonal matrix, as in Example 3.

12. $\begin{pmatrix} 1 & -1 \\ 1 & 1 \end{pmatrix}$

13. $\begin{pmatrix} 3 & 0 & 4 \\ -4 & 0 & 3 \\ 0 & 1 & 0 \end{pmatrix}$

14. $\begin{pmatrix} 1 & 2 & 2 \\ 2 & 1 & -2 \\ 2 & -2 & 1 \end{pmatrix}$

15. $\begin{pmatrix} 1 & -1 & 1 & 1 \\ -1 & 1 & 1 & 1 \\ 1 & 1 & -1 & 1 \\ 1 & 1 & 1 & -1 \end{pmatrix}$

16. Prove the preservation-of-dot-product property stated in Theorem 5.5.

17. Prove the preservation-of-length property stated in Theorem 5.5.

18. Prove the preservation-of-angle property stated in Theorem 5.5.

In Exercises 19–22, use the technique illustrated in Examples 4 and 5 to solve the least-squares problem.

19. Find the least-squares linear fit to the data points $(-4, -2)$, $(-2, 0)$, $(0, 1)$, $(2, 4)$, $(4, 5)$.

20. Find the least-squares linear fit to the data points $(0, 1)$, $(1, 4)$, $(2, 6)$, $(3, 8)$, $(4, 9)$.

21. The number of gallons of maple syrup made from the sugar bush of a Vermont farmer over the past five years was:

    80 gallons five years ago,
    70 gallons four years ago,
    75 gallons three years ago,
    65 gallons two years ago,
    60 gallons last year.

    Use a least-squares linear fit of these data to project the number of gallons that will be produced this year. (Does this problem make practical sense? Why?)

22. The number of minutes required for a rat to find its way out of a maze on repeated attempts were 8 minutes, 8 minutes, 6 minutes, 5 minutes, and 6 minutes on its first five tries. Use a least-squares linear fit of these data to project the time the rat will take on its sixth try.

23. Let $(a_1, b_1), (a_2, b_2), \ldots, (a_m, b_m)$ be data points. If $\sum_{i=1}^{m} a_i = 0$, show that the line that best fits the data in the least-squares sense is given by $r_0 + r_1 x$, where

$$r_0 = \left(\sum_{i=1}^{m} b_i \right) \bigg/ m$$

and

$$r_1 = \left(\sum_{i=1}^{m} a_i b_i \right) \bigg/ \left(\sum_{i=1}^{m} a_i^2 \right).$$

24. Repeat Exercise 23 but do not assume $\sum_{i=1}^{m} a_i = 0$. Show that the least-squares linear fit of the data is given by $y = r_0 + r_1(x - c)$, where $c = (\sum_{i=1}^{m} a_i)/m$ and r_0 and r_1 have the values given in Exercise 23.

5.4

The Gram–Schmidt Process

In Section 5.3, we saw the advantage of projecting a vector onto a subspace S of $\mathbb{R}^n$ using an orthonormal basis for S in the computations (see Theorem 5.3). It is our purpose in this section to show how we can transform any basis of a subspace of $\mathbb{R}^n$ into one that is orthonormal.

ORTHOGONAL SETS AND ORTHONORMAL BASES

A set $\{\mathbf{v}_1, \mathbf{v}_2, \ldots, \mathbf{v}_k\}$ of nonzero vectors in $\mathbb{R}^n$ is **orthogonal** if the vectors $\mathbf{v}_j$ are mutually perpendicular. The problem of finding an orthonormal basis for a subspace S of $\mathbb{R}^n$ will be solved if we can find an orthogonal generating set for S, as our next theorem shows.

Theorem 5.6 Orthogonal Sets and Orthonormal Bases

Let $\{\mathbf{v}_1, \mathbf{v}_2, \ldots, \mathbf{v}_k\}$ be an orthogonal set of nonzero vectors in $\mathbb{R}^n$. Then the vectors $\mathbf{v}_j$ in this set are independent and

$$\left\{ \frac{\mathbf{v}_1}{\|\mathbf{v}_1\|}, \frac{\mathbf{v}_2}{\|\mathbf{v}_2\|}, \ldots, \frac{\mathbf{v}_k}{\|\mathbf{v}_k\|} \right\}$$

is an orthonormal basis for $\mathrm{sp}(\mathbf{v}_1, \mathbf{v}_2, \ldots, \mathbf{v}_k)$.

Proof To show that the orthogonal set $\{\mathbf{v}_1, \mathbf{v}_2, \ldots, \mathbf{v}_k\}$ is independent, let us suppose that

$$\mathbf{v}_j = s_1\mathbf{v}_1 + s_2\mathbf{v}_2 + \cdots + s_{j-1}\mathbf{v}_{j-1}. \tag{1}$$

Taking the dot product of both sides of Eq. (1) with $\mathbf{v}_j$ yields $\mathbf{v}_j \cdot \mathbf{v}_j = 0$, which contradicts $\mathbf{v}_j \neq \mathbf{0}$. Thus no $\mathbf{v}_j$ can be equal to a linear combination of its predecessors, so the $\mathbf{v}_j$ are independent and thus form a basis for S. We leave it to the student to verify that the vectors $\mathbf{v}_j/\|\mathbf{v}_j\|$ also form a basis for S; of course, this basis is orthonormal. ■

Example 1   Show that $\mathbf{v}_1 = (1, 1, 1, 1)$, $\mathbf{v}_2 = (-1, 1, -1, 1)$, and $\mathbf{v}_3 = (1, -1, -1, 1)$ are orthogonal, and find an orthonormal basis for $S = \text{sp}(\mathbf{v}_1, \mathbf{v}_2, \mathbf{v}_3)$ in $\mathbb{R}^4$.

Solution   We easily find that $\mathbf{v}_1 \cdot \mathbf{v}_2 = \mathbf{v}_1 \cdot \mathbf{v}_3 = \mathbf{v}_2 \cdot \mathbf{v}_3 = 0$ so $\{\mathbf{v}_1, \mathbf{v}_2, \mathbf{v}_3\}$ is an orthogonal set. Since $\|\mathbf{v}_1\| = \|\mathbf{v}_2\| = \|\mathbf{v}_3\| = \sqrt{4} = 2$, we see that

$$\{(1/2, 1/2, 1/2, 1/2), (-1/2, 1/2, -1/2, 1/2), (1/2, -1/2, -1/2, 1/2)\}$$

is an orthonormal basis for S. $\triangleleft$

THE GRAM—SCHMIDT PROCESS

We now describe a computational technique for creating an orthonormal basis from a given basis of a subspace S of $\mathbb{R}^n$. The following theorem asserts the existence of such a basis, and its proof is constructive in nature. That is, the proof shows how an orthonormal basis can be constructed.

Theorem 5.7   Orthonormal Basis Theorem (Gram—Schmidt)

> Every nonzero subspace S of $\mathbb{R}^n$ has an orthonormal basis.

Proof   Let $\{\mathbf{a}_1, \mathbf{a}_2, \ldots, \mathbf{a}_k\}$ be any basis for S, and let

$$T_j = \text{sp}(\mathbf{a}_1, \mathbf{a}_2, \ldots, \mathbf{a}_j) \quad \text{for } j = 1, 2, \ldots, k.$$

Let $\mathbf{v}_1 = \mathbf{a}_1$. For $j = 2, \ldots, k$, let $\mathbf{p}_j$ be the projection of $\mathbf{a}_j$ on T_{j-1}, and let $\mathbf{v}_j = \mathbf{a}_j - \mathbf{p}_j$. That is, $\mathbf{v}_j$ is obtained by subtracting from $\mathbf{a}_j$ its projection on the subspace generated by its predecessors. Figure 5.11 gives a symbolic illustration. We note that $\mathbf{v}_j = \mathbf{a}_j - \mathbf{p}_j$ lies in the subspace T_j since $\mathbf{a}_j$ lies in T_j and $\mathbf{p}_j$ lies in T_{j-1}, which is contained in T_j. Now $\mathbf{v}_j$ is perpendicular to every vector in T_{j-1} by number (2) of the boxed properties of projections on page 248. Therefore, $\mathbf{v}_j$ is perpendicular to $\mathbf{v}_1, \mathbf{v}_2, \ldots, \mathbf{v}_{j-1}$. Taking $j = k$, we conclude that each vector in the set

$$\{\mathbf{v}_1, \mathbf{v}_2, \ldots, \mathbf{v}_k\} \tag{2}$$

THE GRAM-SCHMIDT PROCESS is named for the Danish mathematician Jorgen P. Gram (1850–1916) and the German Erhard Schmidt (1876–1959). It was first published by Gram in 1883 in a paper titled "Series Development Using the Method of Least Squares." It was published again with a careful proof by Schmidt in 1907 in a work on integral equations. In fact, Schmidt even referred to Gram's result. For Schmidt, as for Gram, the vectors were continuous functions defined on an interval $[a, b]$ with the inner product of two such functions ϕ, ψ being given as $\int_a^b \phi(x)\psi(x)\, dx$. Schmidt was more explicit than Gram, however, writing out the process in great detail and proving that the set of functions ψ_i derived from his original set ϕ_i was in fact an orthonormal set.

Schmidt, who was at the University of Berlin from 1917 until his death, is best known for his definitive work on Hilbert spaces, spaces of square summable sequences of complex numbers. In fact, he applied the Gram—Schmidt process to sets of vectors in these spaces to help develop necessary and sufficient conditions for such sets to be linearly independent.

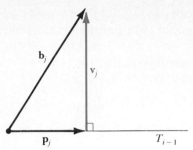

Figure 5.11
The vector v_j in the Gram–Schmidt construction.

is perpendicular to each of its predecessors. Thus the set of vectors (2) consists of k mutually perpendicular vectors in the k-dimensional subspace S, and so is an *orthogonal basis* for S. It follows that if we set $\mathbf{q}_i = (1/\|\mathbf{v}_i\|)\mathbf{v}_i$ for $i = 1, 2, \ldots, k$, then

$$\{\mathbf{q}_1, \mathbf{q}_2, \ldots, \mathbf{q}_k\}$$

is an orthonormal basis for S. ■

The proof of Theorem 5.7 was computational, providing us with a technique for constructing an orthonormal basis for a subspace S of $\mathbb{R}^n$. The technique is known as the "Gram–Schmidt Process," and we have boxed it for easy reference.

Gram–Schmidt Process

To find an orthonormal basis for a subspace S of $\mathbb{R}^n$:

1. Find a basis $\{\mathbf{a}_1, \mathbf{a}_2, \ldots, \mathbf{a}_k\}$ for S.
2. Let $\mathbf{v}_1 = \mathbf{a}_1$. For $j = 2, \ldots, k$, let $\mathbf{v}_j$ be obtained by subtracting from $\mathbf{a}_j$ its projection on the subspace generated by its predecessors.
3. The $\mathbf{v}_j$ so obtained form an orthogonal basis for S, and can be normalized to yield an orthonormal basis.

Corollary   ***Expansion of an Orthogonal Set to an Orthogonal Basis***

Every orthogonal set of vectors in a subspace S of $\mathbb{R}^n$ can be expanded if necessary to an orthogonal basis for S.

Proof   An orthogonal set $\{\mathbf{v}_1, \mathbf{v}_2, \ldots, \mathbf{v}_r\}$ of vectors in S is an independent set by Theorem 5.6, and can be expanded to a basis $\{\mathbf{v}_1, \ldots, \mathbf{v}_r, \mathbf{a}_1, \ldots, \mathbf{a}_s\}$

of S by Theorem 3.14 on page 174. We apply the Gram–Schmidt process to this basis for S. Since the $\mathbf{v}_j$ are already mutually perpendicular, none of them will be changed by the Gram–Schmidt process, which thus yields an orthogonal basis containing the given vectors $\mathbf{v}_j$ for $j = 1, \ldots, r$. ■

When actually executing the Gram–Schmidt process, we project vectors onto subspaces, as described by number (2) in the box above. Now we know that it is best to work with an orthogonal or orthonormal basis for a subspace when projecting on it. Since the subspace T_j generated by $\mathbf{a}_1, \mathbf{a}_2, \ldots, \mathbf{a}_j$ is also the subspace generated by the orthogonal set $\mathbf{v}_1, \mathbf{v}_2, \ldots, \mathbf{v}_j$, it is surely best to work with this latter basis for T_j when computing the desired projection. One may normalize the $\mathbf{v}_j$, forming the vector $\mathbf{q}_j = (1/\|\mathbf{v}_j\|)\mathbf{v}_j$ to obtain a vector of length 1 at each step of the construction. In that case, Theorem 5.3 shows that the specific formula for $\mathbf{v}_j$ is as follows.

Projections on Orthonormal $\mathbf{q}_i$

$$\mathbf{v}_j = \mathbf{a}_j - [(\mathbf{a}_j \cdot \mathbf{q}_1)\mathbf{q}_1 + (\mathbf{a}_j \cdot \mathbf{q}_2)\mathbf{q}_2 + \cdots + (\mathbf{a}_j \cdot \mathbf{q}_{j-1})\mathbf{q}_{j-1}] \tag{3}$$

If the $\mathbf{v}_j$ are not normalized at each step, then replacing each $\mathbf{q}_i$ by $(1/\|\mathbf{v}_i\|)\mathbf{v}_i$ in Eq. (3), we obtain the following formula.

Projections on Orthogonal $\mathbf{v}_i$

$$\mathbf{v}_j = \mathbf{a}_j - \left[\frac{\mathbf{a}_j \cdot \mathbf{v}_1}{\mathbf{v}_1 \cdot \mathbf{v}_1} \mathbf{v}_1 + \frac{\mathbf{a}_j \cdot \mathbf{v}_2}{\mathbf{v}_2 \cdot \mathbf{v}_2} \mathbf{v}_2 + \cdots + \frac{\mathbf{a}_j \cdot \mathbf{v}_{j-1}}{\mathbf{v}_{j-1} \cdot \mathbf{v}_{j-1}} \mathbf{v}_{j-1} \right] \tag{4}$$

The arithmetic using (3) and using (4) is similar, but (4) postpones the introduction of the radicals from normalizing until the entire orthogonal basis is obtained. We will use (4) in our work.

Example 2 Find an orthonormal basis for the subspace $S = \text{sp}((1, 0, 1), (1, 1, 1))$ of $\mathbb{R}^3$.

Solution We use the Gram–Schmidt process with formula (4), finding first an orthogonal basis for S. We take $\mathbf{v}_1 = (1, 0, 1)$. From formula (4) with $\mathbf{v}_1 = (1, 0, 1)$ and $\mathbf{a}_2 = (1, 1, 1)$, we have

$$\mathbf{v}_2 = \mathbf{a}_2 - \frac{\mathbf{a}_2 \cdot \mathbf{v}_1}{\mathbf{v}_1 \cdot \mathbf{v}_1} \mathbf{v}_1 = (1, 1, 1) - (2/2)(1, 0, 1) = (0, 1, 0).$$

An orthogonal basis for S is $\{(1, 0, 1), (0, 1, 0)\}$ and an orthonormal basis is $\{(1/\sqrt{2}, 0, 1/\sqrt{2}), (0, 1, 0)\}$. ◁

Example 3 Expand $\{(1, 1, 0), (1, -1, 1)\}$ to an orthogonal basis for $\mathbb{R}^3$ and then transform this to an orthonormal basis for $\mathbb{R}^3$.

Solution   First we expand the given set to a basis $\{\mathbf{a}_1, \mathbf{a}_2, \mathbf{a}_3\}$ for $\mathbb{R}^3$. We take $\mathbf{a}_1 = (1, 1, 0)$, $\mathbf{a}_2 = (1, -1, 1)$, and $\mathbf{a}_3 = (1, 0, 0)$, which we easily see form a basis for $\mathbb{R}^3$.

Now we use the Gram–Schmidt process with formula (4). Since $\mathbf{a}_1$ and $\mathbf{a}_2$ are perpendicular, we let $\mathbf{v}_1 = \mathbf{a}_1 = (1, 1, 0)$ and $\mathbf{v}_2 = \mathbf{a}_2 = (1, -1, 1)$. From formula (4), we have

$$\mathbf{v}_3 = \mathbf{a}_3 - \left[\frac{\mathbf{a}_3 \cdot \mathbf{v}_1}{\mathbf{v}_1 \cdot \mathbf{v}_1} \mathbf{v}_1 + \frac{\mathbf{a}_3 \cdot \mathbf{v}_2}{\mathbf{v}_2 \cdot \mathbf{v}_2} \mathbf{v}_2 \right]$$

$$= (1, 0, 0) - [(\tfrac{1}{2})(1, 1, 0) + (\tfrac{1}{3})(1, -1, 1)]$$

$$= (1, 0, 0) - (\tfrac{5}{6}, \tfrac{1}{6}, \tfrac{1}{3}) = (\tfrac{1}{6}, -\tfrac{1}{6}, -\tfrac{1}{3}).$$

Multiplying this vector by -6, we replace $\mathbf{v}_3$ by $(-1, 1, 2)$. Thus we have expanded the given set to an orthogonal basis

$$\{(1, 1, 0), (1, -1, 1), (-1, 1, 2)\}$$

of $\mathbb{R}^3$. Normalizing these vectors to unit length, we obtain

$$\{(1/\sqrt{2}, 1/\sqrt{2}, 0), (1/\sqrt{3}, -1/\sqrt{3}, 1/\sqrt{3}), (-1/\sqrt{6}, 1/\sqrt{6}, 2/\sqrt{6})\}$$

as an orthonormal basis.   ◁

Example 4   Transform the basis $\{(1, 2, 1), (2, 1, 1), (1, 1, 2)\}$ of $\mathbb{R}^3$ to an orthonormal one using the Gram–Schmidt process.

Solution   First we find an orthogonal basis, using formula (4). We take $\mathbf{v}_1 = (1, 2, 1)$, and we compute $\mathbf{v}_2$ by subtracting from $\mathbf{a}_2 = (2, 1, 1)$ its projection onto $\mathbf{v}_1$:

$$\mathbf{v}_2 = \mathbf{a}_2 - \frac{\mathbf{v}_1 \cdot \mathbf{a}_2}{\mathbf{v}_1 \cdot \mathbf{v}_1} \mathbf{v}_1 = (2, 1, 1) - (\tfrac{5}{6})(1, 2, 1)$$

$$= (\tfrac{7}{6}, -\tfrac{4}{6}, \tfrac{1}{6}).$$

To ease computations, we replace $\mathbf{v}_2$ by $6\mathbf{v}_2$, obtaining $\mathbf{v}_2 = (7, -4, 1)$. Finally we subtract from $\mathbf{a}_3 = (1, 1, 2)$ its projection on the subspace sp($\mathbf{v}_1$, $\mathbf{v}_2$), obtaining

$$\mathbf{v}_3 = \mathbf{a}_3 - \left[\frac{\mathbf{a}_3 \cdot \mathbf{v}_1}{\mathbf{v}_1 \cdot \mathbf{v}_1} \mathbf{v}_1 + \frac{\mathbf{a}_3 \cdot \mathbf{v}_2}{\mathbf{v}_2 \cdot \mathbf{v}_2} \mathbf{v}_2 \right]$$

$$= (1, 1, 2) - [(\tfrac{5}{6})(1, 2, 1) + (\tfrac{5}{66})(7, -4, 1)]$$

$$= (1, 1, 2) - (\tfrac{1}{66})(90, 90, 60)$$

$$= (-\tfrac{24}{66}, -\tfrac{24}{66}, \tfrac{72}{66}) = (\tfrac{4}{11})(-1, -1, 3).$$

Replacing $\mathbf{v}_3$ by $(\tfrac{11}{4})\mathbf{v}_3$, we see that the basis $\{(1, 2, 1), (7, -4, 1), (-1, -1, 3)\}$ is also orthogonal. An orthonormal basis is

$$\{(1/\sqrt{6})(1, 2, 1), (1/\sqrt{66})(7, -4, 1), (1/\sqrt{11})(-1, -1, 3)\}.   ◁$$

The arithmetic involved in the Gram–Schmidt process is a bit tedious with pencil and paper, but it is very easy to implement the construction on a computer.

ORTHOGONAL COMPLEMENTS

A vector $\mathbf{v}$ in $\mathbb{R}^n$ is **orthogonal to a subspace** S of $\mathbb{R}^n$ if it is orthogonal to each vector in S. Two subspaces S and T of $\mathbb{R}^n$ are said to be **orthogonal** if each vector of one is orthogonal to each vector of the other. If S is a subspace of $\mathbb{R}^n$, the set of *all* vectors in $\mathbb{R}^n$ that are orthogonal to S is the **orthogonal complement** of S and is denoted by $S^\perp$.

Example 5   If A is an $m \times n$ matrix, show that the orthogonal complement of the row space of A is the nullspace of A.

Solution   The nullspace of A is precisely the set of all vectors $\mathbf{x}$ in $\mathbb{R}^n$ that are solutions of the homogeneous system $A\mathbf{x} = \mathbf{0}$. Thus the nullspace of A is the set of all vectors $\mathbf{x}$ in $\mathbb{R}^n$ that are orthogonal to each of the rows of A, and hence to the row space of A.   ◁

Example 5 did not show that, conversely, the row space of A is the orthogonal complement of the nullspace of A. This is true, and follows from part (3) of the next theorem.

Theorem 5.8   Properties of Orthogonal Complements

Let S be a subspace of $\mathbb{R}^n$ and let $S^\perp$ be its orthogonal complement.

  i) $S^\perp$ is a subspace of $\mathbb{R}^n$.
  ii) $\dim(S^\perp) = n - \dim(S)$.
  iii) $(S^\perp)^\perp = S$.
  iv) Each vector $\mathbf{b}$ in $\mathbb{R}^n$ is the sum $\mathbf{b} = \mathbf{b}_S + \mathbf{b}_{S^\perp}$ of its projections onto S and onto $S^\perp$. This is the only representation of $\mathbf{b}$ as the sum of vectors in S and $S^\perp$.

Proof   (i) Recall that a subspace must contain $\mathbf{0}$, which is orthogonal to all vectors, and must be closed under addition and scalar multiplication. We leave the verification of these closure properties for $S^\perp$ to Exercise 23.

(ii)–(iii) If $S = \{\mathbf{0}\}$, then $S^\perp = \mathbb{R}^n$ and the theorem is trivial. We assume $S \neq \{\mathbf{0}\}$. By Theorem 5.7, we can find an orthonormal basis $\{\mathbf{q}_1, \ldots , \mathbf{q}_k\}$ for S. By the corollary to Theorem 5.7, this set can be enlarged to an orthonormal basis $\{\mathbf{q}_1, \ldots , \mathbf{q}_k, \mathbf{q}_{k+1}, \ldots , \mathbf{q}_n\}$ for $\mathbb{R}^n$. By definition, $S^\perp$ is the set of all vectors in $\mathbb{R}^n$ that are orthogonal to all vectors in S. It is clear that a vector in $\mathbb{R}^n$ is orthogonal to every vector in S if and only if it is orthogonal to each of the vectors $\mathbf{q}_1, \ldots , \mathbf{q}_k$ in our basis for S. But since

$$(r_1\mathbf{q}_1 + \cdots + r_n\mathbf{q}_n) \cdot \mathbf{q}_j = r_j,$$

we see that a vector $\mathbf{v} = r_1\mathbf{q}_1 + \cdots + r_n\mathbf{q}_n$ in $\mathbb{R}^n$ is in $S^\perp$ if and only if $r_1, \ldots , r_k$ are all zero, that is, if and only if $\mathbf{v}$ is in $\mathrm{sp}(\mathbf{q}_{k+1}, \ldots , \mathbf{q}_n)$. This shows that a basis for $S^\perp$ is $\mathbf{q}_{k+1}, \ldots, \mathbf{q}_n$, and assertions (ii) and (iii) of the theorem then follow easily.

(iv) Suppose $\mathbf{b} = \mathbf{u} + \mathbf{v} = \bar{\mathbf{u}} + \bar{\mathbf{v}}$, where $\mathbf{u}$ and $\bar{\mathbf{u}}$ are in S while $\mathbf{v}$ and $\bar{\mathbf{v}}$ are both in $S^\perp$. Then $\mathbf{u} - \bar{\mathbf{u}} = \bar{\mathbf{v}} - \mathbf{v}$ is in both S and $S^\perp$, so this vector must be perpendicular to itself. Clearly the only vector perpendicular to itself is $\mathbf{0}$, so $\mathbf{u} = \bar{\mathbf{u}}$ and $\mathbf{v} = \bar{\mathbf{v}}$. This shows that there is at most one way of writing $\mathbf{b}$ as a sum of vectors in S and in $S^\perp$. Now Theorem 5.1 tells us that

$$\mathbf{b} = \mathbf{b}_S + (\mathbf{b} - \mathbf{b}_S)$$

is one way of writing $\mathbf{b}$ as a sum of vectors in S and $S^\perp$. The same theorem and (iii) above tell us that another way is

$$\mathbf{b} = (\mathbf{b} - \mathbf{b}_{S^\perp}) + \mathbf{b}_{S^\perp}.$$

By the uniqueness just shown, we must have $(\mathbf{b} - \mathbf{b}_S) = \mathbf{b}_{S^\perp}$, which proves (iv). ∎

It follows from Example 5 and Theorem 5.8 (iii) that the nullspace N_A and the row space of an $m \times n$ matrix A are orthogonal complements of each other. In particular, assertion (ii) of the theorem provides another proof that the rank of A plus the nullity of A equals the number of columns of A.

Example 6   Find a nonzero vector in $\mathbb{R}^4$ orthogonal to $\mathbf{v}_1 = (1, 2, 2, 1)$, $\mathbf{v}_2 = (3, 4, 2, 3)$, and $\mathbf{v}_3 = (0, 1, 3, 2)$.

Solution   We wish to find the orthogonal complement of $S = \text{sp}(\mathbf{v}_1, \mathbf{v}_2, \mathbf{v}_3)$ in $\mathbb{R}^4$. Following Example 5, we find the nullspace of the matrix

$$A = \begin{pmatrix} 1 & 2 & 2 & 1 \\ 3 & 4 & 2 & 3 \\ 0 & 1 & 3 & 2 \end{pmatrix}.$$

Reducing A, we have

$$\begin{pmatrix} 1 & 2 & 2 & 1 \\ 3 & 4 & 2 & 3 \\ 0 & 1 & 3 & 2 \end{pmatrix} \sim \begin{pmatrix} 1 & 2 & 2 & 1 \\ 0 & -2 & -4 & 0 \\ 0 & 1 & 3 & 2 \end{pmatrix} \sim \begin{pmatrix} 1 & 0 & -2 & 1 \\ 0 & 1 & 2 & 0 \\ 0 & 0 & 1 & 2 \end{pmatrix}.$$

Therefore, the nullspace of A is the set of vectors of the form

$$(-5r, 4r, -2r, r) \quad \text{for any scalar } r,$$

and any such vector for $r \neq 0$ satisfies our requirements. ◁

* THE *QR*-FACTORIZATION

Let $\{\mathbf{a}_1, \mathbf{a}_2, \ldots, \mathbf{a}_k\}$ be a set of independent vectors in $\mathbb{R}^n$ generating a subspace S. Let $\mathbf{v}_1, \mathbf{v}_2, \ldots, \mathbf{v}_k$ be the orthogonal basis for S obtained by using

---

* The *QR*-factorization can be omitted.

Eq. (4) in the Gram–Schmidt process. If

$$\mathbf{q}_j = (1/\|\mathbf{v}_j\|)\mathbf{v}_j, \tag{5}$$

then $\{\mathbf{q}_1, \mathbf{q}_2, \dots, \mathbf{q}_k\}$ is an orthonormal basis for S.

We rewrite Eq. (3) in the form

$$\mathbf{a}_j = (\mathbf{a}_j \cdot \mathbf{q}_1)\mathbf{q}_1 + (\mathbf{a}_j \cdot \mathbf{q}_2)\mathbf{q}_2 + \dots + (\mathbf{a}_j \cdot \mathbf{q}_{j-1})\mathbf{q}_{j-1} + \mathbf{v}_j. \tag{6}$$

Equation (6) expresses $\mathbf{a}_j$ as a linear combination of $\mathbf{q}_1, \mathbf{q}_2, \dots, \mathbf{q}_{j-1}, \mathbf{v}_j$. Since $\mathbf{v}_j = \|\mathbf{v}_j\|\mathbf{q}_j$, we can use Eq. (6) to express $\mathbf{a}_j$ as a linear combination of $\mathbf{q}_1, \mathbf{q}_2, \dots, \mathbf{q}_j$. Namely, if we let

$$r_{1j} = \mathbf{q}_1 \cdot \mathbf{a}_j, \; r_{2j} = \mathbf{q}_2 \cdot \mathbf{a}_j, \dots, r_{j-1,j} = \mathbf{q}_{j-1} \cdot \mathbf{a}_j, \; r_{jj} = \|\mathbf{v}_j\|, \tag{7}$$

then Eq. (6) takes the simple form

$$\mathbf{a}_j = r_{1j}\mathbf{q}_1 + r_{2j}\mathbf{q}_2 + \dots + r_{jj}\mathbf{q}_j \tag{8}$$

for $j = 1, 2, \dots, k$. Equation (8) can be summarized for $j = 1, 2, \dots, k$ by the single matrix equation

$$\underbrace{\begin{pmatrix} | & | & & | \\ \mathbf{a}_1 & \mathbf{a}_2 & \cdots & \mathbf{a}_k \\ | & | & & | \end{pmatrix}}_{A} = \underbrace{\begin{pmatrix} | & | & & | \\ \mathbf{q}_1 & \mathbf{q}_2 & \cdots & \mathbf{q}_k \\ | & | & & | \end{pmatrix}}_{Q} \underbrace{\begin{pmatrix} r_{11} & r_{12} & \cdots & r_{1k} \\ & r_{22} & \cdots & r_{2k} \\ & & \ddots & \vdots \\ & & & r_{kk} \end{pmatrix}}_{R}.$$

If we call the matrices in the equation A, Q, and R as indicated, then we see that we have a factorization $A = QR$, where A is an $n \times k$ matrix with k independent column vectors, Q is an $n \times k$ matrix with orthonormal column vectors, and R is an upper-triangular invertible $k \times k$ matrix. We summarize this in a theorem.

Theorem 5.9 The *QR*-factorization

Let A be a $n \times k$ matrix with independent column vectors in $\mathbb{R}^n$. There exists an $n \times k$ matrix Q with orthonormal column vectors and an upper-triangular invertible $k \times k$ matrix R such that $A = QR$.

We state the special case of Theorem 5.9 for square matrices as a corollary.

Corollary *QR-factorization for Invertible Matrices*

An $n \times n$ matrix A of rank n can be factored as $A = QR$, where Q and R are $n \times n$ matrices, Q is orthogonal, and R is upper-triangular and invertible.

Example 7   Let

$$A = \begin{pmatrix} 1 & 1 \\ 0 & 1 \\ 1 & 1 \end{pmatrix}.$$

Factor A in the form $A = QR$ described in Theorem 5.9, using the computations in Example 2.

Solution   It is clear from our discussion that the matrix R could be formed as we perform the Gram–Schmidt process on the column vectors in A, using Eq. (3) and normalizing as we go. The entries r_{ij} are obtained from Eq. (7). For the present example, it will be more efficient for us to simply take

$$Q = \begin{pmatrix} 1/\sqrt{2} & 0 \\ 0 & 1 \\ 1/\sqrt{2} & 0 \end{pmatrix}$$

from Example 2 and solve $QR = A$ for the matrix R. That is, we solve the matrix equation

$$\begin{pmatrix} 1/\sqrt{2} & 0 \\ 0 & 1 \\ 1/\sqrt{2} & 0 \end{pmatrix} \begin{pmatrix} r_{11} & r_{12} \\ 0 & r_{22} \end{pmatrix} = \begin{pmatrix} 1 & 1 \\ 0 & 1 \\ 1 & 1 \end{pmatrix}$$

for the entries r_{11}, r_{12}, and r_{22}. This corresponds to two linear systems of three equations each, and by inspection we easily see that $r_{11} = \sqrt{2}, r_{12} = \sqrt{2}$, and $r_{22} = 1$. Thus

$$A = \begin{pmatrix} 1 & 1 \\ 0 & 1 \\ 1 & 1 \end{pmatrix} = \begin{pmatrix} 1/\sqrt{2} & 0 \\ 0 & 1 \\ 1/\sqrt{2} & 0 \end{pmatrix} \begin{pmatrix} \sqrt{2} & \sqrt{2} \\ 0 & 1 \end{pmatrix} = QR. \quad \triangleleft$$

The QR-factorization has application to the method of least squares. Recall that the least-squares solution of an overdetermined system $A\mathbf{x} = \mathbf{b}$ is the vector

$$(A^TA)^{-1}A^T\mathbf{b},$$

where the matrix A has independent column vectors. We factor A as QR and remember that $Q^TQ = I$ since Q has orthonormal column vectors. Then we obtain

$$\begin{aligned} (A^TA)^{-1}A^T\mathbf{b} &= [(QR)^TQR]^{-1}(QR)^T\mathbf{b} = (R^TQ^TQR)^{-1}R^TQ^T\mathbf{b} \\ &= (R^TR)^{-1}R^TQ^T\mathbf{b} = R^{-1}(R^T)^{-1}R^TQ^T\mathbf{b} \\ &= R^{-1}Q^T\mathbf{b}. \end{aligned} \tag{9}$$

Once Q and R have been found, the least-squares solution vector $\bar{\mathbf{x}} = R^{-1}Q^T\mathbf{b}$ can be found by multiplying $\mathbf{b}$ by Q^T and solving the upper-triangular system

$$R\bar{\mathbf{x}} = Q^T\mathbf{b} \tag{10}$$

by back substitution. Since R is already upper triangular, no matrix reduction is needed! We should regard R as a record of the projection operations in the Gram–Schmidt process, just as we regarded L in the LU-factorization in Section 2.2 as a record of the row operations in reduction of a matrix to upper-triangular form.

Suppose we believe that the output y of a process should be a linear function of time t. We measure the y-values at specified times during the process. We can then find the least-squares linear fit $y = r_1 + s_1 t$ from the measured data by finding the QR-factorization of an appropriate matrix A and finding the solution vector (9) in the way we just described. If we then repeat the process and make measurements again at the same times, the matrix A will not change, so we can find another least-squares linear fit using (10) with the same matrices Q and R. We can repeat this many times, obtaining linear fits $r_1 + s_1 t$, $r_2 + s_2 t$, . . . , $r_m + s_m t$. For our final estimate of the proper linear fit for this process, we might take $r + st$, where r is the average of the m-values r_i and s is the average of the m-values s_i. Hopefully, errors made in measurements will roughly cancel each other out over repetitions of the process, and our estimated linear fit $r + st$ will be fairly accurate.

Section 7.4 indicates that QR-factorization is also useful for finding eigenvalues of a square matrix.

SUMMARY

1. Any orthogonal set of vectors in $\mathbb{R}^n$ is a basis for the subspace it generates.

2. Every subspace S of $\mathbb{R}^n$ has an orthonormal basis. Any basis can be transformed into an orthogonal basis using the Gram–Schmidt process, in which each vector $\mathbf{a}_j$ of the given basis is replaced by the vector $\mathbf{v}_j$ obtained by subtracting from $\mathbf{a}_j$ its projection on the subspace generated by its predecessors.

3. Any orthogonal set of vectors in a subspace S of $\mathbb{R}^n$ can be expanded, if necessary, to an orthogonal basis for S.

4. Two subspaces of $\mathbb{R}^n$ are orthogonal if every vector in one is perpendicular to every vector in the other.

5. The orthogonal complement $S^\perp$ of a subspace S of $\mathbb{R}^n$ is the set of all vectors in $\mathbb{R}^n$ orthogonal to every vector in S. Also, $S^\perp$ is a subspace of $\mathbb{R}^n$, and $(S^\perp)^\perp = S$.

6. The row space and the nullspace of an $m \times n$ matrix A are orthogonal complements of each other.

*7. Let A be an $n \times k$ matrix of rank k. Then A can be factored as QR, where Q is an $n \times k$ matrix with orthonormal column vectors and R is a $k \times k$ upper-triangular invertible matrix. The matrix R can be viewed as a record of the Gram–Schmidt process on the column vectors of A to obtain the column vectors of Q. The least-squares solution of an overdetermined system $A\mathbf{x} = \mathbf{b}$ is the solution of the square system $R\mathbf{x} = Q^T\mathbf{b}$.

EXERCISES

1. Find an orthonormal basis for the subspace $\mathrm{sp}((0, 1, 0), (1, 1, 1))$ of $\mathbb{R}^3$.

2. Find an orthonormal basis for the subspace $\mathrm{sp}((1, 1, 0), (-1, 2, 1))$ of $\mathbb{R}^3$.

3. Transform the basis $\{(1, 0, 1), (0, 1, 2), (2, 1, 0)\}$ for $\mathbb{R}^3$ into an orthonormal basis using the Gram–Schmidt process.

4. Repeat Exercise 3 using the basis $\{(1, 1, 1), (1, 0, 1), (0, 1, 1)\}$ for $\mathbb{R}^3$.

5. Find an orthonormal basis for the subspace of $\mathbb{R}^4$ spanned by $(1, 0, 1, 0)$, $(1, 1, 1, 0)$, and $(1, -1, 0, 1)$.

6. Find an orthonormal basis for the subspace of $\mathbb{R}^5$ spanned by $(1, -1, 1, 0, 0)$, $(-1, 0, 0, 0, 1)$, $(0, 0, 1, 0, 1)$, and $(1, 0, 0, 1, 1)$.

7. Find an orthonormal basis for $\mathbb{R}^4$ that contains an orthonormal basis for the subspace $\mathrm{sp}((1, 0, 1, 0), (0, 1, 1, 0))$.

8. Find a basis for the orthogonal complement of $\mathrm{sp}((1, -1, 3))$ in $\mathbb{R}^3$.

9. Find a nonzero vector in $\mathbb{R}^3$ orthogonal to both $(2, 1, 1)$ and $(-1, 3, 2)$.

10. Find a nonzero vector in $\mathbb{R}^4$ orthogonal to the vectors $(2, 1, 1, 0)$, $(2, -1, 3, 1)$, and $(2, 4, 0, 1)$.

11. Find a nonzero vector in $\mathbb{R}^5$ orthogonal to the vectors $(1, 2, 3, 2, 1)$, $(1, 2, 0, 1, 2)$, and $(0, 1, 2, 0, 1)$.

12. Let A be an $m \times n$ matrix.
   a) Show that the set S of row vectors $\mathbf{x}$ in $\mathbb{R}^m$ such that $\mathbf{x}A = \mathbf{0}$ is a subspace of $\mathbb{R}^m$.
   b) Show that the subspace S in part (a) and the row space of A are orthogonal complements.

13. Let S and T be orthogonal subspaces of $\mathbb{R}^n$ such that $\dim(S) + \dim(T) = n$. Why must S and T be orthogonal complements?

14. Let S be a subspace of $\mathbb{R}^n$ with orthogonal complement $S^\perp$. Writing $\mathbf{a} = \mathbf{a}_S + \mathbf{a}_{S^\perp}$ as in Theorem 5.8, prove that

$$\|\mathbf{a}\| = \sqrt{\|\mathbf{a}_S\|^2 + \|\mathbf{a}_{S^\perp}\|^2}.$$

[*Hint:* Use the formula $\|\mathbf{a}\|^2 = \mathbf{a} \cdot \mathbf{a}$.]

15. (*Distance from a point to a subspace*). Let S be a subspace of $\mathbb{R}^n$. Figure 5.12 suggests that the distance from a point $\mathbf{a}$ in $\mathbb{R}^n$ to the subspace S is equal to the magnitude of the projection of the vector $\mathbf{a}$ on the orthogonal complement of S. Find the distance from the point $(1, 2, 3)$ in $\mathbb{R}^3$ to the subspace (plane) $\mathrm{sp}((2, 2, 1), (1, 2, 1))$.

16. Find the distance from the point $(2, 1, 3, 1)$ in $\mathbb{R}^4$ to the subspace (plane) $\mathrm{sp}((1, 2, 1, 0), (1, 1, 2, 1))$. [*Hint:* See Exercise 13.]

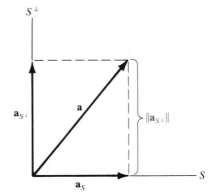

Figure 5.12
The distance from a to S is $\|\mathbf{a}_{S^\perp}\|$.

In Exercises 17–22, use the idea in Exercise 15 to find the distance from the given point $\mathbf{a}$ to the given line (one-dimensional subspace) S. [Note: *To calculate $\|\mathbf{a}_{S^\perp}\|$, first calculate $\|\mathbf{a}_S\|$ and then use Exercise 14.*]

17. $\mathbf{a} = (1, 2, 1)$,
   $S = \mathrm{sp}((2, 1, 0))$ in $\mathbb{R}^3$

18. $\mathbf{a} = (2, -1, 3)$,
   $S = \mathrm{sp}((1, 2, 4))$ in $\mathbb{R}^3$

19. $\mathbf{a} = (1, 2, -1, 0)$,
$S = \text{sp}((3, 1, 4, -1))$ in $\mathbb{R}^4$

20. $\mathbf{a} = (2, 1, 1, 2)$,
$S = \text{sp}((1, 2, 1, 3))$ in $\mathbb{R}^4$

21. $\mathbf{a} = (1, 2, 3, 4, 5)$,
$S = \text{sp}((1, 1, 1, 1, 1))$ in $\mathbb{R}^5$

22. $\mathbf{a} = (1, 0, 1, 0, 1, 0, 1)$,
$S = \text{sp}((1, 2, 3, 4, 3, 2, 1))$ in $\mathbb{R}^7$

23. Let S be a subspace of $\mathbb{R}^n$ and let

$$S^\perp = \{\mathbf{v} \text{ in } \mathbb{R}^n \mid \mathbf{v} \cdot \mathbf{w} = 0 \text{ for each } \mathbf{w} \text{ in } S\}$$

be the orthogonal complement of S. Prove that $S^\perp$ is a subspace of $\mathbb{R}^n$.

In Exercises 24–29, use the work in the indicated exercise to find the QR-factorization of the matrix having as column vectors the transposes of the row vectors given in that exercise.

* 24. Exercise 2

* 25. Exercise 1

* 26. Exercise 4

* 27. Exercise 3

* 28. Exercise 6

* 29. Exercise 5

🔲 *The supplied program QRFACTOR allows the user to enter k independent vectors in $\mathbb{R}^n$ for n and k at most 10. The program can then be used to find an orthonormal set of vectors spanning the same subspace. It will also exhibit the QR-factorization of the $n \times k$ matrix A having the entered vectors as column vectors. Finally, it will use the QR-factorization to give the least-squares solution of a linear system $A\mathbf{x} = \mathbf{b}$. Students who have not studied the QR-factorization can still use the program to execute the Gram–Schmidt process and to find least-squares solutions. Use this program for the remaining exercises.*

30. Check answers to those of Exercises 1–6 that were attempted.

* 31. Check answers to those of Exercises 24–29 that were attempted.

32. Find the least-squares linear fit for the data points $(-3, 10)$, $(-2, 8)$, $(-1, 7)$, $(0, 6)$, $(1, 4)$, $(2, 5)$, $(3, 6)$.

33. Find the least-squares quadratic fit for the data points in Exercise 32.

34. Find the least-squares cubic fit for the data points in Exercise 32.

35. Find the least-squares quartic fit for the data points in Exercise 32.

36. Find the quadratic polynomial function whose graph passes through the points $(1, 4)$, $(2, 15)$, and $(3, 32)$.

37. Find the cubic polynomial function whose graph passes through the points $(-1, 12)$ $(0, -5)$, $(2, 15)$, and $(3, 52)$.

5.5
Lines, Planes, and Hyperplanes

In discussing the subspace of $\mathbb{R}^n$ spanned by nonzero vectors $\mathbf{d}_1, \mathbf{d}_2, \ldots, \mathbf{d}_k$, we saw in Section 3.3 that $\text{sp}(\mathbf{d}_1)$ is a *line* containing $\mathbf{0}$ and $\text{sp}(\mathbf{d}_1, \mathbf{d}_2)$ is a plane containing $\mathbf{0}$ in $\mathbb{R}^n$ if $\mathbf{d}_1$ and $\mathbf{d}_2$ are not parallel. Lines and planes in general need not contain the origin, and so need not be subspaces.

LINES IN $\mathbb{R}^n$

Let $\mathbf{a}$ be a point and $\mathbf{d}$ be a vector in $\mathbb{R}^n$. We wish to *translate* the line $\text{sp}(\mathbf{d})$ to obtain a parallel line through the point $\mathbf{a}$, as shown in Fig. 5.13. Geometrically, translation of a set by a vector $\mathbf{a}$ is accomplished by sliding the set in the direction determined by $\mathbf{a}$ through a distance $\|\mathbf{a}\|$. Analytically, translation by $\mathbf{a}$ corresponds to adding $\mathbf{a}$ to every vector in the set.

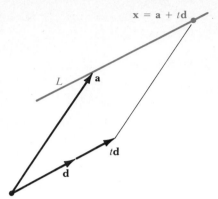

Figure 5.13
The line L through the point a with parallel vector d.

For the line L in Fig. 5.13, the vector **d** parallel to the line determines the *direction* of L, while the point **a** is a specific *point* lying on the line. That is, we think of the line as being determined by

Point **a**,   Parallel vector **d**.

Every point x on the line can be expressed in the form

$$\mathbf{x} = \mathbf{a} + t\mathbf{d},$$

as in Fig. 5.13.

Definition 5.3   A Line in $\mathbb{R}^n$

The **line in $\mathbb{R}^n$ through** $\mathbf{a} = (a_1, a_2, \ldots, a_n)$ **with nonzero parallel vector d** $= (d_1, d_2, \ldots, d_n)$ is the set of points $\mathbf{x} = (x_1, x_2, \ldots, x_n)$ satisfying

$$\mathbf{x} = \mathbf{a} + t\mathbf{d} \quad \text{for some scalar } t. \tag{1}$$

Equation (1) is a **vector equation of the line** and **parametric equations** are the corresponding component equations

$$x_i = a_i + td_i \quad \text{for } i = 1, 2, \ldots, n. \tag{2}$$

Example 1   Find parametric equations of the line in $\mathbb{R}^2$ through $(2, 1)$ having parallel vector $(3, 4)$.

Solution   A vector equation of the line is

$$(x_1, x_2) = (2, 1) + t(3, 4)$$

and parametric equations are $x_1 = 2 + 3t$, $x_2 = 1 + 4t$.   ◁

Example 2   Find parametric equations of the line in $\mathbb{R}^3$ that passes through the points $\mathbf{a} = (1, 2, -1)$ and $\mathbf{b} = (3, 1, 4)$.

Solution   In order to find a suitable parallel vector for the line, we compute $\mathbf{d} = \mathbf{b} - \mathbf{a} = (2, -1, 5)$, which is the vector that starts at $\mathbf{a}$ and terminates at $\mathbf{b}$. (See Fig. 5.14) Therefore, a vector equation of the line is

$$(x_1, x_2, x_3) = (1, 2, -1) + t(2, -1, 5)$$

and parametric equations are

$$x_1 = 1 + 2t, \qquad x_2 = 2 - t, \qquad x_3 = -1 + 5t. \quad \lhd$$

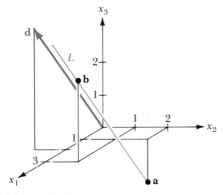

Figure 5.14
The line L passing through a and b.

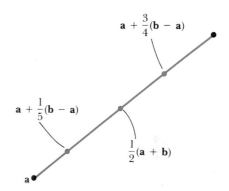

Figure 5.15
Points on the line segment between a and b corresponding to $t = 1/5$, $1/2$, and $3/4$.

LINE SEGMENTS

The last example illustrates that a suitable parallel vector for the line passing through two points $\mathbf{a}$ and $\mathbf{b}$ in $\mathbb{R}^n$ is $\mathbf{d} = \mathbf{b} - \mathbf{a}$, and a vector equation of the line is the following.

Line Passing Through a and b in $\mathbb{R}^n$

$$\mathbf{x} = \mathbf{a} + t(\mathbf{b} - \mathbf{a}) \quad \text{for scalars } t \qquad\qquad (3)$$

Equations (2) and (3) really present the line as a t-axis with origin at $\mathbf{a}$ and where one unit on the t-axis equals $\|\mathbf{d}\|$ units in $\mathbb{R}^n$.

We see that each point in $\mathbb{R}^n$ on the **line segment** between $\mathbf{a}$ and $\mathbf{b}$ consists of all points $\mathbf{x}$ in Eq. (3) obtained for some value t from 0 to 1. By choosing t appropriately, we can locate any point on the line (3) that we please, as illustrated in Fig. 5.15. In particular, the **midpoint of the line segment** between $\mathbf{a}$ and $\mathbf{b}$ is given by

$$\mathbf{a} + (\tfrac{1}{2})(\mathbf{b} - \mathbf{a}) = (\tfrac{1}{2})(\mathbf{a} + \mathbf{b}).$$

Example 3   Find the points that divide the line segment between $\mathbf{a} = (1, 2, 1, 3)$ and $\mathbf{b} = (2, 1, 4, 2)$ in $\mathbb{R}^4$ into five equal parts.

Solution   The line segment between $\mathbf{a}$ and $\mathbf{b}$ is given by Eq. (3) for values of t from 0 to 1. Equation (3) becomes

$$(x_1, x_2, x_3, x_4) = (1, 2, 1, 3) + t(1, -1, 3, -1), \quad 0 \le t \le 1.$$

By choosing $t = 0, \frac{1}{5}, \frac{2}{5}, \frac{3}{5}, \frac{4}{5}$, and 1 we obtain the division required, shown in Table 5.9.   ◁

Table 5.9

t	Equally spaced points from $\mathbf{a}$ to $\mathbf{b}$
0	(1, 2, 1, 3)
1/5	(1.2, 1.8, 1.6, 2.8)
2/5	(1.4, 1.6, 2.2, 2.6)
3/5	(1.6, 1.4, 2.8, 2.4)
4/5	(1.8, 1.2, 3.4, 2.2)
1	(2, 1, 4, 2)

PLANES IN $\mathbb{R}^n$

Just as a line is a translation of a one-dimensional subspace in $\mathbb{R}^n$, a plane in $\mathbb{R}^n$ is a translation of a two-dimensional subspace $\text{sp}(\mathbf{a}_1, \mathbf{a}_2)$, where $\mathbf{a}_1$ and $\mathbf{a}_2$ are nonzero and nonparallel vectors in $\mathbb{R}^n$. Initially, we will restrict our attention to the case of a plane in $\mathbb{R}^3$. For each point $\mathbf{b}$ and nonzero vector $\mathbf{d}$ in $\mathbb{R}^3$, there is a unique plane in $\mathbb{R}^3$ through $\mathbf{b}$ and perpendicular to $\mathbf{d}$. That is, $\mathbf{d}$ is perpendicular to every vector in the plane, as illustrated in Fig. 5.16.

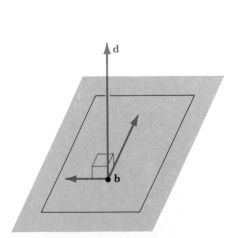

Figure 5.16
d is perpendicular to every vector in the plane.

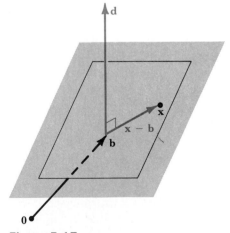

Figure 5.17
x − b is perpendicular to the vector d.

We think of $\mathbf{d}$ as establishing a *direction* for the plane, just as a vector parallel to a line establishes a direction for the line. A perpendicular vector to a plane can be found by taking the cross product of two vectors lying in the plane. Referring to Fig. 5.17, we see that a vector $\mathbf{x}$ in the plane passing through a point $\mathbf{b}$ and having orthogonal vector $\mathbf{d}$ must satisfy $(\mathbf{x} - \mathbf{b}) \perp \mathbf{d}$, that is,

$$\mathbf{d} \cdot (\mathbf{x} - \mathbf{b}) = 0. \tag{4}$$

This is the vector equation of the plane, and is determined by a direction vector $\mathbf{d}$ perpendicular to the plane together with a point $\mathbf{b}$ in the plane.

Definition 5.4 A Plane in $\mathbb{R}^3$

The **plane in $\mathbb{R}^3$ passing through $\mathbf{b} = (b_1, b_2, b_3)$ with nonzero orthogonal vector $\mathbf{d} = (d_1, d_2, d_3)$** is the set of points $\mathbf{x} = (x_1, x_2, x_3)$ satisfying $\mathbf{d} \cdot (\mathbf{x} - \mathbf{b}) = 0$. This vector equation can be written as the single linear scalar equation

$$d_1 x_1 + d_2 x_2 + d_3 x_3 = c,$$

where $c = b_1 d_1 + b_2 d_2 + b_3 d_3$.

Example 4   Find a scalar equation of the plane in $\mathbb{R}^3$ that passes through $\mathbf{b} = (3, 2, -4)$ and is orthogonal to the line with parametric equations $x_1 = 2 + 3t$, $x_2 = 1 - 2t$, $x_3 = 4 + t$.

THE EQUATION OF A PLANE IN $\mathbb{R}^3$ presented in Definition 5.3 appears as early as 1732 in a paper of Jacob Hermann (1678–1733). He was able to determine the plane's position by using intercepts and also noted that the sine of the angle between the plane and the one coordinate plane he dealt with (what we call the x_1, x_2-plane) was

$$\frac{\sqrt{d_1{}^2 + d_2{}^2}}{\sqrt{d_1{}^2 + d_2{}^2 + d_3{}^2}}.$$

In his 1748 *Introduction to Infinitesimal Analysis*, Leonhard Euler (1707–1783) used instead the cosine of this angle, $d_3 / \sqrt{d_1{}^2 + d_2{}^2 + d_3{}^2}$.

At the end of the eighteenth century it was Gaspard Monge (1746–1818), however, who, in his notes for a course on solid analytic geometry at the Ecole Polytechnique, related the equation of a plane to all three coordinate planes and gave the cosines of the angles the plane made with each of these (the so-called direction cosines). He also presented many of the standard problems of solid analytic geometry, examples of which appear in the exercises. For example, he showed how to find the plane passing through three given points, the line through a point perpendicular to a plane, the distance between two parallel planes, and the angle between a line and a plane.

Known as "the greatest geometer of the eighteenth century," Monge developed new graphical geometric techniques as a student and later professor at a military school. The first problem he solved involved a procedure enabling soldiers to make quickly a fortification capable of shielding a position from both the view and the firepower of the enemy. Monge served the French revolutionary government as minister of the navy and later served Napoleon in various scientific offices. Ultimately, he was appointed a senator for life by the emperor.

Solution   The parallel vector $\mathbf{d} = (3, -2, 1)$ for the given line is orthogonal to the desired plane. Thus the plane has vector equation

$$(3, -2, 1) \cdot [(x_1, x_2, x_3) - (3, 2, -4)] = 0.$$

Computing the dot product, we obtain the single linear equation $3x_1 - 2x_2 + x_3 - (9 - 4 - 4) = 0$, or

$$3x_1 - 2x_2 + x_3 = 1. \quad \triangleleft$$

Example 5   Find a scalar equation of the plane in $\mathbb{R}^3$ that passes through $\mathbf{b} = (2, 3, 4)$ and is parallel to the plane $\mathrm{sp}(\mathbf{a}_1, \mathbf{a}_2)$, where $\mathbf{a}_1 = (2, -2, 3)$ and $\mathbf{a}_2 = (2, -3, 1)$.

Solution   An orthogonal vector for the given plane is the cross product

$$\mathbf{d} = \mathbf{a}_1 \times \mathbf{a}_2 = \begin{vmatrix} \mathbf{i} & \mathbf{j} & \mathbf{k} \\ 2 & -2 & 3 \\ 2 & -3 & 1 \end{vmatrix} = (7, 4, -2).$$

Therefore, a vector equation of the desired plane is

$$(7, 4, -2) \cdot [(x_1, x_2, x_3) - (2, 3, 4)] = 0.$$

We obtain $7x_1 + 4x_2 - 2x_3 - (14 + 12 - 8) = 0$ or $7x_1 + 4x_2 - 2x_3 = 18$ as a scalar equation. $\triangleleft$

We know that every plane in $\mathbb{R}^3$ can be described by a linear equation of the form

$$d_1x_1 + d_2x_2 + d_3x_3 = c, \quad \text{not all } d_i = 0. \tag{5}$$

Conversely, any equation of this type describes a plane in $\mathbb{R}^3$. For example, if $d_1 \neq 0$ in Eq. (5), then choosing any numbers a_2, a_3 in $\mathbb{R}$, we let

$$a_1 = (c - d_2a_2 - d_3a_3)/d_1.$$

This produces a point $\mathbf{a} = (a_1, a_2, a_3)$ satisfying

$$d_1a_1 + d_2a_2 + d_3a_3 = c.$$

Therefore, any vector $\mathbf{x} = (x_1, x_2, x_3)$ satisfying Eq. (5) must also satisfy the difference

$$d_1(x_1 - a_1) + d_2(x_2 - a_2) + d_3(x_3 - a_3) = 0$$

of the two equations, and so is in the plane passing through $\mathbf{a}$ with orthogonal vector $\mathbf{d}$. We summarize in a theorem.

Theorem 5.10   Equation of a Plane

If d_1, d_2, d_3 are scalars not all of which are zero, then for any number c the equation $d_1x_1 + d_2x_2 + d_3x_3 = c$ is the equation of a plane in $\mathbb{R}^3$ with orthogonal vector $\mathbf{d} = (d_1, d_2, d_3)$.

HYPERPLANES

In trying to generalize the previous discussion to $\mathbb{R}^n$, we are naturally led to consider an $(n - 1)$-dimensional *flat piece* of $\mathbb{R}^n$, which is known as a *hyperplane*. The term "hyperplane" is used to distinguish such $(n - 1)$-dimensional flat pieces from two-dimensional planes in $\mathbb{R}^n$.

Definition 5.5   A Hyperplane in $\mathbb{R}^n$

> The **hyperplane in $\mathbb{R}^n$ through b** $= (b_1, b_2, \ldots, b_n)$ **with nonzero orthogonal vector d** $= (d_1, d_2, \ldots, d_n)$ is the set of all points $\mathbf{x} = (x_1, x_2, \ldots, x_n)$ satisfying
>
> $$\mathbf{d} \cdot (\mathbf{x} - \mathbf{b}) = 0. \tag{6}$$

Equation (6) can be written without the dot product as

$$d_1 x_1 + d_2 x_2 + \cdots + d_n x_n = c, \tag{7}$$

where $c = d_1 b_1 + d_2 b_2 + \cdots + d_n b_n$.

We comment that any equation of the form (7) in which not all d_i are zero is the equation of a hyperplane with orthogonal vector $\mathbf{d} = (d_1, d_2, \ldots, d_n)$. The proof is analogous to that of Theorem 5.10, and is requested in Exercise 40.

Example 6   Find the equation of the hyperplane in $\mathbb{R}^5$ through the point $(2, 1, -1, 3, 4)$ and orthogonal to the vector $(2, 1, 5, 1, 2)$.

Solution   The vector equation of the hyperplane is

$$(2, 1, 5, 1, 2) \cdot [(x_1, x_2, x_3, x_4, x_5) - (2, 1, -1, 3, 4)] = 0,$$

which can be written as

$$2x_1 + x_2 + 5x_3 + x_4 + 2x_5 = 11. \quad \triangleleft$$

A hyperplane in $\mathbb{R}^2$ is a line with equation $d_1 x_1 + d_2 x_2 = c$, which is orthogonal to the vector (d_1, d_2). Exercise 3 requests parametric equations of this line. A hyperplane in $\mathbb{R}^3$ is the plane in Definition 5.5. Let us see why a hyperplane in $\mathbb{R}^n$ is an $(n - 1)$-dimensional "flat piece." Equation (7) corresponds to a system of one equation in n unknowns, and the corresponding homogeneous equation

$$d_1 x_1 + d_2 x_2 + \cdots + d_n x_n = 0$$

has as solutions all vectors in a subspace S of dimension $n - 1$. Let

$$\{\mathbf{a}_1, \mathbf{a}_2, \ldots, \mathbf{a}_{n-1}\}$$

be a basis for S. Our hyperplane can be described as the *translation* $\mathbf{b} + S$ of S, that is, the vectors $\mathbf{x}$ in the hyperplane satisfy

$$\mathbf{x} = \mathbf{b} + t_1\mathbf{a}_1 + t_2\mathbf{a}_2 + \cdots + t_{n-1}\mathbf{a}_{n-1} \tag{8}$$

for scalars t_i in $\mathbb{R}$. See Theorem 1.9 on page 73. An orthogonal vector $\mathbf{d}$ for the hyperplane is $\mathbf{d} = \mathbf{b} - \mathbf{b}_S$, where $\mathbf{b}_S$ is the projection of $\mathbf{b}$ on S as described in Theorem 5.1. That is, if A is the $n \times (n-1)$ matrix with column vectors $\mathbf{a}_1, \mathbf{a}_2, \ldots, \mathbf{a}_{n-1}$, then $\mathbf{b}_S = A(A^TA)^{-1}A^T\mathbf{b}$. Alternatively, an orthogonal vector $\mathbf{d}$ to the hyperplane is any nontrivial solution of the linear system $A^T\mathbf{x} = \mathbf{0}$, for the dot product of any such solution with each $\mathbf{a}_i$ is zero. For still another alternative, $\mathbf{d}$ can be found by forming a symbolic determinant similar to a cross product, but with first row $(\mathbf{e}_1, \mathbf{e}_2, \ldots, \mathbf{e}_n)$, and with subsequent rows the vectors $\mathbf{a}_1, \mathbf{a}_2, \ldots, \mathbf{a}_{n-1}$. Properties of determinants show that the vector obtained by expanding this determinant by minors across the first row is orthogonal to every other row, and hence is orthogonal to S.

FLATS IN $\mathbb{R}^n$

We have studied translations of one-dimensional and $(n-1)$-dimensional subspaces of $\mathbb{R}^n$. We turn to the translation of a general k-dimensional subspace.

Definition 5.6   A *k*-flat in $\mathbb{R}^n$

Let $\{\mathbf{a}_1, \mathbf{a}_2, \ldots, \mathbf{a}_k\}$ be a basis for a subspace S of $\mathbb{R}^n$ and let $\mathbf{b}$ be a point in $\mathbb{R}^n$. The translated set $\mathbf{b} + S$ consisting of all vectors $\mathbf{x}$ of the form

$$\mathbf{x} = \mathbf{b} + t_1\mathbf{a}_1 + t_2\mathbf{a}_2 + \cdots + t_k\mathbf{a}_k \tag{9}$$

for scalars t_i in $\mathbb{R}$ is called a **k-flat** containing $\mathbf{b}$ in $\mathbb{R}^n$.

In $\mathbb{R}^n$ we see that a 1-flat is a line, a 2-flat is a plane, and an $(n-1)$-flat is a hyperplane.

Let $A\mathbf{x} = \mathbf{b}$ be any system of m equations in n unknowns that has at least one solution. See Theorem 1.9 on page 73. Then the solution set of the system is a k-flat in $\mathbb{R}^n$, where k is the dimension of the nullspace of A. The next example illustrates this. If a system of equations has a unique solution, then its solution set is a 0-*flat*.

Example 7   Solve the system of equations

$$
\begin{aligned}
x_1 + 2x_2 - 2x_3 + x_4 + 3x_5 &= 1 \\
2x_1 + 5x_2 - 3x_3 - x_4 + 2x_5 &= 2 \\
-3x_1 - 8x_2 + 6x_3 - x_4 - 5x_5 &= 1
\end{aligned}
$$

and write the solution set as a k-flat.

Solution Reducing the corresponding partitioned matrix, we have

$$
\left(\begin{array}{ccccc|c}
1 & 2 & -2 & 1 & 3 & 1 \\
2 & 5 & -3 & -1 & 2 & 2 \\
-3 & -8 & 6 & -1 & -5 & 1
\end{array}\right)
\sim
\left(\begin{array}{ccccc|c}
1 & 2 & -2 & 1 & 3 & 1 \\
0 & 1 & 1 & -3 & -4 & 0 \\
0 & -2 & 0 & 2 & 4 & 4
\end{array}\right)
$$

$$
\sim
\left(\begin{array}{ccccc|c}
1 & 0 & -4 & 7 & 11 & 1 \\
0 & 1 & 1 & -3 & -4 & 0 \\
0 & 0 & 2 & -4 & -4 & 4
\end{array}\right)
$$

$$
\sim
\left(\begin{array}{ccccc|c}
1 & 0 & 0 & -1 & 3 & 9 \\
0 & 1 & 0 & -1 & -2 & -2 \\
0 & 0 & 1 & -2 & -2 & 2
\end{array}\right).
$$

Thus $\mathbf{b} = (9, -2, 2, 0, 0)$ is a particular solution to the given system, and $\mathbf{a}_1 = (1, 1, 2, 1, 0)$ and $\mathbf{a}_2 = (-3, 2, 2, 0, 1)$ form a basis for the solution space of the corresponding homogeneous system. The solution set of the given system is the 2-flat in $\mathbb{R}^5$ with vector equation $\mathbf{x} = \mathbf{b} + t_1\mathbf{a}_1 + t_2\mathbf{a}_2$, which can be written in the form

$$
\begin{pmatrix} x_1 \\ x_2 \\ x_3 \\ x_4 \\ x_5 \end{pmatrix}
=
\begin{pmatrix} 9 \\ -2 \\ 2 \\ 0 \\ 0 \end{pmatrix}
+ t_1 \begin{pmatrix} 1 \\ 1 \\ 2 \\ 1 \\ 0 \end{pmatrix}
+ t_2 \begin{pmatrix} -3 \\ 2 \\ 2 \\ 0 \\ 1 \end{pmatrix} \quad \triangleleft
$$

SUMMARY

1. A k-flat in $\mathbb{R}^n$ is a translation of a k-dimensional subspace and has the form $\mathbf{b} + \mathrm{sp}(\mathbf{a}_1, \mathbf{a}_2, \ldots, \mathbf{a}_k)$, where $\mathbf{b}$ is a vector in $\mathbb{R}^n$ and $\mathbf{a}_1, \mathbf{a}_2, \ldots, \mathbf{a}_k$ are independent vectors in $\mathbb{R}^n$. A vector equation of the k-flat is
$$\mathbf{x} = \mathbf{b} + t_1\mathbf{a}_1 + t_2\mathbf{a}_2 + \cdots + t_k\mathbf{a}_k$$
for scalars t_i in $\mathbb{R}$.

2. A line in $\mathbb{R}^n$ is a 1-flat. The line passing through the point $\mathbf{b}$ with parallel vector $\mathbf{d}$ is given by $\mathbf{x} = \mathbf{b} + t\mathbf{d}$, where t runs through all scalars. Parametric equations of the line are the component equations
$$x_1 = b_1 + td_1, \, x_2 = b_2 + td_2, \ldots, x_n = b_n + td_n.$$

3. A plane in $\mathbb{R}^n$ is a 2-flat. In $\mathbb{R}^3$, the plane passing through the point $\mathbf{b} = (b_1, b_2, b_3)$ with orthogonal vector $\mathbf{d} = (d_1, d_2, d_3)$ has scalar equation
$$d_1(x_1 - b_1) + d_2(x_2 - b_2) + d_3(x_3 - b_3) = 0$$
as well as the vector equation described in number (1) above.

4. A hyperplane in $\mathbb{R}^n$ is an $(n - 1)$-flat. The hyperplane in $\mathbb{R}^n$ passing through $\mathbf{b} = (b_1, b_2, \ldots, b_n)$ with orthogonal vector $\mathbf{d} =$

$(d_1, d_2, \ldots, d_n)$ has scalar equation

$$d_1(x_1 - b_1) + d_2(x_2 - b_2) + \cdots + d_n(x_n - b_n) = 0$$

as well as the vector equation described in number (1) above.

EXERCISES

1. Give parametric equations for the line in $\mathbb{R}^2$ through $(3, -3)$ with parallel vector $\mathbf{d} = (-8, 4)$. Sketch the line in an appropriate figure.

2. Give parametric equations for the line in $\mathbb{R}^3$ through $(-1, 3, 0)$ with parallel vector $\mathbf{d} = (-2, -1, 4)$. Sketch the line in an appropriate figure.

3. Consider the line in $\mathbb{R}^2$ given by the equation $d_1 x_1 + d_2 x_2 = c$ for numbers d_1, d_2, and c in $\mathbb{R}$, where d_1 and d_2 are not both zero. Find parametric equations of the line.

4. Find parametric equations for the line in $\mathbb{R}^2$ through $(5, -1)$ and orthogonal to the line with parametric equations $x_1 = 4 - 2t$, $x_2 = 7 + t$.

5. For each pair of points, find parametric equations of the line containing them.
   a) $(-2, 4)$ and $(3, -1)$ in $\mathbb{R}^2$
   b) $(3, -1, 6)$ and $(0, -3, -1)$ in $\mathbb{R}^3$
   c) $(2, 0, 4)$ and $(-1, 5, -8)$ in $\mathbb{R}^3$

6. For each of the given pairs of lines in $\mathbb{R}^3$, determine whether the lines intersect. If they intersect, find the point of intersection, and determine whether the lines are orthogonal.
   a) $x_1 = 4 + t$, $x_2 = 2 - 3t$, $x_3 = -3 + 5t$
   and
   $x_1 = 11 + 3s$, $x_2 = -9 - 4s$, $x_3 = -4 - 3s$
   b) $x_1 = 11 + 3t$, $x_2 = -3 - t$, $x_3 = 4 + 3t$
   and
   $x_1 = 6 - 2s$, $x_2 = -2 + s$, $x_3 = -15 + 7s$

7. Find all points in common to the lines in $\mathbb{R}^2$ given by

$$x_1 = 5 - 3t, x_2 = -1 + t, \quad \text{and}$$
$$x_1 = -7 + 6s, x_2 = 3 - 2s.$$

8. Find parametric equations for the line in $\mathbb{R}^3$ through $(-1, 2, 3)$ that is orthogonal to each of the two lines having parametric equations

$$x_1 = -2 + 3t, x_2 = 4, x_3 = 1 - t \quad \text{and}$$
$$x_1 = 7 - t, x_2 = 2 + 3t, x_3 = 4 + t.$$

9. Find the midpoint of the line segment joining the pair of points.
   a) $(-2, 4)$ and $(3, -1)$ in $\mathbb{R}^2$
   b) $(3, -1, 6)$ and $(0, -3, -1)$ in $\mathbb{R}^3$
   c) $(0, 4, 8)$ and $(-4, 5, 9)$ in $\mathbb{R}^3$

10. Find the point in $\mathbb{R}^2$ on the line segment joining $(-1, 3)$ and $(2, 5)$ that is twice as close to $(-1, 3)$ as to $(2, 5)$.

11. Find the point in $\mathbb{R}^3$ on the line segment joining $(-2, 1, 3)$ and $(0, -5, 6)$ that is one fourth of the way from $(-2, 1, 3)$ to $(0, -5, 6)$.

12. Find an equation for the plane in $\mathbb{R}^3$ through $(1, -3, 0)$ and with orthogonal vector $\mathbf{d} = (0, -1, 4)$.

13. Find an equation of the plane in $\mathbb{R}^3$ passing through $(3, -1, 4)$ and parallel to the plane with equation $3x_1 - 2x_2 + 7x_3 = 14$.

14. Find an equation of the plane in $\mathbb{R}^3$ passing through $(-1, 4, -3)$ and orthogonal to the line with parametric equations $x_1 = 3 - 7t$, $x_2 = 4 + t$, $x_3 = 2t$.

15. Find an equation of the plane in $\mathbb{R}^3$ passing through the origin and orthogonal to the line through the points $(-1, 3, 0)$ and $(2, -4, 3)$.

16. Classify the planes given by the following pairs of equations as parallel, orthogonal, or neither. Also, find the angle between the planes.
  a) $x_1 + 4x_2 = 7$ and $x_2 - x_3 = -3$
  b) $3x_1 + 4x_2 - x_3 = 1$ and
      $-6x_1 - 8x_2 + 2x_3 = 4$
  c) $x_1 - 3x_2 - x_3 = 7$ and $2x_1 + 4x_3 = 1$
  d) $x_1 = 8$ and $x_2 - 4x_3 = 2$

17. Find parametric equations of the line through $(-3, 2, 7)$ in $\mathbb{R}^3$ that is orthogonal to the plane with equation

$$x_1 - 2x_2 + x_3 = 3.$$

18. Find the intersection in $\mathbb{R}^3$ of the line given by

$$x_1 = 5 + t, \quad x_2 = -3t, \quad x_3 = -2 + 4t$$

and the plane with equation $x_1 - 3x_2 + 2x_3 = -25$.

19. Find the intersection in $\mathbb{R}^3$ of the line given by
$$x_1 = 2, \quad x_2 = 5 - t, \quad x_3 = 2t$$

and the plane with equation $x_1 + 2x_3 = 10$.

20. Find an equation of the plane that passes through the unit coordinate points $(1, 0, 0)$, $(0, 1, 0)$, $(0, 0, 1)$ in $\mathbb{R}^3$.

21. Find an equation of the plane in $\mathbb{R}^3$ that passes through $(1, 0, 0)$, $(0, 1, -1)$, and $(1, 1, 1)$.

22. Show that the distance from the point $\mathbf{a} = (a_1, a_2, a_3)$ in $\mathbb{R}^3$ to the plane $d_1x_1 + d_2x_2 + d_3x_3 = c$ is

$$\frac{|d_1a_1 + d_2a_2 + d_3a_3 - c|}{\sqrt{d_1^2 + d_2^2 + d_3^2}}.$$

[*Hint:* For $\mathbf{x}$ in the plane, find the magnitude of the projection of $\mathbf{a} - \mathbf{x}$ on a vector $\mathbf{d}$ perpendicular to the plane.]

23. Use the formula in Exercise 22 to find the distance from the point $(1, 3, -1)$ to the plane with equation $2x_1 - x_2 + x_3 = 4$ in $\mathbb{R}^3$.

24. Find a formula like the one in Exercise 22 for the distance from the point $\mathbf{a}$ in $\mathbb{R}^n$ to the hyperplane $d_1x_1 + d_2x_2 + \cdots + d_nx_n = c$.

25. Use the formula found in Exercise 24 to find the distance from the point $(-2, 3, 1, 4, 2)$ to the hyperplane $3x_1 - x_2 + 2x_3 + x_4 - 4x_5 = 1$ in $\mathbb{R}^5$.

26. Find the midpoint of the line segment between $(2, 1, 3, 4, 0)$ and $(1, 2, -1, 3, -1)$ in $\mathbb{R}^5$.

27. Find the points that divide the line segment between $(2, 1, 3, 4)$ and $(-1, 2, 1, 3)$ in $\mathbb{R}^4$ into three equal parts.

28. Find an equation of the hyperplane that passes through the points $(1, 2, 1, 2, 3)$, $(0, 1, 2, 1, 3)$, $(0, 0, 3, 1, 2)$, $(0, 0, 0, 1, 4)$, and $(0, 0, 0, 0, 2)$ in $\mathbb{R}^5$. [*Hint:* Solve an appropriate linear system.]

29. Find an equation of the plane that passes through the points $(1, 2, 1)$, $(-1, 2, 3)$, and $(2, 1, 4)$ in $\mathbb{R}^3$.

30. Find a vector equation for the plane in $\mathbb{R}^4$ that passes through the points $(1, 2, 1, 3)$, $(4, 1, 2, 1)$, and $(3, 1, 2, 0)$. [*Hint:* Since the plane is a 2-flat in $\mathbb{R}^4$, use Eq. (9).]

In Exercises 31–38, solve the given system of linear equations and write the solution set as a k-flat.

31. $\begin{aligned} x_1 - 2x_2 &= 3 \\ 3x_1 - x_2 &= 14 \end{aligned}$

32. $\begin{aligned} x_1 + 2x_2 - x_3 &= -3 \\ 3x_1 + 7x_2 + 2x_3 &= 1 \\ 4x_1 - 2x_2 + x_3 &= -2 \end{aligned}$

33. $x_1 + 4x_2 - 2x_3 = 4$
$2x_1 + 7x_2 - x_3 = -2$
$x_1 + 3x_2 + x_3 = -6$

34. $x_1 - 3x_2 + x_3 = 2$
$3x_1 - 8x_2 + 2x_3 = 5$
$3x_1 - 7x_2 + x_3 = 4$

35. $x_1 - 3x_2 + 2x_3 - x_4 = 8$
$3x_1 - 7x_2 + x_4 = 0$

36. $x_1 \quad - 2x_3 + x_4 = 6$
$2x_1 - x_2 + x_3 - 3x_4 = 0$
$9x_1 - 3x_2 - x_3 - 7x_4 = 4$

37. $x_1 + 2x_2 - 3x_3 + x_4 = 2$
$3x_1 + 6x_2 - 8x_3 - 2x_4 = 1$

38. $x_1 - 3x_2 + x_3 + 2x_4 = 2$
$x_1 - 2x_2 + 2x_3 + 4x_4 = -1$
$2x_1 - 8x_2 - x_3 \quad = 3$
$3x_1 - 9x_2 + 4x_3 \quad = 7$

39. For $n \geq 2$ and $1 \leq j \leq n$, let $\mathbf{a}_j = (a_{1j}, a_{2j}, \ldots, a_{nj})$ be n points in $\mathbb{R}^n$ that do not all lie on a single $(n - 2)$-flat. Consider the determinant

$$f(x_1, x_2, \ldots, x_n) = \begin{vmatrix} a_{11} & a_{12} & \cdots & a_{1n} & x_1 \\ \vdots & \vdots & & \vdots & \vdots \\ a_{n1} & a_{n2} & \cdots & a_{nn} & x_n \\ 1 & 1 & \cdots & 1 & 1 \end{vmatrix}.$$

In the expansion of this determinant on the last column, it can be shown that at least one cofactor x_i' is nonzero. Using this fact, show that the hyperplane in $\mathbb{R}^n$ passing through $\mathbf{a}_1, \mathbf{a}_2 \ldots, \mathbf{a}_n$ has equation $f(x_1, x_2, \ldots, x_n) = 0$.

40. State and prove the analog of Theorem 5.10 for a hyperplane in $\mathbb{R}^n$.

41. Find the distance between the planes in $\mathbb{R}^3$ given by

$$x_1 + 3x_2 - 2x_3 = 1 \quad \text{and} \quad 2x_1 + 6x_2 - 4x_3 = 8.$$

42. Find the distance between the lines in $\mathbb{R}^3$ given by

$$\begin{aligned} x_1 &= 3 - 2t & x_1 &= 2 - s \\ x_2 &= 4 + 3t \quad \text{and} & x_2 &= 5 + 2s \\ x_3 &= -1 + t & x_3 &= -3 + 4s \end{aligned}$$

[*Hint:* Let **a** be a point on one line and **b** be a point on the other line. Find the projection of $\mathbf{b} - \mathbf{a}$ on the orthogonal complement of an appropriate subspace.]

43. Repeat Exercise 42 for the lines in $\mathbb{R}^4$ given by $\mathbf{x} = (2, 1, 4, 0) + t(1, 0, 0, 1)$ and $\mathbf{x} = (1, 2, 0, -1) + s(1, -1, 0, 1)$. [*Hint:* See the hint for Exercise 42.]

5.6
The Volume of an *n*-box in $\mathbb{R}^m$

In Section 4.1, we saw that the area of the parallelogram (or *2-box*) in $\mathbb{R}^2$ determined by vectors $\mathbf{a}_1$ and $\mathbf{a}_2$ is the absolute value $|\det(A)|$ of the determinant of the 2×2 matrix A having $\mathbf{a}_1$ and $\mathbf{a}_2$ as column vectors.† We also saw that the volume of the parallelepiped (or *3-box*) in $\mathbb{R}^3$ determined by vectors

---

† We choose to put the vectors $\mathbf{a}_i$ as columns rather than rows of A to be consistent with earlier sections of this chapter.

$\mathbf{a}_1$, $\mathbf{a}_2$, and $\mathbf{a}_3$ is $|\det(A)|$ for the 3×3 matrix A whose columns are $\mathbf{a}_1$, $\mathbf{a}_2$, $\mathbf{a}_3$. We wish to extend these notions by defining an *n-box* in $\mathbb{R}^m$ for $m \geq n$ and finding its "volume."

Definition 5.7   An *n*-box in $\mathbb{R}^m$

Let $\mathbf{a}_1, \mathbf{a}_2, \ldots, \mathbf{a}_n$ be n independent vectors in $\mathbb{R}^m$ for $n \leq m$. The **n-box in $\mathbb{R}^m$ determined by these vectors** is the set of all vectors $\mathbf{x}$ satisfying

$$x = t_1\mathbf{a}_1 + t_2\mathbf{a}_2 + \cdots + t_n\mathbf{a}_n \quad \text{for} \quad 0 \leq t_i \leq 1$$

and $i = 1, 2, \ldots, n$.

Example 1   Describe geometrically the 1-box determined by the "vector" 2 in $\mathbb{R}$ and the 1-box determined by a nonzero vector $\mathbf{a}$ in $\mathbb{R}^m$.

Solution   The 1-box determined by the "vector" 2 in $\mathbb{R}$ consists of all numbers $t(2)$ for $0 \leq t \leq 1$, which is simply the closed interval $0 \leq x \leq 2$. Similarly, the 1-box in $\mathbb{R}^m$ determined by a nonzero vector $\mathbf{a}$ is the line segment joining the origin to the point $\mathbf{a}$. $\triangleleft$

Example 2   Draw symbolic sketches of a 2-box in $\mathbb{R}^m$ and a 3-box in $\mathbb{R}^m$.

Solution   A 2-box in $\mathbb{R}^m$ is a parallelogram with a vertex at the origin, as shown in Fig. 5.18. Similarly, a 3-box in $\mathbb{R}^m$ is a parallelepiped with a vertex at the origin, as illustrated in Fig. 5.19. $\triangleleft$

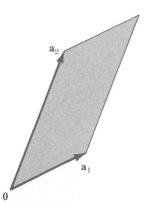

Figure 5.18
A 2-box in $\mathbb{R}^m$.

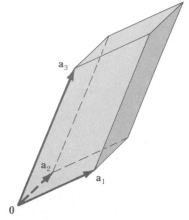

Figure 5.19
A 3-box in $\mathbb{R}^m$.

Note that our boxes need not be rectangular boxes; perhaps we should have used the term "skew box" to make this clear.

We are accustomed to speaking of the *length* of a line segment, of the *area* of a piece of the plane, and of the *volume* of a piece of space. To avoid introducing new terms when discussing a general n-box, we will use the three-dimensional term *volume* when speaking of its size. Note that we already used the three-dimensional term *box* as the name for the object! Thus by the volume of a 1-box, we simply mean its length, and the volume of a 2-box its area, and so on.

The volume of the 1-box determined by $\mathbf{a}_1$ in $\mathbb{R}^m$ is its length $\|\mathbf{a}_1\|$. If we regard $(\mathbf{a}_1 \cdot \mathbf{a}_1)$ as a 1×1 matrix, Exercise 18 shows that this length can be written as

$$\text{Length} = \|\mathbf{a}_1\| = \sqrt{\det(\mathbf{a}_1 \cdot \mathbf{a}_1)}. \tag{1}$$

Let us turn to a 2-box in $\mathbb{R}^m$ determined by nonzero and nonparallel vectors $\mathbf{a}_1$ and $\mathbf{a}_2$. We repeat an argument in Section 4.1, using slightly different notation. From Fig. 5.20, the area of this parallelogram is given by

$$\text{Area} = \|\mathbf{b}\|\|\mathbf{a}_2\|$$

where, for the angle θ between $\mathbf{a}_1$ and $\mathbf{a}_2$, we have $\|\mathbf{b}\| = \|\mathbf{a}_1\| \sin \theta$. We then have

$$\begin{aligned}
(\text{Area})^2 &= \|\mathbf{a}_1\|^2\|\mathbf{a}_2\|^2 \sin^2\theta \\
&= \|\mathbf{a}_1\|^2\|\mathbf{a}_2\|^2[1 - \cos^2\theta] \\
&= (\mathbf{a}_1 \cdot \mathbf{a}_1)(\mathbf{a}_2 \cdot \mathbf{a}_2)\left[1 - \frac{(\mathbf{a}_1 \cdot \mathbf{a}_2)^2}{(\mathbf{a}_1 \cdot \mathbf{a}_1)(\mathbf{a}_2 \cdot \mathbf{a}_2)}\right] \\
&= (\mathbf{a}_1 \cdot \mathbf{a}_1)(\mathbf{a}_2 \cdot \mathbf{a}_2) - (\mathbf{a}_1 \cdot \mathbf{a}_2)(\mathbf{a}_2 \cdot \mathbf{a}_1) \\
&= \begin{vmatrix} \mathbf{a}_1 \cdot \mathbf{a}_1 & \mathbf{a}_1 \cdot \mathbf{a}_2 \\ \mathbf{a}_2 \cdot \mathbf{a}_1 & \mathbf{a}_2 \cdot \mathbf{a}_2 \end{vmatrix} = \det(\mathbf{a}_i \cdot \mathbf{a}_j).
\end{aligned} \tag{2}$$

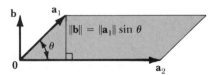

Figure 5.20
The volume of a 2-box.

From Eqs. (1) and (2), one guesses that the square of the volume of an n-box in $\mathbb{R}^m$ might be $\det(\mathbf{a}_i \cdot \mathbf{a}_j)$. Of course, we must define what we mean by the volume of such a box, but with the natural definition, this conjecture is true. Note that if A is the matrix with jth column vector $\mathbf{a}_j$, then A^T is the matrix with ith row vector $\mathbf{a}_i$, and the $n \times n$ matrix $(\mathbf{a}_i \cdot \mathbf{a}_j)$ is A^TA, so that Eq. (2) can be written as

$$(\text{Area})^2 = \det(\mathbf{a}_i \cdot \mathbf{a}_j) = \det(A^TA). \tag{3}$$

We now proceed to define the volume of an n-box in $\mathbb{R}^m$ in an inductive fashion. That is, we will define the volume of a 1-box directly, and then define the volume of an n-box in terms of the volume of an $(n-1)$-box. Our definition of volume is a natural one, essentially taking the product of the measure of the base of a box and the altitude of the box. We will think of the base of the n-box determined by $\mathbf{a}_1, \mathbf{a}_2, \ldots, \mathbf{a}_n$ as an $(n-1)$-box determined by some $n-1$ of the vectors $\mathbf{a}_i$. In general, such a base can be selected in several different ways. We find it convenient to work with boxes determined by *ordered* sequences of vectors. We will choose one special base for the box and give a definition of its volume in terms of this order of the vectors. Once we obtain the expression $\det(A^T A)$ in Eq. (3) for the square of the volume, it becomes a simple matter to show that the volume does not change if the order of the vectors in the sequence is changed. We will show that $\det(A^T A)$ remains unchanged if the order of the columns of A is changed.

Observe that if an n-box is determined by $\mathbf{a}_1, \mathbf{a}_2, \ldots, \mathbf{a}_n$, then $\mathbf{a}_1$ can be uniquely expressed in the form

$$\mathbf{a}_1 = \mathbf{b} + \mathbf{p}, \tag{4}$$

where $\mathbf{p}$ is the projection of $\mathbf{a}_1$ on $S = \mathrm{sp}(\mathbf{a}_2, \ldots, \mathbf{a}_n)$ and $\mathbf{b} = \mathbf{a}_1 - \mathbf{p}$ is orthogonal to S. This follows from Theorem 5.1 and is illustrated in Fig. 5.21.

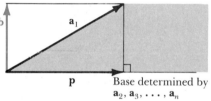

Figure 5.21
The altitude vector $\mathbf{b}$ perpendicular to the base box.

Definition 5.8 Volume of an n-box

The **volume** of the 1-box determined by a nonzero vector $\mathbf{a}_1$ in $\mathbb{R}^m$ is $\|\mathbf{a}_1\|$. Let $\mathbf{a}_1, \mathbf{a}_2, \ldots, \mathbf{a}_n$ be an ordered sequence of n independent vectors, and suppose the volume of an r-box determined by an ordered sequence of r independent vectors has been defined for $r < n$. The **volume of the n-box** in $\mathbb{R}^m$ determined by the ordered sequence of $\mathbf{a}_i$ is the product of the volume of the "base" determined by the ordered sequence $\mathbf{a}_2, \ldots, \mathbf{a}_n$ and the length of the vector $\mathbf{b}$ given in (4). That is,

$$\text{Volume} = (\text{altitude } \|\mathbf{b}\|)(\text{volume of the base}).$$

As a first step in finding a formula for the volume of an n-box, we establish a preliminary result on determinants.

Theorem 5.11 Property of det($A^T A$)

Let $\mathbf{a}_1, \mathbf{a}_2, \ldots, \mathbf{a}_n$ be vectors in $\mathbb{R}^m$ and let A be the $m \times n$ matrix with jth column vector $\mathbf{a}_j$. Let B be the $m \times n$ matrix obtained from A by replacing the first column of A by the vector

$$\mathbf{b} = \mathbf{a}_1 - r_2 \mathbf{a}_2 - \cdots - r_n \mathbf{a}_n$$

for scalars $r_2, \ldots, r_n$. Then

$$\det(A^T A) = \det(B^T B). \tag{5}$$

Proof The matrix B can be obtained from the matrix A by a sequence of $n - 1$ elementary *column-addition* operations. Each of the elementary column operations can be performed on A by multiplying A on the *right* by an elementary matrix formed by executing the same elementary column-addition operation on the $n \times n$ identity matrix I. Each elementary column-addition matrix therefore has the same determinant 1 as the identity matrix I. Their product is an $n \times n$ matrix E such that $B = AE$, and $\det(E) = 1$. Using properties of determinants and the transpose operation, we have

$$\det(B^T B) = \det((AE)^T (AE)) = \det(E^T (A^T A)E)$$
$$= 1 \cdot \det(A^T A) \cdot 1 = \det(A^T A). \ \blacksquare$$

We can now prove our volume formula.

Theorem 5.12 The Volume of a Box

The volume of the n-box in $\mathbb{R}^m$ determined by the ordered sequence $\mathbf{a}_1, \mathbf{a}_2, \ldots, \mathbf{a}_n$ of n independent vectors is given by

$$\text{Volume} = \sqrt{\det(A^T A)},$$

where A is the $m \times n$ matrix with $\mathbf{a}_j$ as jth column vector.

Proof Since our volume was defined inductively, we give an inductive proof. The theorem is valid if $n = 1$ or 2 by Eqs. (1) and (3), respectively. Let $n > 2$ and suppose that the theorem is proved for all k-boxes for $k \le n - 1$. If we write $\mathbf{a}_1 = \mathbf{b} + \mathbf{p}$ as in (4), then since $\mathbf{p}$ lies in $\text{sp}(\mathbf{a}_2, \ldots, \mathbf{a}_n)$, we have

$$\mathbf{p} = r_2 \mathbf{a}_2 + \cdots + r_n \mathbf{a}_n$$

for some scalars $r_2, \ldots, r_n$, so that

$$\mathbf{b} = \mathbf{a}_1 - \mathbf{p} = \mathbf{a}_1 - r_2 \mathbf{a}_2 - \cdots - r_n \mathbf{a}_n.$$

Let B be the matrix obtained from A by replacing the first column vector $\mathbf{a}_1$ of A by the vector $\mathbf{b}$, as in the preceding theorem. Since $\mathbf{b}$ is orthogonal to each of the vectors $\mathbf{a}_2, \ldots, \mathbf{a}_n$, which determine the base of our box, we obtain

$$B^TB = \begin{pmatrix} \mathbf{b} \cdot \mathbf{b} & 0 & \cdots & 0 \\ 0 & \mathbf{a}_2 \cdot \mathbf{a}_2 & \cdots & \mathbf{a}_2 \cdot \mathbf{a}_n \\ \vdots & \vdots & & \vdots \\ 0 & \mathbf{a}_n \cdot \mathbf{a}_2 & \cdots & \mathbf{a}_n \cdot \mathbf{a}_n \end{pmatrix}. \tag{6}$$

From (6) we see at once that

$$\det(B^TB) = \|\mathbf{b}\|^2 \begin{vmatrix} \mathbf{a}_2 \cdot \mathbf{a}_2 & \cdots & \mathbf{a}_2 \cdot \mathbf{a}_n \\ \vdots & & \vdots \\ \mathbf{a}_n \cdot \mathbf{a}_2 & \cdots & \mathbf{a}_n \cdot \mathbf{a}_n \end{vmatrix}.$$

By our induction assumption, the square of the volume of the $(n-1)$-box in $\mathbb{R}^m$ determined by the ordered sequence $\mathbf{a}_2, \ldots, \mathbf{a}_n$ is

$$\begin{vmatrix} \mathbf{a}_2 \cdot \mathbf{a}_2 & \cdots & \mathbf{a}_2 \cdot \mathbf{a}_n \\ \vdots & & \vdots \\ \mathbf{a}_n \cdot \mathbf{a}_2 & \cdots & \mathbf{a}_n \cdot \mathbf{a}_n \end{vmatrix}.$$

Applying Eq. (5) of Theorem 5.11, we then obtain

$$\det(A^TA) = \det(B^TB) = \|\mathbf{b}\|^2[\text{Volume of the base}]^2$$
$$= (\text{Volume})^2.$$

This proves our theorem. ■

Corollary 1   *Independence of Order*

The volume of a box determined by independent vectors $\mathbf{a}_1, \mathbf{a}_2, \ldots, \mathbf{a}_n$ and defined in Definition 5.8 is independent of the order of the vectors; in particular, the volume is independent of a choice of base for the box.

Proof A rearrangement of the sequence $\mathbf{a}_1, \mathbf{a}_2, \ldots, \mathbf{a}_n$ of vectors corresponds to the same rearrangement of the columns of the matrix A. Such a rearrangement of the columns of A can be accomplished by multiplying A on the right by a product of $n \times n$ elementary *column-interchange* matrices all having determinant 1. As in the proof of Theorem 5.11, we see that for the resulting matrix $B = AE$, we have $\det(B^TB) = \det(A^TA)$. ■

The volume of an n-box in $\mathbb{R}^n$ is of such importance that we restate Theorem 5.12 for this special case as a second corollary.

Corollary 2   *Volume of an n-box in $\mathbb{R}^n$*

If A is an $n \times n$ matrix with independent column vectors $\mathbf{a}_1, \mathbf{a}_2, \ldots, \mathbf{a}_n$, then $|\det(A)|$ is the volume of the n-box in $\mathbb{R}^n$ determined by these n vectors.

Proof By Theorem 5.12 and Corollary 1, the square of the volume of the n-box is $\det(A^TA)$. But since A is an $n \times n$ matrix, we have

$$\det(A^TA) = \det(A^T) \cdot \det(A) = (\det(A))^2.$$

The conclusion of the corollary then follows at once. ∎

Comparing Theorem 5.12 and this second corollary, we see that the formula for the volume of a box of dimension n in a space of larger dimension m involves a square root, whereas the formula for the volume of a box in a space of its own dimension does not involve a square root. The student of calculus discovers that the calculus formulas used to find length of a curve (which is one-dimensional) in the plane or in space involve a square root. The same is true of the formulas to find the area of a surface (two-dimensional) in space. However, the calculus formulas for finding the *area* of part of the plane or the *volume* of some part of space do not involve square roots. Theorem 5.12 and its second corollary lie at the heart of this difference in the calculus formulas. We will say more about this in Chapter 8, designed for those who have studied some calculus.

Example 3 Find the area of the parallelogram in $\mathbb{R}^4$ that is determined by the vectors $(2, 1, -1, 3)$ and $(0, 2, 4, -1)$.

Solution If

$$A = \begin{pmatrix} 2 & 0 \\ 1 & 2 \\ -1 & 4 \\ 3 & -1 \end{pmatrix},$$

then

$$A^TA = \begin{pmatrix} 2 & 1 & -1 & 3 \\ 0 & 2 & 4 & -1 \end{pmatrix} \begin{pmatrix} 2 & 0 \\ 1 & 2 \\ -1 & 4 \\ 3 & -1 \end{pmatrix} = \begin{pmatrix} 15 & -5 \\ -5 & 21 \end{pmatrix}.$$

By Theorem 5.12 we have

$$(\text{Area})^2 = \begin{vmatrix} 15 & -5 \\ -5 & 21 \end{vmatrix} = 290.$$

Thus the area of the parallelogram is $\sqrt{290}$. ◁

Example 4 Find the volume of the parallelepiped in $\mathbb{R}^3$ that is determined by the vectors $(1, 0, -1)$, $(-1, 1, 3)$, and $(2, 4, 1)$.

Solution   We compute the determinant

$$\begin{vmatrix} 1 & -1 & 2 \\ 0 & 1 & 4 \\ -1 & 3 & 1 \end{vmatrix} = \begin{vmatrix} 1 & -1 & 2 \\ 0 & 1 & 4 \\ 0 & 2 & 3 \end{vmatrix} = \begin{vmatrix} 1 & 4 \\ 2 & 3 \end{vmatrix} = -5.$$

By Corollary 2 of Theorem 5.12, the volume of the parallelepiped is therefore 5. ◁

SUMMARY

1. A n-box in $\mathbb{R}^m$, where $m \geq n$, is determined by n independent vectors $\mathbf{a}_1, \mathbf{a}_2, \ldots, \mathbf{a}_n$ and consists of all vectors $\mathbf{x}$ in $\mathbb{R}^m$ such that

$$\mathbf{x} = t_1\mathbf{a}_1 + t_2\mathbf{a}_2 + \cdots + t_n\mathbf{a}_n,$$

   where $0 \leq t_i \leq 1$ for $i = 1, 2, \ldots, n$.

2. A 1-box in $\mathbb{R}^m$ is a line segment and its "volume" is its length.

3. A 2-box in $\mathbb{R}^m$ is a parallelogram determined by two independent vectors, and the "volume" of the 2-box is the area of the parallelogram.

4. A 3-box in $\mathbb{R}^m$ is a "skewed" box (parallelepiped) in the usual sense, and its volume is the usual volume.

5. Let $\mathbf{a}_1, \mathbf{a}_2, \ldots, \mathbf{a}_n$ be independent vectors in $\mathbb{R}^m$ for $m \geq n$, and let A be the $m \times n$ matrix with jth column vector $\mathbf{a}_j$. The volume of the n-box in $\mathbb{R}^m$ determined by the n vectors is $\sqrt{\det(A^T A)}$.

6. For the case of an n-box in the space $\mathbb{R}^n$ of the same dimension, the formula for its volume in number (5) reduces to $|\det(A)|$.

EXERCISES

1. Find the area of the parallelogram in $\mathbb{R}^3$ determined by the vectors $(0, 1, 4)$ and $(-1, 3, -2)$.

2. Find the area of the parallelogram in $\mathbb{R}^5$ determined by the vectors $(1, 0, 1, 2, -1)$ and $(0, 1, -1, 1, 3)$.

3. Find the volume of the 3-box in $\mathbb{R}^4$ determined by the vectors $(-1, 2, 0, 1)$, $(0, 1, 3, 0)$, and $(0, 0, 2, -1)$.

4. Find the volume of the 4-box in $\mathbb{R}^5$ determined by the vectors $(1, 0, 1, 0, 1)$, $(0, 1, 1, 0, 0)$, $(3, 0, 1, 0, 0)$, and $(1, -1, 0, 0, 1)$.

In Exercises 5–10, find the volume of the n-box determined by the given vectors in $\mathbb{R}^n$.

5. $(-1, 4), (2, 3)$ in $\mathbb{R}^2$

6. $(-5, 3), (1, 7)$ in $\mathbb{R}^2$

7. $(1, 3, -5), (2, 4, -1), (3, 1, 2)$ in $\mathbb{R}^3$

8. $(-1, 4, 7), (3, -2, -1), (4, 0, 2)$ in $\mathbb{R}^3$

9. $(1, 0, 0, 1), (2, -1, 3, 0), (0, 1, 3, 4),$ $(-1, 1, -2, 1)$ in $\mathbb{R}^4$

10. $(1, -1, 0, 1), (2, -1, 3, 1), (-1, 4, 2, -1),$ $(0, 1, 0, 2)$ in $\mathbb{R}^4$

11. Find the area of the triangle in $\mathbb{R}^3$ with vertices $(-1, 2, 3)$, $(0, 1, 4)$, and $(2, 1, 5)$. [*Hint:* Think of vectors emanating from $(-1, 2, 3)$. The triangle may be viewed as half a parallelogram.]

12. Find the volume of the tetrahedron in $\mathbb{R}^3$ with vertices $(1, 0, 3)$, $(-1, 2, 4)$, $(3, -1, 2)$, and $(2, 0, -1)$. [*Hint:* Think of vectors emanating from $(1, 0, 3)$, and use the formula

$$(\tfrac{1}{3})(\text{area of base})(\text{altitude})$$

for the volume of a tetrahedron.

13. Find the volume of the tetrahedron in $\mathbb{R}^4$ with vertices $(1, 0, 0, 1)$, $(-1, 2, 0, 1)$, $(3, 0, 1, 1)$, and $(-1, 4, 0, 1)$. [*Hint:* See the hint for Exercise 12.]

14. Give a geometric interpretation of the fact that an $n \times n$ matrix with two equal rows has determinant zero.

15. Using the results of this section, give a criterion that four points **a**, **b**, **c**, and **d** in $\mathbb{R}^n$ lie in a plane.

16. Determine whether the points $(1, 0, 1, 0)$, $(-1, 1, 0, 1)$, $(0, 1, -1, 1)$, and $(1, -1, 4, -1)$ lie in a plane in $\mathbb{R}^4$. (See Exercise 15.)

17. Determine whether the points $(2, 0, 1, 3)$, $(3, 1, 0, 1)$, $(-1, 2, 0, 4)$, and $(3, 1, 2, 4)$ lie in a plane in $\mathbb{R}^4$. (See Exercise 15.)

18. Prove Eq. (1); that is, prove that the square of the length of the line segment determined by $\mathbf{a}_1$ in $\mathbb{R}^n$ is $\|\mathbf{a}_1\|^2 = \det(\mathbf{a}_1 \cdot \mathbf{a}_1)$.

19. **a)** If one attempts to define an n-box in $\mathbb{R}^m$ for $n > m$, what will its volume as an n-box be?

    b) Let A be an $m \times n$ matrix with $n > m$. Find $\det(A^T A)$.

6

6.1 – 6.5.

Linear Transformations

We have spent quite a bit of time working with vector spaces, in particular the space $\mathbb{R}^n$. It is time to turn to relationships between two vector spaces.

Relationships between two sets X and Y are usually studied by means of functions. The notion of a **function,** a **mapping,** or simply a **map** $f: X \to Y$ that maps the set X into the set Y is already familiar to us. Recall that such a function is nothing more than a rule that assigns to each element x of X a single element y of Y. We write $y = f(x)$ to express this assignment. The set X is the **domain** of the function. Figure 6.1 gives a visual illustration of the concept.

Let V and W be vector spaces. This chapter is concerned with a certain type of function

$$T: V \to W.$$

Both V and W are endowed with an *algebraic structure* arising from the two operations of vector addition and scalar multiplication. We will be working with functions $T: V \to W$ which, in a sense, preserve this algebraic structure. These functions are known as *linear transformations* or *linear maps*. In Section 6.1 we define linear transformations that map one vector space into another, give many examples, and establish some elementary properties of linear transformations. Subsequent sections deal with computation of linear transformations and their relationship to matrix concepts that we have studied.

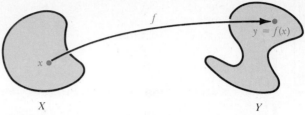

Figure 6.1
A function f that maps X into Y.

6.1

The Notion of a Linear Transformation

EXAMPLES OF FUNCTIONS

To get us thinking in terms of functions, we begin by mentioning some functions that are familiar from past work, and some that we have used before in this text.

The squaring function that maps $\mathbb{R}$ into $\mathbb{R}$. Mathematics instructors are always reaching for the function f, where $f(x) = x^2$, as a simple example of a function whose graph is not a straight line. We have $f(2) = 4$, $f(3) = 9$, $f(-\frac{5}{2}) = \frac{25}{4}$, and so on.

The sine function that maps $\mathbb{R}$ into $\mathbb{R}$. The trigonometric function $h(x) = \sin x$ that maps $\mathbb{R}$ into $\mathbb{R}$ is very important in the study of wave motion. Scientific calculators make it easy to find values of $\sin x$, which always lie in the interval $[-1, 1]$.

The norm or length function that maps $\mathbb{R}^n$ into $\mathbb{R}$. This function assigns to each vector $\mathbf{x}$ in $\mathbb{R}^n$ its length $\|\mathbf{x}\|$ in $\mathbb{R}$.

The dot-product mapping of $\mathbb{R}^n \times \mathbb{R}^n$ into $\mathbb{R}$.† The dot-product function assigns to the pair $(\mathbf{v}_1, \mathbf{v}_2)$ the dot product $\mathbf{v}_1 \cdot \mathbf{v}_2$ in $\mathbb{R}$. The dot-product function makes it easy for us to determine whether two vectors in $\mathbb{R}^n$ are orthogonal. More generally, it can be combined with the norm function to find the angle between two nonzero vectors in $\mathbb{R}^n$.

A projection mapping of $\mathbb{R}^2$ into $\mathbb{R}$. The function $g \colon \mathbb{R}^2 \to \mathbb{R}$ defined by $g(x_1, x_2) = x_1$ projects each vector (x_1, x_2) in $\mathbb{R}^2$ onto its first coordinate.

An evaluation mapping of F into $\mathbb{R}$. Let F be the vector space of all functions f mapping $\mathbb{R}$ into $\mathbb{R}$. For each number c in $\mathbb{R}$, we have the evaluation function $E_c \colon F \to \mathbb{R}$ defined by $E_c(f) = f(c)$, the value of the function f at c. Note that these evaluation mappings act on other functions.

---

† For sets S and T, $S \times T$ is the set of all *ordered* pairs (s, t), where s and t are elements of S and T, respectively.

LINEAR TRANSFORMATIONS

Let V and W be vector spaces. As stated in the introduction to this chapter, we will be dealing mainly with functions $T: V \rightarrow W$ that *preserve* the vector operations of addition and scalar multiplication.

Definition 6.1 **Linear Transformation**

A function T that maps a vector space V into a vector space W is a **linear transformation** if it satisfies:

1. $T(\mathbf{v}_1 + \mathbf{v}_2) = T(\mathbf{v}_1) + T(\mathbf{v}_2)$,   Preservation of addition

2. $T(r\mathbf{v}_1) = rT(\mathbf{v}_1)$,   Preservation of scalar multiplication

for all vectors $\mathbf{v}_1$ and $\mathbf{v}_2$ in V and scalars r in $\mathbb{R}$.

Exercise 20 indicates that the two conditions of Definition 6.1 may be combined into the single condition

$$T(r_1\mathbf{v}_1 + r_2\mathbf{v}_2) = r_1 T(\mathbf{v}_1) + r_2 T(\mathbf{v}_2) \tag{1}$$

for all vectors $\mathbf{v}_1$ and $\mathbf{v}_2$ in V and all scalars r_1 and r_2 in $\mathbb{R}$.

Example 1 Determine whether or not $h: \mathbb{R} \rightarrow \mathbb{R}$, where $h(x) = \sin x$, is a linear transformation.

Solution We know that

$$\sin(\pi/4 + \pi/4) \neq \sin(\pi/4) + \sin(\pi/4),$$

for $\sin(\pi/4 + \pi/4) = \sin(\pi/2) = 1$, whereas we have $\sin(\pi/4) + \sin(\pi/4) = 1/\sqrt{2} + 1/\sqrt{2} = 2/\sqrt{2}$. Thus $\sin x$ is not a linear transformation, for it does not preserve addition. ◁

THE CONCEPT OF A LINEAR SUBSTITUTION dates back to the eighteenth century. But it was only after physicists became used to dealing with vectors that the idea of a function of vectors became explicit. One of the founders of vector analysis, Oliver Heaviside (1850–1925), introduced the idea of a linear vector operator in one of his works on electromagnetism in 1885. He defined it using coordinates: $\vec{B}$ comes from $\vec{H}$ by a linear vector operator if, when $\vec{B}$ has components B_1, B_2, B_3 and $\vec{H}$ has components H_1, H_2, H_3, there are numbers μ_{ij} for $i,j = 1, 2, 3$, where

$$B_1 = \mu_{11}H_1 + \mu_{12}H_2 + \mu_{13}H_3$$
$$B_2 = \mu_{21}H_1 + \mu_{22}H_2 + \mu_{23}H_3$$
$$B_3 = \mu_{31}H_1 + \mu_{32}H_2 + \mu_{33}H_3.$$

In his lectures at Yale which were published in 1901, J. Willard Gibbs called this same transformation a linear vector function. But he also defined this more abstractly as a continuous function f such that $f(\vec{v} + \vec{w}) = f(\vec{v}) + f(\vec{w})$. A fully abstract definition, exactly like Definition 6.1, was given by Hermann Weyl in *Space–Time–Matter* (1918).

Oliver Heaviside was a self-taught expert on mathematical physics who played an important role in the development of electromagnetic theory and especially its practical applications. In 1901 he predicted the existence of a reflecting ionized region surrounding the earth; the existence of this layer, now called the ionosphere, was soon confirmed.

The next example describes all the linear transformations $T: \mathbb{R} \to \mathbb{R}$. We see that they are all very simple functions.

Example 2   Show that if $T: \mathbb{R} \to \mathbb{R}$ is a linear transformation, then there is a number a in $\mathbb{R}$ such that

$$T(x) = ax \tag{2}$$

for all x in $\mathbb{R}$. Show conversely that Eq. (2) defines a linear transformation for each number a in $\mathbb{R}$.

Solution   Let $T: \mathbb{R} \to \mathbb{R}$ be a linear transformation, and let $T(1) = a$. By the preservation of scalar multiplication, for any "vector" x in $\mathbb{R}$ we then have

$$T(x) = T(x\,1) = xT(1) = xa = ax$$

for all x in $\mathbb{R}$.

Conversely, let a be any number in $\mathbb{R}$ and let $T: \mathbb{R} \to \mathbb{R}$ be defined by $T(x) = ax$. Then for any "vectors" x and y in $\mathbb{R}$, we have

$$T(x + y) = a(x + y) = ax + ay$$

and for any scalar r

$$T(rx) = a(rx) = r(ax) = rT(x).$$

Thus T preserves the vector operations, so it is a linear transformation. $\triangleleft$

Example 3   Show that the projection function $g: \mathbb{R}^2 \to \mathbb{R}$ defined by $g(x_1, x_2) = x_1$ is a linear transformation.

Solution   If $\mathbf{a} = (a_1, a_2)$ and $\mathbf{b} = (b_1, b_2)$ are any two vectors in $\mathbb{R}^2$, then $\mathbf{a} + \mathbf{b} = (a_1 + b_1, a_2 + b_2)$ so that

$$g(\mathbf{a} + \mathbf{b}) = a_1 + b_1 = g(\mathbf{a}) + g(\mathbf{b}).$$

Thus g preserves addition.

If r is any scalar, then $r\mathbf{a} = (ra_1, ra_2)$ so that

$$g(r\mathbf{a}) = ra_1 = rg(\mathbf{a})$$

and g preserves multiplication by scalars. $\triangleleft$

Example 4 gives the obvious generalization of Example 3.

Example 4   Show that the function $g: \mathbb{R}^n \to \mathbb{R}$ defined by $f(a_1, a_2, \ldots , a_n) = a_k$ is a linear transformation.

Solution   The solution is analogous to that in Example 3. $\triangleleft$

If we identify the point $(0, \ldots , 0, a_k, 0, \ldots , 0)$ in $\mathbb{R}^n$ with the kth coordinate a_k of $\mathbf{a}$ in $\mathbb{R}^n$ we recognize that Example 4 is a special case of projection of $\mathbb{R}^n$ onto a subspace.

Example 5   *(Projection transformations).* Let S be a subspace of $\mathbb{R}^n$. Show that the function $g: \mathbb{R}^n \to S$ defined by $g(\mathbf{v}) = \mathbf{v}_S$, the projection of the vector $\mathbf{v}$ in $\mathbb{R}^n$ onto the subspace S, is a linear transformation.

Solution   We recall from Section 5.1 that if $\mathbb{R}^n$ is regarded as a space of column vectors and if $\dim(S) = m$, then the projection of $\mathbf{v}$ on S is given by $\mathbf{v}_S = P_S\mathbf{v}$, where P_S is the $n \times n$ projection matrix for S. Thus our function g has the form

$$g(\mathbf{v}) = P_S\mathbf{v}.$$

The condition $g(\mathbf{v}_1 + \mathbf{v}_2) = g(\mathbf{v}_1) + g(\mathbf{v}_2)$ becomes nothing more than the familiar distributive property $P_S(\mathbf{v}_1 + \mathbf{v}_2) = P_S\mathbf{v}_1 + P_S\mathbf{v}_2$ for matrix multiplication. For any scalar r in $\mathbb{R}$, the condition $g(r\mathbf{v}) = rg(\mathbf{v})$ takes the form $P_S(r\mathbf{v}) = r(P_S\mathbf{v})$, which is again a familiar property of matrix multiplication. Therefore, the projection transformation g is a linear transformation.   ◁

In the preceding example, the fact that P_S is a projection matrix was not used; any $m \times n$ matrix A can be used to give a linear transformation of $\mathbb{R}^n$ into $\mathbb{R}^m$ as follows.

Example 6   *(Transformation by a matrix).* Let A be an $m \times n$ matrix and let $\mathbb{R}^n$ and $\mathbb{R}^m$ be viewed as spaces of *column* vectors. Show that the function $T_A: \mathbb{R}^n \to \mathbb{R}^m$ defined by

$$T_A(\mathbf{x}) = A\mathbf{x}$$

is a linear transformation.

Solution   The argument is identical to that of Example 5 if P_S is replaced by A.   ◁

We will show in Section 6.2 that the linear transformation of $\mathbb{R}^n$ into $\mathbb{R}^m$ given in Example 6 is the most general type possible. That is, we will see that every linear transformation of $\mathbb{R}^n$ into $\mathbb{R}^m$ can be computed by multiplying the column vectors in $\mathbb{R}^n$ on the left by some $m \times n$ matrix A. Note that we have already established this for the case $n = m = 1$ in Example 2. We illustrate it for an important type of linear transformation of $\mathbb{R}^2$ into $\mathbb{R}^2$.

Example 7   *(Rotation of the plane).* Show that the function R that maps $\mathbb{R}^2$ into $\mathbb{R}^2$ by rotating the plane counterclockwise through a positive angle α is a linear transformation.

Solution   We view $\mathbb{R}^2$ as a space of column vectors. Referring to Fig. 6.2, we compute a formula for $R(\mathbf{v})$, where $\mathbf{v}$ is any vector in $\mathbb{R}^2$. If $\mathbf{v}$ makes an angle of θ with the x-axis and if $r = \|\mathbf{v}\|$, then

$$\mathbf{v} = \begin{pmatrix} r \cos \theta \\ r \sin \theta \end{pmatrix}$$

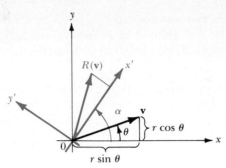

Figure 6.2
A rotation R of the plane.

and

$$R(\mathbf{v}) = \begin{pmatrix} r\cos(\alpha + \theta) \\ r\sin(\alpha + \theta) \end{pmatrix} = \begin{pmatrix} r(\cos\alpha\cos\theta - \sin\alpha\sin\theta) \\ r(\sin\alpha\cos\theta + \cos\alpha\sin\theta) \end{pmatrix}.$$

This last equation can be written as

$$R(\mathbf{v}) = \begin{pmatrix} \cos\alpha & -\sin\alpha \\ \sin\alpha & \cos\alpha \end{pmatrix} \begin{pmatrix} r\cos\theta \\ r\sin\theta \end{pmatrix}. \qquad (3)$$

Thus the function R has the form $R(\mathbf{v}) = A\mathbf{v}$, where A is the 2×2 matrix in Eq. (3), and therefore R is the linear transformation T_A that we described in Example 6.   ◁

Example 8   Determine if $f: \mathbb{R}^3 \to \mathbb{R}^2$ defined by $f(x_1, x_2, x_3) = (x_1 - 2x_2, x_2 + 3x_3)$ is a linear transformation.

Solution   Let $\mathbf{a} = (a_1, a_2, a_3)$ and $\mathbf{b} = (b_1, b_2, b_3)$. Then $\mathbf{a} + \mathbf{b} = (a_1 + b_1, a_2 + b_2, a_3 + b_3)$ so that

$$f(\mathbf{a} + \mathbf{b}) = (a_1 + b_1 - 2(a_2 + b_2), a_2 + b_2 + 3(a_3 + b_3)),$$

while

$$f(\mathbf{a}) + f(\mathbf{b}) = (a_1 - 2a_2, a_2 + 3a_3) + (b_1 - 2b_2, b_2 + 3b_3)$$
$$= (a_1 - 2a_2 + b_1 - 2b_2, a_2 + 3a_3 + b_2 + 3b_3).$$

Thus we see that $f(\mathbf{a} + \mathbf{b}) = f(\mathbf{a}) + f(\mathbf{b})$, so f preserves addition.
    If r is any scalar, then

$$r\mathbf{a} = (ra_1, ra_2, ra_3)$$

and

$$f(r\mathbf{a}) = (ra_1 - 2ra_2, ra_2 + 3ra_3) = r(a_1 - 2a_2, a_2 + 3a_3) = rf(\mathbf{a}),$$

so f preserves scalar multiplication also.   ◁

Example 9   Let F be the vector space of all functions $f: \mathbb{R} \to \mathbb{R}$ and let c be in $\mathbb{R}$. Show that the evaluation function $E_c: F \to \mathbb{R}$ defined by $E_c(f) = f(c)$, which maps each function f in F into its value at c, is a linear transformation.

Solution   We show that E_c preserves addition and scalar multiplication. If f and g are functions in the vector space F, then evaluating $f + g$ at c, we obtain

$$(f + g)(c) = f(c) + g(c).$$

Therefore,

$$
\begin{aligned}
E_c(f + g) &= (f + g)(c) && \text{Definition of } E_c \\
&= f(c) + g(c) && \text{Definition of } f + g \text{ in } F \\
&= E_c(f) + E_c(g). && \text{Definition of } E_c
\end{aligned}
$$

This shows that E_c preserves addition. Since $(rf)(c)$ is defined to be $r[f(c)]$, the computation

$$
\begin{aligned}
E_c(rf) &= (rf)(c) && \text{Definition of } E_c \\
&= r[f(c)] && \text{Definition of } rf \text{ in } F \\
&= r[E_c(f)] && \text{Definition of } E_c
\end{aligned}
$$

shows that E_c preserves scalar multiplication.   ◁

PROPERTIES OF LINEAR TRANSFORMATIONS

The two properties of linear transformations in our next theorem are useful and easy to prove.

Theorem 6.1   Preservation of Zero and Subtraction

Let V and W be vector spaces and let $T: V \to W$ be a linear transformation. Then:

i) $T(\mathbf{0}) = \mathbf{0}$,   Preservation of zero

ii) $T(\mathbf{v}_1 - \mathbf{v}_2) = T(\mathbf{v}_1) - T(\mathbf{v}_2)$,   Preservation of subtraction

for any vectors $\mathbf{v}_1$ and $\mathbf{v}_2$ in V.

Proof   (i) Of course when we write $T(\mathbf{0}) = \mathbf{0}$, the first zero vector is in V and the second is in W. Let us use the notations $\mathbf{0}_V$ and $\mathbf{0}_W$ for the moment to make this clear. Preservation of addition for T shows that

$$T(\mathbf{0}_V) + T(\mathbf{0}_V) = T(\mathbf{0}_V + \mathbf{0}_V) = T(\mathbf{0}_V). \tag{4}$$

Subtracting $T(\mathbf{0}_V)$ in the vector space W from both sides of Eq. (4), we obtain $T(\mathbf{0}_V) = T(\mathbf{0}_V) - T(\mathbf{0}_V) = \mathbf{0}_W$, which is (i).

(ii) For vectors $\mathbf{v}_1$ and $\mathbf{v}_2$ in V we have

$$
\begin{aligned}
T(\mathbf{v}_1 - \mathbf{v}_2) &= T(\mathbf{v}_1 + (-1)\mathbf{v}_2) = T(\mathbf{v}_1) + T((-1)\mathbf{v}_2) \\
&= T(\mathbf{v}_1) + (-1)T(\mathbf{v}_2) = T(\mathbf{v}_1) - T(\mathbf{v}_2). \blacksquare
\end{aligned}
$$

Example 10 Determine if the function $f: \mathbb{R}^2 \to \mathbb{R}^2$ defined by $f(x_1, x_2) = (x_1 + x_2, x_1 + 1)$ is a linear transformation.

Solution Since $f(0, 0) = (0, 1)$, we see that f does not preserve zero, and so cannot be a linear transformation. ◁

Let $f: X \to Y$ be a function, and let Q be a subset of X and S a subset of Y. Recall that

$$f[Q] = \{f(q) \mid q \text{ in } Q\} \qquad \text{Image of } Q \text{ under } f$$

and

$$f^{-1}[S] = \{x \text{ in } X \mid f(x) \text{ is in } S\}. \qquad \text{Inverse image of } S \text{ under } f$$

For example, if $f: \mathbb{R} \to \mathbb{R}$ is the squaring function so that $f(x) = x^2$, then

$$f[\{1, 2, 3\}] = \{1, 4, 9\},$$

while

$$f^{-1}[\{1, 4, 9\}] = \{1, -1, 2, -2, 3, -3\}.$$

Recall too that for $f: X \to Y$, the set $f[X]$ is called the **range** of f. We will be concerned with a linear transformation T of a vector space V into a vector space W, where Q is a subspace of V and S is a subspace of W.

Theorem 6.2 Preservation of Subspaces

Let V and W be vector spaces and let $T: V \to W$ be a linear transformation.

i) If Q is a subspace of V, then $T[Q]$ is a subspace of W.
ii) If S is a subspace of W, then $T^{-1}[S]$ is a subspace of V.

Proof (i) We need only show that $T[Q]$ is closed under vector addition and multiplication by scalars. Let $T(\mathbf{v}_1)$ and $T(\mathbf{v}_2)$ be any vectors in $T[Q]$, where $\mathbf{v}_1$ and $\mathbf{v}_2$ are vectors in Q. Then

$$T(\mathbf{v}_1) + T(\mathbf{v}_2) = T(\mathbf{v}_1 + \mathbf{v}_2)$$

by preservation of addition. Now $\mathbf{v}_1 + \mathbf{v}_2$ is in Q since Q is itself closed under addition, so $T(\mathbf{v}_1 + \mathbf{v}_2)$ is in $T[Q]$. This shows that $T[Q]$ is closed under vector addition. Also, if r is any scalar, then $r\mathbf{v}_1$ is in Q, and of course

$$rT(\mathbf{v}_1) = T(r\mathbf{v}_1).$$

This shows that $rT(\mathbf{v}_1)$ is in $T[Q]$, so $T[Q]$ is closed under multiplication by scalars. Thus $T[Q]$ is a subspace of W.

(ii) Let $\mathbf{x}$ and $\mathbf{y}$ be any vectors in $T^{-1}[S]$, so that $T(\mathbf{x})$ and $T(\mathbf{y})$ are in S. Then

$$T(\mathbf{x} + \mathbf{y}) = T(\mathbf{x}) + T(\mathbf{y})$$

is also in the subspace S, so $\mathbf{x} + \mathbf{y}$ is in $T^{-1}[S]$. For any scalar r, we know that

$$rT(\mathbf{x}) = T(r\mathbf{x})$$

and $rT(\mathbf{x})$ is in S. Thus $r\mathbf{x}$ is also in $T^{-1}[S]$. This shows that $T^{-1}[S]$ is also closed under addition and multiplication by scalars, so $T^{-1}[S]$ is a subspace of V. ∎

The proof of the first part of Theorem 6.2 illustrates a fundamental analogy between the defining condition of a subspace, namely *closure* under addition and scalar multiplication, and the defining condition of a linear transformation, namely *preservation* of addition and scalar multiplication.

Example 11   Following the notation of Example 9, let F be the vector space of all functions $f: \mathbb{R} \to \mathbb{R}$, and let c be a scalar in $\mathbb{R}$. Determine if the mapping $T: F \to \mathbb{R}$ defined by $T(f) = |f(c)|$ is a linear transformation.

Solution   Since $T[F]$ contains positive numbers but no negative numbers, it is not a subspace of the one-dimensional space $\mathbb{R}$. Therefore, T is not a linear transformation, as shown by Theorem 6.2.   ◁

Two important special cases of Theorem 6.2 are basic to the study of linear transformations.

Definition 6.2 The Subspaces range (T) and ker(T)

Let V and W be vector spaces, and let $T: V \to W$ be a linear transformation. The subspace $T[V]$ of W is the **range** of T, and is denoted by range(T). The subspace $T^{-1}[\{0\}]$, where $\{0\}$ is the zero subspace of W, is called the **kernel** or **nullspace** of T, and is denoted by ker(T).

As mentioned earlier, Section 6.2 shows that every linear transformation $T: \mathbb{R}^n \to \mathbb{R}^m$ is actually a linear transformation T_A, where $T_A(\mathbf{x}) = A\mathbf{x}$ for some $m \times n$ matrix A. The relationship of the range and kernel of T_A to the matrix A (which we box) follows easily (see Theorem 3.8).

Properties of T_A: $V \to W$

1. The range of T_A is the column space of A.
2. The kernel of T_A is the nullspace of A.

We illustrate with two examples.

Example 12   Find the kernel of the linear transformation of Example 8.

Solution   We must find all vectors (x_1, x_2, x_3) such that $f(x_1, x_2, x_3) = (x_1 - 2x_2, x_2 + 3x_3)$ is the zero vector. Thus we must solve the system of equations

$$x_1 - 2x_2 \qquad\quad = 0$$
$$x_2 + 3x_3 = 0,$$

which can be written in matrix form as

$$A \begin{pmatrix} x_1 \\ x_2 \\ x_3 \end{pmatrix} = \begin{pmatrix} 0 \\ 0 \end{pmatrix}, \quad \text{where} \quad A = \begin{pmatrix} 1 & -2 & 0 \\ 0 & 1 & 3 \end{pmatrix}.$$

The solution to the system is $(x_1, x_2, x_3) = (-6s, -3s, s)$ for any scalar s. Thus the kernel of f is the one-dimensional subspace of $\mathbb{R}^3$ generated by $(-6, -3, 1)$. ◁

Example 13   Find the range of the linear transformation of Example 8.

Solution   We must find the set of all $f[\mathbf{v}]$ for $\mathbf{v}$ in $\mathbb{R}^3$. Writing vectors as column vectors, we must find all vectors of the form

$$f \begin{pmatrix} x_1 \\ x_2 \\ x_3 \end{pmatrix} = \begin{pmatrix} x_1 - 2x_2 \\ x_2 + 3x_3 \end{pmatrix} = x_1 \begin{pmatrix} 1 \\ 0 \end{pmatrix} + x_2 \begin{pmatrix} -2 \\ 1 \end{pmatrix} + x_3 \begin{pmatrix} 0 \\ 3 \end{pmatrix}.$$

Therefore, the range of f is the column space of the matrix

$$A = \begin{pmatrix} 1 & -2 & 0 \\ 0 & 1 & 3 \end{pmatrix}.$$

Since the first two columns of A are independent, we conclude that range(f) = $\mathbb{R}^2$. ◁

Recall that a function $f: X \rightarrow Y$ is **one to one** if $f(x_1) \neq f(x_2)$ unless $x_1 = x_2$ in X. When f is one to one, each y_1 in the range $f[X]$ is equal to $f(x_1)$ for a *unique* x_1 in X. The **inverse function** $f^{-1}: f[X] \rightarrow X$ is then defined by $f^{-1}(y_1) = x_1$. For a linear transformation $T: V \rightarrow W$, the kernel of T plays a crucial role in determining whether T is one to one, so that T^{-1} is well defined.

Theorem 6.3   Role of the Kernel

Let V and W be vector spaces, and let $T: V \rightarrow W$ be a linear transformation. Then T is one to one if and only if ker$(T) = \{\mathbf{0}\}$, the zero subspace of V. When this is the case, the inverse map $T^{-1}: T[V] \rightarrow V$ is also a linear transformation.

Proof   Suppose ker$(T) = \{\mathbf{0}\}$ and let $T(\mathbf{v}_1) = T(\mathbf{v}_2)$ for $\mathbf{v}_1$ and $\mathbf{v}_2$ in V. Then $T(\mathbf{v}_1 - \mathbf{v}_2) = T(\mathbf{v}_1) - T(\mathbf{v}_2) = \mathbf{0}$, so $\mathbf{v}_1 - \mathbf{v}_2$ is in ker(T) and hence $\mathbf{v}_1 - \mathbf{v}_2 = \mathbf{0}$. We conclude that $\mathbf{v}_1 = \mathbf{v}_2$, showing that T is one to one.

Conversely, suppose that T is one to one, and that $\mathbf{v}$ lies in ker(T) so that $T(\mathbf{v}) = \mathbf{0}$. Now $T(\mathbf{0}) = \mathbf{0}$ by Theorem 6.1, so $T(\mathbf{v}) = T(\mathbf{0})$. Since T is one to one, we must have $\mathbf{v} = \mathbf{0}$. This shows that ker$(T) = \{\mathbf{0}\}$.

We suppose now that ker$(T) = \{\mathbf{0}\}$ so that T is one to one and $T^{-1}: T[V] \rightarrow V$ is well defined. We show that T^{-1} is also a linear transformation. Let

$\mathbf{w}_1 = T(\mathbf{v}_1)$ and $\mathbf{w}_2 = T(\mathbf{v}_2)$ be two vectors in $T[V]$ so that $T^{-1}(\mathbf{w}_1) = \mathbf{v}_1$ and $T^{-1}(\mathbf{w}_2) = \mathbf{v}_2$. We also have $T(\mathbf{v}_1 + \mathbf{v}_2) = \mathbf{w}_1 + \mathbf{w}_2$, which shows that $T^{-1}(\mathbf{w}_1 + \mathbf{w}_2) = \mathbf{v}_1 + \mathbf{v}_2$. Therefore, $T^{-1}(\mathbf{w}_1 + \mathbf{w}_2) = T^{-1}(\mathbf{w}_1) + T^{-1}(\mathbf{w}_2)$. Similarly, the equation $T(r\mathbf{v}_1) = rT(\mathbf{v}_1) = r\mathbf{w}_1$ shows that $r\mathbf{v}_1 = T^{-1}(r\mathbf{w}_1)$, that is, $rT^{-1}(\mathbf{w}_1) = T^{-1}(r\mathbf{w}_1)$. ∎

Example 14 Show that the linear transformation $T: \mathbb{R}^2 \to \mathbb{R}^2$ defined by $T(x, y) = (x - 2y, x + y)$ is one to one, and find a formula for the inverse transformation T^{-1}.

Solution To show that T is one to one, we show that $\ker(T)$ is the zero subspace. As in Example 12, we find $\ker(T)$ by solving the system

$$x - 2y = 0$$
$$x + y = 0.$$

We easily find that $(x, y) = (0, 0)$ is the only solution. Thus $\ker(T)$ is the zero subspace and T^{-1} is a well-defined linear transformation by Theorem 6.3.

To find a formula for T^{-1}, we take a vector (u, v) in the range of T and find a vector (x, y) such that $T(x, y) = (u, v)$. This equation can be written as

$$(x - 2y, x + y) = (u, v),$$

which gives rise to the system of equations

$$x - 2y = u$$
$$x + y = v.$$

Reducing the corresponding partitioned matrix, we obtain

$$\begin{pmatrix} 1 & -2 & | & u \\ 1 & 1 & | & v \end{pmatrix} \sim \begin{pmatrix} 1 & -2 & | & u \\ 0 & 3 & | & v - u \end{pmatrix} \sim \begin{pmatrix} 1 & 0 & | & u + 2(v - u)/3 \\ 0 & 1 & | & (v - u)/3 \end{pmatrix}$$
$$= \begin{pmatrix} 1 & 0 & | & (u + 2v)/3 \\ 0 & 1 & | & (v - u)/3 \end{pmatrix}.$$

Thus the solution to the system of equation is

$$(x, y) = ((u + 2v)/3, (v - u)/3).$$

It follows that $T^{-1}(u, v) = ((u + 2v)/3, (v - u)/3)$ and is defined for all vectors (u, v) in $\mathbb{R}^2$. ◁

SUMMARY

Let V and W be vector spaces, and let T be a function mapping V into W.

1. The function T is a linear transformation if it preserves addition and scalar multiplication, that is, if

$$T(\mathbf{v}_1 + \mathbf{v}_2) = T(\mathbf{v}_1) + T(\mathbf{v}_2)$$

and

$$T(r\mathbf{v}_1) = rT(\mathbf{v}_1)$$

for all vectors $\mathbf{v}_1$ and $\mathbf{v}_2$ in V and scalars r in $\mathbb{R}$.

2. If T is a linear transformation, then $T(\mathbf{0}) = \mathbf{0}$ and $T(\mathbf{v}_1 - \mathbf{v}_2) = T(\mathbf{v}_1) - T(\mathbf{v}_2)$.

3. If T is a linear transformation and Q is a subspace of V, then $T[Q]$ is a subspace of W. If S is a subspace of W, then $T^{-1}[S]$ is a subspace of V.

4. If T is a linear transformation, then range(T) is the subspace $\{T(\mathbf{v}) \mid \mathbf{v}$ in $V\}$ of W, and ker(T) is the subspace $\{\mathbf{v}$ in $V \mid T(\mathbf{v}) = \mathbf{0}\}$ of V.

5. If T is a linear transformation, then T is one to one if and only if ker$(T) = \{\mathbf{0}\}$. When this is the case, then $T^{-1}: T[V] \to V$ is well defined, and is a linear transformation.

EXERCISES

In Exercises 1–7, determine whether the given function is a linear transformation.

1. $f: \mathbb{R}^2 \to \mathbb{R}^2$ defined by $f(x, y) = (x + y, 2y)$
2. $f: \mathbb{R}^2 \to \mathbb{R}^2$ defined by $f(x, y) = (x + 4, x - y)$
3. $f: \mathbb{R}^2 \to \mathbb{R}^3$ defined by $f(x, y) = (2x, 3x + y, x + y)$
4. $f: \mathbb{R}^3 \to \mathbb{R}^3$ defined by $f(x_1, x_2, x_3) = (x_3, 0, x_1 + x_2)$
5. $f: \mathbb{R}^3 \to \mathbb{R}^2$ defined by $f(x_1, x_2, x_3) = (x_1 + x_2 + x_3, 2x_1)$
6. $f: \mathbb{R}^3 \to \mathbb{R}^4$ defined by $f(x_1, x_2, x_3) = (x_1 + x_2, x_2 + x_3, x_3 + x_1, x_1 - x_3)$
7. $f: \mathbb{R}^4 \to \mathbb{R}^4$ defined by $f(x_1, x_2, x_3, x_4) = (x_1, 1, x_3, 1)$

In Exercises 8–12, let F be the vector space of all functions $f: \mathbb{R} \to \mathbb{R}$. Determine whether the given function T is a linear transformation.

8. $T: F \to \mathbb{R}$ defined by $T(f) = f(-4)$
9. $T: F \to \mathbb{R}$ defined by $T(f) = [f(5)]^2$
10. $T: F \to F$ defined by $T(f) = f + f$
11. $T: F \to F$ defined by $T(f) = f + 3$, where 3 is the constant function with value 3 for all x in $\mathbb{R}$
12. $T: F \to F$ defined by $T(f) = -f$

In Exercises 13–19, find the kernel of the given linear transformation. Use Theorem 6.3 to determine whether the linear transformation is one to one. For those that are one to one, find a formula for the inverse transformation, as in Example 14.

13. $T: \mathbb{R}^2 \to \mathbb{R}^2$ defined by $T(x, y) = (x - y, x + y)$
14. $T: \mathbb{R}^2 \to \mathbb{R}^3$ defined by $T(x, y) = (x + y, 2x + y, 3x + y)$
15. $T: \mathbb{R}^3 \to \mathbb{R}^3$ defined by $T(x_1, x_2, x_3) = (2x_1 + x_3, 3x_1 + x_2, x_1 + x_2 - x_3)$
16. $T: \mathbb{R}^3 \to \mathbb{R}^3$ defined by $T(x_1, x_2, x_3) = (x_1 - x_2, x_1 + x_2, x_1 + x_2 + x_3)$

17. $T: \mathbb{R}^3 \to \mathbb{R}^2$ defined by $T(x_1, x_2, x_3) = (x_1 + x_2, x_2 + x_3)$

18. $T: \mathbb{R}^4 \to \mathbb{R}^4$ defined by $T(x_1, x_2, x_3, x_4) = (x_4, x_3, x_2, x_1)$

19. $T: F \to \mathbb{R}$ defined by $T(f) = E_3(f) = f(3)$ as in Example 9

20. Prove that the two conditions in Definition 6.1 for a linear transformation are equivalent to the single condition in Eq. (1).

In Exercises 21–27, describe the range of the linear transformation given in the indicated exercise.

21. Exercise 13   **22.** Exercise 14   **23.** Exercise 15   **24.** Exercise 16

25. Exercise 17   **26.** Exercise 18   **27.** Exercise 19

28. Let V, V', and V'' be vector spaces, and let $T_1: V \to V'$ and $T_2: V' \to V''$ be linear transformations. Show that the composite function $(T_2 \circ T_1): V \to V''$ defined by $(T_2 \circ T_1)(\mathbf{v}) = T_2(T_1(\mathbf{v}))$ for each $\mathbf{v}$ in V is again a linear transformation.

The remaining exercises are concerned with the algebra of all linear maps of a vector space V into a vector space W. The exercises show that this collection of linear maps has a natural vector-space structure. We let $L(V, W)$ be the collection of all linear maps of V into W.

29. Let T_1 and T_2 be in $L(V, W)$, and let $(T_1 + T_2): V \to W$ be defined by

$$(T_1 + T_2)(\mathbf{v}) = T_1(\mathbf{v}) + T_2(\mathbf{v})$$

for each vector $\mathbf{v}$ in V. Show that $T_1 + T_2$ is again a linear transformation of V into W.

30. Let T be in $L(V, W)$, let r be any scalar in $\mathbb{R}$, and let $rT: V \to W$ be defined by $(rT)(\mathbf{v}) = r(T(\mathbf{v}))$ for each vector $\mathbf{v}$ in V. Show that rT is again a linear transformation of V into W.

31. Verify the properties required for addition and scalar multiplication in a vector space for the addition and scalar multiplication defined on $L(V, W)$ in Exercises 29 and 30. Careful proofs of all the properties need not be written out but sufficient thought should be given to see why they are true. What is the zero vector in $L(V, W)$?

32.   Work with Topic 3 of the available program VECTGRPH until a score of at least 80 percent can be achieved consistently.

6.2
Matrix Representations of Linear Transformations

As Examples 5, 6, and 7 in Section 6.1 suggested, every linear transformation T of $\mathbb{R}^n$ into $\mathbb{R}^m$ can be computed by multiplying column vectors in $\mathbb{R}^n$ on the left by an $m \times n$ matrix A_T. In this section we show how to find such a *matrix representation* A_T for a linear transformation T. Results from our study of matrices are then used to study linear transformations.

STANDARD MATRIX REPRESENTATIONS

Let $T: \mathbb{R}^n \to \mathbb{R}^m$ be a linear transformation. The following theorem shows that if we can find a matrix A_T such that $A_T\mathbf{b}_i = T(\mathbf{b}_i)$ for each vector $\mathbf{b}_i$ in some basis for $\mathbb{R}^n$, then $A_T\mathbf{x} = T(\mathbf{x})$ for every vector $\mathbf{x}$ in $\mathbb{R}^n$. The theorem is as easy to state and prove for general finite-dimensional vector spaces V and W as it is for the Euclidean spaces $\mathbb{R}^n$ and $\mathbb{R}^m$.

Theorem 6.4   Role of a Basis in the Determination of T

Let V and W be finite-dimensional vector spaces, and let $\{\mathbf{b}_1, \mathbf{b}_2, \ldots, \mathbf{b}_n\}$ be a basis for V. Each linear transformation $T: V \to W$ is completely determined by its effect on this basis. That is, if the n vectors

$$T(\mathbf{b}_1), T(\mathbf{b}_2), \ldots, T(\mathbf{b}_n)$$

are known, then $T(\mathbf{x})$ can be computed in terms of them for any vector $\mathbf{x}$ in V.

Proof   Let $\mathbf{x}$ be any vector in V. We know that $\mathbf{x}$ can be expressed *uniquely* in the form

$$\mathbf{x} = x_1\mathbf{b}_1 + x_2\mathbf{b}_2 + \cdots + x_n\mathbf{b}_n, \tag{1}$$

where the x_i are scalars in $\mathbb{R}$. By preservation of addition and scalar multiplication for T, we obtain from Eq. (1)

$$T(\mathbf{x}) = x_1 T(\mathbf{b}_1) + x_2 T(\mathbf{b}_2) + \cdots + x_n T(\mathbf{b}_n). \tag{2}$$

Equation (2) shows that $T(\mathbf{x})$ is completely determined if the vector values of T at vectors in the basis are known.   ■

Now let $T: \mathbb{R}^n \to \mathbb{R}^m$ be a linear transformation, and let

$$E = \{\mathbf{e}_1, \mathbf{e}_2, \ldots, \mathbf{e}_n\}$$

be the standard basis for $\mathbb{R}^n$. By Theorem 6.4, T is completely determined by its vector values $T(\mathbf{e}_j)$ for $j = 1, 2, \ldots, n$. If we regard $\mathbf{e}_j$ as a *column* vector in $\mathbb{R}^n$, then for any $m \times n$ matrix A, the column vector $A\mathbf{e}_j$ is the jth column vector of A. This follows from the equation $AI = A$, where I is the $n \times n$ identity matrix. Therefore, if A_T is to be a matrix such that $A_T\mathbf{e}_j = T(\mathbf{e}_j)$, then A_T must be the $m \times n$ matrix with jth column vector $T(\mathbf{e}_j)$. Taking $\mathbf{b}_j = \mathbf{e}_j$ in Eqs. (1) and (2), we see that $A_T\mathbf{x} = T(\mathbf{x})$ for every vector $\mathbf{x}$ in $\mathbb{R}^n$. We summarize our work in a definition and a theorem.

Definition 6.3   Standard Matrix Representation

Let $T: \mathbb{R}^n \to \mathbb{R}^m$ be a linear transformation, and let $E = \{\mathbf{e}_1, \mathbf{e}_2, \ldots, \mathbf{e}_n\}$ be the standard basis for $\mathbb{R}^n$. The $m \times n$ matrix A_T having $T(\mathbf{e}_j)$ as jth column vector is the **standard matrix representation** of T.

Theorem 6.5   Role of the Standard Matrix Representation

If A_T is the standard matrix representation of a linear transformation $T: \mathbb{R}^n \to \mathbb{R}^m$, then $A_T\mathbf{x} = T(\mathbf{x})$ for every column vector $\mathbf{x}$ in $\mathbb{R}^n$.

Example 1 Find the standard matrix representation A_T for the linear transformation $T: \mathbb{R}^4 \to \mathbb{R}^3$ defined by

$$T(x_1, x_2, x_3, x_4) = (x_1 - 2x_2 + 2x_4, \; x_1 + x_3 - x_4, \; 4x_1 - 2x_2 + 3x_3 - x_4).$$

Solution Writing vectors as column vectors, we see that A_T is the 3×4 matrix whose columns are

$$T\begin{pmatrix} 1 \\ 0 \\ 0 \\ 0 \end{pmatrix} = \begin{pmatrix} 1 \\ 1 \\ 4 \end{pmatrix}, \quad T\begin{pmatrix} 0 \\ 1 \\ 0 \\ 0 \end{pmatrix} = \begin{pmatrix} -2 \\ 0 \\ -2 \end{pmatrix}, \quad T\begin{pmatrix} 0 \\ 0 \\ 1 \\ 0 \end{pmatrix} = \begin{pmatrix} 0 \\ 1 \\ 3 \end{pmatrix}, \quad \text{and} \quad T\begin{pmatrix} 0 \\ 0 \\ 0 \\ 1 \end{pmatrix} = \begin{pmatrix} 2 \\ -1 \\ -1 \end{pmatrix},$$

so that

$$A_T = \begin{pmatrix} 1 & -2 & 0 & 2 \\ 1 & 0 & 1 & -1 \\ 4 & -2 & 3 & -1 \end{pmatrix}. \quad \triangleleft$$

Note in Example 1 that the first row of the matrix A_T contains precisely the coefficients in the first-component formula

$$x_1 - 2x_2 + 2x_4$$

used to define T. Similarly, the second row of A_T consists of the coefficients in the second-component formula, and the third row has the coefficients in the third-component formula. One can easily see why this is true in general, and this observation makes it very easy for us to write down the standard matrix representation when T is defined by such formulas.

We mentioned that matrix representations allow us to derive properties of linear transformations from known properties of matrices. Here is an illustration. Let A_T be the standard matrix representation of $T: \mathbb{R}^n \to \mathbb{R}^m$. We saw on page 317 that

$$\text{range}(T) = \text{column space of } A_T \tag{3}$$

and

$$\ker(T) = \text{nullspace of } A_T. \tag{4}$$

The dimension of the $\ker(T)$ is called the **nullity** of T, denoted by $\text{null}(T)$, and the dimension of $\text{range}(T)$ is called the **rank** of T, denoted by $\text{rank}(T)$. In view of Eqs. (3) and (4), we have

$$\text{rank}(T) = \text{rank}(A_T) \quad \text{and} \quad \text{null}(T) = \text{null}(A_T).$$

Theorem 3.16 showed that for any $m \times n$ matrix A, we have

$$\text{rank}(A) + \text{null}(A) = n. \tag{5}$$

Translating Eq. (5) to the context of linear transformations, we obtain the following result.

Theorem 6.6 Rank Equation for Linear Transformations

Let $T: \mathbb{R}^n \to \mathbb{R}^m$ be a linear transformation. Then

$$\text{rank}(T) + \text{null}(T) = n.$$

Example 2 Find the rank and nullity of the linear transformation T of Example 1.

Solution We find the rank of the matrix representation A_T of T as follows:

$$A_T = \begin{pmatrix} 1 & -2 & 0 & 2 \\ 1 & 0 & 1 & -1 \\ 4 & -2 & 3 & -1 \end{pmatrix} \sim \begin{pmatrix} 1 & -2 & 0 & 2 \\ 0 & 2 & 1 & -3 \\ 0 & 6 & 3 & -9 \end{pmatrix}$$

$$\sim \begin{pmatrix} 1 & 0 & -2 & 2 \\ 0 & 1 & 2 & -3 \\ 0 & 3 & 6 & -9 \end{pmatrix} \sim \begin{pmatrix} 1 & 0 & -2 & 2 \\ 0 & 1 & 2 & -3 \\ 0 & 0 & 0 & 0 \end{pmatrix}.$$

Thus $\text{rank}(T) = \text{rank}(A_T) = 2$. Using Theorem 6.6, we have

$$\text{null}(T) = 4 - \text{rank}(T) = 4 - 2 = 2. \quad \lhd$$

We give another illustration of the relationship between properties of matrices and properties of linear transformations. Let T be a linear transformation of $\mathbb{R}^n$ into itself. From Theorem 6.3, we know that the inverse function T^{-1} of T is well defined if and only if $\ker(T) = \{\mathbf{0}\}$. On the other hand, we know that the matrix representation A_T is an invertible matrix if and only if its nullspace is $\{\mathbf{0}\}$. Applying Eq. (4), we see that T is well defined if and only if A_T is invertible. In this case Theorem 6.6 indicates that $\text{rank}(T) = n$ so that T maps $\mathbb{R}^n$ *onto all* of $\mathbb{R}^n$. We arrive at the following criterion for the inverse of T to be well defined.

Theorem 6.7 Criterion for T^{-1} to be Well Defined

Let $T: \mathbb{R}^n \to \mathbb{R}^n$ be a linear transformation. Then T is one to one with well-defined inverse transformation $T^{-1}: \mathbb{R}^n \to \mathbb{R}^n$ if and only if its standard matrix representation A_T is invertible, in which case

$$(A_T)^{-1} = A_{(T^{-1})}.$$

The proof of the equation in Theorem 6.7 is left to Exercise 27.

OTHER MATRIX REPRESENTATIONS OF T

We computed the standard matrix representation A_T of a linear transformation $T: \mathbb{R}^n \to \mathbb{R}^m$ by working with the standard basis E. We could work with any basis B of $\mathbb{R}^n$ and any basis B' of $\mathbb{R}^m$. We will obtain another matrix representation of T, which we will refer to as the *matrix representation A_T relative to bases B,B'*. The standard matrix representation thus becomes the matrix representation A_T relative to bases E,E', where E and E' are the standard bases for $\mathbb{R}^n$ and $\mathbb{R}^m$, respectively.

The standard matrix representation is certainly the easiest one to compute. However, a matrix representation relative to other bases B,B' may have a much simpler form than the standard matrix representation, and this simpler form may provide very useful information about the linear transformation T. This feature will be illustrated in Section 6.4 when we study eigenvalues and eigenvectors of a linear transformation T.

We turn now to the discussion of the matrix representation of T relative to bases B,B'. The construction of the matrix will be analogous to the construction of the standard matrix representation relative to E,E'. However, the vectors in the standard bases E and E' have a natural *order*. The vector having 1 in the first component is the first basis vector, the vector having 1 in the second component is the second basis vector, and so on. There is no such natural order for vectors in a general basis B of $\mathbb{R}^n$. We will need to consider an order for them because rearranging the order of the basis vectors corresponds to rearranging the order of the column vectors in the matrix representation, as we will see. Thus we introduce the notion of an **ordered basis** $B = (\mathbf{b}_1, \mathbf{b}_2, \ldots, \mathbf{b}_n)$ for $\mathbb{R}^n$, where we use parentheses rather than braces to indicate that the vectors have a specific order. We will henceforth consider E to be the ordered basis $(\mathbf{e}_1, \mathbf{e}_2, \ldots, \mathbf{e}_n)$ of $\mathbb{R}^n$.

Let $T: \mathbb{R}^n \to \mathbb{R}^m$ be a linear transformation, let $B = (\mathbf{b}_1, \mathbf{b}_2, \ldots, \mathbf{b}_n)$ be an ordered basis for $\mathbb{R}^n$, and let $B' = (\mathbf{b}'_1, \mathbf{b}'_2, \ldots, \mathbf{b}'_m)$ be an ordered basis for $\mathbb{R}^m$. We want to find a matrix representation A_T relative to B,B' so that it can be used to compute $T(\mathbf{x})$ for $\mathbf{x}$ in $\mathbb{R}^n$ by following the boxed steps.

Computation of $T(\mathbf{x})$ Using the Matrix Representation A_T relative to B,B'

Step 1 Express $\mathbf{x}$ in the form

$$\mathbf{x} = r_1\mathbf{b}_1 + r_2\mathbf{b}_2 + \cdots + r_n\mathbf{b}_n.$$

Step 2 Using A_T, compute the $m \times 1$ column vector

$$A_T \begin{pmatrix} r_1 \\ r_2 \\ \vdots \\ r_n \end{pmatrix} \quad \text{and denote it by} \quad \begin{pmatrix} r'_1 \\ r'_2 \\ \vdots \\ r'_m \end{pmatrix}.$$

Step 3 Then we have

$$T(\mathbf{x}) = r'_1\mathbf{b}'_1 + r'_2\mathbf{b}'_2 + \cdots + r'_m\mathbf{b}'_m.$$

In step 1, the vector

$$\mathbf{r} = (r_1, r_2, \ldots, r_n)$$

is called the **coordinate vector of x relative to the ordered basis** B and is denoted by $\mathbf{x}_B$. Of course, the usual components $x_1, x_2, \ldots, x_n$ of $\mathbf{x}$ are its coordinates relative to the standard basis E.

Let us determine the structure of A_T that makes these steps valid. The coordinate vector of $\mathbf{b}_1$ relative to the ordered basis B is $(1, 0, \ldots, 0)$ since $\mathbf{b}_1 = 1\mathbf{b}_1 + 0\mathbf{b}_2 + \cdots + 0\mathbf{b}_n$. Similarly, the coordinate vector of $\mathbf{b}_j$ relative to the ordered basis B is given by the vector $\mathbf{e}_j$ in $\mathbb{R}^n$. Recall that for any $m \times n$ matrix A, the product $A\mathbf{e}_j$ is the jth column of A, where $\mathbf{e}_j$ is taken as a column vector. The steps then require that

$$T(\mathbf{x})_{B'} = A_T \mathbf{x}_B.$$

Taking $\mathbf{x} = \mathbf{b}_j$, we see that the matrix A_T must have as jth column vector the coordinate vector of $T(\mathbf{b}_j)$ relative to the ordered basis B' for $\mathbb{R}^m$.

Definition 6.4   Matrix Representation of *T* relative to *B*, *B'*

Let $T: \mathbb{R}^n \rightarrow \mathbb{R}^m$ be a linear transformation, and let B and B' be ordered bases for $\mathbb{R}^n$ and $\mathbb{R}^m$, respectively. Let A_T be the $m \times n$ matrix whose jth column vector is the column coordinate vector of $T(\mathbf{b}_j)$ relative to the ordered basis B'. This matrix A_T is the **matrix representation of *T* relative to bases *B*, *B'***.

Writing out A_T we see that if $T(\mathbf{b}_j) = a_{1j}\mathbf{b}'_1 + a_{2j}\mathbf{b}'_2 + \cdots + a_{mj}\mathbf{b}'_m$, then

$$A_T = \begin{pmatrix} a_{11} & \cdots & a_{1j} & \cdots & a_{1n} \\ a_{21} & \cdots & a_{2j} & \cdots & a_{2n} \\ \vdots & & \vdots & & \vdots \\ a_{m1} & \cdots & a_{mj} & \cdots & a_{mn} \end{pmatrix}.$$

$$T(\mathbf{b}_1)_{B'} \qquad T(\mathbf{b}_j)_{B'} \qquad T(\mathbf{b}_n)_{B'}$$

Theorem 6.8   Role of the Matrix Representation Relative to *B*, *B'*

Let A_T be the matrix representation of $T: \mathbb{R}^n \rightarrow \mathbb{R}^m$ relative to bases B, B'. For each $\mathbf{x}$ in $\mathbb{R}^n$, we have

$$A_T \mathbf{x}_B = T(\mathbf{x})_{B'},$$

where $\mathbf{x}_B$ and $T(\mathbf{x})_{B'}$ are column coordinate vectors for $\mathbf{x}$ relative to B and B', respectively.

Proof Let A_T be the matrix representation of a linear transformation T as in Definition 6.4. If we follow steps 1 and 2 listed above using the matrix A_T, then it is easy to see that the right side of the equation in step 3 does describe a linear transformation of $\mathbb{R}^n$ into $\mathbb{R}^m$. Since this transformation has the same action as T on the basis vectors in B, we know by Theorem 6.1 that it has the same action as T on each $\mathbf{x}$ in $\mathbb{R}^n$. That is, the equation in step 3 is valid. ■

We start our illustrations with two that find coordinate vectors relative to nonstandard bases in order to be sure that the concept is well in mind. Once again, we will find ourselves faced with linear systems to solve. Our study of linear systems in Chapter 1 is the foundation for most of our work.

Example 3 Find the coordinate vectors of $(1, -1)$ and of $(-1, -8)$ relative to the ordered basis $B = ((1, -1), (1, 2))$ of $\mathbb{R}^2$.

Solution Clearly, $(1, -1)_B = (1, 0)$ since

$$(1, -1) = 1(1, -1) + 0(1, 2).$$

To find $(-1, -8)_B$, we must find r_1 and r_2 such that $(-1, -8) = r_1(1, -1) + r_2(1, 2)$. Equating components of this vector equation, we obtain the linear system

$$
\begin{aligned}
r_1 + r_2 &= -1 \\
-r_1 + 2r_2 &= -8.
\end{aligned}
$$

The solution of this system is easily found to be $r_1 = 2$, $r_2 = -3$, so we have $(-1, 8)_B = (2, -3)$. ◁

Example 4 Find the coordinate vector of $(1, 2, -2)$ relative to the ordered basis $B = ((1, 1, 1), (1, 2, 0), (1, 0, 1))$ in $\mathbb{R}^3$.

Solution We must express $(1, 2, -2)$ as a linear combination of the basis vectors in B. The equation

$$(1, 2, -2) = r_1(1, 1, 1) + r_2(1, 2, 0) + r_3(1, 0, 1)$$

gives rise to the linear system

$$
\begin{aligned}
r_1 + r_2 + r_3 &= 1 \\
r_1 + 2r_2 \phantom{{}+ r_3} &= 2 \\
r_1 \phantom{{}+ 2r_2} + r_3 &= -2.
\end{aligned}
$$

We find the unique solution by a Gauss–Jordan reduction of the corresponding partitioned matrix, obtaining

$$\begin{pmatrix} 1 & 1 & 1 & | & 1 \\ 1 & 2 & 0 & | & 2 \\ 1 & 0 & 1 & | & -2 \end{pmatrix} \sim \begin{pmatrix} 1 & 1 & 1 & | & 1 \\ 0 & 1 & -1 & | & 1 \\ 0 & -1 & 0 & | & -3 \end{pmatrix}$$

$$\sim \begin{pmatrix} 1 & 0 & 2 & | & 0 \\ 0 & 1 & -1 & | & 1 \\ 0 & 0 & -1 & | & -2 \end{pmatrix} \sim \begin{pmatrix} 1 & 0 & 0 & | & -4 \\ 0 & 1 & 0 & | & 3 \\ 0 & 0 & 1 & | & 2 \end{pmatrix}.$$

Therefore, $(1, 2, -2)_B = (-4, 3, 2)$.   ◁

We box the procedure illustrated by the last two examples.

Finding the Coordinate Vector of a in $\mathbb{R}^n$ Relative to an Ordered Basis B

Step 1 Writing vectors as column vectors, form the partitioned matrix $(\mathbf{b}_1, \mathbf{b}_2, \ldots, \mathbf{b}_n \mid \mathbf{a})$.

Step 2 Use a Gauss–Jordan reduction to obtain the partitioned matrix $(I \mid \mathbf{a}_B)$, where I is the $n \times n$ identity matrix and $\mathbf{a}_B$ is the desired coordinate vector.

We note that the Gauss–Jordan reduction in step 2 is always possible because B is a basis; consequently, the matrix to the left of the partition in step 1 has rank n.

Of course, one may find at the same time coordinates of several different vectors relative to the ordered basis B by just lining up all the vectors as column vectors to the right of the partition in step 1. This is illustrated in the next example, where we find a matrix representation with respect to nonstandard bases.

Example 5   Let $T: \mathbb{R}^3 \to \mathbb{R}^3$ be the linear transformation defined by $T(x_1, x_2, x_3) = (2x_1 - x_2 + 2x_3, x_1 + x_2, 3x_1 - x_2 - x_3)$. Find the matrix representation A_T relative to ordered bases B, B', where

$$B = ((1, 0, 0), (1, 1, 1), (1, -1, 1))$$

and

$$B' = ((1, 1, 2), (1, 2, 1), (2, 1, 1)).$$

Solution   We have

$$T(1, 0, 0) = (2, 1, 3), \qquad T(1, 1, 1) = (3, 2, 1), \quad \text{and} \quad T(1, -1, 1) = (5, 0, 3).$$

We must find the coordinate vector of each of these relative to the ordered basis B'. Following the steps in the preceding box for all three vectors at once, we reduce a partitioned matrix as follows:

$$(\mathbf{b}_1', \mathbf{b}_2', \mathbf{b}_3', \mid T(\mathbf{b}_1), T(\mathbf{b}_2), T(\mathbf{b}_3)) = \begin{pmatrix} 1 & 1 & 2 & \mid & 2 & 3 & 5 \\ 1 & 2 & 1 & \mid & 1 & 2 & 0 \\ 2 & 1 & 1 & \mid & 3 & 1 & 3 \end{pmatrix}$$

$$\sim \begin{pmatrix} 1 & 1 & 2 & \mid & 2 & 3 & 5 \\ 0 & 1 & -1 & \mid & -1 & -1 & -5 \\ 0 & -1 & -3 & \mid & -1 & -5 & -7 \end{pmatrix}$$

$$\sim \begin{pmatrix} 1 & 0 & 3 & \mid & 3 & 4 & 10 \\ 0 & 1 & -1 & \mid & -1 & -1 & -5 \\ 0 & 0 & -4 & \mid & -2 & -6 & -12 \end{pmatrix}$$

$$\sim \begin{pmatrix} 1 & 0 & 0 & \mid & \frac{3}{2} & -\frac{1}{2} & 1 \\ 0 & 1 & 0 & \mid & -\frac{1}{2} & \frac{1}{2} & -2 \\ 0 & 0 & 1 & \mid & \frac{1}{2} & \frac{3}{2} & 3 \end{pmatrix}.$$

Thus

$$A_T = \begin{pmatrix} \frac{3}{2} & -\frac{1}{2} & 1 \\ -\frac{1}{2} & \frac{1}{2} & -2 \\ \frac{1}{2} & \frac{3}{2} & 3 \end{pmatrix}. \quad \triangleleft$$

We box the technique illustrated in the preceding example.

Finding the Matrix Representation of T: $\mathbb{R}^n \to \mathbb{R}^m$
Relative to Ordered Bases B, B'

Step 1 Form the partitioned matrix

$$(\mathbf{b}_1', \mathbf{b}_2', \ldots, \mathbf{b}_m' \mid T(\mathbf{b}_1), \ldots, T(\mathbf{b}_n)).$$

Step 2 Use a Gauss–Jordan reduction to obtain the
partitioned matrix $(I \mid A_T)$, where I is the $m \times m$
identity matrix and A_T is the desired matrix
representation.

We note that the Gauss–Jordan reduction in step 2 is always possible because B' is a basis; consequently, the matrix to the left of the partition in step 1 has rank m.

SUMMARY

Let T: $\mathbb{R}^n \to \mathbb{R}^m$ be a linear transformation, and let $B = (\mathbf{b}_1, \mathbf{b}_2, \ldots, \mathbf{b}_n)$ and $B' = (\mathbf{b}_1', \mathbf{b}_2', \ldots, \mathbf{b}_m')$ be ordered bases for $\mathbb{R}^n$ and $\mathbb{R}^m$, respectively.
1. The coordinate vector of $\mathbf{a}$ in $\mathbb{R}^n$ relative to B is the vector $\mathbf{a}_B = (r_1, r_2, \ldots, r_n)$ such that

$$\mathbf{a} = r_1\mathbf{b}_1 + r_2\mathbf{b}_2 + \cdots + r_n\mathbf{b}_n.$$

See the box on page 328 for its computation.

2. The standard matrix representation of T is the $m \times n$ matrix A_T whose jth column vector is $T(\mathbf{e}_j)$, where $\mathbf{e}_j$ is the jth vector in the standard basis E for $\mathbb{R}^n$.

3. The matrix representation of T relative to bases B, B' is the $m \times n$ matrix A_T whose jth column is the coordinate vector of $T(\mathbf{b}_j)$ relative to the basis B'. See the box on page 329 for its computation.

EXERCISES

In Exercises 1–6, find the coordinate vector of the given vector relative to the indicated ordered basis.

1. $(-1, 1)$ in $\mathbb{R}^2$ relative to $((0, 1), (1, 0))$ 2. $(-2, 4)$ in $\mathbb{R}^2$ relative to $((0, -2), (-\frac{1}{2}, 0))$

3. $(4, 6, 2)$ in $\mathbb{R}^3$ relative to $((2, 0, 0), (0, 1, 1), (0, 0, 1))$

4. $(4, -2, 1)$ in $\mathbb{R}^3$ relative to $((0, 1, 1), (2, 0, 0), (0, 3, 0))$

5. $(3, 13, -1)$ in $\mathbb{R}^3$ relative to $((1, 3, -2), (4, 1, 3), (-1, 2, 0))$

6. $(9, 6, 11, 0)$ in $\mathbb{R}^4$ relative to $((1, 0, 1, 0), (2, 1, 1, -1), (0, 1, 1, -1), (2, 1, 3, 1))$

In Exercises 7–12, find (a) the standard matrix representation of the given linear transformation T and (b) the matrix representation of T relative to B, B'.

7. $T: \mathbb{R}^2 \to \mathbb{R}^2$ defined by $T(x, y) = (2x - 3y, x + y)$,
   $B = ((1, 1), (1, 0))$, $B' = ((0, 1), (1, 1))$

8. $T: \mathbb{R}^2 \to \mathbb{R}^3$ defined by $T(x, y) = (2x + y, x + 2y, x - 3y)$,
   $B = ((1, 0), (0, 1))$, $B' = ((1, 1, 1), (1, 1, 0), (1, 0, 0))$

9. $T: \mathbb{R}^3 \to \mathbb{R}^3$ defined by $T(x_1, x_2, x_3) = (x_1 + x_2 + x_3, x_1 - x_2 - x_3, -x_1 - x_2 + x_3)$,
   $B = ((2, 3, 1), (1, 2, 0), (2, 0, 3))$,
   $B' = ((1, 0, 0), (0, 1, 0), (0, 0, 1))$

10. $T: \mathbb{R}^3 \to \mathbb{R}^3$ defined by $T(x_1, x_2, x_3) = (x_1 + 2x_2 + x_3, x_1, x_2 + x_3)$,
    $B = ((1, 0, 1), (1, 1, 0), (0, 1, 1))$,
    $B' = ((0, 1, 1), (1, 1, 0), (1, 0, 1))$

11. $T: \mathbb{R}^3 \to \mathbb{R}^4$ defined by $T(x_1, x_2, x_3) = (x_1 + x_2, x_2 + x_3, x_3 + x_1, x_1 + x_2 + x_3)$,
    $B = ((1, 1, 1), (1, 1, 0), (1, 0, 0))$,
    $B' = ((1, 1, 1, 1), (1, 1, 1, 0), (1, 1, 0, 0), (1, 0, 0, 0))$

12. $T: \mathbb{R}^3 \to \mathbb{R}^2$ defined by $T(x_1, x_2, x_3) = (x_1 + 2x_2 + 3x_3, x_1)$,
    $B = ((1, 1, 1), (1, 1, 0), (1, 0, 0))$,
    $B' = ((1, 1), (1, 2))$

In Exercises 13–18, find the rank and nullity of the linear transformation in the indicated exercise.

13. Exercise 7   14. Exercise 8   15. Exercise 9

16. Exercise 10   17. Exercise 11   18. Exercise 12

*Exercises 19–24 deal with reflections of $\mathbb{R}^2$ and of $\mathbb{R}^3$. Let L be a line in the plane $\mathbb{R}^2$. The map $f: \mathbb{R}^2 \to \mathbb{R}^2$ which carries a point in $\mathbb{R}^2$ into its mirror image in L is the **reflection** of the plane in L. (See Fig. 6.3.) One has the obvious analogous notion of the reflection of $\mathbb{R}^3$ in a plane of $\mathbb{R}^3$.*

19. Argue geometrically that if the line L in $\mathbb{R}^2$ passes through the origin, then reflection in L is a linear transformation of $\mathbb{R}^2$ into itself.

20. Find the standard matrix representation of the reflection of $\mathbb{R}^2$ in
 a) the x-axis;
 b) the y-axis.

21. Find the standard matrix representation of the reflection of $\mathbb{R}^2$ in
 a) the line $y = x$;
 b) the line $y = -x$.

22. Argue geometrically that reflection in a plane through the origin is a linear transformation of $\mathbb{R}^3$ into $\mathbb{R}^3$.

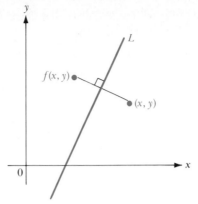

Figure 6.3
The reflection f in the line L.

23. Find the standard matrix representation of the reflection of $\mathbb{R}^3$ in
 a) the plane $x_2 = 0$;
 b) the plane $x_3 = 0$.

24. Find the standard matrix representation of the reflection of $\mathbb{R}^3$ in
 a) the plane $x_2 = x_1$;
 b) the plane $x_1 = x_3$.

In Exercises 25–27, let $T: \mathbb{R}^n \to \mathbb{R}^n$ be a linear transformation with matrix representation A_T relative to bases B,B where B is an ordered basis of $\mathbb{R}^n$.

25. If A_T is a diagonal matrix, describe the effect of T on the basis vectors in B.

26. If A_T is an invertible matrix, describe in terms of T the linear transformation $T'(\mathbf{x}) = A^{-1}\mathbf{x}$.

27. If A_T is an invertible matrix, show that $(A_T)^{-1} = A_{(T^{-1})}$. (See Exercise 26.)

28. Let A be an $m \times n$ matrix and $T_A: \mathbb{R}^n \to \mathbb{R}^m$ be the matrix transformation given by $T_A(\mathbf{x}) = A\mathbf{x}$ for any column vector $\mathbf{x}$ in $\mathbb{R}^n$. Show that the standard matrix representation $A_{(T_A)}$ of T_A is A, that is, $A_{(T_A)} = A$.

29. Following the idea in Exercise 28, explain the meaning of the equation $T_{(A_T)} = T$ for the standard matrix representation A_T of T.

30. Let V be a vector space of dimension n, and let $B = (\mathbf{b}_1, \mathbf{b}_2, \ldots, \mathbf{b}_n)$ be an ordered basis for V. Let $T: V \to \mathbb{R}^n$ be defined by $T(\mathbf{v}) = \mathbf{v}_B$, the coordinate vector of $\mathbf{v}$ relative to the ordered basis B, for each vector $\mathbf{v}$ in V.
 a) Show that T is a linear transformation.
 b) Show that T is one to one.
 c) Show that range$(T) = \mathbb{R}^n$.

[*Comment:* In the introduction to Chapter 3 we justified our concentration on the Euclidean spaces $\mathbb{R}^n$ by saying that every sufficiently small vector space is algebraically indistinguishable from some $\mathbb{R}^n$. We can now say that "sufficiently small" means finite dimensional. If $T: V \to W$ is a one-to-one linear transformation for vector spaces V and W, then we can rename each vector $\mathbf{v}$ in V using the name of the vector $T(\mathbf{v})$ in W, giving different vectors in V different names. Assuming now also that range$(T) = W$, all the names of vectors in W are used up. The conditions

$$T(\mathbf{v}_1 + \mathbf{v}_2) = T(\mathbf{v}_1) + T(\mathbf{v}_2) \quad \text{and} \quad T(r\mathbf{v}) = rT(\mathbf{v})$$

for a linear transformation then show that so far as the vector-space operations of addition and scalar multiplication are concerned, we can regard V and W as the same vector space just discussed in two different languages: the language using the names in V and the language using the names in W. The map T performs the translation from one language to the other. The vector spaces V and W are called **isomorphic**. Exercise 30 shows that every finite-dimensional real vector space V is isomorphic to $\mathbb{R}^n$, where n is dim(V). This demonstrates our assertion in the introduction to Chapter 3.]

31. ▪ Work with Topic 4 of the supplied program VECTGRPH until a score of at least 82 percent can be obtained reliably.

32. ▪ Repeat Exercise 31 using Topic 5 of VECTGRPH.

33. ▪ Repeat Exercise 31 using Topic 6 of VECTGRPH.

6.3
Change of Basis and Similar Matrices

In this section we will be interested primarily in linear transformations of a finite-dimensional vector space V into itself and their matrix representations relative to B,B where B is an ordered basis of V. We will show the relationship between two matrix representations of the same linear transformation T, one representation relative to B,B and the other relative to B',B'. We will start by finding a matrix that can be used to change coordinates of a vector relative to one ordered basis into coordinates relative to another.

CHANGE-OF-BASIS MATRIX

Let V be a finite-dimensional vector space and let $B = (\mathbf{b}_1, \mathbf{b}_2, \ldots, \mathbf{b}_n)$ and $B' = (\mathbf{b}'_1, \mathbf{b}'_2, \ldots, \mathbf{b}'_n)$ be two ordered bases for V. As an application of our work in Section 6.2, we show how to find a matrix C that carries the column coordinate vector $\mathbf{v}_B$ of the vector $\mathbf{v}$ in V to the column coordinate vector $\mathbf{v}_{B'}$. That is, the matrix C satisfies

$$C\mathbf{v}_B = \mathbf{v}_{B'}. \tag{1}$$

Our work in Section 6.2 shows that C is the matrix representation relative to B,B' of the linear transformation

$$\iota\colon V \to V, \qquad \text{(the identity transformation)}$$

where $\iota(\mathbf{v}) = \mathbf{v}$ for all $\mathbf{v}$ in V. The boxed steps on page 329 show precisely how to find C. Since $\iota(\mathbf{b}_j) = \mathbf{b}_j$, one simply forms the partitioned matrix

$$(\mathbf{b}'_1, \mathbf{b}'_2, \ldots, \mathbf{b}'_n \mid \mathbf{b}_1, \mathbf{b}_2, \ldots, \mathbf{b}_n) \tag{2}$$

and uses Gauss–Jordan reduction to obtain the partitioned matrix $(I \mid C)$. This Gauss–Jordan reduction is possible since the matrix to the left of the partition in the expression (2) has rank n. Indeed, the matrix to the right of the partition also has rank n, so the matrix C we obtain is invertible. We call C the **change-of-basis matrix relative to bases B,B'**.

Example 1   Consider the ordered basis

$$B = ((1,\ 1,\ 0),\ (1,\ 0,\ 1),\ (0,\ 1,\ 1))$$

for $\mathbb{R}^3$. Find the change-of-basis matrix C relative to bases E,B where E is the standard ordered basis for $\mathbb{R}^3$. Use C to find the coordinates of

$$\begin{pmatrix} 2 \\ -2 \\ 4 \end{pmatrix}$$

in $\mathbb{R}^n$ relative to the basis B.

Solution   Reducing the partitioned matrix having column vectors in B to the left of the partition and column vectors in E to the right, we have

$$\left(\begin{array}{ccc|ccc} 1 & 1 & 0 & 1 & 0 & 0 \\ 1 & 0 & 1 & 0 & 1 & 0 \\ 0 & 1 & 1 & 0 & 0 & 1 \end{array}\right) \sim \left(\begin{array}{ccc|ccc} 1 & 1 & 0 & 1 & 0 & 0 \\ 0 & -1 & 1 & -1 & 1 & 0 \\ 0 & 1 & 1 & 0 & 0 & 1 \end{array}\right)$$

$$\sim \left(\begin{array}{ccc|ccc} 1 & 0 & 1 & 0 & 1 & 0 \\ 0 & 1 & -1 & 1 & -1 & 0 \\ 0 & 0 & 2 & -1 & 1 & 1 \end{array}\right)$$

$$\sim \left(\begin{array}{ccc|ccc} 1 & 0 & 0 & \frac{1}{2} & \frac{1}{2} & -\frac{1}{2} \\ 0 & 1 & 0 & \frac{1}{2} & -\frac{1}{2} & \frac{1}{2} \\ 0 & 0 & 1 & -\frac{1}{2} & \frac{1}{2} & \frac{1}{2} \end{array}\right).$$

Therefore, the change-of-basis matrix C relative to E,B is

$$C = \begin{pmatrix} \frac{1}{2} & \frac{1}{2} & -\frac{1}{2} \\ \frac{1}{2} & -\frac{1}{2} & \frac{1}{2} \\ -\frac{1}{2} & \frac{1}{2} & \frac{1}{2} \end{pmatrix}.$$

To find the coordinates of

$$\begin{pmatrix} 2 \\ -2 \\ 4 \end{pmatrix}$$

relative to B, we compute

$$C\begin{pmatrix} 2 \\ -2 \\ 4 \end{pmatrix} = \begin{pmatrix} \frac{1}{2} & \frac{1}{2} & -\frac{1}{2} \\ \frac{1}{2} & -\frac{1}{2} & \frac{1}{2} \\ -\frac{1}{2} & \frac{1}{2} & \frac{1}{2} \end{pmatrix}\begin{pmatrix} 2 \\ -2 \\ 4 \end{pmatrix} = \begin{pmatrix} -2 \\ 4 \\ 0 \end{pmatrix}.$$

As a check, we verify that

$$-2\begin{pmatrix}1\\1\\0\end{pmatrix} + 4\begin{pmatrix}1\\0\\1\end{pmatrix} + 0\begin{pmatrix}0\\1\\1\end{pmatrix} = \begin{pmatrix}2\\-2\\4\end{pmatrix}. \quad \triangleleft$$

Following Eq. (2) we observed that a change-of-basis matrix C is invertible. From the equation

$$C\mathbf{v}_B = \mathbf{v}_{B'}$$

for the change-of-basis matrix C relative to B,B', we obtain

$$C^{-1}\mathbf{v}_{B'} = \mathbf{v}_B.$$

That is, the change-of-basis matrix relative to bases B',B, going in the other direction, is the inverse of the matrix C. We summarize in a theorem.

Theorem 6.9 Change-of-basis Matrix

Let $B = (\mathbf{b}_1, \mathbf{b}_2, \ldots, \mathbf{b}_n)$ and $B' = (\mathbf{b}_1', \mathbf{b}_2', \ldots, \mathbf{b}_n')$ be ordered bases of a vector space V. The change-of-basis matrix C relative to bases B,B' satisfying Eq. (1) is found by reducing the augmented matrix

$$(\mathbf{b}_1', \mathbf{b}_2', \ldots, \mathbf{b}_n' \mid \mathbf{b}_1, \mathbf{b}_2, \ldots, \mathbf{b}_n)$$

to $(I \mid C)$. This matrix C is invertible, and its inverse is the change-of-basis matrix relative to B',B.

COMPOSITION OF LINEAR TRANSFORMATIONS

Recall that the composite function $g \circ f$ of two functions $f \colon U \to V$ and $g \colon V \to W$ is defined by the formula

$$(g \circ f)(u) = g(f(u)) \tag{3}$$

for u in U. This action of f followed by g can be visualized as shown in Fig. 6.4. Note the left-to-right execution in the figure as opposed to the right-to-left computation of $(g \circ f)(u)$ as given in Eq. (3). It is an easy exercise to show that if $T \colon U \to V$ and $T' \colon V \to W$ are linear transformations of vector spaces, then $T' \circ T \colon U \to W$ is a linear transformation (see Exercise 11).

Example 2   Let T and T' be the linear transformations of $\mathbb{R}^3$ into itself defined by

$$T(x_1, x_2, x_3) = (x_1 + x_2 + x_3, x_1 + x_2, x_3)$$

and

$$T'(x_1, x_2, x_3) = (2x_1 - x_2, x_2 + x_3, 2x_3 - x_2).$$

Find formulas for $T' \circ T$ and for $T \circ T'$, and then find standard matrix representations for T, T', $T' \circ T$, and $T \circ T'$.

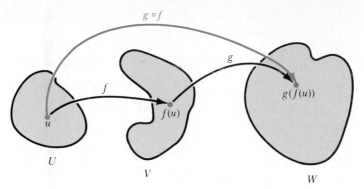

Figure 6.4
The composite function $g \circ f$.

Solution We have

$$(T' \circ T)(a_1, a_2, a_3)$$
$$= T'\big(T(a_1, a_2, a_3)\big)$$
$$= T'(a_1 + a_2 + a_3, a_1 + a_2, a_3)$$
$$= \big(2(a_1 + a_2 + a_3) - (a_1 + a_2), a_1 + a_2 + a_3, 2a_3 - (a_1 + a_2)\big)$$
$$= (a_1 + a_2 + 2a_3, a_1 + a_2 + a_3, -a_1 - a_2 + 2a_3)$$

and

$$(T \circ T')(a_1, a_2, a_3)$$
$$= T\big(T'(a_1, a_2, a_3)\big)$$
$$= T(2a_1 - a_2, a_2 + a_3, 2a_3 - a_2)$$
$$= (2a_1 - a_2 + a_2 + a_3 + 2a_3 - a_2, 2a_1 - a_2 + a_2 + a_3, 2a_3 - a_2)$$
$$= (2a_1 - a_2 + 3a_3, 2a_1 + a_3, -a_2 + 2a_3).$$

As standard matrix representations, we obtain

$$A_T = \begin{pmatrix} 1 & 1 & 1 \\ 1 & 1 & 0 \\ 0 & 0 & 1 \end{pmatrix}, \qquad A_{T'} = \begin{pmatrix} 2 & -1 & 0 \\ 0 & 1 & 1 \\ 0 & -1 & 2 \end{pmatrix},$$

$$A_{T' \circ T} = \begin{pmatrix} 1 & 1 & 2 \\ 1 & 1 & 1 \\ -1 & -1 & 2 \end{pmatrix}, \qquad A_{T \circ T'} = \begin{pmatrix} 2 & -1 & 3 \\ 2 & 0 & 1 \\ 0 & -1 & 2 \end{pmatrix}. \quad \triangleleft$$

MATRIX REPRESENTATION OF $T' \circ T$

Note that for the matrices found in Example 2, we have

$$A_{T' \circ T} = A_{T'} A_T \quad \text{and} \quad A_{T \circ T'} = A_T A_{T'}.$$

This multiplicative property of matrix representations is true in general as we now show. While the theorem and proof are valid for any finite-dimensional vector spaces, we state and prove the theorem for spaces $\mathbb{R}^n$, $\mathbb{R}^m$, and $\mathbb{R}^s$ so that the dimensions of the spaces are apparent from the notations.

Theorem 6.10   Multiplicative Property of Matrix Representations

Let B, B', and B'' be ordered bases of $\mathbb{R}^n$, $\mathbb{R}^m$, and $\mathbb{R}^s$, respectively, and let $T: \mathbb{R}^n \to \mathbb{R}^m$ and $T': \mathbb{R}^m \to \mathbb{R}^s$ be linear transformations, so that we have the diagram

$$\mathbb{R}^n \xrightarrow{\quad T \quad} \mathbb{R}^m \xrightarrow{\quad T' \quad} \mathbb{R}^s.$$
$$\phantom{\mathbb{R}^n}B \phantom{\xrightarrow{\quad T \quad}} B' \phantom{\xrightarrow{\quad T' \quad}} B''$$

Let A_T, $A_{T'}$, and $A_{T' \circ T}$ be matrix representations relative to B,B' and B',B'' and B,B'', respectively. Then

$$A_{T' \circ T} = A_{T'} A_T.$$

Proof   Let $B = (\mathbf{b}_1, \mathbf{b}_2, \ldots, \mathbf{b}_n)$, $B' = (\mathbf{b}'_1, \mathbf{b}'_2, \ldots, \mathbf{b}'_m)$, and $B'' = (\mathbf{b}''_1, \mathbf{b}''_2, \ldots, \mathbf{b}''_s)$. Then $A_T = (a_{kj})$ is an $m \times n$ matrix, and $D = A_{T'} = (d_{ik})$ is an $s \times m$ matrix. To find the matrix representation of $A_{T' \circ T}$ relative to B,B'' we must compute $(T' \circ T)(\mathbf{b}_j)$ for each basis vector $\mathbf{b}_j$ in B and find its coordinate vector relative to B''. Using the matrix representation A_T, we have

$$T(\mathbf{b}_j) = \sum_{k=1}^{m} a_{kj} \mathbf{b}'_k.$$

Since T' is a linear transformation, we obtain

$$T'(T(\mathbf{b}_j)) = T' \left(\sum_{k=1}^{m} a_{kj} \mathbf{b}'_k \right) = \sum_{k=1}^{m} a_{kj} T'(\mathbf{b}'_k).$$

Now using the matrix representation D for T', we can write

$$T'(\mathbf{b}'_k) = \sum_{i=1}^{s} d_{ik} \mathbf{b}''_i.$$

Therefore,

$$T'(T(\mathbf{b}_j)) = \sum_{k=1}^{m} a_{kj} \left(\sum_{i=1}^{s} d_{ik} \mathbf{b}''_i \right) = \sum_{i=1}^{s} \left(\sum_{k=1}^{m} d_{ik} a_{kj} \right) \mathbf{b}''_i,$$

and the matrix representation $A_{T' \circ T}$ is the $s \times n$ matrix whose entry in the ith row and jth column is $\sum_{k=1}^{m} d_{ik} a_{kj}$. However, this is precisely the entry in row i and column j of the matrix product DA. Therefore, $A_{T' \circ T} = DA$, which is what we wished to prove. ∎

We never did prove that matrix multiplication is associative. The straightforward computational proof is not difficult, but is a bit complicated with a lot of summation signs. It is very easy to show that function composition is associative, that is, that $h \circ (g \circ f) = (h \circ g) \circ f$ for suitable functions f, g, and h (see Exercise 16). We use this fact and apply Theorem 6.10 to give a proof that matrix multiplication is associative.

Theorem 6.11   Associativity of Matrix Multiplication

Let B, C, and D be matrices of sizes such that DC and CB are both defined. Then $(DC)B = D(CB)$.

Proof   Let B, C, and D have sizes $m \times n$, $r \times m$, and $s \times r$, respectively, and let $T(\mathbf{x}) = B\mathbf{x}$, $T'(\mathbf{y}) = C\mathbf{y}$, and $T''(\mathbf{z}) = D\mathbf{z}$ be corresponding linear transformations as shown by the diagram

$$\mathbb{R}^n \xrightarrow[\;B\;]{\;T\;} \mathbb{R}^m \xrightarrow[\;C\;]{\;T'\;} \mathbb{R}^r \xrightarrow[\;D\;]{\;T''\;} \mathbb{R}^s.$$

We know that composition of functions is associative, so that

$$(T'' \circ T') \circ T = T'' \circ (T' \circ T).$$

Using Theorem 6.10, we have

$$(DC)B = A_{T'' \circ T'} A_T = A_{(T'' \circ T') \circ T}$$

$$= A_{T'' \circ (T' \circ T)} = A_{T''} A_{T' \circ T} = D(CB),$$

which shows that matrix multiplication is associative. ◼

Let B and B' be ordered bases of a finite-dimensional vector space V. As another application of Theorem 6.10, we give another proof that the change-of-basis matrix C relative to B,B' is invertible with inverse the change-of-basis matrix D relative to B',B. Recall that a change-of-basis matrix is a matrix representation of the identity map. We consider the following diagram of identity maps ι, bases, and matrices:

$$V \xrightarrow[C]{\iota} V \xrightarrow[D]{\iota} V.$$
$$B \quad\quad B' \quad\quad B$$

Now the matrix representation of the identity map ι relative to B,B is the identity matrix I. Applying Theorem 6.10, we have $I = DC$, which proves that these change-of-basis matrices are inverses of each other.

SIMILAR MATRICES

We apply Theorem 6.10 to prove the main result of this section.

Theorem 6.12   Similarity of Matrix Representations of *T*

Let T be a linear transformation of a finite-dimensional vector space V into itself, and let B and B' be ordered bases of V. If A and D are the matrix representations of T *relative to* B,B and B',B', respectively, then

$$D = C^{-1}AC,$$

where C is the change-of-basis matrix relative to B',B.

Proof   Recall that the change-of-basis matrix relative to B',B is the matrix representation C of the identity transformation ι relative to B',B while its inverse C^{-1} is the change-of-basis matrix relative to B,B'. We form the following diagram of linear transformations, bases, and matrices:

$$V \xrightarrow{\;\;\iota\;\;} V \xrightarrow{\;\;T\;\;} V \xrightarrow{\;\;\iota\;\;} V \; . \qquad (4)$$
$$B' \quad C \quad B \quad A \quad B \quad C^{-1} \quad B'$$

Since the identity transformation leaves vectors fixed, we have $\iota \circ T \circ \iota = T$, so the matrix representation of $\iota \circ T \circ \iota$ relative to B',B' is also the matrix representation D of T relative to B',B'. Applying Theorem 6.10 to the diagram, we see that the matrix representation of $\iota \circ T \circ \iota$ relative to B',B' can also be written as $C^{-1}AC$. Thus $D = C^{-1}AC$ as asserted. ◼

Example 3   Consider the linear transformation $T: \mathbb{R}^3 \to \mathbb{R}^3$ defined by

$$T(x_1, x_2, x_3) = (x_1 + x_2 + x_3, x_1 + x_2, x_3).$$

Find the standard matrix representation A of T and the matrix representation D of T relative to B,B where

$$B = ((1, 1, 0), (1, 0, 1), (0, 1, 1)).$$

Also find the invertible matrix C such that $D = C^{-1}AC$.

Solution   The diagram (4) above becomes

$$\mathbb{R}^3 \xrightarrow{\;\;\iota\;\;} \mathbb{R}^3 \xrightarrow{\;\;T\;\;} \mathbb{R}^3 \xrightarrow{\;\;\iota\;\;} \mathbb{R}^3.$$
$$B \quad C \quad E \quad A \quad E \quad C^{-1} \quad B$$

Since

$$T(\mathbf{e}_1) = (1, 1, 0) = \mathbf{e}_1 + \mathbf{e}_2,$$
$$T(\mathbf{e}_2) = (1, 1, 0) = \mathbf{e}_1 + \mathbf{e}_2,$$
$$T(\mathbf{e}_3) = (1, 0, 1) = \mathbf{e}_1 + \mathbf{e}_3,$$

we find that

$$A = \begin{pmatrix} 1 & 1 & 1 \\ 1 & 1 & 0 \\ 0 & 0 & 1 \end{pmatrix}.$$

Since

$$T(\mathbf{b}_1) = (2, 2, 0) = 2\mathbf{b}_1,$$
$$T(\mathbf{b}_2) = (2, 1, 1) = \mathbf{b}_1 + \mathbf{b}_2,$$
$$T(\mathbf{b}_3) = (2, 1, 1) = \mathbf{b}_1 + \mathbf{b}_2,$$

we find that

$$D = \begin{pmatrix} 2 & 1 & 1 \\ 0 & 1 & 1 \\ 0 & 0 & 0 \end{pmatrix}.$$

Finally we have

$$C = \begin{pmatrix} 1 & 1 & 0 \\ 1 & 0 & 1 \\ 0 & 1 & 1 \end{pmatrix},$$

which is the change-of-basis matrix relative to B,E. As a partial check of our work, we can verify the equation $CD = AC$. ◁

We repeat here Definition 4.3 of Section 4.6, which formalizes the relationship of the matrices A and D in Theorem 6.12.

Definition 6.5 **Similar Matrices**

Let A and D be two $n \times n$ matrices. Then A is **similar** to D if $D = C^{-1}AC$ for some invertible $n \times n$ matrix C.

Exercise 13 of Section 4.6 shows that the definition of similarity is symmetric. That is, if D is similar to A, then A is similar to D. Thus we may simply speak of similar matrices.

We have just shown that any two matrix representations of the same linear transformation $T: V \to V$ relative to different bases for V are similar. We now show that, conversely, two different but similar $n \times n$ matrices can be regarded as representing the same linear transformation relative to different bases. Suppose that $D = C^{-1}AC$ for $n \times n$ matrices. Let T_A be the linear transformation of $\mathbb{R}^n$ into itself given by $T_A(\mathbf{x}) = A\mathbf{x}$. Clearly A is the matrix representation of T_A relative to the standard bases E,E. If we let B be the ordered basis consisting of the column vectors of C, then applying the proof of Theorem 6.12, we see that $D = C^{-1}AC$ is the matrix representation of T_A relative to B,B.

THE IDEA OF SIMILARITY, like many matrix notions, appears without a definition in works as early as the 1820s. In fact, in his 1826 work on quadratic forms (see note on page 364) Cauchy showed that if two quadratic forms are related by a change of variables—that is, if their matrices are similar—then their characteristic equations are the same. But like the concept of orthogonality, that of similarity was first formally defined and discussed by Georg Frobenius in 1878. Frobenius began by discussing the general case: He called two matrices A, D *equivalent* if there exist invertible matrices P, Q such that $D = PAQ$. These latter matrices were called the substitutions through which A was transformed into D.

Frobenius then dealt with the special cases where $P = Q^T$ (the two matrices are then called *congruent*) and where $P = Q^{-1}$ (the similarity case of this section). Frobenius went on to prove many results on similarity including the useful theorem that if A is similar to D, then $f(A)$ is similar to $f(D)$, where f is any polynomial matrix function.

Since we have stated theorems in the context of general finite-dimensional vector spaces, we should give one example involving a vector space that is not one of the spaces $\mathbb{R}^n$. (See, however, the comment following Exercise 30 of Section 6.2.) No knowledge of calculus is needed in the example that follows, but a little experience with calculus certainly makes it more intriguing.

Example 4 Let P_3 be the vector space of polynomials of degree at most 3, and let B be the ordered basis $(1, x, x^2, x^3)$ for P_3. The calculus operation of differentiation, applied to polynomials in P_3, is a linear transformation T of P_3 into itself such that $T(1) = 0$, $T(x) = 1$, $T(x^2) = 2x$, and $T(x^3) = 3x^2$. Find the matrix A_T representing T relative to the bases B,B and use A_T to differentiate the polynomial $4x^3 - x^2 + 7x - 5$. Then find the matrix representation of $T^2 = T \circ T$ for B,B and use it to differentiate twice (find the second derivative of) the polynomial $2x^3 - 8x^2 + 4x + 3$.

Solution From the values given for T on the basis vectors in B, we find that

$$A_T = \begin{pmatrix} 0 & 1 & 0 & 0 \\ 0 & 0 & 2 & 0 \\ 0 & 0 & 0 & 3 \\ 0 & 0 & 0 & 0 \end{pmatrix}.$$

The coordinate vector of $4x^3 - x^2 + 7x - 5$ relative to B is

$$\begin{pmatrix} -5 \\ 7 \\ -1 \\ 4 \end{pmatrix}.$$

We compute $T(4x^3 - x^2 + 7x - 5)$ using A_T:

$$A_T \begin{pmatrix} -5 \\ 7 \\ -1 \\ 4 \end{pmatrix} = \begin{pmatrix} 0 & 1 & 0 & 0 \\ 0 & 0 & 2 & 0 \\ 0 & 0 & 0 & 3 \\ 0 & 0 & 0 & 0 \end{pmatrix} \begin{pmatrix} -5 \\ 7 \\ -1 \\ 4 \end{pmatrix} = \begin{pmatrix} 7 \\ -2 \\ 12 \\ 0 \end{pmatrix}.$$

Thus $T(4x^3 - x^2 + 7x - 5) = 7(1) + (-2)x + 12x^2 + 0x^3 = 12x^2 - 2x + 7$. The matrix representation of T^2 relative to B,B is

$$A_T A_T = \begin{pmatrix} 0 & 1 & 0 & 0 \\ 0 & 0 & 2 & 0 \\ 0 & 0 & 0 & 3 \\ 0 & 0 & 0 & 0 \end{pmatrix} \begin{pmatrix} 0 & 1 & 0 & 0 \\ 0 & 0 & 2 & 0 \\ 0 & 0 & 0 & 3 \\ 0 & 0 & 0 & 0 \end{pmatrix} = \begin{pmatrix} 0 & 0 & 2 & 0 \\ 0 & 0 & 0 & 6 \\ 0 & 0 & 0 & 0 \\ 0 & 0 & 0 & 0 \end{pmatrix}.$$

Thus we can find the second derivative of $2x^3 - 8x^2 + 4x + 3$ by computing

$$\begin{pmatrix} 0 & 0 & 2 & 0 \\ 0 & 0 & 0 & 6 \\ 0 & 0 & 0 & 0 \\ 0 & 0 & 0 & 0 \end{pmatrix} \begin{pmatrix} 3 \\ 4 \\ -8 \\ 2 \end{pmatrix} = \begin{pmatrix} -16 \\ 12 \\ 0 \\ 0 \end{pmatrix}$$

and we obtain the answer $12x - 16$. ◁

SUMMARY

Let V be a vector space of dimension n.

1. Let B and B' be ordered bases of V, and let $\iota: V \to V$ be the identity transformation, so that $\iota(\mathbf{v}) = \mathbf{v}$ for all $\mathbf{v}$ in V. The change-of-basis matrix C relative to B,B' is the matrix representation of ι relative to B,B'. We have $C\mathbf{v}_B = \mathbf{v}_{B'}$, for all $\mathbf{v}$ in V. The change-of-basis matrix relative to B',B is C^{-1}.

2. If T and T' are linear transformations of V into itself and B is any ordered basis for V, then the matrix representations relative to B,B satisfy $A_{T'\circ T} = A_{T'}A_T$.

3. Two $n \times n$ matrices A and D are similar if there exists an invertible $n \times n$ matrix C such that $D = C^{-1}AC$.

4. Let $T: V \to V$ be a linear transformation, and let B and B' be ordered bases for V. If A is the matrix representation of T relative to B,B and D is the matrix representation of T relative to B',B', then A and D are similar matrices and $D = C^{-1}AC$, where C is the change-of-base matrix relative to B',B.

EXERCISES

In Exercises 1–4, find the change-of-basis matrix (a) relative to B,B' and (b) relative to B',B. Verify that these matrices are inverses of each other.

1. $B = ((1, 1), (1, 0))$ and $B' = ((0, 1), (1, 1))$ in $\mathbb{R}^2$
2. $B = ((2, 3, 1), (1, 2, 0), (2, 0, 3))$ and $B' = ((1, 0, 0), (0, 1, 0), (0, 0, 1))$ in $\mathbb{R}^3$
3. $B = ((1, 0, 1), (1, 1, 0), (0, 1, 1))$ and $B' = ((0, 1, 1), (1, 1, 0), (1, 0, 1))$ in $\mathbb{R}^3$
4. $B = ((1, 1, 1, 1), (1, 1, 1, 0), (1, 1, 0, 0), (1, 0, 0, 0))$ and the standard ordered basis $B' = E$ for $\mathbb{R}^4$

In Exercises 5–10, find (a) the standard matrix representation A of the given linear transformation T, (b) the matrix representation D of T relative to B,B, where B is the given ordered basis, and (c) an invertible matrix C such that $D = C^{-1}AC$.

5. $T: \mathbb{R}^2 \to \mathbb{R}^2$ defined by $T(x, y) = (2x - 3y, x + y)$, $B = ((1, 1), (1, -1))$
6. $T: \mathbb{R}^2 \to \mathbb{R}^2$ defined by $T(x, y) = (x - y, 2x)$, $B = ((2, 1), (1, 2))$

7. $T: \mathbb{R}^3 \rightarrow \mathbb{R}^3$ defined by $T(x_1, x_2, x_3) = (x_1 + x_2, x_1 - x_2, x_1 + x_3)$, $B = ((1, 1, 1), (1, 0, 1), (0, 1, 1))$

8. $T: \mathbb{R}^3 \rightarrow \mathbb{R}^3$ defined by $T(x_1, x_2, x_3) = (x_3, x_2 + x_3, x_1 + x_2 + x_3)$, $B = ((0, 0, 1), (0, 1, 1), (1, 1, 1))$

9. $T: \mathbb{R}^4 \rightarrow \mathbb{R}^4$ defined by $T(x_1, x_2, x_3, x_4) = (x_4, x_3, x_2, x_1)$, $B = ((1, 1, 1, 1), (1, 1, 1, 0), (1, 1, 0, 0), (1, 0, 0, 0))$

10. $T: \mathbb{R}^5 \rightarrow \mathbb{R}^5$ defined by $T(x_1, x_2, x_3, x_4, x_5) = (x_1 + x_2, x_2 + x_3, x_3 + x_4, x_4 + x_5, x_1 + x_5)$, $B = (\mathbf{e}_5, \mathbf{e}_4, \mathbf{e}_3, \mathbf{e}_2, \mathbf{e}_1)$, where $\mathbf{e}_i$ is the ith unit coordinate vector.

11. Show that if $T: U \rightarrow V$ and $T': V \rightarrow W$ are linear transformations of vector spaces, then so is $T' \circ T: U \rightarrow W$.

In Exercises 12–14, find a formula for $T' \circ T$ in two ways: (a) by computing $T'(T(\mathbf{x}))$ and (b) by finding and multiplying appropriate matrix representations for T' and T.

12. T and T' defined on $\mathbb{R}^2$ by $T(x, y) = (x + y, 2x + y)$ and $T'(x, y) = (x + 3y, x - y)$

13. T and T' defined on $\mathbb{R}^3$ by $T(x_1, x_2, x_3) = (x_1 + 2x_2, x_2 + 2x_3, x_1 - x_3)$ and $T'(x_1, x_2, x_3) = (x_1 - x_2 + x_3, x_1 + x_2, 3x_1 - x_2 + 2x_3)$

14. T and T' defined on $\mathbb{R}^3$ by $T(x_1, x_2, x_3) = (x_1, -x_2, x_2)$ and $T'(x_1, x_2, x_3) = (x_1 + x_2, 0, x_1 + x_2 + x_3)$

15. Let T be a linear transformation of $\mathbb{R}^n$ into itself and assume that the standard matrix representation A_T of T is invertible. Use Theorem 6.10 to prove that the standard matrix representation of the inverse transformation T^{-1} is A_T^{-1}. That is, show that $A_{T^{-1}} = A_T^{-1}$.

16. Prove that the composition of functions is associative, that is, if $f: U \rightarrow V$, $g: V \rightarrow W$, and $h: W \rightarrow S$ are three functions, then $(h \circ g) \circ f = h \circ (g \circ f)$.

17. Review the notion of a *reflection* discussed following Exercise 18 in Section 6.2. Let T be the reflection of the plane in the line $ax + by = 0$, as shown in Fig. 6.5. This exercise shows how a change of basis can be used to find easily the standard matrix representation of T relative to the bases E,E.

    a) Find basis vectors $\mathbf{b}_1$ and $\mathbf{b}_2$ for $\mathbb{R}^2$ that are perpendicular and parallel, respectively, to the line $ax + by = 0$, as shown in Fig. 6.5.

    b) Let B be the ordered basis $(\mathbf{b}_1, \mathbf{b}_2)$. Find the matrix representation of T relative to B,B.

    c) Use the ideas of this section to find the standard matrix representation of T relative to E,E.

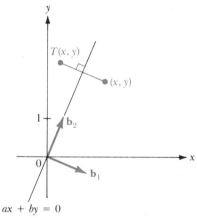

Figure 6.5
The reflection T of $\mathbb{R}^2$ in $ax + by = 0$.

18. Let T be the linear transformation of $\mathbb{R}^3$ giving the reflection in the plane $ax_1 + bx_2 + cx_3 = 0$. Proceed as in Exercise 17 to find the standard matrix representation of T relative to the bases E,E.

19. (This exercise shows that if A and B are $n \times n$ matrices such that $AB = I$, then $BA = I$, as stated in Section 1.5.) Let A and B be $n \times n$ matrices such that $AB = I$, the $n \times n$ identity matrix. Let T_A and T_B be the linear transformations of $\mathbb{R}^n$ into itself, having A and B, respectively, as matrix representations relative to E,E.

    a) Show that $T_A T_B$ is the identity map of $\mathbb{R}^n$ into $\mathbb{R}^n$.

    b) Deduce from (a) that T_B is a one-to-one transformation and that $T_A[\mathbb{R}^n] = \mathbb{R}^n$.

c) Show from (b) and the rank equation for linear transformations that $T_B[\mathbb{R}^n] = \mathbb{R}^n$ and T_A is a one-to-one transformation, so that both T_A and T_B are one-to-one maps having $\mathbb{R}^n$ as range.

d) Deduce from (c) that $T_B T_A$ is the identity map of $\mathbb{R}^n$ into $\mathbb{R}^n$ and that therefore $BA = I$.

6.4
Eigenvalues and Transformations

We have seen that linear transformations and matrices are closely related: $T(\mathbf{x})$ can be computed as $A\mathbf{x}$. In Section 4.6, we studied $A^k\mathbf{x}$ using eigenvalues and eigenvectors of a square matrix A. This section is devoted to eigenvalues and eigenvectors in the context of linear transformations.

Let $T: V \to V$ be a linear transformation of a vector space of finite dimension n *into itself*. Let $B = (\mathbf{b}_1, \mathbf{b}_2, \ldots, \mathbf{b}_n)$ be an ordered basis for V, and let A be the matrix representation of T relative to bases B,B. Recall from Section 4.5 that the $n \times n$ matrix A has eigenvalues that are solutions of the *characteristic equation* $\det(A - \lambda I) = 0$. For any eigenvalue λ of A, there exists a nonzero column eigenvector $\mathbf{x}$ in $\mathbb{R}^n$ such that $A\mathbf{x} = \lambda\mathbf{x}$. Consider now the vector

$$\mathbf{v} = x_1\mathbf{b}_1 + x_2\mathbf{b}_2 + \cdots + x_n\mathbf{b}_n$$

in V having $\mathbf{x}$ as column coordinate vector relative to the basis B. Since A is the matrix representation of T relative to bases B,B and $A\mathbf{x} = \lambda\mathbf{x}$, it must be the case that $T(\mathbf{v}) = \lambda\mathbf{v}$. We will call λ an *eigenvalue* and $\mathbf{v}$ a corresponding *eigenvector* of T.

It is significant that we can define eigenvalues and eigenvectors for a linear transformation $T: V \to V$ without any reference to a matrix representation, and without even assuming that V is finite-dimensional.

Definition 6.6   **Eigenvalues and Eigenvectors**

Let T be a linear transformation of a vector space V into itself. A scalar λ is an **eigenvalue** of T if there is a *nonzero* vector $\mathbf{v}$ in V such that $T(\mathbf{v}) = \lambda\mathbf{v}$. The vector $\mathbf{v}$ is then a corresponding **eigenvector** of T.

As shown by the discussion preceding the definition, we can find eigenvalues and eigenvectors of a linear transformation by finding the eigenvalues and eigenvectors of a matrix representation. We first give an example in $\mathbb{R}^3$, which will serve also to review the computation of eigenvalues and eigenvectors.

Example 1   Find the eigenvalues and corresponding eigenvectors of the linear transformation of $\mathbb{R}^3$ into itself defined by

$$T(x_1, x_2, x_3) = (3x_1 + 2x_2, \, 2x_1, \, x_1 + 2x_3).$$

Solution The standard matrix representation of T is

$$A = \begin{pmatrix} 3 & 2 & 0 \\ 2 & 0 & 0 \\ 1 & 0 & 2 \end{pmatrix}.$$

The *characteristic polynomial* of A is

$$p(\lambda) = \begin{vmatrix} 3 - \lambda & 2 & 0 \\ 2 & -\lambda & 0 \\ 1 & 0 & 2 - \lambda \end{vmatrix} = (2 - \lambda) \begin{vmatrix} 3 - \lambda & 2 \\ 2 & -\lambda \end{vmatrix}$$

$$= (2 - \lambda)(\lambda^2 - 3\lambda - 4) = (2 - \lambda)(\lambda - 4)(\lambda + 1).$$

Thus the eigenvalues of A are $\lambda = -1, 2,$ and 4.

The eigenvectors are obtained by reducing the coefficient matrix $A - \lambda I$. For $\lambda_1 = -1$, we have

$$A - \lambda_1 I = A + I = \begin{pmatrix} 4 & 2 & 0 \\ 2 & 1 & 0 \\ 1 & 0 & 3 \end{pmatrix}$$

$$\sim \begin{pmatrix} 1 & 0 & 3 \\ 0 & 1 & -6 \\ 0 & 2 & -12 \end{pmatrix} \sim \begin{pmatrix} 1 & 0 & 3 \\ 0 & 1 & -6 \\ 0 & 0 & 0 \end{pmatrix}.$$

The solution of the homogeneous system $(A + I)\mathbf{x} = \mathbf{0}$ is given by

$$\mathbf{v} = \begin{pmatrix} -3r \\ 6r \\ r \end{pmatrix} \quad \text{for any scalar } r.$$

An eigenvector corresponding to $\lambda_1 = -1$ is then

$$\mathbf{v}_1 = \begin{pmatrix} -3r \\ 6r \\ r \end{pmatrix} \quad \text{for any nonzero scalar } r.$$

For $\lambda_2 = 2$ we have

$$A - \lambda_2 I = A - 2I = \begin{pmatrix} 1 & 2 & 0 \\ 2 & -2 & 0 \\ 1 & 0 & 0 \end{pmatrix}$$

$$\sim \begin{pmatrix} 1 & 2 & 0 \\ 0 & -6 & 0 \\ 0 & -2 & 0 \end{pmatrix} \sim \begin{pmatrix} 1 & 2 & 0 \\ 0 & 1 & 0 \\ 0 & 0 & 0 \end{pmatrix},$$

which yields an eigenvector

$$\mathbf{v}_2 = \begin{pmatrix} 0 \\ 0 \\ r \end{pmatrix} \quad \text{for any nonzero scalar } r.$$

For $\lambda_3 = 4$, we have

$$A - \lambda_3 I = A - 4I = \begin{pmatrix} -1 & 2 & 0 \\ 2 & -4 & 0 \\ 1 & 0 & -2 \end{pmatrix}$$

$$\sim \begin{pmatrix} 1 & -2 & 0 \\ 0 & 0 & 0 \\ 0 & 2 & -2 \end{pmatrix} \sim \begin{pmatrix} 1 & -2 & 0 \\ 0 & 1 & -1 \\ 0 & 0 & 0 \end{pmatrix},$$

which yields an eigenvector

$$\mathbf{v}_3 = \begin{pmatrix} 2r \\ r \\ r \end{pmatrix} \quad \text{for any nonzero scalar } r.$$

In the statement of the example, we considered T to be acting on row vectors. Writing our results above in row notation, we see that $\lambda_1 = -1$, $\lambda_2 = 2$, and $\lambda_3 = 4$ are eigenvalues for T with corresponding eigenvectors

$$\mathbf{w}_1 = (-3r, 6r, r), \qquad \mathbf{w}_2 = (0, 0, s), \quad \text{and} \quad \mathbf{w}_3 = (2t, t, t)$$

for any choice of nonzero scalars r, s, and t. As a check of our work, we can verify that

$$T(\mathbf{w}_1) = -\mathbf{w}_1, \qquad T(\mathbf{w}_2) = 2\mathbf{w}_2, \quad \text{and} \quad T(\mathbf{w}_3) = 4\mathbf{w}_3. \quad \triangleleft$$

We now give an example involving a vector space that is not one of the spaces $\mathbb{R}^n$.

Example 2   Find all polynomials $p(x)$ of degree at most 2 such that $p(1 - x) = -p(x)$.

Solution   Let P_2 be the vector space of all polynomials of degree at most 2. With a moment of thought, we see that replacing x by $1 - x$ gives a linear transformation T of P_2 into itself.

Let B be the ordered basis $(1, x, x^2)$ of P_2. The values of T on the polynomials in this basis are given by

$$T(1) = 1, \qquad T(x) = 1 - x, \quad \text{and} \quad T(x^2) = 1 - 2x + x^2.$$

Thus the matrix representation of T relative to bases B, B is

$$A = \begin{pmatrix} 1 & 1 & 1 \\ 0 & -1 & -2 \\ 0 & 0 & 1 \end{pmatrix}.$$

The eigenvalues of this upper-triangular matrix are of course the values that appear on the diagonal. Since $\lambda_1 = -1$ is one of these eigenvalues, we see that there will be some nonzero polynomials $p(x)$ such that T maps $p(x)$ to $-p(x)$, that is, such that $p(1 - x) = -p(x)$. Computing these polynomials (eigenvectors), we find that

$$A - \lambda_1 I = \begin{pmatrix} 2 & 1 & 1 \\ 0 & 0 & -2 \\ 0 & 0 & 2 \end{pmatrix} \sim \begin{pmatrix} 2 & 1 & 1 \\ 0 & 0 & 1 \\ 0 & 0 & 0 \end{pmatrix}.$$

Eigenvectors of A corresponding to $\lambda_1 = -1$ are thus

$$\begin{pmatrix} -r \\ 2r \\ 0 \end{pmatrix}, \quad \text{where } r \neq 0.$$

Thus the polynomials satisfying $p(1 - x) = -p(x)$ in P_2 are 0 and the eigenvectors of T, namely, the polynomials of the form $-r1 + 2rx + 0x^2 = r(2x - 1)$. $\lhd$

EIGENVALUES AND EIGENVECTORS OF SIMILAR MATRICES

Let T be a linear transformation of a vector space V of finite dimension n. Our opening remarks indicate that λ is an eigenvalue of $T: V \to V$ in the sense of Definition 6.6 if and only if λ is an eigenvalue of the matrix representation A of T relative to bases B,B. But B may be chosen to be any basis of V. Thus all matrix representations of T must have the same eigenvalues, namely, the eigenvalues of T in the sense of Definition 6.6. This shows the significance of the remark preceding the definition that the eigenvalues of T can be defined without reference to a basis.

We know from the preceding section that matrix representations of T relative to different bases are similar matrices. That is, if A and D are both matrix representations of T relative to bases B,B and B',B' respectively, then we must have $D = C^{-1}AC$ for an invertible $n \times n$ matrix C. We also saw that any two similar matrices can be regarded as representing the same linear transformation relative to different bases. In conjunction with the preceding paragraph, we obtain this nice result:

Similar matrices have the same eigenvalues.

We take a moment to expand on the argument we have just given. Let V be a vector space of finite dimension n. The study of linear transformations $T: V \to V$ is essentially the same as the study of products $A\mathbf{x}$ of vectors $\mathbf{x}$ in $\mathbb{R}^n$ by $n \times n$ matrices A. We should understand this relationship thoroughly, and be able to bounce back and forth at will from $T(\mathbf{v})$ for $\mathbf{v}$ in V to $A\mathbf{x}$ for $\mathbf{x}$ in $\mathbb{R}^n$. This is illustrated in Example 2. Sometimes a theorem that is not immediately obvious from one point of view is quite easy from the other viewpoint. For example, it is not immediately apparent that similar matrices have the same eigenvalues from the matrix point of view; we ask for a matrix proof in Exercise 14. However, this becomes obvious from the linear-transformation point of view as explained above. On the other hand, it is not obvious from Definition 6.6 that a linear transformation of V can have at most n eigenvalues. However, this is easy to establish from the matrix point of view, for $A\mathbf{x} = \lambda\mathbf{x}$ has a nontrivial solution if and only if the coefficient matrix $A - \lambda I$ of the system $(A - \lambda I)\mathbf{x} = \mathbf{0}$ has determinant zero, and $\det(A - \lambda I) = 0$ is a polynomial equation of degree n having at most n solutions. We state a theorem that includes the displayed result above. Exercises 15 and 16 ask for proofs of the second and third statements in the theorem.

Theorem 6.13 Eigenvalues and Eigenvectors of Similar Matrices

Let A and D be similar $n \times n$ matrices, so that $D = C^{-1}AC$ for some invertible $n \times n$ matrix C. Let the eigenvalues of A be the (not necessarily distinct) numbers $\lambda_1, \lambda_2, \ldots, \lambda_n$.

i) The eigenvalues of D are also $\lambda_1, \lambda_2, \ldots, \lambda_n$.

ii) The algebraic and geometric multiplicity of each λ_i as an eigenvalue of A is the same as when viewed as an eigenvalue of D.

iii) If $\mathbf{v}_i$ in $\mathbb{R}^n$ is an eigenvector of the matrix A corresponding to λ_i, then $C^{-1}\mathbf{v}_i$ is an eigenvector of the matrix D corresponding to λ_i.

If λ is an eigenvalue of $T: V \to V$, then Exercise 17 shows that the subset of V consisting of the zero vector and all eigenvectors for λ is a subspace of V, the **eigenspace** E_λ of λ. Of course, this concept parallels the eigenspace E_λ in $\mathbb{R}^n$ of an $n \times n$ matrix A.

DIAGONALIZATION

Let $T: V \to V$ be a linear transformation of a vector space V of finite dimension n. Suppose that there exists an ordered basis $B = (\mathbf{b}_1, \mathbf{b}_2, \ldots, \mathbf{b}_n)$ of V comprised of eigenvectors of T. Let the eigenvalue corresponding to $\mathbf{b}_i$ be λ_i. Then the matrix representation of T relative to bases B,B has the simple diagonal form

$$D = \begin{pmatrix} \lambda_1 & & & \\ & \lambda_2 & & \bigcirc \\ & & \ddots & \\ & \bigcirc & & \lambda_n \end{pmatrix}.$$

We make a definition for linear transformations that clearly parallels one for matrices.

Definition 6.7 Diagonalizable Transformation

A linear transformation T of a finite-dimensional vector space V into itself is **diagonalizable** if there exists an ordered basis B of V such that the matrix representation D of T relative to B,B is a diagonal matrix.

From the work of Section 4.6 with matrices, we realize that $T: V \to V$ is diagonalizable if and only if V has a basis B consisting of eigenvectors of T, and that this will be the case if and only if the geometric multiplicity of each eigenvalue coincides with its algebraic multiplicity. Note that while the geometric multiplicity of an eigenvalue λ of T can be defined directly in terms of T and V, namely, as the dimension of the eigenspace E_λ, we must take a matrix representation of T to define the algebraic multiplicity of λ. By Theorem 6.13, it doesn't matter which matrix representation we take.

If V has a basis B of eigenvectors, then it becomes easy to compute $T^k(\mathbf{v})$ for any positive integer k and any vector $\mathbf{v}$ in V. We need only find the coordinate vector $\mathbf{d}$ in $\mathbb{R}^n$ of $\mathbf{v}$ relative to the ordered basis B, so that

$$\mathbf{v} = d_1\mathbf{b}_1 + d_2\mathbf{b}_2 + \cdots + d_n\mathbf{b}_n.$$

Then

$$T^k(\mathbf{v}) = d_1\lambda_1{}^k\mathbf{b}_1 + d_2\lambda_2{}^k\mathbf{b}_2 + \cdots d_n\lambda_n{}^k\mathbf{b}_n.$$

Of course, this is the transformation analogue of the computation of $A^k\mathbf{x}$ in Section 4.6. We illustrate with an example.

Example 3 Consider the vector space P_2 of polynomials of degree at most 2, and let B' be the ordered basis $(1, x, x^2)$ for P_2. Let $T: P_2 \to P_2$ be the linear transformation such that

$$T(1) = 3 + 2x + x^2, \quad T(x) = 2, \quad \text{and} \quad T(x^2) = 2x^2.$$

Find $T^4(x + 2)$ and discuss the behavior of $T^k(p(x))$ for any polynomial $p(x)$ in P_2 and large positive integers k.

Solution The matrix representation of T relative to B',B' is

$$A = \begin{pmatrix} 3 & 2 & 0 \\ 2 & 0 & 0 \\ 1 & 0 & 2 \end{pmatrix},$$

which is the matrix of Example 1. From Example 1, we find that we have as eigenvalues and corresponding eigenvectors

Eigenvalues	Eigenvectors of A	Eigenvectors of T
$\lambda_1 = -1,$	$\mathbf{w}_1 = (-3, 6, 1),$	$p_1(x) = -3 + 6x + x^2,$
$\lambda_2 = 2,$	$\mathbf{w}_2 = (0, 0, 1),$	$p_2(x) = x^2,$
$\lambda_3 = 4,$	$\mathbf{w}_3 = (2, 1, 1),$	$p_3(x) = 2 + x + x^2.$

Let B be the ordered basis $(-3 + 6x + x^2, x^2, 2 + x + x^2)$ consisting of these eigenvectors. We can find the coordinate vector (d_1, d_2, d_3) of $x + 2$ relative to the basis B by solving the linear system corresponding to the augmented matrix.

$$\begin{pmatrix} -3 & 0 & 2 & | & 2 \\ 6 & 0 & 1 & | & 1 \\ 1 & 1 & 1 & | & 0 \end{pmatrix} \sim \begin{pmatrix} 1 & 1 & 1 & | & 0 \\ -3 & 0 & 2 & | & 2 \\ 6 & 0 & 1 & | & 1 \end{pmatrix} \sim \begin{pmatrix} 1 & 1 & 1 & | & 0 \\ 0 & 3 & 5 & | & 2 \\ 0 & -6 & -5 & | & 1 \end{pmatrix}$$

$$\sim \begin{pmatrix} 1 & 1 & 1 & | & 0 \\ 0 & 3 & 5 & | & 2 \\ 0 & 0 & 5 & | & 5 \end{pmatrix}.$$

We see that $d_3 = 1$, $d_2 = -1$, and $d_1 = 0$. Thus

$$x + 2 = 0(x^2 + 6x - 3) + (-1)x^2 + (1)(x^2 + x + 2)$$

so

$$T^k(x + 2) = 2^k(-1)x^2 + 4^k(1)(x^2 + x + 2).$$

In particular,

$$T^4(x + 2) = -16x^2 + 256(x^2 + x + 2) = 240x^2 + 256x + 512.$$

For a general polynomial $p(x)$ in P_2 with coordinate vector $\mathbf{d}$ relative to B, we have

$$T^k(p(x)) = d_1(-1)^k(x^2 + 6x - 3) + d_2 2^k x^2 + d_3 4^k(x^2 + x + 2).$$

Now let k take on larger and larger integer values. The exponential term 4^k dominates 2^k, which in turn dominates $(-1)^k$. Thus if $d_3 \neq 0$ we see that $T^k(p(x))$ for large k is approximately $4^k(x^2 + x + 2)$. This is illustrated in our computation of $T^4(x + 2)$ above, even though 4 cannot be regarded as a very large value of k. If $d_3 = 0$ but $d_2 \neq 0$, then $T^k(p(x))$ is approximately $2^k x^2$. Finally, if $d_2 = d_3 = 0$ and $d_1 \neq 0$, then $T^k(p(x))$ is $d_1(-1)^k(x^2 + 6x - 3)$.

$\triangleleft$

Topic 8 of the available program VECTGRPH is designed for graphic illustration and drill on the dominance of an eigenvalue of maximum magnitude, which Example 3 illustrates.

Sometimes one wants to find a formula for a transformation that has certain eigenvector and eigenvalues. Example 4 illustrates this.

Example 4 Consider a line through the origin in the x,y-plane $\mathbb{R}^2$. Mapping each point (x, y) in the plane into its mirror image in the line is a linear transformation $T: \mathbb{R}^2 \to \mathbb{R}^2$ known as the **reflection** of the plane in the line. (See the note after Exercises 13–18 of Section 6.2.) Figure 6.6 illustrates the reflection of the plane in the line $x + 2y = 0$. Find a formula for $T(x, y)$ if T is the reflection of $\mathbb{R}^2$ in this line.

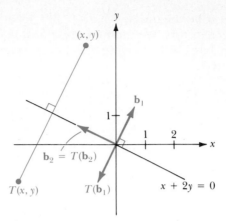

Figure 6.6
The reflection T of $\mathbb{R}^2$ in $x + 2y = 0$.

Solution From Fig. 6.6 we see that for this reflection a vector $\mathbf{b}_1$ perpendicular to $x + 2y = 0$ should be an eigenvector with eigenvalue -1, while a vector $\mathbf{b}_2$ along the line should be an eigenvector with eigenvalue 1. Viewing $x + 2y$ as the equation of a hyperplane, we see that $\mathbf{b}_1 = (1, 2)$ is a vector perpendicular to the line. Of course $\mathbf{b}_2 = (2, -1)$ may be taken as a vector along the line perpendicular to $\mathbf{b}_1$, because $\mathbf{b}_1 \cdot \mathbf{b}_2 = 0$. Thus the matrix representation of T relative to $B = (\mathbf{b}_1, \mathbf{b}_2)$ is

$$D = \begin{pmatrix} -1 & 0 \\ 0 & 1 \end{pmatrix}.$$

We want to find the standard matrix representation A of T relative to E,E where $E = (\mathbf{e}_1, \mathbf{e}_2)$. We draw the diagram

$$\mathbb{R}^2 \underset{E}{\xrightarrow[\;C^{-1}\;]{\iota}} \mathbb{R}^2 \underset{B}{\xrightarrow[\;D\;]{T}} \mathbb{R}^2 \underset{B}{\xrightarrow[\;C\;]{\iota}} \mathbb{R}^2$$

so we can keep track of what is happening. Here C is the matrix representation of the identity transformation ι relative to B,E and so

$$C = \begin{pmatrix} 1 & 2 \\ 2 & -1 \end{pmatrix}.$$

Then

$$C^{-1} = \begin{pmatrix} \frac{1}{5} & \frac{2}{5} \\ \frac{2}{5} & -\frac{1}{5} \end{pmatrix}$$

$$A = CDC^{-1} = \begin{pmatrix} 1 & 2 \\ 2 & -1 \end{pmatrix} \begin{pmatrix} -1 & 0 \\ 0 & 1 \end{pmatrix} \begin{pmatrix} \frac{1}{5} & \frac{2}{5} \\ \frac{2}{5} & -\frac{1}{5} \end{pmatrix}$$

$$= \begin{pmatrix} -1 & 2 \\ -2 & -1 \end{pmatrix} \begin{pmatrix} \frac{1}{5} & \frac{2}{5} \\ \frac{2}{5} & -\frac{1}{5} \end{pmatrix} = \begin{pmatrix} \frac{3}{5} & -\frac{4}{5} \\ -\frac{4}{5} & -\frac{3}{5} \end{pmatrix}.$$

Thus a formula for T is $T(x, y) = (\frac{3}{5}x - \frac{4}{5}y, -\frac{4}{5}x - \frac{3}{5}y)$. As a check, we easily compute that $T(1, 2) = (-1, -2)$ and $T(2, -1) = (2, -1)$. ◁

SUMMARY

1. Let T be a linear transformation of a vector space V into itself. A nonzero $\mathbf{v}$ in V is an eigenvector of T if $T(\mathbf{v}) = \lambda\mathbf{v}$ for some scalar λ, called the eigenvalue of T corresponding to $\mathbf{v}$.

2. If T is a linear transformation of a vector space V of finite dimension n into itself, then the eigenvalues of T are the eigenvalues of every matrix representation of T.

3. If A and D are similar $n \times n$ matrices so that $D = C^{-1}AC$, where C is an invertible $n \times n$ matrix, then A and D have the same eigenvalues with the same geometric and algebraic multiplicities. If $\mathbf{v}$ is an eigenvector of A, then $C^{-1}\mathbf{v}$ is an eigenvector of D.

4. A linear transformation T of a finite-dimensional vector space V into itself is diagonalizable if and only if V has a basis of eigenvectors of T. This is the case if and only if the geometric multiplicity of each eigenvalue is equal to its algebraic multiplicity.

EXERCISES

In Exercises 1–6, find the eigenvalues λ_i and corresponding eigenvectors $\mathbf{v}_i$ of the linear transformation T. Determine whether the linear transformation is diagonalizable.

1. T defined on $\mathbb{R}^2$ by $T(x, y) = (2x - 3y, -3x + 2y)$

2. T defined on $\mathbb{R}^2$ by $T(x, y) = (x - y, -x + y)$

3. T defined on $\mathbb{R}^3$ by $T(x_1, x_2, x_3) = (x_1 + x_3, x_2, x_1 + x_3)$

4. T defined on $\mathbb{R}^3$ by $T(x_1, x_2, x_3) = (x_1, 4x_2 + 7x_3, 2x_2 - x_3)$

5. T defined on $\mathbb{R}^3$ by $T(x_1, x_2, x_3) = (x_1, -5x_1 + 3x_2 - 5x_3, -3x_1 - 2x_2)$

6. T defined on $\mathbb{R}^3$ by $T(x_1, x_2, x_3) = (3x_1 - x_2 + x_3, -2x_1 + 2x_2 - x_3, 2x_1 + x_2 + 4x_3)$

In Exercises 7–12, find all polynomials $p(x)$ in the indicated vector space that satisfy the given condition, if any such polynomials exist.

7. $p(2x + 1) = 4p(x)$ in P_2

8. $p(2x + 1) = 2p(x)$ in P_2

9. $p(2x + 1) = 3p(x)$ in P_2

10. $p(1 - 3x) = p(x)$ in P_3

11. $p(1 - 3x) = 3p(x)$ in P_3

12. $p(1 - 3x) = 9p(x)$ in P_3

13. Find in terms of a and b all numbers λ such that there exists a nonzero polynomial $p(x)$ for which $p(ax + b) = \lambda p(x)$ for $p(x)$ in (a) P_2; (b) P_3.

14. Give a matrix proof that similar matrices have the same eigenvalues.

15. Prove statement (ii) of Theorem 6.13. [*Hint:* Note that $C^{-1}AC - \lambda I = C^{-1}AC - C^{-1}IC$.]

16. Prove statement (iii) of Theorem 6.13.

17. Let λ be an eigenvalue of a linear transformation $T: V \to V$. Show from Definition 6.6 that the zero vector together with all eigenvectors for λ in V form a subspace of V.

19. Let $T: \mathbb{R}^2 \to \mathbb{R}^2$ be the reflection in the line $ax + by = 0$, where a and b are not both zero. Find a formula for $T(x, y)$ in terms of a, b, x, and y. [*Hint:* See Example 4.]

21. Let $T: \mathbb{R}^3 \to \mathbb{R}^3$ be the reflection of $\mathbb{R}^3$ in the plane $x_1 + 2x_2 - x_3 = 0$. Find $T(x_1, x_2, x_3)$.

23. Let $T: \mathbb{R}^n \to \mathbb{R}^n$ be a linear transformation. Give conditions on the eigenvalues and eigenvectors of T for the transformation to be a reflection in a hyperplane of $\mathbb{R}^n$.

25. Let A be an $n \times n$ matrix whose characteristic polynomial $p(\lambda)$ factors into linear factors, so that A has n (not necessarily distinct) eigenvalues $\lambda_1, \lambda_2, \ldots, \lambda_n$. Show that

$$\det(A) = \text{Constant term of } p(\lambda)$$
$$= \lambda_1 \lambda_2 \ldots \lambda_n.$$

18. Let $T: \mathbb{R}^2 \to \mathbb{R}^2$ be the reflection in the line $2x + 3y = 0$. Find a formula for $T(x, y)$. [*Hint:* See Example 4.]

20. Consider a line $ax + by = 0$ through the origin in the plane $\mathbb{R}^2$. For each point (x, y) in the plane but not on the line, let $T(x, y)$ be the point on the same side of the line twice as far from the line as (x, y) is and such that the line through (x, y) and $T(x, y)$ is perpendicular to $ax + by = 0$. If (x, y) is on the line, then $T(x, y) = (x, y)$. Argue that T is a linear transformation of $\mathbb{R}^2$. Find a formula for $T(x, y)$ in terms of a, b, x, and y.

22. Let $T: \mathbb{R}^3 \to \mathbb{R}^3$ be the reflection of $\mathbb{R}^3$ in the plane $ax_1 + bx_2 + cx_3 = 0$, where a, b, c are not all zero. Find $T(x_1, x_2, x_3)$ in terms of a, b, c, x_1, x_2, and x_3.

24. Show that if $p(\lambda)$ is the characteristic polynomial of a square matrix A, then the constant term of $p(\lambda)$ is $\det(A)$.

26. The **trace** of an $n \times n$ matrix A is defined by

$$\text{tr}(A) = a_{11} + a_{22} + \cdots + a_{nn}.$$

Let the characteristic polynomial $p(\lambda)$ factor into linear factors, so that A has n (not necessarily distinct) eigenvalues $\lambda_1, \lambda_2, \ldots, \lambda_n$. Show that

$$\text{tr}(A) = (-1)^{n-1} [\text{Coefficient of } \lambda^{n-1} \text{ in } p(\lambda)]$$
$$= \lambda_1 + \lambda_2 + \cdots + \lambda_n.$$

27. ▣ Work with Topic 8 of the available program VECTGRPH until a grade of at least 80 percent can be achieved consistently.

6.5
Symmetric and Orthogonal Matrices

Recall that a symmetric matrix is an $n \times n$ square matrix A in which the kth row vector is equal to the kth column vector for each $k = 1, 2, \ldots, n$. Equivalently, A is symmetric if and only if it is equal to its transpose A^T. The problem of diagonalizing a symmetric matrix arises in many applications (see Section 7.1, for example). As we stated in Theorem 4.12 this diagonalization can always be achieved. In fact, we will see that diagonalization of a symmetric matrix A can always be achieved in a particularly nice way: One can find an *orthogonal* matrix C such that $C^{-1}AC = D$ is a diagonal matrix. We begin by proving the perpendicularity of two eigenvectors of a symmetric matrix that correspond to distinct eigenvalues.

Theorem 6.14 Orthogonality of Eigenspaces of a Symmetric Matrix

Eigenvectors of a symmetric matrix that correspond to different eigenvalues are orthogonal. That is, the eigenspaces of a symmetric matrix are orthogonal.

Proof Let A be an $n \times n$ symmetric matrix, and let $\mathbf{v}_1$ and $\mathbf{v}_2$ be eigenvectors corresponding to distinct eigenvalues λ_1 and λ_2, respectively. Writing vectors as column vectors, we have

$$A\mathbf{v}_1 = \lambda_1\mathbf{v}_1 \quad \text{and} \quad A\mathbf{v}_2 = \lambda_2\mathbf{v}_2.$$

To show that $\mathbf{v}_1$ and $\mathbf{v}_2$ are orthogonal, we compute

$$\lambda_1(\mathbf{v}_1 \cdot \mathbf{v}_2) = (\lambda_1\mathbf{v}_1) \cdot \mathbf{v}_2 = (A\mathbf{v}_1) \cdot \mathbf{v}_2.$$

The final dot product can be written in matrix form as

$$(A\mathbf{v}_1)^T\mathbf{v}_2 = (\mathbf{v}_1^T A^T)\mathbf{v}_2.$$

Therefore,

$$\lambda_1(\mathbf{v}_1 \cdot \mathbf{v}_2) = \mathbf{v}_1^T A^T \mathbf{v}_2. \tag{1}$$

In a similar way (see Exercise 9), we find that

$$\lambda_2(\mathbf{v}_1 \cdot \mathbf{v}_2) = \mathbf{v}_1^T A \mathbf{v}_2. \tag{2}$$

Since $A = A^T$, Eqs. (1) and (2) show that

$$\lambda_1(\mathbf{v}_1 \cdot \mathbf{v}_2) = \lambda_2(\mathbf{v}_1 \cdot \mathbf{v}_2) \quad \text{or} \quad (\lambda_1 - \lambda_2)(\mathbf{v}_1 \cdot \mathbf{v}_2) = 0.$$

Since $\lambda_1 - \lambda_2 \neq 0$, we conclude that $\mathbf{v}_1 \cdot \mathbf{v}_2 = 0$, which shows that $\mathbf{v}_1$ is orthogonal to $\mathbf{v}_2$. ■

Theorem 6.14 and the results stated for symmetric matrices in Section 4.6 tell us that an $n \times n$ real symmetric matrix A has real numbers for all its eigenvalues and has n mutually perpendicular real eigenvectors. These n eigenvectors can be normalized to length 1, and provide an orthonormal basis for $\mathbb{R}^n$ consisting of n eigenvectors of A. Thus A is diagonalizable and a diagonalizing matrix C such that $D = C^{-1}AC$ can be chosen with columns consisting of the vectors in this basis for $\mathbb{R}^n$. This matrix C is an *orthogonal matrix* as defined in Section 5.3. We summarize our discussion in a theorem.

Theorem 6.15 Fundamental Theorem of Real Symmetric Matrices

Every real symmetric matrix A is diagonalizable. The diagonalization $D = C^{-1}AC$ can be achieved using a real orthogonal matrix C.

The orthogonal property of the $n \times n$ change-of-basis matrix C in the equation $D = C^{-1}AC$ of Theorem 6.15 is very important in applications. We will see in a moment that a dot product of two vectors in $\mathbb{R}^n$ can be computed equally well by the familiar formula using either its usual coordinates relative to the standard basis E or its coordinates relative to the basis consisting of the column vectors in C.

The converse of Theorem 6.15 is also true. If $D = C^{-1}AC$ is a diagonal matrix and C is an orthogonal matrix, then A is symmetric (see Exercise 7). The equation $D = C^{-1}AC$ is said to be an **orthogonal diagonalization** of A.

Example 1 Find an orthogonal diagonalization of the matrix

$$A = \begin{pmatrix} 1 & 2 \\ 2 & 4 \end{pmatrix}.$$

Solution The eigenvalues of A are the roots of the characteristic equation

$$\begin{vmatrix} 1 - \lambda & 2 \\ 2 & 4 - \lambda \end{vmatrix} = \lambda^2 - 5\lambda = 0.$$

They are $\lambda_1 = 0$ and $\lambda_2 = 5$. We proceed to find the corresponding eigenvectors of A.

For $\lambda_1 = 0$, we have

$$A - \lambda_1 I = A = \begin{pmatrix} 1 & 2 \\ 2 & 4 \end{pmatrix} \sim \begin{pmatrix} 1 & 2 \\ 0 & 0 \end{pmatrix},$$

which yields the eigenspace

$$E_0 = \text{sp}\begin{pmatrix} -2 \\ 1 \end{pmatrix}.$$

For $\lambda_2 = 5$, we have

$$A - \lambda_2 I = A - 5I = \begin{pmatrix} -4 & 2 \\ 2 & -1 \end{pmatrix} \sim \begin{pmatrix} 2 & -1 \\ 0 & 0 \end{pmatrix},$$

which yields the eigenspace

$$E_5 = \text{sp}\begin{pmatrix} 1 \\ 2 \end{pmatrix}.$$

Thus

$$\begin{pmatrix} 0 & 0 \\ 0 & 5 \end{pmatrix} = \begin{pmatrix} -2 & 1 \\ 1 & 2 \end{pmatrix}^{-1} \begin{pmatrix} 1 & 2 \\ 2 & 4 \end{pmatrix} \begin{pmatrix} -2 & 1 \\ 1 & 2 \end{pmatrix}$$

is a diagonalization of A, and

$$\begin{pmatrix} 0 & 0 \\ 0 & 5 \end{pmatrix} = \begin{pmatrix} -2/\sqrt{5} & 1/\sqrt{5} \\ 1\sqrt{5} & 2/\sqrt{5} \end{pmatrix}^{-1} \begin{pmatrix} 1 & 2 \\ 2 & 4 \end{pmatrix} \begin{pmatrix} -2/\sqrt{5} & 1/\sqrt{5} \\ 1/\sqrt{5} & 2/\sqrt{5} \end{pmatrix}$$

is an orthogonal diagonalization of A. ◁

Example 2   Show that an orthogonal diagonalization of the matrix

$$A = \begin{pmatrix} 1 & -1 & -1 \\ -1 & 1 & -1 \\ -1 & -1 & 1 \end{pmatrix}$$

is provided by the matrix

$$C = \begin{pmatrix} 1/\sqrt{3} & -1/\sqrt{2} & -1/\sqrt{6} \\ 1/\sqrt{3} & 1/\sqrt{2} & -1/\sqrt{6} \\ 1/\sqrt{3} & 0 & 2/\sqrt{6} \end{pmatrix}.$$

Solution   Clearly C is an orthogonal matrix so $C^{-1} = C^T$. The obvious thing to do is simply to compute $C^{-1}AC = C^TAC$ and observe that the result is the diagonal matrix

$$D = \begin{pmatrix} -1 & 0 & 0 \\ 0 & 2 & 0 \\ 0 & 0 & 2 \end{pmatrix}.$$

An alternative and slightly easier approach is to consider the column vectors $\mathbf{c}_1$, $\mathbf{c}_2$, and $\mathbf{c}_3$ of C, and simply compute $A\mathbf{c}_1$, $A\mathbf{c}_2$, and $A\mathbf{c}_3$. We easily find that

$$A\mathbf{c}_1 = -\mathbf{c}_1, \qquad A\mathbf{c}_2 = 2\mathbf{c}_2, \quad \text{and} \quad A\mathbf{c}_3 = 2\mathbf{c}_3.$$

Thus $(\mathbf{c}_1, \mathbf{c}_2, \mathbf{c}_3)$ is an ordered basis of mutually perpendicular eigenvectors of A, so $C^{-1}AC$ must be a diagonal matrix, and in fact must be the matrix D given above. ◁

ORTHOGONAL MATRICES

Recall that an $n \times n$ orthogonal matrix is one whose columns form an orthonormal basis for $\mathbb{R}^n$. Equivalently, A is an orthogonal matrix if and only if $A^TA = I$. We conclude this section with a brief discussion of linear transformations that can be represented by an orthogonal matrix.

First we show that any orthogonal matrix A gives rise to a linear transformation $T(\mathbf{x}) = A\mathbf{x}$ that preserves the dot product. Writing vectors as column vectors, we have

$$T(\mathbf{v}) \cdot T(\mathbf{w}) = (A\mathbf{v}) \cdot (A\mathbf{w}),$$

which can be written in matrix form as $(A\mathbf{v})^T(A\mathbf{w}) = \mathbf{v}^T(A^TA)\mathbf{w} = \mathbf{v}^T\mathbf{w}$, that is,

$$T(\mathbf{v}) \cdot T(\mathbf{v}) = \mathbf{v} \cdot \mathbf{w}. \tag{3}$$

Definition 6.8   Orthogonal Linear Transformation

A linear transformation T of $\mathbb{R}^n$ into itself is **orthogonal** if it satisfies $T(\mathbf{v}) \cdot T(\mathbf{w}) = \mathbf{v} \cdot \mathbf{w}$ for all vectors $\mathbf{v}$ and $\mathbf{w}$ in $\mathbb{R}^n$.

We just showed that every orthogonal matrix gives rise to an orthogonal linear transformation. The converse is also true.

Theorem 6.16   Orthogonality

A linear transformation T of $\mathbb{R}^n$ into itself is orthogonal if and only if its standard matrix representation A_T is an orthogonal matrix.

Proof   It remains to show that A_T is an orthogonal matrix if T preserves the dot product. The columns of A_T are the vectors $T(\mathbf{e}_1)$, $T(\mathbf{e}_2)$, . . . , $T(\mathbf{e}_n)$, where $\mathbf{e}_j$ is the jth unit coordinate vector of $\mathbb{R}^n$. We have

$$T(\mathbf{e}_i) \cdot T(\mathbf{e}_j) = \mathbf{e}_i \cdot \mathbf{e}_j = \begin{cases} 0 & \text{if } i = j, \\ 1 & \text{if } i \neq j, \end{cases}$$

showing that the columns of A_T form an orthonormal basis of $\mathbb{R}^n$. Thus A_T is an orthogonal matrix. ∎

Example 3   Show that the linear transformation $T: \mathbb{R}^3 \to \mathbb{R}^3$ defined by $T(x_1, x_2, x_3) = (x_1/\sqrt{2} + x_3/\sqrt{2}, x_2, -x_1/\sqrt{2} + x_3/\sqrt{2})$ is orthogonal.

Solution   This follows from the fact that the standard matrix representation

$$A_T = \begin{pmatrix} 1/\sqrt{2} & 0 & -1/\sqrt{2} \\ 0 & 1 & 0 \\ 1/\sqrt{2} & 0 & 1/\sqrt{2} \end{pmatrix}$$

is an orthogonal matrix.   ◁

Our next theorem explains why an orthogonal linear transformation is sometimes referred to as *unitary*.

Theorem 6.17   Preservation of Unit Vectors

A linear transformation $T: \mathbb{R}^n \to \mathbb{R}^n$ is orthogonal if and only if T maps unit vectors into unit vectors, so that $\|\mathbf{u}\| = 1$ implies $\|T(\mathbf{u})\| = 1$ for all unit vectors $\mathbf{u}$ in $\mathbb{R}^n$.

Proof   First assume that T is an orthogonal matrix. Let $\mathbf{u}$ be a vector in $\mathbb{R}^n$ such that $\|\mathbf{u}\| = 1$. Then $\mathbf{u} \cdot \mathbf{u} = \|\mathbf{u}\|^2 = 1$ and therefore $\|T(\mathbf{u})\|^2 = T(\mathbf{u}) \cdot T(\mathbf{u}) = 1$ so that $T(\mathbf{u})$ is a unit vector.

Conversely, let T be a linear transformation that preserves unit vectors. If $\mathbf{v}$ is any nonzero vector in $\mathbb{R}^n$, then $\mathbf{v}/\|\mathbf{v}\|$ is a unit vector, so $T(\mathbf{v}/\|\mathbf{v}\|) = (1/\|\mathbf{v}\|)T(\mathbf{v})$ is also a unit vector. This means that $1 = \|(1/\|\mathbf{v}\|)T(\mathbf{v})\| = (1/\|\mathbf{v}\|)\|T(\mathbf{v})\|$ from which we conclude that $\|T(\mathbf{v})\| = \|\mathbf{v}\|$. Thus T preserves the norm. That T preserves the dot product and hence is orthogonal follows immediately from the identity

$$\mathbf{v} \cdot \mathbf{w} = (1/4)[\|\mathbf{v} + \mathbf{w}\|^2 - \|\mathbf{v} - \mathbf{w}\|^2] \tag{4}$$

whose proof is left as an exercise (see Exercise 11). ∎

Example 4   Show that the linear transformation that rotates the plane counterclockwise through any angle is an orthogonal linear transformation.

Solution   Clearly a rotation of the plane preserves the lengths of all vectors, in particular unit vectors. Example 7 of Section 6.1 shows that a rotation is a linear transformation, and Theorem 6.17 then shows that the transformation is orthogonal. ◁

Finally, we note that besides preserving lengths of vectors, an orthogonal transformation T of $\mathbb{R}^n$ into itself preserves angles between vectors. That is, the angle between $\mathbf{v}_1$ and $\mathbf{v}_2$ is the same as the angle between $T(\mathbf{v}_1)$ and $T(\mathbf{v}_2)$ for all choices of $\mathbf{v}_1$ and $\mathbf{v}_2$ in $\mathbb{R}^n$. This follows at once from the fact that T preserves dot products and lengths, and the angle between $\mathbf{v}_1$ and $\mathbf{v}_2$ can be computed using the formula $\cos^{-1}[(\mathbf{v}_1 \cdot \mathbf{v}_2)/(\|\mathbf{v}_1\|\|\mathbf{v}_2\|)]$.

SUMMARY

Let A be an $n \times n$ real symmetric matrix.

1. All eigenvalues of A are real numbers.

2. The eigenspaces of A are mutually orthogonal and A has n mutually perpendicular eigenvectors.

3. The matrix A is diagonalizable by an orthogonal matrix C. That is, there is an orthogonal matrix C such that $D = C^{-1}AC$ is a diagonal matrix.

4. A linear transformation of $\mathbb{R}^n$ into itself is said to be orthogonal if it preserves the dot product, or equivalently, if its standard matrix is orthogonal, or equivalently, if it maps unit vectors into unit vectors.

EXERCISES

In Exercises 1–6, find an orthogonal matrix C such that $D = C^{-1}AC$ is a diagonalization of the given symmetric matrix A.

1. $\begin{pmatrix} 1 & 2 \\ 2 & 1 \end{pmatrix}$

2. $\begin{pmatrix} 3 & 2 \\ 2 & 0 \end{pmatrix}$

3. $\begin{pmatrix} 2 & 1 & 0 \\ 1 & 2 & 1 \\ 0 & 1 & 2 \end{pmatrix}$

4. $\begin{pmatrix} 0 & 1 & 1 \\ 1 & 2 & 1 \\ 1 & 1 & 0 \end{pmatrix}$

5. $\begin{pmatrix} 0 & 1 & 1 & 0 \\ 1 & -2 & 2 & 1 \\ 1 & 2 & -2 & 1 \\ 0 & 1 & 1 & 0 \end{pmatrix}$

6. $\begin{pmatrix} 1 & -1 & 0 & 0 \\ -1 & 1 & 0 & 0 \\ 0 & 0 & 1 & 3 \\ 0 & 0 & 3 & 1 \end{pmatrix}$

7. Let $D = C^{-1}AC$ be a diagonal matrix with C an orthogonal matrix. Prove that A must be symmetric.

8. Prove that the product of two $n \times n$ symmetric matrices is also symmetric.

9. Verify Eq. (2) in the text.

10. Show that the columns of an $n \times n$ matrix A form an orthonormal basis for $\mathbb{R}^n$ if and only if the rows of A form an orthonormal basis for $\mathbb{R}^n$.

11. Prove Eq. (4) of the text.

12. Show that if A is an orthogonal matrix, then $\det(A) = \pm 1$.

13. Find a 2×2 matrix with determinant 1 that is not an orthogonal matrix.

In Exercises 14–19, determine whether the given linear transformation is orthogonal.

14. $T: \mathbb{R}^2 \to \mathbb{R}^2$ defined by $T(x,y) = (y, x)$

15. $T: \mathbb{R}^3 \to \mathbb{R}^3$ defined by $T(x, y, z) = (x, y, 0)$

16. $T: \mathbb{R}^2 \to \mathbb{R}^2$ defined by $T(x, y) = (2x, y)$

17. $T: \mathbb{R}^2 \to \mathbb{R}^2$ defined by $T(x, y) = (x, -y)$

18. $T: \mathbb{R}^2 \to \mathbb{R}^2$ defined by $T(x, y) = (x/2 - \sqrt{3}y/2, -\sqrt{3}x/2 + y/2)$

19. $T: \mathbb{R}^3 \to \mathbb{R}^3$ defined by $T(x, y, z) = (x/3 + 2y/3 + 2z/3, -2x/3 - y/3 + 2z/3, -2x/3 + 2y/3 - z/3)$

20. Show that the eigenvalues of an orthogonal matrix must be 1 or -1. [*Hint:* Think in terms of linear transformations.]

21. For the orthogonal matrix C found in Exercise 1, verify directly that the linear map $T(\mathbf{x}) = C\mathbf{x}$ preserves the dot product.

22. Repeat Exercise 21 for the matrix C found in Exercise 3.

6.6
Volume-change Factor for a Linear Transformation

In this section, we will be working with the volume of *sufficiently nice regions* in $\mathbb{R}^n$. To define such a notion carefully would require an excursion into calculus of several variables. We will simply assume that we all have an intuitive notion of such regions and their volume.

Let $T: \mathbb{R}^n \to \mathbb{R}^n$ be a linear transformation and let A be its standard matrix representation. Recall that the jth column vector of A is $T(\mathbf{e}_j)$. We will show that if a region G in $\mathbb{R}^n$ has volume V, then the image of G under T has volume

$$|\det(A)| \cdot V.$$

That is, the volume of a region is multiplied by $|\det(A)|$ when the region is transformed by T. This result has important applications to integral calculus, as we indicate in Chapter 8.

As a first step toward this result, we examine a special case. The volume of the *unit n-cube* determined by $\mathbf{e}_1, \mathbf{e}_2, \ldots, \mathbf{e}_n$ in $\mathbb{R}^n$ is 1. To find the volume of the box in $\mathbb{R}^n$ determined by the vectors $T(\mathbf{e}_1), T(\mathbf{e}_2), \ldots, T(\mathbf{e}_n)$, we should compute the determinant of the matrix having $T(\mathbf{e}_j)$ as jth column vector. (See Theorem 5.12, Corollary 2 on page 305.) Thus the volume of the image under T of our unit n-cube is $|\det(A)|$.

Now for $c > 0$, the vectors $c\mathbf{e}_1, c\mathbf{e}_2, \ldots, c\mathbf{e}_n$ in $\mathbb{R}^n$ determine an n-cube of volume c^n. Since T is a linear transformation, we have $T(c\mathbf{e}_j) = cT(\mathbf{e}_j)$. Using the scalar-multiplication property of determinants, we see that the volume of the n-box determined by the vectors

$$cT(\mathbf{e}_1), \; cT(\mathbf{e}_2), \; . \; . \; . \; , \; cT(\mathbf{e}_n)$$

is $c^n|\det(A)|$. Given any cube in $\mathbb{R}^n$, we can choose a vertex of the cube as origin and let the edges from that vertex determine coordinate axes. Thus we see that every n-cube in $\mathbb{R}^n$ having edges of length c parallel to the coordinate axes is mapped by T into an n-box of volume $c^n|\det(A)|$.

The volume of a sufficiently nice region G in $\mathbb{R}^n$ may be approximated by adding the volumes of small n-cubes contained in G and having edges of length c parallel to the coordinate axes. Figure 6.7(a) illustrates this situation for a plane region in $\mathbb{R}^2$, where a grid of small squares (2-cubes) is placed on the region. As the length c of the edges of the squares approaches zero, the sum of the areas of the colored squares inside the region approaches the area of the region. These squares inside G are mapped by T into parallelograms of area $c^2|\det(A)|$ inside the image of G under T. (See the colored parallelograms in Fig. 6.7b.) As c approaches zero, the sum of the areas of these parallelograms approaches the area of the image of G under T, which thus must be the area of G multiplied by $|\det(A)|$. A similar construction can be made with a grid of n-cubes for a region G in $\mathbb{R}^n$. Each such cube is mapped by T into an n-box of volume $c^n|\det(A)|$. Adding the volumes of these n-boxes and taking the limiting value of the sum as c approaches zero, it is clear that the volume of the image under T of the region G is given by

$$\text{Volume of image of } G = |\det(A)| \cdot (\text{volume of } G). \tag{1}$$

We summarize this work in a theorem.

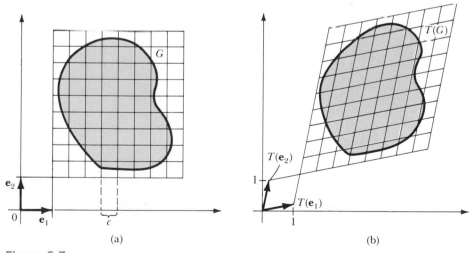

Figure 6.7
(a) A region G in the domain of T; (b) the image of G under T.

Theorem 6.18   Volume-change Factor for $T: \mathbb{R}^n \rightarrow \mathbb{R}^n$

Let G be a region in $\mathbb{R}^n$ of volume V, and let $T: \mathbb{R}^n \rightarrow \mathbb{R}^n$ be a linear transformation with standard matrix representation A. Then the volume of the image of G under T is $|\det(A)| \cdot V$.

As the title of Theorem 6.18 indicates, $|\det(A)|$ is the **volume-change factor** for T.

Example 1   Let $T: \mathbb{R}^2 \rightarrow \mathbb{R}^2$ be the linear transformation given by $T(x, y) = (2x - y, x + 3y)$. Find the area under T of the image of the disk $x^2 + y^2 \leq 4$ in the domain of T.

Solution   The disk $x^2 + y^2 \leq 4$ has radius 2 and area 4π. It is mapped by T into a region (actually bounded by an ellipse) of area

$$|\det(A)| \cdot (4\pi) = \begin{vmatrix} 2 & -1 \\ 1 & 3 \end{vmatrix} \cdot (4\pi) = (6 + 1)(4\pi) = 28\pi. \quad \triangleleft$$

We can generalize Theorem 6.18 to a linear transformation $T: \mathbb{R}^n \rightarrow \mathbb{R}^m$, where $m \geq n$ and T has rank n. This time the standard matrix representation A is an $m \times n$ matrix. The image under T of the unit n-cube in $\mathbb{R}^n$ outlined by $\mathbf{e}_1, \mathbf{e}_2, \ldots, \mathbf{e}_n$ is the n-box in $\mathbb{R}^m$ outlined by $T(\mathbf{e}_1), T(\mathbf{e}_2), \ldots, T(\mathbf{e}_n)$. According to Theorem 5.12 on page 304, the volume of this box in $\mathbb{R}^m$ is

$$\sqrt{\det(A^T A)}.$$

The same grid argument used above and illustrated in Fig. 6.7 shows that a region G in $\mathbb{R}^n$ of volume V is mapped by T into a region of $\mathbb{R}^m$ of volume

$$\sqrt{\det(A^T A)} \cdot V.$$

We summarize this generalization in a theorem.

Theorem 6.19   Volume-change Factor for $T: \mathbb{R}^n \rightarrow \mathbb{R}^m$

Let G be a region in $\mathbb{R}^n$ of volume V. Let $m \geq n$ and let $T: \mathbb{R}^n \rightarrow \mathbb{R}^m$ be a linear transformation of rank n. Let A be the standard matrix representation of T. Then the volume of the image of G in $\mathbb{R}^m$ under T is

$$\sqrt{\det(A^T A)} \cdot V.$$

Example 2   Let $T: \mathbb{R}^2 \rightarrow \mathbb{R}^3$ be given by $T(x, y) = (2x + 3y, x - y, 2y)$. Find the area of the image under T of the disk $x^2 + y^2 \leq 4$ in $\mathbb{R}^3$.

Solution   The standard matrix representation A of T is

$$A = \begin{pmatrix} 2 & 3 \\ 1 & -1 \\ 0 & 2 \end{pmatrix}$$

and

$$A^T A = \begin{pmatrix} 2 & 1 & 0 \\ 3 & -1 & 2 \end{pmatrix} \begin{pmatrix} 2 & 3 \\ 1 & -1 \\ 0 & 2 \end{pmatrix} = \begin{pmatrix} 5 & 5 \\ 5 & 14 \end{pmatrix}.$$

Thus

$$\sqrt{\det(A^T A)} = \sqrt{70 - 25} = \sqrt{45} = 3\sqrt{5}.$$

A region G of $\mathbb{R}^2$ having area V is mapped by T into a plane region of area $3\sqrt{5} \cdot V$ in $\mathbb{R}^3$. Thus the disk $x^2 + y^2 \leq 4$ of area 4π is mapped into a plane region in $\mathbb{R}^3$ of area

$$(3\sqrt{5})(4\pi) = 12\pi\sqrt{5}. \quad \lhd$$

SUMMARY

1. If $T: \mathbb{R}^n \to \mathbb{R}^n$ is a linear transformation with standard matrix representation A, then T maps a region in its domain of volume V into a region of volume $|\det(A)|V$.

2. If $T: \mathbb{R}^n \to \mathbb{R}^m$ is a linear transformation of rank n with standard matrix representation A, then T maps a region in its domain of volume V into a piece of $\mathbb{R}^m$ of volume $\sqrt{\det(A^T A)} \cdot V$.

EXERCISES

In Exercises 1–4, let $T: \mathbb{R}^2 \to \mathbb{R}^2$ be the linear transformation defined by $T(x, y) = (4x - 2y, 2x + 3y)$. Find the area of the image under T of each of the given regions in $\mathbb{R}^2$.

1. The square $0 \leq x \leq 1, 0 \leq y \leq 1$

2. The rectangle $-1 \leq x \leq 1, 1 \leq y \leq 2$

3. The parallelogram determined by $2\mathbf{e}_1 + 3\mathbf{e}_2$ and $4\mathbf{e}_1 - \mathbf{e}_2$

4. The disk $(x - 1)^2 + (y + 2)^2 \leq 9$

In Exercises 5–8, let $T: \mathbb{R}^3 \to \mathbb{R}^3$ be defined by $T(x, y, z) = (x - 2y, 3x + z, 4x + 3y)$. Find the volume of the image under T of each of the given regions in $\mathbb{R}^3$.

5. The cube $0 \leq x \leq 1, 0 \leq y \leq 1, 0 \leq z \leq 1$

6. The box $0 \leq x \leq 2, -1 \leq y \leq 3, 2 \leq z \leq 5$

7. The box determined by $2\mathbf{e}_1 + 3\mathbf{e}_2 - \mathbf{e}_3$, $4\mathbf{e}_1 - 2\mathbf{e}_3$, and $\mathbf{e}_1 - \mathbf{e}_2 + 2\mathbf{e}_3$

8. The ball $x^2 + (y - 3)^2 + (z + 2)^2 \leq 16$

In Exercises 9–12, let $T: \mathbb{R}^2 \to \mathbb{R}^3$ be the linear transformation defined by $T(x, y) = (y, x, x + y)$. Find the area of the image under T of each of the following regions in $\mathbb{R}^2$.

9. The square $0 \leq x \leq 1, 0 \leq y \leq 1$

10. The rectangle $2 \leq x \leq 3, -1 \leq y \leq 4$

11. The triangle with vertices $(0, 0), (6, 0), (0, 3)$

12. The disk $x^2 + y^2 \leq 25$

In Exercises 13–15, let $T: \mathbb{R}^2 \to \mathbb{R}^4$ *be defined by* $T(x, y) = (x - y, x, - y, 2x + y)$. *Find the area of the image under* T *of each of the following regions in* $\mathbb{R}^2$.

13. The square $0 \le x \le 1, 0 \le y \le 1$ **14.** The square $-1 \le x \le 3, -1 \le y \le 3$

15. The disk $x^2 + y^2 \le 9$

16. We have seen that for $n \times n$ matrices A and B, we have $\det(AB) = \det(A) \cdot \det(B)$, but the proof was not really intuitive. Give an intuitive geometric argument showing that at least $|\det(AB)| = |\det(A)| \cdot |\det(B)|$. [*Hint:* Use the fact that if A is the standard matrix representation of $T: \mathbb{R}^n \to \mathbb{R}^n$ and B is the standard matrix representation of $T': \mathbb{R}^n \to \mathbb{R}^n$, then AB is the standard matrix representation of $T \circ T'$.]

Eigenvalues: Applications and Computations

This chapter deals with further applications of eigenvalues and with the computation of eigenvalues. In Section 7.1, we discuss quadratic forms and their diagonalization. The principal axis theorem (Theorem 7.1) asserts that every quadratic form can be diagonalized. This is probably the most important result in the chapter, having applications to vibration of elastic bodies, quantum mechanics, and electric circuits. Presentations of such applications are beyond the scope of this text, and we have chosen to present a more familiar application in Section 7.2: classification of conic sections and quadric surfaces. Although the facts about conic sections may be familiar to the reader, their easy derivation from the principal axis theorem should be seen and enjoyed.

Section 7.3 applies the parallel axis theorem to local extrema of functions, and discusses maximization and minimization of quadratic forms on unit spheres. The latter topic is again important to vibration problems; it indicates that eigenvalues of maximum and of minimum magnitudes can be found for symmetric matrices using techniques from advanced calculus for finding extrema of quadratic forms on unit spheres.

In Section 7.4, we sketch three methods for computing eigenvalues: the power method, Jacobi's method, and the QR method. We attempt to make as intuitively clear as we can how and why each method works, but proofs and discussions of efficiency are omitted.

7.1

Diagonalization of Quadratic Forms

QUADRATIC FORMS

A quadratic form in one variable x is a polynomial $f(x) = ax^2$, where $a \neq 0$. A quadratic form in two variables x and y is a polynomial $f(x, y) = ax^2 + bxy + cy^2$, where at least one of a, b, or c is nonzero. The term *quadratic* means *degree* 2. The term *form* means *homogeneous;* that is, each summand contains a product of the *same number* of variables, namely 2 for a quadratic form. Thus $3x^2 - 4xy$ is a quadratic form in x and y, but $x^2 + y^2 - 4$ and $x - 3y^2$ are not quadratic forms.

Turning to the general case, a **quadratic form** $f(\mathbf{x}) = f(x_1, x_2, \ldots, x_n)$ in n variables is a polynomial that can be written using summation notation as

$$f(\mathbf{x}) = \sum_{\substack{i \leq j \\ i,j=1}}^{n} u_{ij} x_i x_j, \tag{1}$$

where not all u_{ij} are zero. To illustrate, the general quadratic form in x_1, x_2, and x_3 is

$$u_{11} x_1{}^2 + u_{12} x_1 x_2 + u_{13} x_1 x_3 + u_{22} x_2{}^2 + u_{23} x_2 x_3 + u_{33} x_3{}^2. \tag{2}$$

A computation shows that

$$(x_1, x_2, x_3) \begin{pmatrix} u_{11} & u_{12} & u_{13} \\ 0 & u_{22} & u_{23} \\ 0 & 0 & u_{33} \end{pmatrix} \begin{pmatrix} x_1 \\ x_2 \\ x_3 \end{pmatrix}$$

$$= (x_1, x_2, x_3) \begin{pmatrix} u_{11} x_1 + u_{12} x_2 + u_{13} x_3 \\ u_{22} x_2 + u_{23} x_3 \\ u_{33} x_3 \end{pmatrix}$$

$$= x_1(u_{11} x_1 + u_{12} x_2 + u_{13} x_3) + x_2(u_{22} x_2 + u_{23} x_3) + x_3(u_{33} x_3)$$

$$= u_{11} x_1{}^2 + u_{12} x_1 x_2 + u_{13} x_1 x_3 + u_{22} x_2{}^2 + u_{23} x_2 x_3 + u_{33} x_3{}^2,$$

IN 1826 CAUCHY DISCUSSED QUADRATIC FORMS IN THREE VARIABLES, that is, forms of the type $Ax^2 + By^2 + Cz^2 + 2Dxy + 2Exz + 2Fyz$. He showed that the characteristic equation formed from the determinant

$$\begin{vmatrix} A & D & E \\ D & B & F \\ E & F & C \end{vmatrix}$$

remains the same under any change of rectangular axes, what we would call an orthogonal coordinate change. Furthermore, he demonstrated a result already proved by Euler (see note on page 380), that one could find axes such that the new form had only the square terms. Three years later, Cauchy generalized the result to quadratic forms in n variables. In fact, he demonstrated that the roots of the characteristic equation $\lambda_1, \lambda_2, \ldots, \lambda_n$ are all real and showed how to find the linear substitution that converts the original form to the form $\lambda_1 x_1{}^2 + \lambda_2 x_2{}^2 + \cdots + \lambda_n x_n{}^2$.

In 1833 Jacobi proved the same result by a somewhat different method (see Section 7.4).

which is again the form (2). A moment of thought convinces us that the term involving u_{ij} for $i \leq j$ in the expansion of

$$(x_1, x_2, \ldots, x_n) \begin{pmatrix} u_{11} & u_{12} & \cdots & u_{1n} \\ 0 & u_{22} & \cdots & u_{2n} \\ \vdots & \vdots & & \vdots \\ 0 & 0 & \cdots & u_{nn} \end{pmatrix} \begin{pmatrix} x_1 \\ x_2 \\ \vdots \\ x_n \end{pmatrix} \tag{3}$$

is precisely $u_{ij}x_ix_j$. Thus the matrix product (3) is equal to the sum (1).

> Every quadratic form in n variables x_i can be written as $\mathbf{x}^T U \mathbf{x}$, where $\mathbf{x}$ is the column vector of variables and U is a nonzero upper-triangular matrix.

We will call the matrix U the **upper-triangular coefficient matrix** of the quadratic form.

Example 1   Write $x^2 - 2xy + 6xz + z^2$ in the matrix form (3).

Solution   We easily obtain the matrix expression

$$(x, y, z) \begin{pmatrix} 1 & -2 & 6 \\ 0 & 0 & 0 \\ 0 & 0 & 1 \end{pmatrix} \begin{pmatrix} x \\ y \\ z \end{pmatrix}.$$

Just think of x as the first variable, y the second, and z the third. The summand $-2xy$, for example, gives the coefficient -2 in the row 1, column 2 position. ◁

A matrix expression for a given quadratic form is by no means unique. Clearly $\mathbf{x}^T A \mathbf{x}$ gives a quadratic form for any nonzero $n \times n$ matrix A, and of course the form can be rewritten as $\mathbf{x}^T U \mathbf{x}$, where U is upper-triangular.

Example 2   Expand

$$(x, y, z) \begin{pmatrix} -1 & 3 & 1 \\ 2 & 1 & 0 \\ -2 & 2 & 4 \end{pmatrix} \begin{pmatrix} x \\ y \\ z \end{pmatrix},$$

and find the upper-triangular coefficient matrix for the quadratic form.

Solution   We find that

$$(x, y, z) \begin{pmatrix} -1 & 3 & 1 \\ 2 & 1 & 0 \\ -2 & 2 & 4 \end{pmatrix} \begin{pmatrix} x \\ y \\ z \end{pmatrix} = (x, y, z) \begin{pmatrix} -x + 3y + z \\ 2x + y \\ -2x + 2y + 4z \end{pmatrix}$$

$$= x(-x + 3y + z) + y(2x + y) + z(-2x + 2y + 4z)$$
$$= -x^2 + 3xy + xz + 2xy + y^2 - 2xz + 2yz + 4z^2$$
$$= -x^2 + 5xy + y^2 - xz + 2yz + 4z^2.$$

The upper-triangular coefficient matrix is

$$U = \begin{pmatrix} -1 & 5 & -1 \\ 0 & 1 & 2 \\ 0 & 0 & 4 \end{pmatrix}. \quad \triangleleft$$

All the nice things that we will prove about quadratic forms come from the fact that any quadratic form $f(\mathbf{x})$ can be expressed as $f(\mathbf{x}) = \mathbf{x}^T A \mathbf{x}$, where A is a *symmetric* matrix.

Example 3   Find the symmetric coefficient matrix for the form $x^2 - 2xy + 6xz + z^2$ in Example 1.

Solution   Rewriting the form as

$$x^2 - xy - yx + 3xz + 3zx + z^2,$$

we obtain the symmetric coefficient matrix

$$A = \begin{pmatrix} 1 & -1 & 3 \\ -1 & 0 & 0 \\ 3 & 0 & 1 \end{pmatrix}. \quad \triangleleft$$

As illustrated in Example 3, we obtain the symmetric matrix A for the form

$$f(\mathbf{x}) = \sum_{\substack{i \le j \\ i,j=1}}^{n} u_{ij} x_i x_j \tag{4}$$

as follows: Write each *cross term* $u_{ij}x_i x_j$ as $(u_{ij}/2)x_i x_j + (u_{ij}/2)x_j x_i$, and take $a_{ij} = a_{ji} = u_{ij}/2$. Clearly the resulting matrix A is symmetric. We call it the **symmetric coefficient matrix** of the form.

Every quadratic form in n variables x_i can be written as $\mathbf{x}^T A \mathbf{x}$, where $\mathbf{x}$ is the column vector of variables and A is a symmetric matrix.

DIAGONALIZATION OF QUADRATIC FORMS

We saw in Section 6.5 that a symmetric matrix can be diagonalized by an orthogonal matrix. That is, if A is an $n \times n$ symmetric matrix, then there exists an $n \times n$ orthogonal change-of-basis matrix C such that $C^{-1}AC = D$, where D is a diagonal matrix. Recall that the diagonal entries of D are $\lambda_1, \lambda_2, \ldots, \lambda_n$, where the λ_j are the (not necessarily distinct) eigenvalues of the matrix A, the jth column of C is a unit eigenvector corresponding to λ_j, and the column vectors of C form an orthonormal basis for $\mathbb{R}^n$.

Consider now a quadratic form $\mathbf{x}^T A \mathbf{x}$, where A is symmetric, and let C be an orthogonal diagonalizing matrix for A. Since $C^{-1} = C^T$, the substitution

$$\mathbf{x} = C\mathbf{t} \qquad \text{Diagonalizing substitution}$$

changes our quadratic form as follows:

$$\mathbf{x}^T A \mathbf{x} = (C\mathbf{t})^T A (C\mathbf{t}) = \mathbf{t}^T C^T A C \mathbf{t} = \mathbf{t}^T C^{-1} A C \mathbf{t}$$
$$= \mathbf{t}^T D \mathbf{t} = \lambda_1 t_1^2 + \lambda_2 t_2^2 + \cdots + \lambda_n t_n^2,$$

where the λ_j are the eigenvalues of A. We have thus diagonalized the quadratic form. The value of $\mathbf{x}^T A \mathbf{x}$ for any $\mathbf{x}$ in $\mathbb{R}^n$ is the same as the value of $\mathbf{t}^T D \mathbf{t}$ for $\mathbf{t} = C^{-1}\mathbf{x}$.

Example 4   Find a substitution $\mathbf{x} = C\mathbf{t}$ that diagonalizes the form $3x_1^2 + 10x_1x_2 + 3x_2^2$, and find the corresponding diagonalized form.

Solution   The symmetric coefficient matrix for the quadratic form is

$$A = \begin{pmatrix} 3 & 5 \\ 5 & 3 \end{pmatrix}.$$

We need to find the eigenvalues and eigenvectors for A. We have

$$|A - \lambda I| = \begin{vmatrix} 3 - \lambda & 5 \\ 5 & 3 - \lambda \end{vmatrix} - \lambda^2 - 6\lambda - 16 = (\lambda + 2)(\lambda - 8).$$

Thus $\lambda_1 = -2$ and $\lambda_2 = 8$ are eigenvalues of A. Finding the eigenvectors to become the columns of the substitution matrix C, we have

$$A - \lambda_1 I = A + 2I = \begin{pmatrix} 5 & 5 \\ 5 & 5 \end{pmatrix} \sim \begin{pmatrix} 1 & 1 \\ 0 & 0 \end{pmatrix}$$

and

$$A - \lambda_2 I = A - 8I = \begin{pmatrix} -5 & 5 \\ 5 & -5 \end{pmatrix} \sim \begin{pmatrix} 1 & -1 \\ 0 & 0 \end{pmatrix}.$$

Thus eigenvectors are $\mathbf{v}_1 = (-1, 1)$ and $\mathbf{v}_2 = (1, 1)$. Normalizing them to length 1 and placing them in the columns of the substitution matrix C, we obtain the diagonalizing substitution

$$\begin{pmatrix} x_1 \\ x_2 \end{pmatrix} = C \begin{pmatrix} t_1 \\ t_2 \end{pmatrix} = \begin{pmatrix} -1/\sqrt{2} & 1/\sqrt{2} \\ 1/\sqrt{2} & 1/\sqrt{2} \end{pmatrix} \begin{pmatrix} t_1 \\ t_2 \end{pmatrix}.$$

Our theory then tells us that making the variable substitution

$$x_1 = (1/\sqrt{2})(-t_1 + t_2),$$
$$x_2 = (1/\sqrt{2})(t_1 + t_2),$$ (5)

in the form $3x_1{}^2 + 10x_1x_2 + 3x_2{}^2$ will give the diagonal form

$$\lambda_1 t_1{}^2 + \lambda_2 t_2{}^2 = -2t_1{}^2 + 8t_2{}^2.$$

This can, of course, be checked by actually substituting the expression for x_1 and x_2 in Eqs. (5) into the form $3x_1{}^2 + 10x_1x_2 + 3x_2{}^2$. ◁

In the preceding example, we might like to clear denominators in our substitution (5) by using $\mathbf{x} = (\sqrt{2}C)\mathbf{t}$, so that $x_1 = -t_1 + t_2$ and $x_2 = t_1 + t_2$. Using the substitution $\mathbf{x} = kC\mathbf{t}$ with a quadratic form $\mathbf{x}^T A\mathbf{x}$, we obtain

$$\mathbf{x}^T A\mathbf{x} = (kC\mathbf{t})^T A(kC\mathbf{t}) = k^2(\mathbf{t}^T C^T AC\mathbf{t}) = k^2(\mathbf{t}^T D\mathbf{t}).$$

Thus with $k = \sqrt{2}$, the substitution $x_1 = -t_1 + t_2$, $x_2 = t_1 + t_2$ in Example 1 results in the diagonal form

$$2(-2t_1{}^2 + 8t_2{}^2) = -4t_1{}^2 + 16t_2{}^2.$$

In any particular situation, one would have to balance the desire for arithmetic simplicity against the desire to have the new orthogonal basis actually be orthonormal.

As an orthogonal matrix, C has determinant ± 1 (Exercise 12 of Section 6.5). We state without proof the significance of the sign of $\det(C)$. If $\det(C) = 1$, the new ordered orthonormal basis B given by the column vectors of C has the **same orientation** in $\mathbb{R}^n$ as the standard ordered basis E, while if $\det(C) = -1$, then B has the **opposite orientation** to E. In order for $B = (\mathbf{b}_1, \mathbf{b}_2)$ in $\mathbb{R}^2$ to have the same orientation as $(\mathbf{e}_1, \mathbf{e}_2)$, there must be a *rotation* of the plane, given by a matrix transformation $\mathbf{x} = C\mathbf{t}$, which carries both $\mathbf{e}_1$ to $\mathbf{b}_1$ and $\mathbf{e}_2$ to $\mathbf{b}_2$. The same interpretation in terms of rotation is true in $\mathbb{R}^3$, with the additional condition that $\mathbf{e}_3$ be carried to $\mathbf{b}_3$. To illustrate for $\mathbb{R}^2$, Fig. 7.1(a) shows an ordered orthonormal basis $B = (\mathbf{b}_1, \mathbf{b}_2)$, where $\mathbf{b}_1 = (-1/\sqrt{2}, 1/\sqrt{2})$ and $\mathbf{b}_2 = (-1/\sqrt{2}, -1/\sqrt{2})$, having the same orientation as E. Note that

$$\det(C) = \begin{vmatrix} -1/\sqrt{2} & -1/\sqrt{2} \\ 1/\sqrt{2} & -1/\sqrt{2} \end{vmatrix} = 1/2 + 1/2 = 1.$$

$$\qquad \mathbf{b}_1 \qquad\quad \mathbf{b}_2$$

Clearly, counterclockwise rotation through an angle of 135° carries E into B, preserving the order of the vectors. However, we see that the basis $(\mathbf{b}_1, \mathbf{b}_2) = ((0, -1), (-1, 0))$ shown in Fig. 7.1(b) does not have the same orientation as E, and this time

$$\det(C) = \begin{vmatrix} 0 & -1 \\ -1 & 0 \end{vmatrix} = -1.$$

$$\qquad \mathbf{b}_1 \quad\ \mathbf{b}_2$$

For any orthogonal matrix C having determinant -1, multiplication of any *single* column of C by -1 gives an orthogonal matrix with determinant 1. For

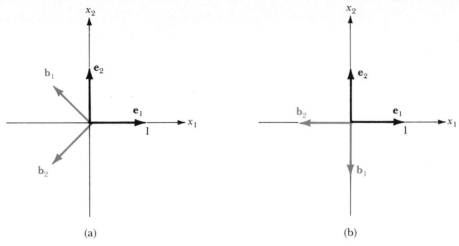

Figure 7.1
(a) Rotation of axes: (b₁, b₂) and (e₁, e₂) have the same orientation. (b) Not a rotation of axes: (b₁, b₂) and (e₁, e₂) have opposite orientation.

the diagonalization we are considering, where the columns are normalized eigenvectors, multiplication by −1 still gives a normalized eigenvector. While there will be some sign changes in the diagonalizing substitution, the final diagonal form remains the same, for the eigenvalues are the same, and their order has not been changed.

We summarize all our work in one main theorem, and then conclude with a final example. Section 7.2 will give an application of the theorem to geometry, and Section 7.3 will give an application to optimization.

Theorem 7.1   Principal Axis Theorem

Every quadratic form $f(\mathbf{x})$ in n variables $x_1, x_2, \ldots, x_n$ can be diagonalized by a substitution $\mathbf{x} = C\mathbf{t}$, where C is an $n \times n$ orthogonal matrix. The diagonalized form appears as

$$\lambda_1 t_1^2 + \lambda_2 t_2^2 + \cdots + \lambda_n t_n^2,$$

where the λ_j are the eigenvalues of the symmetric coefficient matrix A of $f(\mathbf{x})$. The jth column vector of C is a normalized eigenvector $\mathbf{v}_j$ of A corresponding to λ_j. Moreover, C can be chosen so that $\det(C) = 1$.

We box a step-by-step outline for diagonalizing a quadratic form.

Diagonalizing a Quadratic Form $f(\mathbf{x})$

Step 1 Find the symmetric coefficient matrix A of the form $f(\mathbf{x})$.

Step 2 Find the (not necessarily distinct) eigenvalues $\lambda_1, \lambda_2, \ldots, \lambda_n$ of A.

Step 3 Find an orthonormal basis for $\mathbb{R}^n$ consisting of normalized eigenvectors of A.

Step 4 Form the matrix C whose columns are the basis vectors found in step 3, in the order corresponding to the listing of eigenvalues in step 2. The transformation $\mathbf{x} = C\mathbf{t}$ is a **rotation** if $\det(C) = 1$. If a rotation is desired and $\det(C) = -1$, change the sign of all components of just one column vector in C.

Step 5 The substitution $\mathbf{x} = C\mathbf{t}$ transforms $f(\mathbf{x})$ to the diagonal form $\lambda_1 t_1^2 + \lambda_2 t_2^2 + \cdots + \lambda_n t_n^2$.

Example 5   Find a variable substitution that diagonalizes the form $2xy + 2xz$, and give the resulting diagonal form.

Solution   The symmetric coefficient matrix for the given quadratic form is

$$A = \begin{pmatrix} 0 & 1 & 1 \\ 1 & 0 & 0 \\ 1 & 0 & 0 \end{pmatrix}.$$

Expanding the determinant on the last column, we find that

$$|A - \lambda I| = \begin{vmatrix} -\lambda & 1 & 1 \\ 1 & -\lambda & 0 \\ 1 & 0 & -\lambda \end{vmatrix} = 1(\lambda) - \lambda(\lambda^2 - 1)$$

$$= -\lambda(-1 + \lambda^2 - 1) = -\lambda(\lambda^2 - 2).$$

The eigenvalues of A are then $\lambda_1 = 0$, $\lambda_2 = \sqrt{2}$, and $\lambda_3 = -\sqrt{2}$. Computing eigenvectors, we have

$$A - \lambda_1 I = A = \begin{pmatrix} 0 & 1 & 1 \\ 1 & 0 & 0 \\ 1 & 0 & 0 \end{pmatrix} \sim \begin{pmatrix} 1 & 0 & 0 \\ 0 & 1 & 1 \\ 0 & 0 & 0 \end{pmatrix}$$

so $\mathbf{v}_1 = (0, -1, 1)$ is an eigenvector for $\lambda_1 = 0$. Also,

$$A - \lambda_2 I = \lambda - \sqrt{2}I = \begin{pmatrix} -\sqrt{2} & 1 & 1 \\ 1 & -\sqrt{2} & 0 \\ 1 & 0 & -\sqrt{2} \end{pmatrix} \sim \begin{pmatrix} 1 & -\sqrt{2} & 0 \\ -\sqrt{2} & 1 & 1 \\ 1 & 0 & -\sqrt{2} \end{pmatrix}$$

$$\sim \begin{pmatrix} 1 & -\sqrt{2} & 0 \\ 0 & -1 & 1 \\ 0 & \sqrt{2} & -\sqrt{2} \end{pmatrix} \sim \begin{pmatrix} 1 & -\sqrt{2} & 0 \\ 0 & 1 & -1 \\ 0 & 0 & 0 \end{pmatrix}$$

so $\mathbf{v}_2 = (\sqrt{2}, 1, 1)$ is an eigenvector for $\lambda_2 = \sqrt{2}$. In a similar fashion, we find that $\mathbf{v}_3 = (-\sqrt{2}, 1, 1)$ is an eigenvector for $\lambda_3 = -\sqrt{2}$. Thus

$$C = \begin{pmatrix} 0 & \sqrt{2}/2 & -\sqrt{2}/2 \\ -1/\sqrt{2} & 1/2 & 1/2 \\ 1/\sqrt{2} & 1/2 & 1/2 \end{pmatrix}$$

is an orthogonal diagonalizing matrix for A. The substitution

$$x = (1/\sqrt{2})(t_2 - t_3)$$
$$y = (1/2)(-\sqrt{2}t_1 + t_2 + t_3)$$
$$z = (1/2)(\sqrt{2}t_1 + t_2 + t_3)$$

will diagonalize $2xy + 2xz$ to become $\sqrt{2}t_2{}^2 - \sqrt{2}t_3{}^2$. $\triangleleft$

SUMMARY

Let $\mathbf{x}$ and $\mathbf{t}$ be $n \times 1$ column vectors with entries x_i and t_i, respectively.

1. For every nonzero $n \times n$ matrix A, the product $f(\mathbf{x}) = \mathbf{x}^T A \mathbf{x}$ gives a quadratic form in the variables x_i.

2. Given any quadratic form $f(\mathbf{x})$, there exists an upper-triangular coefficient matrix U such that $f(\mathbf{x}) = \mathbf{x}^T U \mathbf{x}$ and a symmetric coefficient matrix A such that $f(\mathbf{x}) = \mathbf{x}^T A \mathbf{x}$.

3. Any quadratic form $f(\mathbf{x})$ can be orthogonally diagonalized using a variable substitution $\mathbf{x} = C\mathbf{t}$ by following the boxed steps on page 369.

EXERCISES

In Exercises 1–8, find the upper-triangular coefficient matrix U and the symmetric coefficient matrix A of the given quadratic form.

1. $3x^2 - 6xy + y^2$

2. $8x^2 + 9xy - 3y^2$

3. $x^2 - y^2 - 4xy + 3xz - 8yz$

4. $x_1{}^2 - 2x_2 + x_3{}^2 + 6x_4{}^2 - 2x_1x_4 + 6x_2x_4 - 8x_1x_3$

5. $(x, y) \begin{pmatrix} -2 & 1 \\ 7 & 3 \end{pmatrix} \begin{pmatrix} x \\ y \end{pmatrix}$

6. $(x, y) \begin{pmatrix} 7 & -10 \\ 15 & 2 \end{pmatrix} \begin{pmatrix} x \\ y \end{pmatrix}$

7. $(x, y, z) \begin{pmatrix} 8 & 3 & 1 \\ 2 & 1 & -4 \\ -5 & 2 & 10 \end{pmatrix} \begin{pmatrix} x \\ y \\ z \end{pmatrix}$

8. $(x_1, x_2, x_3, x_4) \begin{pmatrix} 2 & -1 & 3 & 0 \\ 4 & 2 & -1 & 3 \\ -6 & 3 & 0 & 7 \\ 10 & 2 & 1 & 5 \end{pmatrix} \begin{pmatrix} x_1 \\ x_2 \\ x_3 \\ x_4 \end{pmatrix}$

In Exercises 9–16, find an orthogonal substitution that diagonalizes the given quadratic form, and find the diagonalized form.

9. $2xy$

10. $3x^2 + 4xy$

11. $-6xy + 8y^2$

12. $x^2 + 2xy + y^2$

13. $3x^2 - 4xy + 3y^2$

14. $x_2^2 + 2x_1x_3$

15. $x_1^2 + x_2^2 + x_3^2 - 2x_1x_2 - 2x_1x_3 - 2x_2x_3$ [*Suggestion:* Use Example 2 in Section 6.5.)

16. $x_1^2 + 2x_2^2 - 6x_1x_3 + x_3^2 - 4x_4^2$

17. Find a necessary and sufficient condition on a, b, and c such that the quadratic form $ax^2 + 2bxy + cy^2$ can be orthogonally diagonalized to kt^2.

18. Repeat Exercise 17, but require also that $k = 1$.

In Exercises 19–24, use MATCOMP or similar software to find a diagonal form into which the given form can be transformed by an orthogonal substitution. Do not give the substitution.

19. $3x^2 + 4xy - 5y^2$

20. $x^2 - 8xy + y^2$

21. $3x^2 + y^2 - 2z^2 - 4xy + 6yz$

22. $y^2 - 8z^2 + 3xy - 4xz + 7yz$

23. $x_1^2 - 3x_1x_4 + 5x_4^2 - 8x_2x_3$

24. $x_1^2 - 8x_1x_2 + 6x_2x_3 - 4x_3x_4$

7.2

Applications to Geometry

CONIC SECTIONS IN $\mathbb{R}^2$

Figure 7.2 shows three different types of plane curves obtained when a double right circular cone is cut by a plane. These *conic sections* are the ellipse, the hyperbola, and the parabola. Figure 7.3(a) shows an ellipse in *standard position*, with center at the origin. The equation of this ellipse is

$$\frac{x^2}{a^2} + \frac{y^2}{b^2} = 1. \tag{1}$$

The ellipse in Fig. 7.3(b) with center at (h, k) has equation

$$\frac{(x - h)^2}{a^2} + \frac{(y - k)^2}{b^2} = 1. \tag{2}$$

If we make the translation substitution $\bar{x} = x - h$, $\bar{y} = y - k$, then Eq. (2) becomes $\bar{x}^2/a^2 + \bar{y}^2/b^2 = 1$, which resembles Eq. (1).

By completing the square, any quadratic polynomial equation with no xy-term but where x^2 and y^2 have coefficients of the same sign can be put into the form (2), possibly with 0 or -1 on the right-hand side. The procedure should be familiar from working with circles. We give one illustration for an ellipse.

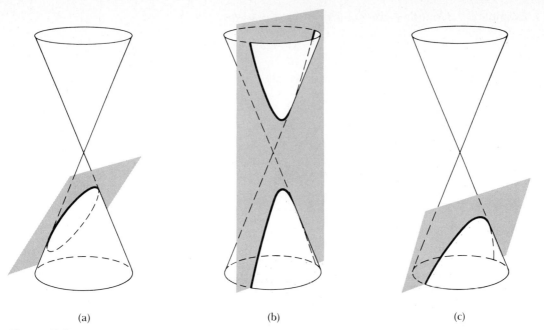

Figure 7.2
Sections of a cone: (a) an elliptic section; (b) a hyperbolic section; (c) a parabolic section.

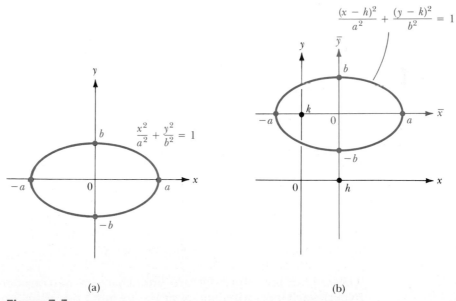

(a)   (b)

Figure 7.3
(a) An ellipse in standard position; (b) an ellipse centered at (h, k).

Example 1   Complete the square in the equation

$$x^2 + 3y^2 - 4x + 6y = -1.$$

Solution   Completing the square, we obtain

$$(x^2 - 4x) + 3(y^2 + 2y) = -1$$
$$(x - 2)^2 + 3(y + 1)^2 = 4 + 3 - 1 = 6$$
$$\frac{(x - 2)^2}{6} + \frac{(y + 1)^2}{2} = 1,$$

which is of the form of Eq. (2). Setting $\bar{x} = x - 2$ and $\bar{y} = y + 1$, we obtain the equation

$$\frac{\bar{x}^2}{6} + \frac{\bar{y}^2}{2} = 1.$$

This is the equation of an ellipse with center $(h, k) = (2, -1)$.   ◁

If the constant on the right-hand side of the initial equation in Example 1 had been -7, we would have obtained $\bar{x}^2 + 3\bar{y}^2 = 0$, which describes the single point $(\bar{x}, \bar{y}) = (0, 0)$. This single point is regarded as a *degenerate* ellipse. If the constant had been -8, we would have obtained $\bar{x}^2 + 3\bar{y}^2 = -1$, which has no real solution; the ellipse is then called *empty*.

Thus every polynomial equation

$$c_1 x^2 + c_2 y^2 + c_3 x + c_4 y = d, \qquad c_1 c_2 > 0, \tag{3}$$

describes an ellipse, possibly degenerate or empty.

In a similar fashion, an equation

$$c_1 x^2 + c_2 y^2 + c_3 x + c_4 y = d, \qquad c_1 c_2 < 0 \tag{4}$$

describes a hyperbola. Note that the coefficients of x^2 and y^2 have *opposite* signs. The standard form of the equation for a hyperbola centered at $(0, 0)$ is

$$\frac{x^2}{a^2} - \frac{y^2}{b^2} = 1 \quad \text{or} \quad \frac{-x^2}{a^2} + \frac{y^2}{b^1} = 1$$

as shown in Fig. 7.4. The dashed diagonal lines $y = \pm(b/a)x$ shown in the figure are the *asymptotes* of the hyperbola. By completing squares and translating axes, we can reduce Eq. (4) to one of the standard forms shown in Fig. 7.4, in variables $\bar{x}$ and $\bar{y}$, unless the constant in the final equation reduces to zero. In that case one obtains

$$\frac{\bar{x}^2}{a^2} - \frac{\bar{y}^2}{b^2} = 0 \quad \text{or} \quad \bar{y} = \pm(b/a)\bar{x}.$$

These equations represent two lines that can be considered a degenerate hyperbola. Thus any equation of the form (4) describes a (possibly degenerate) hyperbola.

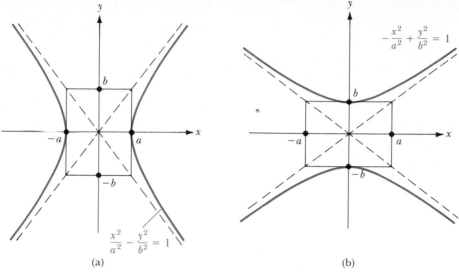

Figure 7.4

(a) The hyperbola $\dfrac{x^2}{a^2} - \dfrac{y^2}{b^2} = 1$; (b) the hyperbola $\dfrac{-x^2}{a^2} + \dfrac{y^2}{b^2} = 1$.

Finally, the equations

$$c_1x^2 + c_2x + c_3y = d \quad \text{and} \quad c_1y^2 + c_2x + c_3y = d, \qquad c_1 \neq 0, \qquad (5)$$

describe parabolas. If $c_3 \neq 0$ in the first equation in (5) and $c_2 \neq 0$ in the second, then these equations can be reduced to the form

$$\bar{x}^2 = a\bar{y} \quad \text{and} \quad \bar{y}^2 = a\bar{x} \qquad (6)$$

by completing the square and translating axes. Figure 7.5 shows two parabolas in standard position. If Eqs. (5) reduce to $c_1x^2 + c_2x = d$ and $c_1y^2 + c_3y = d$, in Eq. (6), then each describes two parallel lines that can be considered degenerate parabolas.

> In summary, every equation of the form
>
> $$c_1x^2 + c_2y^2 + c_3x + c_4y = d \qquad (7)$$
>
> with at least one of c_1 or c_2 nonzero describes a (possibly degenerate or empty) ellipse, hyperbola, or parabola.

CLASSIFICATION OF SECOND-DEGREE CURVES

We can now apply our work in Section 7.1 on diagonalizing quadratic forms to classify the plane curves described by an equation of the type

$$ax^2 + bxy + cy^2 + dx + ey + f = 0 \quad \text{for} \quad a, b, c \text{ not all zero.} \qquad (8)$$

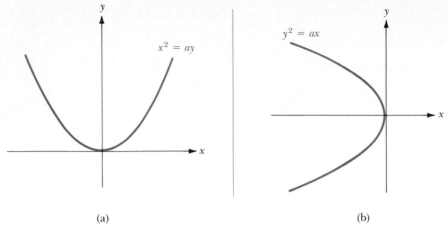

(a)   (b)

Figure 7.5
(a) The parabola $x^2 = ay$, $a > 0$; (b) the parabola $y^2 = ax$, $a < 0$.

We make a substitution

$$\begin{pmatrix} x \\ y \end{pmatrix} = C \begin{pmatrix} t_1 \\ t_2 \end{pmatrix}, \quad \text{where } \det(C) = 1, \tag{9}$$

which orthogonally diagonalizes the quadratic-form part of Eq. (8), which is in color, and we obtain an equation of the form

$$\lambda_1 t_1^2 + \lambda_2 t_2^2 + g t_1 + h t_2 + k = 0. \tag{10}$$

This equation has the form of Eq. (7), and describes an ellipse, hyperbola, or parabola. Remember that Eq. (9) corresponds to a rotation that carries the vector $\mathbf{e}_1$ to the first column vector $\mathbf{b}_1$ of C, and carries $\mathbf{e}_2$ to the second column vector $\mathbf{b}_2$. We think of $\mathbf{b}_1$ as pointing out the t_1-axis and $\mathbf{b}_2$ as pointing out the t_2-axis. We summarize our work in a theorem and give just one illustration, leaving others to the exercises.

ANALYTIC GEOMETRY is generally considered to have been founded by René Descartes (1596–1650) and Pierre Fermat (1601–1665) in the first half of the seventeenth century. But it was not until the appearance of a Latin version of Descartes' *Geometry* in 1661 by Frans van Schooten (1615–1660) that its influence began to be felt. This Latin version was published along with many commentaries; in particular, the *Elements of Curves* by Jan de Witt (1625–1672) gave a systematic treatment of conic sections. De Witt gave canonical forms of these equations similar to those in use today; for example, $y^2 = ax$, $by^2 + x^2 = f^2$, and $x^2 - by^2 = f^2$ represented the parabola, ellipse, and hyperbola, respectively. He then showed how, given an arbitrary second-degree equation in x and y, to find a transformation of axes that reduces the given equation to one of the canonical forms. This is, of course, equivalent to diagonalizing a particular symmetric matrix.

De Witt was a talented mathematician who, because of his family background, could devote but little time to mathematics. In 1653 he became in effect the prime minister of the Netherlands. Over the next decades, he guided the fortunes of the country through a most difficult period, including three wars with England. In 1672 the hostility of one of the Dutch factions culminated in his murder.

Theorem 7.2   **Classification of Second-degree Plane Curves**

Every equation of the form (8) can be reduced to an equation of the form (10) by an orthogonal substitution corresponding to a rotation of the plane. The coefficients λ_1 and λ_2 in Eq. (10) are the eigenvalues of the symmetric coefficient matrix of the quadratic-form portion of Eq. (8). The curve describes a (possibly degenerate or empty)

$$\text{ellipse} \quad \text{if } \lambda_1\lambda_2 > 0,$$
$$\text{hyperbola} \quad \text{if } \lambda_1\lambda_2 < 0,$$
$$\text{parabola} \quad \text{if } \lambda_1\lambda_2 = 0.$$

Example 2   Use rotation and translation of axes to sketch the plane curve $2xy + 2\sqrt{2}x = 1$.

Solution   The symmetric coefficient matrix of the quadratic form $2xy$ is

$$A = \begin{pmatrix} 0 & 1 \\ 1 & 0 \end{pmatrix}.$$

We easily find that the eigenvalues are $\lambda_1 = 1$ and $\lambda_2 = -1$, and that

$$C = \begin{pmatrix} 1/\sqrt{2} & -1/\sqrt{2} \\ 1/\sqrt{2} & 1/\sqrt{2} \end{pmatrix}$$

is an orthogonal diagonalizing matrix with determinant 1. The substitution

$$x = (1/\sqrt{2})(t_1 - t_2)$$
$$y = (1/\sqrt{2})(t_1 + t_2)$$

then yields

$$t_1{}^2 - t_2{}^2 + 2t_1 - 2t_2 = 1.$$

Completing the square, we obtain

$$(t_1 + 1)^2 - (t_2 + 1)^2 = 1,$$

which describes the hyperbola shown in Fig. 7.6.   ◁

QUADRIC SURFACES

An equation in three variables of the form

$$c_1x^2 + c_2y^2 + c_3z^2 + c_4x + c_5y + c_6z = d, \tag{11}$$

where at least one of c_1, c_2, or c_3 is nonzero, describes a *quadric surface* in space, which again might be degenerate or empty. Figures 7.7–7.15 show some of the quadric surfaces in standard position.

By completing the square in (11), which corresponds to translating axes, we see that (11) can be reduced to an equation involving $\bar{x}$, $\bar{y}$, and $\bar{z}$ in which a variable appearing to the second power does not appear to the first power. Note that this is true for all the equations in Figs. 7.7–7.15.

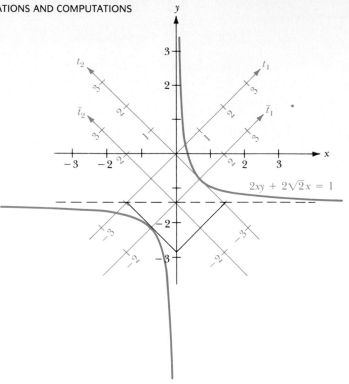

Figure 7.6
The hyperbola $2xy + 2\sqrt{2}x = 1$.

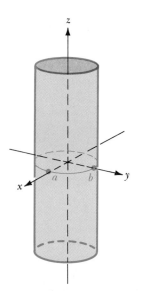

Figure 7.7
The elliptic cylinder
$$\frac{x^2}{a^2} + \frac{y^2}{b^2} = 1.$$

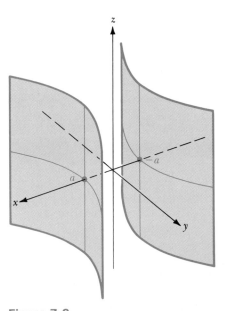

Figure 7.8
The hyperbolic cylinder $\dfrac{x^2}{a^2} - \dfrac{y^2}{b^2} = 1$.

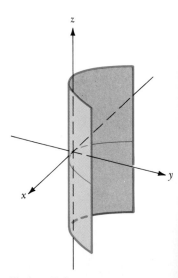

Figure 7.9
The parabolic cylinder $ay = x^2$.

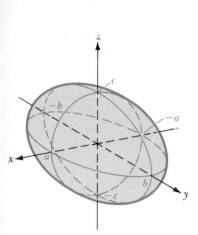

Figure 7.10

The ellipsoid $\dfrac{x^2}{a^2} + \dfrac{y^2}{b^2} + \dfrac{z^2}{c^2} = 1$.

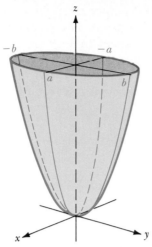

Figure 7.11
The elliptic paraboloid

$$z = \dfrac{x^2}{a^2} + \dfrac{y^2}{b^2}.$$

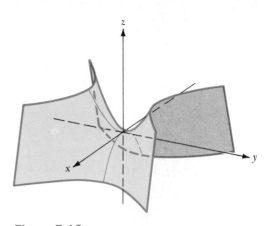

Figure 7.12

The hyperbolic paraboloid $z = \dfrac{y^2}{b^2} - \dfrac{x^2}{a^2}$.

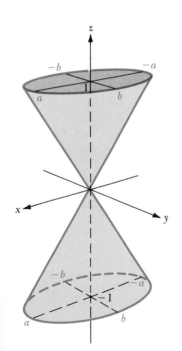

Figure 7.13
The elliptic cone

$$z^2 = \dfrac{x^2}{a^2} + \dfrac{y^2}{b^2}.$$

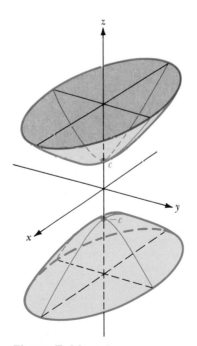

Figure 7.14
The hyperboloid of two sheets

$$\dfrac{z^2}{c^2} - 1 = \dfrac{x^2}{a^2} + \dfrac{y^2}{b^2}.$$

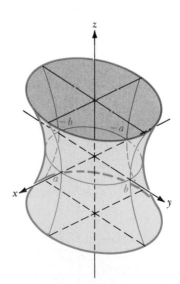

Figure 7.15
The hyperboloid of one sheet

$$\dfrac{z^2}{c^2} + 1 = \dfrac{x^2}{a^2} + \dfrac{y^2}{b^2}.$$

Again, degenerate and empty cases are possible. For example, $x^2 + 2y^2 + z^2 = -4$ gives an empty ellipsoid. The elliptic cone, hyperboloid of two sheets, and hyperboloid of one sheet in Figs. 7.13–7.15 differ only in whether a constant in their equations is zero, negative, or positive; the cone could be considered a degenerate case of each of the other two.

Now consider a general second-degree polynomial equation

$$ax^2 + by^2 + cz^2 + dxy + exz + fyz + px + gy + rz + s = 0, \tag{12}$$

where the coefficient of at least one term of degree two is nonzero. If we make a substitution

$$\begin{pmatrix} x \\ y \\ z \end{pmatrix} = C \begin{pmatrix} t_1 \\ t_2 \\ t_3 \end{pmatrix}, \quad \text{where } \det(C) = 1, \tag{13}$$

that orthogonally diagonalizes the quadratic-form portion of Eq. (12) (in color), we obtain an equation

$$\lambda_1 t_1{}^2 + \lambda_2 t_2{}^2 + \lambda_3 t_3{}^2 + p' t_1 + q' t_2 + r' t_3 + s' = 0, \tag{14}$$

which is of the form of Eq. (11). We state this as a theorem.

Theorem 7.3   **Principal Axis Theorem for $\mathbb{R}^3$**

Every equation of the form (12) can be reduced to an equation of the form (14) by an orthogonal substitution (13) that corresponds to a rotation of axes.

Again, computation of the eigenvalues λ_1, λ_2, and λ_3 of the symmetric coefficient matrix of the quadratic-form portion of Eq. (12) may give us quite a bit of information as to the type of quadric surface. However, actually execut-

---

THE EARLIEST CLASSIFICATION OF QUADRIC SURFACES was given by Leonhard Euler, in his precalculus text *Introduction to Infinitesimal Analysis* (1748). Euler's classification was similar to the conic-section classification of De Witt. Euler considered the second-degree equation in three variables $Ap^2 + Bq^2 + Cr^2 + Dpq + Epr + Fqr + Gp + Hq + Ir + K = 0$ as representing a surface in 3-space. As did De Witt, he gave canonical forms for these surfaces and showed how to rotate and translate the axes to reduce any given equation to a standard form such as $Ap^2 + Bq^2 + Cr^2 + K = 0$. An analysis of the signs of the new coefficients determined whether the given equation represented an ellipsoid, a hyperboloid of one or two sheets, an elliptic or hyperbolic paraboloid, or one of the degenerate cases. Euler did not, however, make explicit use of eigenvalues. He gave a general formula for rotation of axes in 3-space as functions of certain angles and then showed how to choose the angles to make the coefficients D, E, and F all zero.

Euler was the most prolific mathematician of all time. His collected works fill over 70 large volumes. Euler was born in Switzerland, but spent his professional life in St. Petersburg and Berlin. His texts on precalculus, differential calculus, and integral calculus (1748, 1755, 1768) had immense influence and became the bases for such texts up to the present. He standardized much of our current notation, introducing the numbers e, π, and i, as well as giving our current definitions for the trigonometric functions. Even though he was blind for the last 17 years of his life, he continued to produce mathematical papers almost up to the day he died.

ing the substitution and completing squares, which can be tedious, may be necessary to distinguish among certain surfaces. We give a rough classification scheme in Table 7.1. Remember that empty or degenerate cases are possible, even where we do not explicitly give them.

Table 7.1

Eigenvalues λ_1, λ_2, λ_3	Quadric Surface
All same sign	Ellipsoid
Two of one sign and one of the other sign	Elliptic cone, hyperboloid of two sheets, or hyperboloid of one sheet
One zero, two of the same sign	Elliptic paraboloid or elliptic cylinder (degenerate case)
One zero, two of opposite signs	Hyperbolic paraboloid or hyperbolic cylinder (degenerate case)
Two zero, one nonzero	Parabolic cylinder or two parallel planes (degenerate case)

The reader who verifies Table 7.1 will wonder about an equation of the form $ax^2 + by + cz = d$ in the last case given. Exercise 11 indicates that by a rotation of axes, this equation can be reduced to $at_1^2 + rt_2 = d$, which can be written as $at_1^2 + r(t_2 - d/r) = 0$, and consequently describes a parabolic cylinder. We conclude with four examples.

Example 3   Classify the quadric surface $2xy + 2xz = 1$.

Solution   Example 5 of Section 7.1 shows that the orthogonal substitution

$$x = (1/\sqrt{2})(t_2 - t_3)$$
$$y = (1/2)(-\sqrt{2}t_1 + t_2 + t_3)$$
$$z = (1/2)(\sqrt{2}t_1 + t_2 + t_3)$$

transforms $2xy + 2xz$ into $\sqrt{2}t_2^2 - \sqrt{2}t_3^2$. It can be easily checked that the matrix C corresponding to this substitution has determinant 1. Thus the substitution above corresponds to a rotation of axes and transforms the equation $2xy + 2xz = 1$ into $\sqrt{2}t_2^2 - \sqrt{2}t_3^2 = 1$, which we recognize as a hyperbolic cylinder. ◁

Example 4   Classify the quadric surface $2xy + 2xz = y + 1$.

Solution   Using the same substitution as we did in Example 3, we obtain the equation

$$\sqrt{2}t_2^2 - \sqrt{2}t_3^2 = (1/2)(-\sqrt{2}t_1 + t_2 + t_3) + 1.$$

Translation of axes by completing squares will yield an equation of the form

$$\sqrt{2}(t_2 - h)^2 - \sqrt{2}(t_3 - k)^2 = -(1/\sqrt{2})(t_1 - r),$$

which we recognize as a hyperbolic paraboloid. ◁

Example 5   Classify the quadric surface $2xy + 2xz = x + 1$.

Solution   Using again the substitution in Example 3, we obtain

$$\sqrt{2}t_2^2 - \sqrt{2}t_3^2 = (1/\sqrt{2})(t_2 - t_3) + 1$$

or

$$2t_2^2 - t_2 - 2t_3^2 + t_3 = \sqrt{2}.$$

Completing squares yields

$$2(t_2 - 1/4)^2 - 2(t_3 - 1/4)^2 = \sqrt{2},$$

which is a hyperbolic cylinder. ◁

Example 6   Classify the quadric surface

$$2x^2 - 3y^2 + z^2 - 2xy + 4yz - 6x + 8y - 8z = 17$$

as far as possible by finding just the eigenvalues of the symmetric coefficient matrix of the quadratic-form portion of the equation.

Solution   The symmetric matrix of the quadratic-form portion is

$$A = \begin{pmatrix} 2 & -1 & 0 \\ -1 & -3 & 2 \\ 0 & 2 & 1 \end{pmatrix}.$$

We find that

$$\det(A - \lambda I) = \begin{vmatrix} 2 - \lambda & -1 & 0 \\ -1 & -3 - \lambda & 2 \\ 0 & 2 & 1 - \lambda \end{vmatrix}$$

$$= (2 - \lambda) \begin{vmatrix} -3 - \lambda & 2 \\ 2 & 1 - \lambda \end{vmatrix} + \begin{vmatrix} -1 & 2 \\ 0 & 1 - \lambda \end{vmatrix}$$

$$= (2 - \lambda)(\lambda^2 + 2\lambda - 7) + \lambda - 1$$

$$= -\lambda^3 + 12\lambda - 15.$$

Clearly $\lambda = 0$ is not a solution of the characteristic equation $-\lambda^3 + 12\lambda - 15 = 0$. We could plot a rough sketch of the graph of $y = -\lambda^3 + 12\lambda - 15 = 0$, just to determine the signs of the eigenvalues. However, we prefer to use either one of the available software programs MATCOMP with the matrix A or NEWT123 with the characteristic equation. We quickly find that the eigenvalues are approximately

$$\lambda_1 = -3.9720, \qquad \lambda_2 = 1.5765, \qquad \lambda_3 = 2.3954.$$

According to Table 7.1, we have either an elliptic cone, a hyperboloid of two sheets, or a hyperboloid of one sheet. ◁

SUMMARY

1. Given an equation $ax^2 + bxy + cy^2 + dx + ey + f = 0$, let A be the symmetric coefficient matrix of the quadratic-form portion of the equation (in color), let λ_1 and λ_2 be the eigenvalues of A, and let C be an orthogonal matrix of determinant 1 with eigenvectors of A for columns. The substitution corresponding to the matrix C followed by translation of axes reduces the given equation to a standard form for the equation of a conic section. In particular, the equation describes a (possibly degenerate or empty)

   ellipse   if $\lambda_1\lambda_2 > 0$,
   hyperbola if $\lambda_1\lambda_2 < 0$,
   parabola   if $\lambda_1\lambda_2 = 0$.

2. Proceeding in the analogous way to that described in number (1) but for the equation

$$ax^2 + by^2 + cz^2 + dxy + exz + fyz + px + qy + rz + s = 0,$$

   one obtains a standard form for the equation of a (possibly degenerate or empty) quadric surface in space. Table 7.1 lists the information that can be obtained from the three eigenvalues λ_1, λ_2, and λ_3 alone.

EXERCISES

In Exercises 1–8, rotate axes using a substitution

$$\begin{pmatrix} x \\ y \end{pmatrix} = C \begin{pmatrix} t_1 \\ t_2 \end{pmatrix},$$

complete squares if necessary, and sketch the graph (if it is not empty) of the given conic section.

1. $2xy = 1$
2. $2xy - 2\sqrt{2}y = 1$
3. $x^2 + 2xy + y^2 = 4$
4. $x^2 - 2xy + y^2 + 4\sqrt{2}x = 4$
5. $10x^2 + 6xy + 2y^2 = 4$
6. $5x^2 + 4xy + 2y^2 = -1$
7. $3x^2 + 4xy + 6y^2 = 8$
8. $x^2 + 8xy + 7y^2 + 18\sqrt{5}x = -36$

9. Show that the plane curve $ax^2 + bxy + cy^2 + dx + ey + f = 0$ is a (possibly degenerate or empty)

   ellipse   if $b^2 - 4ac < 0$,
   hyperbola if $b^2 - 4ac > 0$,
   parabola   if $b^2 - 4ac = 0$.

   [*Hint:* Diagonalize $ax^2 + bxy + cy^2$, using the quadratic formula, and check the signs of the eigenvalues.]

10. Use Exercise 9 to classify the conic section with the given equation.
    a) $2x^2 + 8xy + 8y^2 - 3x + 2y = 13$
    b) $y^2 + 4xy - 5x^2 - 3x = 12$

c) $-x^2 + 5xy - 7y^2 - 4y + 11 = 0$
d) $xy + 4x - 3y = 8$
e) $2x^2 - 3xy + y^2 - 8x + 5y = 30$
f) $x^2 + 6xy + 9y^2 - 2x + 14y = 10$
g) $4x^2 - 2xy - 3y^2 + 8x - 5y = 17$
h) $8x^2 + 6xy + 2y^2 - 5x = 25$
i) $x^2 - 2xy + 4x - 5y = 6$
j) $2x^2 - 3xy + 2y^2 - 8y = 15$

11. Show that the equation $ax^2 + by + cz = d$ can be transformed into an equation of the form $at_1^2 + rt_2 = d$ by a rotation of axes in space. [*Hint:*

$$\text{If } \begin{pmatrix} x \\ y \\ z \end{pmatrix} = C \begin{pmatrix} t_1 \\ t_2 \\ t_3 \end{pmatrix}, \text{ then } \begin{pmatrix} t_1 \\ t_2 \\ t_3 \end{pmatrix} = C^{-1} \begin{pmatrix} x \\ y \\ z \end{pmatrix} = C^T \begin{pmatrix} x \\ y \\ z \end{pmatrix}.$$

Find an orthogonal matrix C^T such that $\det(C^T) = 1$, $t_1 = x$, and $t_2 = (by + cz)/r$ for some r.]

In Exercises 12–20, classify the quadric surface with the given equation as one of the (possibly empty or degenerate) types illustrated in Figs. 7.7–7.15.

12. $2x^2 + 2y^2 + 6yz + 10z^2 = 9$ **13.** $2xy + z^2 + 1 = 0$

14. $2xz + y^2 + 2y + 1 = 0$ **15.** $x^2 + 2yz + 4x + 1 = 0$

16. $3x^2 + 2y^2 + 6xz + 3z^2 = 1$ **17.** $x^2 - 8xy + 16y^2 - 3z^2 = 8$

18. $x^2 + 4y^2 - 4xz + 4z^2 = 8$ **19.** $-3x^2 + 2y^2 + 8yz + 16z^2 = 10$

20. $x^2 + y^2 + z^2 - 2xy + 2xz - 2yz = 9$

◐ *In Exercises 21–27, use MATCOMP or NEWT123 or similar software to classify the quadric surface according to Table 7.1.*

21. $x^2 + y^2 + z^2 - 2xy - 2xz - 2yz + 3x - 3z = 8$

22. $3x^2 + 2y^2 + 5z^2 + 4xy + 2yz - 3x + 10y = 4$

23. $2x^2 - 8y^2 + 3z^2 - 4xy + yz + 6xz - 3x - 8z = 3$

24. $3x^2 + 7y^2 + 4z^2 + 6xy - 8yz + 16x = 20$

25. $x^2 + 4y^2 + 16z^2 + 4xy + 8xz + 16yz - 8x + 3y = 8$

26. $x^2 + 6y^2 + 4z^2 - xy - 2xz - 3yz - 9x = 20$

27. $4x^2 - 3y^2 + z^2 + 8xz + 6yz + 2x - 3y = 8$

7.3
Applications to Extrema

FINDING EXTREMA OF FUNCTIONS

We turn now to one of the applications of linear algebra to calculus. We simply state the facts from calculus that we need, trying to make them seem reasonable where possible.

There are many situations in which one desires to maximize or minimize a function of one or more variables. Applications involve maximizing profit, minimizing costs, maximizing speed, minimizing time, and so on. Such problems are of great practical importance.

Polynomial functions are especially easy to work with. Many important functions, such as trigonometric, exponential, and logarithmic functions, can't be expressed by a polynomial formula. However, each of those functions, and many others, can be expressed near a point in the domain of the function as a "polynomial" of infinite degree, or an *infinite series* as it is called. For example, it is shown in calculus that

$$\cos x = 1 - \left(\frac{1}{2!}\right) x^2 + \left(\frac{1}{4!}\right) x^4 - \left(\frac{1}{6!}\right) x^6 + \cdots \tag{1}$$

for any number x. [Recall that $2! = 2 \cdot 1$, $3! = 3 \cdot 2 \cdot 1$, and in general, $n! = n(n-1)(n-2) \cdots 3 \cdot 2 \cdot 1$.] We leave to calculus the discussion of the interpretation of the infinite sum

$$1 - \left(\frac{1}{2!}\right) + \left(\frac{1}{4!}\right) - \left(\frac{1}{6!}\right) + \cdots$$

as cos 1. From Eq. (1), it would seem that we should have

$$\cos(2x - y) = 1 - \left(\frac{1}{2!}\right) (2x - y)^2 + \left(\frac{1}{4!}\right) (2x - y)^4 - \left(\frac{1}{6!}\right) (2x - y)^6 + \cdots . \tag{2}$$

Note that the term

$$-\left(\frac{1}{2!}\right) (2x - y)^2 = (-\tfrac{1}{2})(4x^2 - 4xy + y^2)$$

is a quadratic form, that is, a form (homogeneous polynomial) of degree 2. Similarly,

$$\left(\frac{1}{4!}\right) (2x - y)^4 = (\tfrac{1}{24})(16x^4 - 32x^3y + 24x^2y^2 - 8xy^3 + y^4)$$

is a form of degree 4.

Consider a function $g(x_1, x_2, \ldots, x_n)$ of n variables, which we denote as usual by $g(\mathbf{x})$. For many of the most common functions $g(\mathbf{x})$ it can be shown that if $\mathbf{x}$ is near $\mathbf{0}$, so that all x_i are small in magnitude, then

$$g(\mathbf{x}) = c + f_1(\mathbf{x}) + f_2(\mathbf{x}) + f_3(\mathbf{x}) + \cdots + f_n(\mathbf{x}) + \cdots , \tag{3}$$

where each $f_i(\mathbf{x})$ is a form of degree i or is zero. Equation (2) illustrates this; forms of odd degree there are all zero, so we did not write them in Eq. (2).

We wish to determine whether $g(\mathbf{x})$ in Eq. (3) has a *local extremum,* that is, a *local maximum* or *local minimum* at the origin $\mathbf{x} = \mathbf{0}$. A function $g(\mathbf{x})$ has a **local maximum at 0** if $g(\mathbf{x}) \le g(\mathbf{0})$ for all $\mathbf{x}$, where $\|\mathbf{x}\|$ is sufficiently small. The notion of a **local minimum at 0** for $g(\mathbf{x})$ is analogously defined. For example, the function of one variable $g(x) = 1 + x^2$, whose graph is shown in Fig. 7.16, has a local minimum of 1 at $x = 0$.

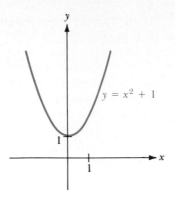

Figure 7.16
The graph of $y = g(x) = x^2 + 1$.

Now if $\mathbf{x}$ is really close to $\mathbf{0}$, all the forms $f_1(\mathbf{x}), f_2(\mathbf{x}), f_3(\mathbf{x}), \ldots$ in Eq. (3) have very small values, so the constant c, if $c \neq 0$, is the dominant term of Eq. (3) near zero. Note that for $g(x) = 1 + x^2$ in Fig. 7.16, the function has values close to the constant 1 for x close to zero.

After a nonzero constant, the form in Eq. (3) that contributes the most to $g(\mathbf{x})$ for $\mathbf{x}$ close to $\mathbf{0}$ is the nonzero form $f_i(\mathbf{x})$ of lowest degree, for the *lower* the degree of a term of a polynomial, the *more* it contributes when the variables are all near zero. For example, x is greater than x^2 near zero; if $x = 1/100$, then x^2 is only $1/10{,}000$. If $f_1(\mathbf{x})$ itself is the zero form, then $f_2(\mathbf{x})$ is the dominant form of Eq. (3) after a nonzero constant c, and so on.

We claim that if $f_1(\mathbf{x}) \neq 0$ in Eq. (3), then $g(\mathbf{x})$ does not have a local maximum or minimum at $\mathbf{x} = \mathbf{0}$. Suppose

$$f_1(\mathbf{x}) = d_1 x_1 + d_2 x_2 + \cdots + d_n x_n$$

with some coefficient, say d_1, nonzero. If k is small but of the same sign as d_1, then near the point $\mathbf{a} = (k, 0, 0, \ldots, 0)$ we see that $g(\mathbf{x}) \approx c + d_1 k > c$. On the other hand, if k is small but of opposite sign to d_1, then near this point we have $g(\mathbf{x}) < c$. Thus $g(\mathbf{x})$ can have a local maximum or minimum of c at $\mathbf{0}$ only if $f_1(\mathbf{x}) = 0$ for all $\mathbf{x}$.

If $f_1(\mathbf{x}) = 0$ in Eq. (3), then for $\mathbf{x}$ near $\mathbf{0}$, the dominant form $f_i(\mathbf{x})$ in the equation after a nonzero constant term is the (nonzero) quadratic form $f_2(\mathbf{x})$; it can be expected to dominate the terms of higher degree for $\mathbf{x}$ close to $\mathbf{0}$. It seems reasonable that if $f_2(\mathbf{x}) > 0$ for *all* $\mathbf{x} \neq \mathbf{0}$ but near $\mathbf{0}$, then for such $\mathbf{x}$

$$g(\mathbf{x}) = c + (\text{Little bit})$$

and $g(\mathbf{x})$ has a local minimum of c at $\mathbf{0}$. On the other hand, if $f_2(\mathbf{x}) < 0$ for *all* such $\mathbf{x}$, then we expect that

$$g(\mathbf{x}) = c - (\text{Little bit})$$

for these values $\mathbf{x}$, and $g(\mathbf{x})$ has a local maximum of c at $\mathbf{0}$. This is proved in an advanced calculus course.

We know that we can orthogonally diagonalize the form $f_2(\mathbf{x})$ with a substitution $\mathbf{x} = C\mathbf{t}$ to become

$$\lambda_1 t_1{}^2 + \lambda_2 t_2{}^2 + \cdots + \lambda_n t_n{}^2, \tag{4}$$

where the λ_i are the eigenvalues of the symmetric coefficient matrix A of $f_2(\mathbf{x})$. Clearly the form (4) is > 0 for *all* nonzero $\mathbf{t}$, and hence $f_2(\mathbf{x}) > 0$ for *all* nonzero $\mathbf{x}$, if and only if we have *all* $\lambda_i > 0$. Similarly, $f_2(\mathbf{x}) < 0$ for *all* nonzero $\mathbf{x}$ if and only if *all* $\lambda_i < 0$. It is also clear that if some λ_i are positive and some are negative, then the form (4), and hence $f_2(\mathbf{x})$, assume both positive and negative values arbitrarily close to zero.

Definition 7.1   Definite Forms

A quadratic form $f(\mathbf{x})$ is **positive definite** if $f(\mathbf{x}) > 0$ for all nonzero $\mathbf{x}$ in $\mathbb{R}^n$, and is **negative definite** if $f(\mathbf{x}) < 0$ for all such nonzero x.

Our work in Section 7.1 and the statement above give us at once the following theorem and corollary.

Theorem 7.4   Definite Forms

A quadratic form is positive definite if and only if all the eigenvalues of its symmetric coefficient matrix A are positive, and is negative definite if and only if all those eigenvalues are negative.

Corollary   *A Test for Local Extrema*

Let $g(\mathbf{x})$ be a function of n variables given by Eq. (3). Suppose that the form $f_1(\mathbf{x})$ in Eq. (3) is zero. If $f_2(\mathbf{x})$ is positive definite, then $g(\mathbf{x})$ has a local minimum of c at $\mathbf{x} = \mathbf{0}$, whereas if $f_2(\mathbf{x})$ is negative definite, then $g(\mathbf{x})$ has a local maximum of c at $\mathbf{x} = \mathbf{0}$. If $f_2(\mathbf{x})$ assumes both positive and negative values, then $g(\mathbf{x})$ has no local extremum at $\mathbf{x} = \mathbf{0}$.

We have discussed local extrema only at the origin. To do similar work at another point $\mathbf{h} = (h_1, h_2, \ldots, h_n)$, we need only translate axes to this point, letting $\bar{x}_i = x_i - h_i$ so that Eq. (3) becomes

$$g(\bar{\mathbf{x}}) = c + f_1(\bar{\mathbf{x}}) + f_2(\bar{\mathbf{x}}) + \cdots + f_n(\bar{\mathbf{x}}) + \cdots . \tag{5}$$

Our discussion indicates a method of attempting to find local extrema of a function, which we box.

Finding an Extremum of $g(x)$

Step 1   Find a point $\mathbf{h}$ where the form $f_1(\bar{\mathbf{x}}) = f_1(\mathbf{x} - \mathbf{h})$ in Eq. (5) becomes the zero form. (This is the province of calculus.)

> **Step 2** Find the quadratic form $f_2(\bar{\mathbf{x}})$ at the point $\mathbf{h}$. (This also requires calculus.)
>
> **Step 3** Find the eigenvalues of the symmetric coefficient matrix of the quadratic form.
>
> **Step 4** If all eigenvalues are positive, then $g(\bar{\mathbf{x}})$ has a local minimum of c at $\mathbf{h}$. If they are all negative, then $g(\bar{\mathbf{x}})$ has a local maximum of c at $\mathbf{h}$. If eigenvalues of both signs occur, then $g(\bar{\mathbf{x}})$ has no local extremum at $\mathbf{h}$.
>
> **Step 5** If any of the eigenvalues is zero, then further study is necessary.

Exercises 17–22 illustrate some of the things that can occur if step 5 is the case. We shall not tackle steps 1 or 2 since they require calculus but will simply start with equations of the form of Eqs. (3) or (5) that meet the requirements in steps 1 and 2.

Example 1   Let

$$g(x, y) = 3 + (2x^2 - 4xy + 4y^2) + (x^3 + 4xy^2 + y^3).$$

Determine whether $g(x, y)$ has a local extremum at the origin.

Solution   The symmetric coefficient matrix for the quadratic-form portion $2x^2 - 4xy + 4y^2$ of $g(x, y)$ is

$$A = \begin{pmatrix} 2 & -2 \\ -2 & 4 \end{pmatrix}.$$

We find that

$$|A - \lambda I| = \begin{vmatrix} 2 - \lambda & -2 \\ -2 & 4 - \lambda \end{vmatrix} = \lambda^2 - 6\lambda + 4.$$

The solutions of the characteristic equation $\lambda^2 - 6\lambda + 4 = 0$ are found by the quadratic formula to be

$$\lambda = \frac{6 \pm \sqrt{36 - 16}}{2} = 3 \pm \sqrt{5}.$$

Clearly $\lambda_1 = 3 + \sqrt{5}$ and $\lambda_2 = 3 - \sqrt{5}$ are both positive, so our form is positive definite. Thus $g(x, y)$ has a minimum of 3 at $(0, 0)$.   ◁

Example 2   Suppose

$$g(x, y, z) = 7 + (2x^2 - 8y^2 + 3z^2 - 4xy + 2yz + 6xz) + (xz^2 - 5y^3)$$
$$+ \text{(higher-degree terms)}.$$

Determine whether $g(x, y, z)$ has a local extremum at the origin.

Solution   The symmetric coefficient matrix of the quadratic-form portion is

$$A = \begin{pmatrix} 2 & -2 & 3 \\ -2 & -8 & 1 \\ 3 & 1 & 3 \end{pmatrix}.$$

We could find $p(\lambda) = |A - \lambda I|$ and attempt to solve the characteristic equation, or try to sketch the graph of $p(\lambda)$ enough to determine the signs of the eigenvalues. However, we prefer to use the available software MATCOMP, or similar software, and find that the eigenvalues are approximately

$$\lambda_1 = -8.605, \qquad \lambda_2 = 0.042, \quad \text{and} \quad \lambda_3 = 5.563.$$

Since both positive and negative eigenvalues appear, there is no local extremum at the origin. ◁

MAXIMIZING OR MINIMIZING A QUADRATIC FORM ON THE UNIT SPHERE

Let $f(\mathbf{x})$ be a quadratic form in the variables $x_1, x_2, \ldots, x_n$. We consider the problem of finding the maximum and minimum values of $f(\mathbf{x})$ for $\mathbf{x}$ on the *unit sphere,* where $\|\mathbf{x}\| = 1$, that is $x_1^2 + x_2^2 + \cdots + x_n^2 = 1$. It is shown in advanced calculus that such extrema of $f(\mathbf{x})$ on the unit sphere always exist.

We need only orthogonally diagonalize the form $f(\mathbf{x})$ using an orthogonal transformation $\mathbf{x} = C\mathbf{t}$, obtaining as usual

$$\lambda_1 t_1^2 + \lambda_2 t_2^2 + \cdots + \lambda_n t_n^2. \tag{6}$$

Since our new basis for $\mathbb{R}^n$ is again orthonormal, the unit sphere has $\mathbf{t}$-equation

$$t_1^2 + t_2^2 + \cdots + t_n^2 = 1; \tag{7}$$

this is a very important point. Suppose the λ_i are arranged so that

$$\lambda_1 \geq \lambda_2 \geq \cdots \geq \lambda_n.$$

On the unit sphere with Eq. (7), we see that formula (6) can be written as

$$\lambda_1(1 - t_2^2 - \cdots - t_n^2) + \lambda_2 t_2^2 + \cdots + \lambda_n t_n^2$$
$$= \lambda_1 - (\lambda_1 - \lambda_2)t_2^2 - \cdots - (\lambda_1 - \lambda_n)t_n^2. \tag{8}$$

Since $\lambda_1 - \lambda_i \geq 0$ for $i > 1$, we see at once that the maximum value assumed by the formula (8) is λ_1 when $t_2 = t_3 = \cdots = t_n = 0$, and $t_1 = \pm 1$. Exercise 32 indicates similarly that the minimum value assumed by the form (6) is λ_n, when $t_n = \pm 1$ and all other $t_i = 0$. We state this as a theorem.

Theorem 7.5   Extrema of a Quadratic Form on the Unit Sphere

Let $f(\mathbf{x})$ be a quadratic form and let $\lambda_1, \lambda_2, \ldots, \lambda_n$ be the eigenvalues of the symmetric coefficient matrix A of $f(\mathbf{x})$. The maximum value assumed by $f(\mathbf{x})$ on the unit sphere $\|\mathbf{x}\| = 1$ is the maximum of the λ_i, and the minimum value assumed is the minimum of the λ_i. Each extremum is assumed at any eigenvector of length 1 corresponding to the eigenvalue that gives the extremum.

The preceding theorem is very important in vibration applications ranging from aerodynamics to particle physics. In such applications, one often needs to know the eigenvalue of maximum or of minimum magnitude for a symmetric matrix. These eigenvalues are frequently found by using advanced-calculus techniques to maximize or minimize the value of a quadratic form on a unit sphere, rather than using algebraic techniques like those presented in this text for finding eigenvalues. The principal axis theorem (Theorem 7.1) and the preceding theorem are the algebraic foundation for such an analytic approach. We illustrate the preceding theorem with an example, which simply maximizes a quadratic form on a unit sphere by finding the eigenvalues, rather than illustrating the more important reverse procedure.

Example 3   Find the maximum and minimum values assumed by $2xy + 2xz$ on the unit sphere $x^2 + y^2 + z^2 = 1$, and find all points where these extrema are assumed.

Solution   From Example 5 of Section 7.1, we see that the eigenvalues of the symmetric coefficient matrix A and an associated eigenvector are given by

$$\lambda_1 = 0, \qquad \mathbf{v}_1 = (0, -1, 1),$$
$$\lambda_2 = \sqrt{2}, \quad \mathbf{v}_2 = (\sqrt{2}, 1, 1),$$
$$\lambda_3 = -\sqrt{2}, \ \mathbf{v}_3 = (-\sqrt{2}, 1, 1).$$

We see that the maximum value assumed by $2xy + 2xz$ on the unit sphere is $\sqrt{2}$, and it is assumed at the points $\pm(\sqrt{2}/2, 1/2, 1/2)$. The minimum value assumed is $-\sqrt{2}$, and it is assumed at the points $\pm(-\sqrt{2}/2, 1/2, 1/2)$. Note that we normalized our eigenvectors to length 1 so that they lie on the unit sphere.   ◁

The extension of Theorem 7.5 to a sphere centered at the origin with radius other than 1 is quite easy, and is left to Exercise 33.

SUMMARY

1. Let $g(\mathbf{x}) = c + f_1(\mathbf{x}) + f_2(\mathbf{x}) + \cdots + f_n(\mathbf{x}) + \cdots$ near $\mathbf{x} = \mathbf{0}$, where $f_i(\mathbf{x})$ is a form of degree i or is zero.

   a) If $f_1(\mathbf{x}) = 0$ and $f_2(\mathbf{x})$ is positive definite, then $g(\mathbf{x})$ has a local minimum of c at $\mathbf{x} = \mathbf{0}$.

   b) If $f_1(\mathbf{x}) = 0$ and $f_2(\mathbf{x})$ is negative definite, then $g(\mathbf{x})$ has a local maximum of c at $\mathbf{x} = \mathbf{0}$.

   c) If $f_1(\mathbf{x}) \neq 0$ or if the symmetric coefficient matrix of $f_2(\mathbf{x})$ has both positive and negative eigenvalues, then $g(\mathbf{x})$ has no local extremum at $\mathbf{x} = \mathbf{0}$.

2. The natural analogue of number (1) holds at $\mathbf{x} = \mathbf{h}$; just translate the axes to the point $\mathbf{h}$ and replace x_i by $\bar{x}_i = x_i - h_i$.

> **3.** A quadratic form in n variables has as maximum (minimum) value on the unit sphere $\|\mathbf{x}\| = 1$ in $\mathbb{R}^n$ the maximum (minimum) of the eigenvalues of the symmetric coefficient matrix of the form. The maximum (minimum) is assumed at each corresponding eigenvector of length 1.

EXERCISES

In Exercises 1–15, assume that $g(\mathbf{x})$, or $g(\bar{\mathbf{x}})$, is described by the given formula for values of $\mathbf{x}$, or $\bar{\mathbf{x}}$, near zero. Draw whatever conclusions are possible concerning local extrema of the function g.

1. $g(x, y) = -7 + (3x^2 - 6xy + 4y^2) + (x^3 - 4y^3)$

2. $g(x, y) = 8 - (2x^2 - 8xy + 3y^2) + (2x^2y - y^3)$

3. $g(x, y) = 4 - 3x + (2x^2 - 2xy + y^2) + (2x^2y + y^3) + \cdots$

4. $g(x, y) = 5 - (8x^2 - 6xy + 2y^2) + (4x^3 - xy^2) + \cdots$

5. $g(\bar{x}, \bar{y}) = 3 - (4\bar{x}^2 - 8\bar{x}\bar{y} + 5\bar{y}^2) + (2\bar{x}^2\bar{y} - \bar{y}^3) + \cdots, \bar{x} = x + 5, \bar{y} = y$

6. $g(x, y) = -2 + (8x^2 + 4xy + y^2) + (x^3 + 5x^2y)$

7. $g(x, y) = 5 + (3x^2 + 10xy + 7y^2) + (7xy^2 - y^3)$

8. $g(x, y) = 4 - (x^2 - 6xy + 9y^2) + (x^3 - y^3) + \cdots$

9. $g(\bar{x}, \bar{y}) = 3 + (2\bar{x}^2 + 8\bar{x}\bar{y} + 8\bar{y}^2) + (4\bar{x}^3 - \bar{x}\bar{y}^2) + \cdots, \bar{x} = x - 3, \bar{y} = y - 1$

10. $g(\bar{x}, \bar{y}) = 4 - (\bar{x}^2 + 3\bar{x}\bar{y} \quad \bar{y}^2) + (4x^2y - 5\bar{x}\bar{y}^2), \bar{x} = x + 1, \bar{y} = y - 7$

11. $g(x, y, z) = 4 - (x^2 + 4xy + 5y^2 + 3z^2) + (x^3 - xyz)$

12. $g(x, y, z) = 3 + (2x^2 + 6xz - y^2 + 5z^2) + (x^2z - y^2z) + \cdots$

13. $g(x, y, z) = 5 + (4x^2 + 2xy + y^2 - z^2) + (xy^2 - 4xyz) + \cdots$

14. $g(\bar{x}, \bar{y}, \bar{z}) = 4 + (\bar{x}^2 + \bar{y}^2 + \bar{z}^2 - 2\bar{x}\bar{z} - 2\bar{x}\bar{y} - 2\bar{y}\bar{z}) + (3\bar{x}^3 - \bar{z}^3) + \cdots, \bar{x} = x + 1, \bar{y} = y - 2, \bar{z} = z + 5$

15. $g(\bar{x}, \bar{y}, \bar{z}) = 4 - 3\bar{z} + (2\bar{x}^2 - 2\bar{x}\bar{y} + 3\bar{x}z + 5\bar{y}^2 + \bar{z}^2) + (2\bar{x}\bar{y}\bar{z} - \bar{z}^3) + \cdots, \bar{x} = x - 7, \bar{y} = y + 6, \bar{z} = z$

16. Define the notion of a local minimum c of a function $f(\mathbf{x})$ of n variables at a point $\mathbf{h}$.

In Exercises 17–22, let

$$g(x, y) = c + f_1(x, y) + f_2(x, y) + f_3(x, y) + \cdots$$

in accordance with the notation we have been using. Let λ_1 and λ_2 be the eigenvalues of the symmetric coefficient matrix of $f_2(x, y)$.

17. Give an example of a polynomial function $g(x, y)$ for which $f_1(x, y) = 0$, $\lambda_1 = 0$, $|\lambda_2| = 1$, and $g(x, y)$ has a local minimum of 10 at the origin.

18. Repeat Exercise 17, but make $g(x, y)$ have a local maximum of -5 at the origin.

19. Repeat Exercise 17, but make $g(x, y)$ have no local extremum at the origin.

20. Give an example of a polynomial function $g(x, y)$ such that $f_1(x, y) = f_2(x, y) = 0$, but which has a local maximum of 1 at the origin.

21. Repeat Exercise 20, but make $g(x, y)$ have a local minimum of 40 at the origin.

22. Repeat Exercise 20, but make $g(x, y)$ have no local extremum at the origin.

In Exercises 23–31, find the maximum and the minimum values of the quadratic form on the unit circle in $\mathbb{R}^2$, or unit sphere in $\mathbb{R}^n$, and find all *points on the unit circle or sphere where the extrema are assumed.*

23. xy, $n = 2$

24. $3x^2 + 4xy$, $n = 2$

25. $-6xy + 8y^2$, $n = 2$

26. $x^2 + 2xy + y^2$, $n = 3$

27. $3x^2 - 6xy - 3y^2$, $n = 2$

28. $y^2 + 2xz$, $n = 3$

29. $x^2 + y^2 + z^2 - 2xy + 2xz - 2yz$, $n = 3$

30. $x^2 + y^2 + z^2 - 2xy - 2xz - 2yz$, $n = 3$ [*Suggestion:* Use Example 2 in Section 6.5.]

31. $x^2 + w^2 + 4yz - 2xw$, $n = 4$

32. Show that the minimum value assumed by a quadratic form in n variables on the unit sphere in $\mathbb{R}^n$ is the minimum eigenvalue of the symmetric coefficient matrix of the form, and that this minimum value is assumed at any corresponding eigenvector of length 1.

33. Let $f(\mathbf{x})$ be a quadratic form in n variables. Describe the maximum and minimum values assumed by $f(\mathbf{x})$ on the sphere $\|\mathbf{x}\| = a^2$ for any $a > 0$. Describe the points of the sphere at which these extrema are assumed. [*Hint:* For any k in $\mathbb{R}$, how does $f(k\mathbf{x})$ compare with $f(\mathbf{x})$?]

🔋 *In Exercises 34–39, use MATCOMP, or similar software, and follow the directions for Exercises 1–15.*

34. $g(x, y, z) = 7 + (3x^2 + 2y^2 + 5z^2 + 4xy + 2yz) + (xyz - x^2y)$

35. $g(x, y, z) = 5 - (x^2 + 6y^2 + 4z^2 - xy - 2xz - 3yz) + (xz^2 - 5z^3) + \cdots$

36. $g(x, y, z) = -1 - (2x^2 - 8y^2 + 3z^2 - 4xy + yz + 6xz) + (8x^3 - 4y^2z + 3z^3)$

37. $g(x, y, z) = 7 + (x^2 + 2y^2 + z^2 - 3xy - 4xz - 3yz) + (x^2y - 4yz^2)$

38. $g(x, y, z) = 5 + (x^2 + 4y^2 + 16z^2 + 4xy + 8xz + 16yz) + (x^3 + y^3 + 7xyz) + \cdots$

7.4
Computing Eigenvalues and Eigenvectors

Computation of the eigenvalues of a matrix is one of the toughest jobs in linear algebra. Many algorithms have been developed, but no one method can be considered the best for all cases. We have used the characteristic equation for computation of eigenvalues in most of our examples and exercises. The available program MATCOMP also uses it. Professionals frown on this method for practical use because small errors made in computing coefficients of the characteristic polynomial can lead to significant errors in eigenvalues. We describe three other methods in this section.

1. The *power method* is especially useful if one wants only the eigenvalue of largest (or of smallest) magnitude, as in many vibration problems.

2. *Jacobi's method* for symmetric matrices is presented without proof. This method is chiefly of historical interest, for the third and most recent of the methods we describe is more general and usually more efficient.

3. The *QR method* was developed by H. Rutishauser in 1958 and J. G. F. Francis in 1961 and is probably the most widely used method today. It finds all eigenvalues, both real and complex, of a real matrix. Details are beyond the scope of this text. We give only a rough idea of the method with no proofs.

The available programs POWER, JACOBI, and QRFACTOR can be used to illustrate the three methods.

THE POWER METHOD

Let A be an $n \times n$ diagonalizable matrix with real eigenvalues. Suppose that one eigenvalue, say λ_1, has greater magnitude than all the others. That is, $|\lambda_1| > |\lambda_i|$ for $i > 1$. We call λ_1 the **dominant eigenvalue** of A. In many vibration problems, one is interested only in computing this dominant eigenvalue as well as the eigenvalue of minimum absolute value.

Since A is diagonalizable, there exists a basis

$$\{\mathbf{b}_1, \mathbf{b}_2, \ldots, \mathbf{b}_n\}$$

for $\mathbb{R}^n$ composed of eigenvectors of A. We assume that $\mathbf{b}_i$ is the eigenvector corresponding to λ_i, and that the numbering is such that $|\lambda_1| > |\lambda_2| \geq |\lambda_3| \geq \cdots \geq |\lambda_n|$.

Let $\mathbf{w}_1$ be any nonzero vector in $\mathbb{R}^n$. Then

$$\mathbf{w}_1 = c_1\mathbf{b}_1 + c_2\mathbf{b}_2 + \cdots + c_n\mathbf{b}_n \tag{1}$$

for some constants c_i in $\mathbb{R}$. Applying A^s to both sides of Eq. (1) and remembering that $A^s\mathbf{b}_i = \lambda_i^s\mathbf{b}_i$, we see that

$$A^s\mathbf{w}_1 = \lambda_1^s c_1\mathbf{b}_1 + \lambda_2^s c_2\mathbf{b}_2 + \cdots + \lambda_n^s c_n\mathbf{b}_n. \tag{2}$$

Since λ_1 is dominant, we see that for large s, the summand $\lambda_1^s c_1\mathbf{b}_1$ dominates the right-hand side of Eq. (2), so long as $c_1 \neq 0$. This is even more evident if we rewrite Eq. (2) in the form

$$A^s\mathbf{w}_1 = \lambda_1^s[c_1\mathbf{b}_1 + (\lambda_2/\lambda_1)^s c_2\mathbf{b}_2 + \cdots + (\lambda_n/\lambda_1)^s c_n\mathbf{b}_n]. \tag{3}$$

If s is large, the quotients $(\lambda_i/\lambda_1)^s$ for $i > 1$ are close to zero, because $|\lambda_i/\lambda_1| < 1$. Thus if $c_1 \neq 0$ and s is large enough, then $A^s\mathbf{w}_1$ is very nearly parallel to $\lambda_1^s c_1\mathbf{b}_1$, which is an eigenvector of A corresponding to the eigenvector λ_1. This suggests that we can approximate an eigenvector of A corresponding to the dominant eigenvalue λ_1 by multiplying an appropriate initial approximation vector $\mathbf{w}_1$ repeatedly by A.

A few comments are in order. In a practical application, one may have a rough idea of an eigenvector for λ_1 and be able to choose a reasonably decent first approximation $\mathbf{w}_1$. In any case, $\mathbf{w}_1$ *should not* be in the subspace of $\mathbb{R}^n$ generated by the eigenvectors corresponding to the λ_j for $j > 1$.

Repeated multiplication of $\mathbf{w}_1$ by A may produce very large (or very small) numbers. It is customary to scale after each multiplication to keep the components of the vectors at a reasonable size. After the first multiplication, we find the maximum d_1 of the magnitudes of all the components of $A\mathbf{w}_1$ and apply A next time to the vector $\mathbf{w}_2 = (1/d_1)A\mathbf{w}_1$. Similarly, we let $\mathbf{w}_3 = (1/d_2)A\mathbf{w}_2$, where d_2 is the maximum of the magnitudes of components of $A\mathbf{w}_2$, and so on. Thus we are always multiplying A times a vector $\mathbf{w}_j$ with components of maximum magnitude 1. This scaling also aids us in estimating the number-of-significant-figure accuracy we have attained in the components of our approximations to an eigenvector.

If $\mathbf{x}$ is an eigenvector corresponding to λ_1, then

$$\frac{A\mathbf{x} \cdot \mathbf{x}}{\mathbf{x} \cdot \mathbf{x}} = \frac{\lambda_1 \mathbf{x} \cdot \mathbf{x}}{\mathbf{x} \cdot \mathbf{x}} = \lambda_1. \tag{4}$$

The quotient $(A\mathbf{x} \cdot \mathbf{x})/(\mathbf{x} \cdot \mathbf{x})$ is called a **Rayleigh quotient.** As we compute the $\mathbf{w}_j$, the Rayleigh quotients $(A\mathbf{w}_j \cdot \mathbf{w}_j)/(\mathbf{w}_j \cdot \mathbf{w}_j)$ should approach λ_1.

This *power method* for finding the dominant eigenvector should, mathematically, break down if we choose the initial approximation $\mathbf{w}_1$ in Eq. (1) in such a way that the coefficient c_1 of $\mathbf{b}_1$ is zero. However, due to roundoff error, it often happens that a nonzero component of $\mathbf{b}_1$ creeps into the $\mathbf{w}_j$ as they are computed, and the $\mathbf{w}_j$ then start swinging toward an eigenvector for λ_1 as desired. This is one case where roundoff error is helpful!

Equation (3) indicates that the ratio $|\lambda_2/\lambda_1|$, which is the maximum of the magnitudes $|\lambda_i/\lambda_1|$ for $i > 1$, should control the speed of convergence of the $\mathbf{w}_j$ to an eigenvector. If $|\lambda_2/\lambda_1|$ is close to 1, then convergence may be quite slow.

We summarize the steps of the power method in the following box.

The Power Method for Finding the Dominant Eigenvalue λ_1 of A

Step 1 Choose an appropriate vector $\mathbf{w}_1$ in $\mathbb{R}^n$ as first approximation to an eigenvector for λ_1.

Step 2 Compute $A\mathbf{w}_1$ and the Rayleigh quotient $(A\mathbf{w}_1 \cdot \mathbf{w}_1)/(\mathbf{w}_1 \cdot \mathbf{w}_1)$.

Step 3 Let $\mathbf{w}_2 = (1/d_1)A\mathbf{w}_1$, where d_1 is the maximum of the magnitudes of components of $A\mathbf{w}_1$.

Step 4 Go to step 2, and repeat with all subscripts increased by 1. The Rayleigh quotients should approach λ_1, and the $\mathbf{w}_j$ should approach an eigenvector of A corresponding to λ_1.

THE RAYLEIGH QUOTIENT is named for John William Strutt, the third Baron Rayleigh (1842–1919). Rayleigh was a hereditary peer who surprised his family by pursuing a scientific career instead of contenting himself with the life of a country gentleman. He set up a laboratory at the family seat in Terling Place, Essex, and spent most of his life there pursuing his research into many aspects of physics, in particular sound and optics. He is particularly famous for his resolution of the long-standing question in optics as to why the sky is blue, as well as for his codiscovery of the element argon, for which he won the Nobel prize in 1904. When he received the British Order of Merit in 1902 he said that "the only merit of which he personally was conscious was that of having pleased himself by his studies, and any results that may have been due to his researches were owing to the fact that it had been a pleasure to him to become a physicist."

Rayleigh used the Rayleigh quotient early in his career in an 1873 work in which he needed to evaluate approximately the normal modes of a complex vibrating system. He subsequently used it and related methods in his classic text *The Theory of Sound* (1877).

Example 1 Illustrate the power method for the matrix

$$A = \begin{pmatrix} 3 & -2 \\ -2 & 0 \end{pmatrix} \quad \text{starting with} \quad \mathbf{w}_1 = \begin{pmatrix} 1 \\ -1 \end{pmatrix},$$

finding $\mathbf{w}_2$ and $\mathbf{w}_3$, and the first two Rayleigh quotients.

Solution We have

$$A\mathbf{w}_1 = \begin{pmatrix} 3 & -2 \\ -2 & 0 \end{pmatrix} \begin{pmatrix} 1 \\ -1 \end{pmatrix} = \begin{pmatrix} 5 \\ -2 \end{pmatrix}.$$

We find as first Rayleigh quotient

$$(A\mathbf{w}_1 \cdot \mathbf{w}_1)/(\mathbf{w}_1 \cdot \mathbf{w}_1) = 7/2 = 3.5.$$

Since 5 is the maximum magnitude of a component of $A\mathbf{w}_1$, we have

$$\mathbf{w}_2 = (1/5)A\mathbf{w}_1 = \begin{pmatrix} 1 \\ -2/5 \end{pmatrix}.$$

Then

$$A\mathbf{w}_2 = \begin{pmatrix} 3 & -2 \\ -2 & 0 \end{pmatrix} \begin{pmatrix} 1 \\ -2/5 \end{pmatrix} = \begin{pmatrix} 19/5 \\ -2 \end{pmatrix}.$$

The next Rayleigh quotient is

$$(A\mathbf{w}_2 \cdot \mathbf{w}_2)/(\mathbf{w}_2 \cdot \mathbf{w}_2) = (23/5)/(29/25) = 115/29 \approx 3.966.$$

Finally,

$$\mathbf{w}_3 = (5/19) \begin{pmatrix} 19/5 \\ -2 \end{pmatrix} = \begin{pmatrix} 1 \\ -10/19 \end{pmatrix}. \quad \triangleleft$$

Example 2 Use the available program POWER or similar software with the data in Example 1, and give the vectors $\mathbf{w}_j$ and the Rayleigh quotients until stabilization to all decimal places printed occurs.

Solution POWER does all computation in double-precision arithmetic but prints the vectors $\mathbf{w}_j$ in single precision to save space. It prints the Rayleigh quotients in double precision. The data obtained using POWER are shown in Table 7.2. It is easy to solve the characteristic equation $\lambda^2 - 3\lambda - 4 = 0$ of A, and see that the eigenvalues are really $\lambda_1 = 4$ and $\lambda_2 = -1$. POWER found the dominant eigenvalue 4, and shows that $(1, -.5)$ is an eigenvector for this eigenvalue. $\quad \triangleleft$

Note that if A has an eigenvalue of nonzero magnitude *smaller* than that of any other eigenvalue, then the power method can be used with A^{-1} to find this smallest eigenvalue. Just recall that the eigenvalues of A^{-1} are the reciprocals of the eigenvalues of A, and the eigenvectors are the same. To illustrate, if 5 is the dominant eigenvalue of A^{-1} and $\mathbf{v}$ is an associated eigenvector, then $1/5$ is the eigenvalue of A of smallest magnitude, and $\mathbf{v}$ is still an associated eigenvector.

Table 7.2 **Power Method for** $A = \begin{pmatrix} 3 & -2 \\ -2 & 0 \end{pmatrix}$

Vector Approximations	Rayleigh Quotients
$(1, -1)$	
$(1, -.4)$	3.5
$(1, -.5263158)$	3.96551724137931
$(1, -.4935065)$	3.997830802603037
$(1, -.5016286)$	3.999864369998644
$(1, -.4995932)$	3.999991522909338
$(1, -.5001018)$	3.999999470180991
$(1, -.4999746)$	3.999999966886309
$(1, -.5000064)$	3.999999997930394
$(1, -.4999984)$	3.99999999987065
$(1, -.5000004)$	3.999999999991916
$(1, -.4999999)$	3.999999999999495
$(1, -.5)$	3.999999999999968
$(1, -.5)$	3.999999999999998
$(1, -.5)$	4
$(1, -.5)$	4

DEFLATION FOR SYMMETRIC MATRICES

The method of deflation gives a way to compute eigenvalues of intermediate magnitude of a *symmetric* matrix by the power method. It is based on an interesting decomposition of a matrix product AB. Let A be an $m \times n$ matrix and B an $n \times s$ matrix. We write AB symbolically as

$$AB = \begin{pmatrix} \mid & \mid & & \mid \\ \mathbf{c}_1 & \mathbf{c}_2 & \cdots & \mathbf{c}_n \\ \mid & \mid & & \mid \end{pmatrix} \begin{pmatrix} -\mathbf{r}_1- \\ -\mathbf{r}_2- \\ \vdots \\ -\mathbf{r}_n- \end{pmatrix},$$

where $\mathbf{c}_j$ is the jth column vector of A and $\mathbf{r}_i$ is the ith row vector of B. We claim that

$$AB = \mathbf{c}_1 \mathbf{r}_1 + \mathbf{c}_2 \mathbf{r}_2 + \cdots + \mathbf{c}_n \mathbf{r}_n, \tag{5}$$

where each $\mathbf{c}_i \mathbf{r}_i$ is the product of an $m \times 1$ matrix with a $1 \times s$ matrix. To see this, remember that the entry in the ith row and jth column of AB is

$$a_{i1}b_{1j} + a_{i2}b_{2j} + \cdots + a_{in}b_{nj}.$$

But $\mathbf{c}_1\mathbf{r}_1$ contributes precisely $a_{i1}b_{1j}$ to the ith row and jth column of the sum in Eq. (5), while $\mathbf{c}_2\mathbf{r}_2$ contributes $a_{i2}b_{2j}$, and so on. This establishes Eq. (5).

Let A be an $n \times n$ *symmetric* matrix with eigenvalues $\lambda_1, \lambda_2, \ldots, \lambda_n$, where

$$|\lambda_1| \geq |\lambda_2| \geq \cdots \geq |\lambda_n|.$$

We know from Section 6.5 that there exists an orthogonal matrix C such that $C^{-1}AC = D$, where D is a diagonal matrix with $d_{ii} = \lambda_i$. Then

$$A = CDC^{-1} = CDC^T$$

$$= \begin{pmatrix} | & | & & | \\ \mathbf{b}_1 & \mathbf{b}_2 & \ldots & \mathbf{b}_n \\ | & | & & | \end{pmatrix} \begin{pmatrix} \lambda_1 & & & \\ & \lambda_2 & & \bigcirc \\ & & \ddots & \\ & \bigcirc & & \lambda_n \end{pmatrix} \begin{pmatrix} -\mathbf{b}_1- \\ -\mathbf{b}_2- \\ \vdots \\ -\mathbf{b}_n- \end{pmatrix}, \tag{6}$$

where $\{\mathbf{b}_1, \mathbf{b}_2, \ldots, \mathbf{b}_n\}$ is an orthonormal basis for $\mathbb{R}^n$ consisting of eigenvectors of A and forming the column vectors of C. From Eq. (6), we see that

$$A = \begin{pmatrix} | & | & & | \\ \mathbf{b}_1 & \mathbf{b}_2 & \ldots & \mathbf{b}_n \\ | & | & & | \end{pmatrix} \begin{pmatrix} -\lambda_1\mathbf{b}_1- \\ -\lambda_2\mathbf{b}_2- \\ \vdots \\ -\lambda_n\mathbf{b}_n- \end{pmatrix}, \tag{7}$$

Using Eq. (5) and writing the $\mathbf{b}_i$ as column vectors, we see that A can be written in the following form:

Spectral Decomposition of A

$$A = \lambda_1\mathbf{b}_1\mathbf{b}_1{}^T + \lambda_2\mathbf{b}_2\mathbf{b}_2{}^T + \cdots + \lambda_n\mathbf{b}_n\mathbf{b}_n{}^T. \tag{8}$$

Equation (8) is known as the **spectral theorem** for symmetric matrices.

Example 3 Illustrate the spectral theorem for the matrix

$$A = \begin{pmatrix} 3 & -2 \\ -2 & 0 \end{pmatrix}$$

in Example 1.

THE TERM "SPECTRUM" was coined around 1905 by David Hilbert (1862–1943) for use in dealing with the eigenvalues of quadratic forms in infinitely many variables. The notion of such forms came out of his study of certain linear operators in spaces of functions. Hilbert was struck by the analogy one could make between these operators and quadratic forms in finitely many variables. Hilbert's approach was greatly expanded and generalized during the next decade by Erhard Schmidt, Frigyes Riesz (1880–1956), and Hermann Weyl. Interestingly enough, in the 1920s physicists called on spectra of certain linear operators to explain optical spectra.

David Hilbert was the most influential mathematician of the early twentieth century. He made major contributions in many fields, including algebraic forms, algebraic number theory, integral equations, foundations of geometry, theoretical physics, and the foundations of mathematics. His speech at the 2nd International Congress of Mathematicians in Paris in 1900 outlining the important mathematical problems of the day proved extremely significant in providing the direction for twentieth-century mathematics.

Solution An easy computation shows that the matrix A has eigenvalues $\lambda_1 = 4$ and $\lambda_2 = -1$, with corresponding eigenvectors

$$\mathbf{v}_1 = \begin{pmatrix} 1 \\ -1/2 \end{pmatrix} \quad \text{and} \quad \mathbf{v}_2 = \begin{pmatrix} 1/2 \\ 1 \end{pmatrix}.$$

Eigenvectors $\mathbf{b}_1$ and $\mathbf{b}_2$ for an orthonormal basis are

$$\mathbf{b}_1 = \begin{pmatrix} 2/\sqrt{5} \\ -1/\sqrt{5} \end{pmatrix} \quad \text{and} \quad \mathbf{b}_2 = \begin{pmatrix} 1/\sqrt{5} \\ 2/\sqrt{5} \end{pmatrix}.$$

We have

$$\lambda_1 \mathbf{b}_1 \mathbf{b}_1^T + \lambda_2 \mathbf{b}_2 \mathbf{b}_2^T = 4 \begin{pmatrix} 2/\sqrt{5} \\ -1/\sqrt{5} \end{pmatrix}(2/\sqrt{5},\, -1/\sqrt{5}) - \begin{pmatrix} 1/\sqrt{5} \\ 2/\sqrt{5} \end{pmatrix}(1/\sqrt{5},\, 2/\sqrt{5})$$

$$= 4 \begin{pmatrix} 4/5 & -2/5 \\ -2/5 & 1/5 \end{pmatrix} - 1 \begin{pmatrix} 1/5 & 2/5 \\ 2/5 & 4/5 \end{pmatrix}$$

$$= \begin{pmatrix} 16/5 & -8/5 \\ -8/5 & 4/5 \end{pmatrix} - \begin{pmatrix} 1/5 & 2/5 \\ 2/5 & 4/5 \end{pmatrix} = \begin{pmatrix} 3 & -2 \\ -2 & 0 \end{pmatrix} = A. \quad \triangleleft$$

Suppose now that we have found the eigenvalue λ_1 of maximum magnitude of A, and a corresponding eigenvector $\mathbf{v}_1$ by the power method. We compute the unit vector $\mathbf{b}_1 = \mathbf{v}_1/\|\mathbf{v}_1\|$. From Eq. (8), we see that

$$A - \lambda_1 \mathbf{b}_1 \mathbf{b}_1^T = 0\mathbf{b}_1 \mathbf{b}_1^T + \lambda_2 \mathbf{b}_2 \mathbf{b}_2^T + \cdots + \lambda_n \mathbf{b}_n \mathbf{b}_n^T \tag{9}$$

is a matrix with eigenvalues $\lambda_2, \lambda_3, \ldots, \lambda_n, 0$ in order of descending magnitude, and corresponding eigenvectors $\mathbf{b}_2, \mathbf{b}_3, \ldots, \mathbf{b}_n, \mathbf{b}_1$. We can now use the power method on this matrix to compute λ_2 and $\mathbf{b}_2$. We then execute this *deflation* again, forming $A - \lambda_1 \mathbf{b}_1 \mathbf{b}_1^T - \lambda_2 \mathbf{b}_2 \mathbf{b}_2^T$ to find λ_3 and $\mathbf{b}_3$, and so on.

The available program POWER has an option to use this method of deflation. For the symmetric matrices of the small size that we use in our examples and exercises, POWER handles deflation well, provided that we compute each eigenvalue until stabilization to all places shown on the screen is achieved. In practice, scientists are wary of using deflation to find more than one or two further eigenvalues, since any error made in computation of an eigenvalue or eigenvector will propagate errors in the computation of subsequent ones.

Example 4 Illustrate Eq. (9) for deflation with the matrix

$$A = \begin{pmatrix} 3 & -2 \\ -2 & 0 \end{pmatrix}$$

of Example 3.

Solution From Example 3, we have

$$A = \lambda_1 \mathbf{b}_1 \mathbf{b}_1^T + \lambda_2 \mathbf{b}_2 \mathbf{b}_2^T = 4 \begin{pmatrix} 4/5 & -2/5 \\ -2/5 & 4/5 \end{pmatrix} - \begin{pmatrix} 1/5 & 2/5 \\ 2/5 & 4/5 \end{pmatrix}.$$

Thus

$$A - \lambda_1 \mathbf{b}_1 \mathbf{b}_1{}^T = -\begin{pmatrix} 1/5 & 2/5 \\ 2/5 & 4/5 \end{pmatrix} = \begin{pmatrix} -1/5 & -2/5 \\ -2/5 & -4/5 \end{pmatrix}.$$

The characteristic polynomial for this matrix is

$$\begin{vmatrix} -1/5 - \lambda & -2/5 \\ -2/5 & -4/5 - \lambda \end{vmatrix} = (\lambda^2 + \lambda) = \lambda(\lambda + 1)$$

and the eigenvalues are indeed $\lambda_2 = -1$ and 0, as claimed after Eq. (9). ◁

Example 5   Use the available program POWER or similar software to find the eigenvalues and eigenvectors, using deflation, for the symmetric matrix

$$\begin{pmatrix} 2 & -8 & 5 \\ -8 & 0 & 10 \\ 5 & 10 & -6 \end{pmatrix}.$$

Solution   Using POWER with deflation and finding eigenvalues as accurately as the printing on the screen permits, we obtain the eigenvectors and eigenvalues

$$\mathbf{v}_1 = (-.6042427, -.8477272, 1), \quad \lambda_1 = -17.49848531152027,$$
$$\mathbf{v}_2 = (-.760632, 1, .3881208), \quad \lambda_2 = 9.96626448890372,$$
$$\mathbf{v}_3 = (1, .3958651, .9398283), \quad \lambda_3 = 3.532220822616553.$$

Using MATCOMP as a check, we find the same eigenvalues and eigenvectors. ◁

JACOBI'S METHOD FOR SYMMETRIC MATRICES

We present Jacobi's method for diagonalizing a symmetric matrix, omitting proofs. Let $A = (a_{ij})$ be an $n \times n$ symmetric matrix, and suppose that a_{pq} is an entry of maximum magnitude among the entries of A that lie above the main diagonal. For example, in the matrix

$$A = \begin{pmatrix} 2 & -8 & 5 \\ -8 & 0 & 10 \\ 5 & 10 & -6 \end{pmatrix} \tag{10}$$

the entry above the diagonal having maximum magnitude is 10. Then form the 2×2 matrix

$$\begin{pmatrix} a_{pp} & a_{pq} \\ a_{qp} & a_{qq} \end{pmatrix}. \tag{11}$$

From the matrix (10), we would form the matrix consisting of the portion shown in color, namely,

$$\begin{pmatrix} 0 & 10 \\ 10 & -6 \end{pmatrix}.$$

Let $C = (c_{ij})$ be a 2×2 orthogonal matrix that diagonalizes the matrix (11). [Recall that C can always be chosen to correspond to a rotation in the plane, although this is not essential for Jacobi's method to work.] Now form an $n \times n$ matrix R, the same size as the matrix A, which looks like the identity matrix except that $r_{pp} = c_{11}, r_{pq} = c_{12}, r_{qp} = c_{21}$, and $r_{qq} = c_{22}$. For the matrix (10) where 10 has maximum magnitude above the diagonal, we would have

$$R = \begin{pmatrix} 1 & 0 & 0 \\ 0 & c_{11} & c_{12} \\ 0 & c_{21} & c_{22} \end{pmatrix}.$$

This matrix R will be an orthogonal matrix, with $\det(R) = \det(C)$. Now form the new symmetric matrix $B_1 = R^T A R$, which it is easy to see has zero entries in the pth row, qth column position and in the qth row, pth column position. Other entries in B_1 can also be changed from those in A, but it can be shown that the maximum magnitude of off-diagonal entries has been reduced, assuming that no other above-diagonal entry in A had the magnitude of a_{pq}. Then repeat this process over again, starting with B_1 instead of A to obtain another symmetric matrix B_2, and so on. It can be shown that the maximum magnitude of off-diagonal entries in the matrices B_i approaches zero as i increases. Thus the sequence of matrices

$$B_1, B_2, B_3, \ldots$$

will approach a diagonal matrix D whose eigenvalues $d_{11}, d_{22}, \ldots, d_{nn}$ are of course the same as those of A.

If one is going to use Jacobi's method much, one should find the 2×2 matrix C that diagonalizes a general 2×2 symmetric matrix

$$\begin{pmatrix} a & b \\ b & c \end{pmatrix};$$

that is, one should find formulas for computing C in terms of the entries a, b, and c. Exercise 23 develops such formulas.

Rather than give a tedious pencil-and-paper example of Jacobi's method, we choose to present data generated by the available program JACOBI for the matrix (10) above, which is the same matrix as that of Example 5 where we used the power method. The reader should note how in each step, the colored entries of maximum magnitude off the diagonal are reduced to "zero." While they may not remain zero in the next step, they never return to their original size.

Example 6 Use the program JACOBI to diagonalize the matrix

$$\begin{pmatrix} 2 & -8 & 5 \\ -8 & 0 & 10 \\ 5 & 10 & -6 \end{pmatrix}.$$

Solution   The program JACOBI gives the following matrices:

$$\begin{pmatrix} 2 & -8 & 5 \\ -8 & 0 & 10 \\ 5 & 10 & -6 \end{pmatrix}$$

$$\begin{pmatrix} 2 & 8.786909 & 3.433691 \\ 8.786909 & -13.44031 & 2.220446\text{E-}16 \\ 3.433691 & 5.551115\text{E-}17 & 7.440307 \end{pmatrix}$$

$$\begin{pmatrix} -17.41676 & -1.110223\text{E-}16 & -1.415677 \\ -1.665335\text{E-}16 & 5.97645 & -3.128273 \\ -1.415677 & -3.128273 & 7.440307 \end{pmatrix}$$

$$\begin{pmatrix} -17.41676 & -.8796473 & -1.109216 \\ -.8796473 & 3.495621 & -1.110223\text{E-}16 \\ -1.109216 & 2.220446\text{E-}16 & 9.921136 \end{pmatrix}$$

$$\begin{pmatrix} -17.46169 & -.8789265 & -8.049117\text{E-}16 \\ -.8789265 & 3.495621 & 3.560333\text{E-}02 \\ -8.673617\text{E-}16 & 3.560333\text{E-}02 & 9.966067 \end{pmatrix}$$

$$\begin{pmatrix} -17.49849 & 7.494005\text{E-}16 & 1.489243\text{E-}03 \\ 5.932754\text{E-}16 & 3.532418 & 3.557217\text{E-}02 \\ 1.489243\text{E-}03 & 3.557217\text{E-}02 & 9.966067 \end{pmatrix}$$

$$\begin{pmatrix} -17.49849 & 8.233765\text{E-}06 & -1.48922\text{E-}03 \\ 8.233765\text{E-}06 & 3.532221 & -8.185293\text{E-}15 \\ -1.48922\text{E-}03 & 8.292846\text{E-}15 & 9.966265 \end{pmatrix}$$

$$\begin{pmatrix} -17.49849 & 8.233765\text{E-}06 & -5.015851\text{E-}13 \\ 8.233765\text{E-}06 & 3.532221 & -4.464509\text{E-}10 \\ -5.015263\text{E-}13 & -4.464508\text{E-}10 & 9.966265 \end{pmatrix}$$

The off-diagonal entries are now quite small, and we obtain the same eigenvalues from the diagonal that we did in Example 5 using the power method. ◁

* *QR* ALGORITHM

At the present time, an algorithm based on the *QR* factorization of an *invertible* matrix, discussed in Section 5.4, is often used by professionals to find eigenvalues of a matrix. A full treatment of the *QR* algorithm is beyond the scope of this text, but we give a brief description of the method.

Let A be a nonsingular matrix. The *QR* algorithm generates a sequence of matrices $A_1, A_2, A_3, A_4 \ldots$, all having the same eigenvalues as A. To generate this sequence, let $A_1 = A$ and factor $A_1 = Q_1 R_1$, where Q_1 is the orthogonal matrix and R_1 the upper-triangle matrix described in Section 5.4. Then let

$A_2 = R_1 Q_1$, factor A_2 into $Q_2 R_2$, and set $A_3 = R_2 Q_2$. Continue in this fashion, factoring A_n into $Q_n R_n$ and setting $A_{n+1} = R_n Q_n$. Under fairly general conditions, the matrices A_i will approach an almost upper-triangular matrix of the form

$$
\begin{pmatrix}
X & X & X & X & X & \cdots & X & X \\
X & X & X & X & X & \cdots & X & X \\
0 & 0 & X & X & X & \cdots & X & X \\
0 & 0 & X & X & X & \cdots & X & X \\
0 & 0 & 0 & 0 & X & \cdots & X & X \\
0 & 0 & 0 & 0 & X & \cdots & X & X \\
& & & & \vdots & & & \\
0 & 0 & 0 & 0 & 0 & \cdots & X & X
\end{pmatrix}.
$$

The colored entries just below the main diagonal may or may not be zero. If one of these entries is nonzero, then the 2×2 submatrix having the entry in its lower left corner, like the matrix shaded above, has a pair of complex conjugate numbers $a \pm bi$ as eigenvalues that are also eigenvalues of the large matrix, and hence of A. Entries on the diagonal that do not lie in such a 2×2 block are real eigenvalues of the matrix and of A.

The available program QRFACTOR can be used to illustrate this procedure. A few comments about the procedure and the program are in order.

From $A_1 = Q_1 R_1$, we have $R_1 = Q_1^{-1} A_1$. Then $A_2 = R_1 Q_1 = Q_1^{-1} A_1 Q_1$, so we see that A_2 is similar to $A_1 = A$ and therefore has the same eigenvalues as A. Continuing in this fashion, we see that each matrix A_i is similar to A. This explains why the eigenvalues don't change as the matrices of the sequence are generated.

Note too that Q_1 and Q_1^{-1} are orthogonal matrices, so that $Q_1^{-1}\mathbf{x}$ and $\mathbf{y}Q_1$ have the same magnitude as the vectors $\mathbf{x}$ and $\mathbf{y}$, respectively. It follows easily that if E is the matrix of errors in the entries of A, then the error matrix $Q_1^{-1} E Q_1$ arising in the computation of $Q_1^{-1} A Q_1$ is of magnitude comparable to that of E. That is, the generation of the sequence of matrices A_i is *stable*. This is highly desirable in numerical computations.

Finally, it is often useful to perform a **shift,** adding a scalar multiple rI of the identity matrix to A_i before generating the next matrix A_{i+1}. Such a shift increases all eigenvalues by r (see Exercise 35 of Section 4.5), but we can keep track of the total change due to such shifts and adjust the eigenvalues of the final matrix found to obtain those of A. To illustrate one use of shifts, suppose we wish to find eigenvalues of a singular matrix B. We can form an initial shift, perhaps taking $A = B + (.001)I$ to obtain an invertible matrix A to start the algorithm. For another illustration, we easily find using the program QRFACTOR that the QR algorithm applied to the matrix

$$
\begin{pmatrix}
0 & 1 \\
1 & 0
\end{pmatrix}
$$

generates this same matrix repeatedly, even though the eigenvalues are 1 and -1 rather than complex numbers. This is an example of a matrix A for which the sequence of matrices A_i does not approach a form described above. However, a shift that adds $(.9)I$ produces the matrix

$$\begin{pmatrix} .9 & 1 \\ 1 & .9 \end{pmatrix},$$

which generates a sequence that quickly converges to

$$\begin{pmatrix} 1.9 & 0 \\ 0 & -.1 \end{pmatrix}.$$

Subtracting the scalar .9 from the eigenvalues 1.9 and $-.1$ of this last matrix, we obtain the eigenvalues 1 and -1 of the original matrix.

Shifts can also be used to speed convergence, which is quite fast when the ratios of magnitudes of eigenvalues is large. The program QRFACTOR displays the matrices A_i as they are generated, and allows shifts. If we notice that we are going to obtain an eigenvalue whose decimal expansion starts with 17.52, then we can speed convergence greatly by adding the shift $(-17.52)I$. The resulting eigenvalue will be near zero and the ratios of the magnitudes of other eigenvalues to its magnitude will probably be large. Using this technique with QRFACTOR, it is quite easy to find all eigenvalues, both real and complex, of most matrices of reasonable size that can be displayed conveniently.

Professional programs make many further improvements in the algorithm we have presented, designed to speed the creation of zeros in the lower part of the matrix.

SUMMARY

1. Let A be an $n \times n$ diagonalizable matrix with real eigenvalues λ_j and with a dominant eigenvalue λ_1 of algebraic multiplicity 1, so that $|\lambda_1| > |\lambda_j|$ for $j = 2, 3, \ldots, n$. Start with any vector $\mathbf{w}_1$ in $\mathbb{R}^n$ that is not in the subspace generated by eigenvectors corresponding to the λ_j for $j > 1$. Form the vectors $\mathbf{w}_2 = (A\mathbf{w}_1)/d_1$, $\mathbf{w}_3 = (A\mathbf{w}_2)/d_2, \ldots$ where d_i is the maximum of the magnitudes of the components of $A\mathbf{w}_i$. The sequence of vectors

$$\mathbf{w}_1, \mathbf{w}_2, \mathbf{w}_3, \ldots$$

approaches an eigenvector of A corresponding to λ_1, and the associated Rayleigh quotients $(A\mathbf{w}_i \cdot \mathbf{w}_i)/(\mathbf{w}_i \cdot \mathbf{w}_i)$ approach λ_1. This is the foundation for the power method, which is summarized in the box on page 394.

2. If A is diagonalizable and invertible, and if $|\lambda_n| < |\lambda_i|$ for $i < n$ with λ_n of algebraic multiplicity 1, then the power method may be used with A^{-1} to find λ_n.

3. Let A be an $n \times n$ symmetric matrix with eigenvalues λ_i such that $|\lambda_1| > |\lambda_2| \geq \cdots \geq |\lambda_n|$. If $\mathbf{b}_1$ is a unit eigenvector corresponding to λ_1,

then $A - \lambda_1 \mathbf{b}_1 \mathbf{b}_1^T$ has eigenvalues $\lambda_2, \lambda_3, \ldots, \lambda_n, 0$ and if $|\lambda_2| > |\lambda_3|$, then λ_2 can be found by applying the power method to $A - \lambda_1 \mathbf{b}_1 \mathbf{b}_1^T$. This deflation can be continued with $A - \lambda_1 \mathbf{b}_1 \mathbf{b}_1^T - \lambda_2 \mathbf{b}_2 \mathbf{b}_2^T$ if $|\lambda_3| > |\lambda_4|$, and so on, to find more eigenvalues.

4. In the Jacobi method for diagonalizing a symmetric matrix A, one generates a sequence of symmetric matrices, starting with A. Each matrix of the sequence is obtained from the preceding one by multiplying it on the left by R^T and on the right by R, where R is an orthogonal "rotation" matrix designed to annihilate the two (symmetrically located) entries of maximum magnitude off the diagonal. The matrices in the sequence approach a diagonal matrix D having the same eigenvalues as A.

5. In the QR method, one generates a sequence of matrices A_i by setting $A_1 = A$ and $A_{n+1} = R_n Q_n$, where the QR factorization of A_n is $Q_n R_n$. The matrices A_i approach almost upper-triangular matrices having the real eigenvalues of A on the diagonal and pairs of complex conjugate eigenvalues of A as eigenvalues of 2×2 blocks appearing along the diagonal. Shifts may be used to speed convergence.

EXERCISES

In Exercises 1–4, use the power method to estimate the eigenvalue of maximum magnitude and a corresponding eigenvector for the given matrix. Start with first estimate

$$\mathbf{w}_1 = \begin{pmatrix} 1 \\ 1 \end{pmatrix},$$

and compute $\mathbf{w}_2$, $\mathbf{w}_3$, and $\mathbf{w}_4$. Also find the three Rayleigh quotients. Then find the exact eigenvalues, for comparison, using the characteristic equation.

1. $\begin{pmatrix} 3 & -3 \\ -5 & 1 \end{pmatrix}$

2. $\begin{pmatrix} 3 & -3 \\ 4 & -5 \end{pmatrix}$

3. $\begin{pmatrix} -3 & 10 \\ -3 & 8 \end{pmatrix}$

4. $\begin{pmatrix} -4 & 9 \\ -2 & 5 \end{pmatrix}$

In Exercises 5–8, find the spectral decomposition (8) of the given symmetric matrix.

5. $\begin{pmatrix} 2 & 3 \\ 3 & 2 \end{pmatrix}$

6. $\begin{pmatrix} 3 & 5 \\ 5 & 3 \end{pmatrix}$

7. $\begin{pmatrix} 1 & 1 & 1 \\ 1 & 0 & 0 \\ 1 & 0 & 0 \end{pmatrix}$

8. $\begin{pmatrix} 1 & -1 & -1 \\ -1 & 1 & 0 \\ -1 & 0 & 1 \end{pmatrix}$

[*Hint:* Use Example 2 in Section 6.5.]

In Exercises 9–12, find the matrix obtained by deflation after the (exact) eigenvalue of maximum magnitude and a corresponding eigenvector are found.

9. The matrix in Exercise 5   **10.** The matrix in Exercise 6

11. The matrix in Exercise 7   **12.** The matrix in Exercise 8

In Exercises 13–15, use the available program POWER or similar software to find the eigenvalue of maximum magnitude and a corresponding eigenvector of the given matrix.

13. $\begin{pmatrix} 1 & -44 & -88 \\ -5 & 55 & 113 \\ 1 & -24 & -48 \end{pmatrix}$ **14.** $\begin{pmatrix} 57 & -2 & 31 \\ -205 & 8 & -113 \\ -130 & 4 & -70 \end{pmatrix}$

15. $\begin{pmatrix} 3 & -22 & -46 \\ -3 & 23 & 47 \\ 1 & -10 & -20 \end{pmatrix}$

16. Use POWER with matrix inversion or similar software to find the eigenvalue of minimum magnitude and a corresponding eigenvector of the matrix in Exercise 13.

17. Repeat Exercise 16 for the matrix

$$\begin{pmatrix} -1 & 3 & 1 \\ 3 & 2 & -11 \\ 1 & -11 & 7 \end{pmatrix}.$$

In Exercises 18–21, use POWER and deflation to find all eigenvalues, and corresponding eigenvectors, for the given symmetric matrix. Always continue the method before deflating until as much stabilization as possible is achieved. Note the relationship between ratios of magnitudes of eigenvalues and the speed of convergence.

18. $\begin{pmatrix} 3 & 5 & -7 \\ 5 & 10 & 11 \\ -7 & 11 & 0 \end{pmatrix}$ **19.** $\begin{pmatrix} 0 & -1 & 4 \\ -1 & 2 & -1 \\ 4 & -1 & 0 \end{pmatrix}$

20. $\begin{pmatrix} 3 & 1 & -2 & 1 \\ 1 & 4 & 0 & -3 \\ -2 & 0 & 0 & 5 \\ 1 & -3 & 5 & 2 \end{pmatrix}$ **21.** $\begin{pmatrix} 5 & 7 & -2 & 6 \\ 7 & 4 & 11 & 3 \\ -2 & 11 & -8 & 0 \\ 6 & 3 & 0 & 6 \end{pmatrix}$

22. The eigenvalue option of the available program VECTGRPH is designed to strengthen geometric understanding of the power method. There is no scaling, but the user has the option of starting again if the numbers get so large that the vectors are off the screen. Read the directions in the program, and then run this eigenvalue option until a score of 85% or better can be reliably achieved.

23. Consider the matrix

$$A = \begin{pmatrix} a & b \\ b & c \end{pmatrix}.$$

The following steps will form an orthogonal diagonalizing matrix C for A.

Step 1 Let $g = (a - c)/2$.

Step 2 Let $h = \sqrt{g^2 + b^2}$.

Step 3 Let $r = \sqrt{b^2 + (g + h)^2}$.

Step 4 Let $s = \sqrt{b^2 + (g - h)^2}$.

Step 5 Let

$$C = \begin{pmatrix} -b/r & -b/s \\ (g + h)/r & (g - h)/s \end{pmatrix}.$$

(If $b < 0$ and a rotation matrix is desired, change the sign of a column vector in A.)
Prove this algorithm by showing the following.

a) The eigenvalues of A are

$$\lambda = \frac{a + c \pm \sqrt{(a - c)^2 + 4b^2}}{2}.$$

[*Hint:* Use the quadratic formula.]

b) The first row vector of $A - \lambda I$ is

$$(g \pm \sqrt{g^2 + b^2}, b).$$

c) Finding an orthogonal matrix whose columns are eigenvectors of A, using part (b), one obtains the matrix C.

d) Prove the parenthetical statement in step 5.

24. Use the algorithm in Exercise 23 to find an orthogonal diagonalizing matrix with determinant 1 for each of the following.

a) $\begin{pmatrix} 0 & 1 \\ 1 & 0 \end{pmatrix}$ 　　　　**b)** $\begin{pmatrix} 3 & -2 \\ -2 & 1 \end{pmatrix}$

In Exercises 25–28, use the available program JACOBI or similar software to find the eigenvalues of the matrix by Jacobi's method.

25. The matrix in Exercise 18 　　　　**26.** The matrix in Exercise 19

27. The matrix in Exercise 20 　　　　**28.** The matrix in Exercise 21

In Exercises 29–32, use the available program QRFACTOR to find all eigenvalues, both real and complex, of the given matrix. Be sure to make use of shifts to speed convergence whenever you can see approximately what an eigenvalue will be.

*29. $\begin{pmatrix} -1 & 5 & 7 \\ 3 & -6 & 8 \\ 9 & 0 & -11 \end{pmatrix}$

*30. $\begin{pmatrix} -3 & 6 & -7 & 2 \\ 13 & -18 & 4 & 6 \\ 21 & 32 & -16 & 9 \\ -4 & 8 & 7 & 11 \end{pmatrix}$

*31. $\begin{pmatrix} -12 & 3 & 15 & 2 & -21 \\ 47 & -34 & 87 & 24 & 7 \\ 35 & 72 & 33 & -57 & 82 \\ -145 & 67 & 32 & 10 & 46 \\ -9 & 22 & 21 & -45 & 8 \end{pmatrix}$

*32. $\begin{pmatrix} -3 & 6 & 4 & -2 & 16 \\ 21 & -33 & -5 & 8 & -12 \\ 15 & -21 & 13 & 4 & 20 \\ -18 & 12 & 4 & 8 & 3 \\ -22 & 31 & 14 & 9 & 10 \end{pmatrix}$

8

The Role of Linear Algebra in Calculus *

We now take a brief look at calculus from the point of view of linear algebra. In Sections 8.1 and 8.2 we assume that the reader has studied calculus of functions of one variable, typically, the first semester of a calculus sequence. In Section 8.1 we observe that differentiation and integration can be viewed as linear transformations of vector spaces. In Section 8.2 we explain how the differential of a differentiable function of one variable provides a local linear approximation of the function at each point in its domain. Newton's method and change of variable in integration are examined from this viewpoint. Section 8.3 requires knowledge of calculus of functions of several variables and develops material for those functions analogous to that in Section 8.2. Section 8.4 indicates how diagonalization of a matrix can be used to solve some systems of differential equations.

8.1
Linearity of Differentiation and Integration

DIFFERENTIATION AS A LINEAR TRANSFORMATION

Let F be the set of all functions $f \colon \mathbb{R} \to \mathbb{R}$. In Section 3.2 we showed that F is a vector space with the usual notions of addition of two functions and of scalar multiplication. Let F_∞ consist of the functions in F that have derivatives of all

---

* For those who have had some calculus.

orders at every point in $\mathbb{R}$. For example, the polynomial functions as well as the functions $\cos x$, $\sin x$, and e^x are in F_∞. We know from calculus that if $f'(a)$ and $g'(a)$ both exist, then

$$(f + g)'(a) = f'(a) + g'(a) \quad \text{and} \quad (cf)'(a) = c[f'(a)]. \tag{1}$$

In particular, $(f + g)'$ and $(cf)'$ both exist. The same is true for derivatives of higher order. Thus F_∞ is closed under addition and scalar multiplication, so it is a vector space, that is, a subspace of F.

Let D denote differentiation, so that $Df = f'$, the derived function of f. We can regard D as a map of F_∞ into itself, that is,

$$D: F_\infty \to F_\infty.$$

It is customary to write Df, as we did above, in place of $D(f)$. Equation (1) can be written

$$D(f + g) = Df + Dg \quad \text{and} \quad D(cf) = c(Df). \tag{2}$$

This shows that the differentiation map D is a *linear transformation* of F_∞ into itself.

We let P be the set of all polynomial functions, which is a subspace of F_∞, and denote by P_n the subspace consisting of all polynomial functions of degree $\leq n$. We will denote the restriction of our linear transformation $D: F_\infty \to F_\infty$ to any subspace of F_∞ again by D, to keep our notation simple. Thus we have a linear transformation $D: P_n \to P_{n-1}$ for each positive integer n. Since P_{n-1} is contained in P_n, we can also write $D: P_n \to P_n$.

Example 1   Let $B = (x^4, x^3, x^2, x, 1)$, which is an ordered basis for P_4. Find the matrix representation A_D of $D: P_4 \to P_4$ relative to B,B and illustrate how A_D can be used to differentiate the polynomial function

$$3x^4 - 5x^3 + 7x^2 - 8x + 2.$$

Solution   Since $D(x^n) = nx^{n-1}$, we see that

$$A_D = \begin{pmatrix} 0 & 0 & 0 & 0 & 0 \\ 4 & 0 & 0 & 0 & 0 \\ 0 & 3 & 0 & 0 & 0 \\ 0 & 0 & 2 & 0 & 0 \\ 0 & 0 & 0 & 1 & 0 \end{pmatrix}.$$

Now the coordinate vector of $p(x) = 3x^4 - 5x^3 + 7x^2 - 8x + 2$ relative to B is

$$p(x)_B = \begin{pmatrix} 3 \\ -5 \\ 7 \\ -8 \\ 2 \end{pmatrix}.$$

Applying Theorem 6.8 on page 326, we have

$$D[p(x)]_B = A_D \cdot p(x)_B = \begin{pmatrix} 0 & 0 & 0 & 0 & 0 \\ 4 & 0 & 0 & 0 & 0 \\ 0 & 3 & 0 & 0 & 0 \\ 0 & 0 & 2 & 0 & 0 \\ 0 & 0 & 0 & 1 & 0 \end{pmatrix} \begin{pmatrix} 3 \\ -5 \\ 7 \\ -8 \\ 2 \end{pmatrix} = \begin{pmatrix} 0 \\ 12 \\ -15 \\ 14 \\ -8 \end{pmatrix},$$

so $p'(x) = 12x^3 - 15x^2 + 14x - 8$. ◁

Computing a second derivative of a function f in F_∞, we obtain $D(Df)$, which we denote by D^2f. Thus D^2 corresponds to the composite transformation $D \circ D$, which we know is again a linear transformation. In general, D^n is a linear transformation of F_∞ into itself for every positive integer n.

Example 2   Let $B = (x^4, x^3, x^2, x, 1)$, which is an ordered basis for P_4. Find A_{D^2} using Example 1 and matrix multiplication.

Solution   We use the fact that $A_{D^2} = A_{D \circ D} = (A_D)^2$. From Example 1, we then obtain

$$A_{D^2} = \begin{pmatrix} 0 & 0 & 0 & 0 & 0 \\ 4 & 0 & 0 & 0 & 0 \\ 0 & 3 & 0 & 0 & 0 \\ 0 & 0 & 2 & 0 & 0 \\ 0 & 0 & 0 & 1 & 0 \end{pmatrix} \begin{pmatrix} 0 & 0 & 0 & 0 & 0 \\ 4 & 0 & 0 & 0 & 0 \\ 0 & 3 & 0 & 0 & 0 \\ 0 & 0 & 2 & 0 & 0 \\ 0 & 0 & 0 & 1 & 0 \end{pmatrix}$$

$$= \begin{pmatrix} 0 & 0 & 0 & 0 & 0 \\ 0 & 0 & 0 & 0 & 0 \\ 12 & 0 & 0 & 0 & 0 \\ 0 & 6 & 0 & 0 & 0 \\ 0 & 0 & 2 & 0 & 0 \end{pmatrix}.$$

Of course, we could also have obtained this matrix by observing that $D^2(x^4) = 12x^2$, $D^2(x^3) = 6x$, $D^2(x^2) = 2$, and $D^2(x) = D^2(1) = 0$. ◁

Example 3   The differentiation transformation D carries the subspace $\mathrm{sp}(\sin x, \cos x)$ of F_∞ into itself. Find the matrix representation A_D relative to B,B where B is the ordered basis $(\sin x, \cos x)$. Find $D(3 \sin x - 7 \cos x)$ using matrix multiplication.

Solution   Since $D(\sin x) = \cos x$ and $D(\cos x) = -\sin x$, we see that

$$A_D = \begin{pmatrix} 0 & -1 \\ 1 & 0 \end{pmatrix}.$$

To find $D(3 \sin x - 7 \cos x)$, we can compute

$$\begin{pmatrix} 0 & -1 \\ 1 & 0 \end{pmatrix} \begin{pmatrix} 3 \\ -7 \end{pmatrix} = \begin{pmatrix} 7 \\ 3 \end{pmatrix},$$

and we obtain the answer $7 \sin x + 3 \cos x$.   ◁

Example 4   Describe the kernel of the differentiation transformation $D: F_\infty \to F_\infty$.

Solution   We know that the solution of the differential equation $Df = 0$ consists of all the constant functions f in F_∞, that is, the functions f for which there exists a constant c in $\mathbb{R}$ such that $f(x) = c$ for all x in $\mathbb{R}$. These constant functions form the subspace $\ker(D)$ of F_∞; it is the subspace P_0 of all polynomial functions of degree zero.   ◁

Now consider $D: F_\infty \to F_\infty$. Note that $D(e^{ax}) = ae^{ax}$. Thus the function e^{ax} is an eigenvector for D with eigenvalue a. In particular, e^x is an eigenvector for D with eigenvalue 1. We know that for each eigenvector $\mathbf{v}$ of a linear transformation, $c\mathbf{v}$ is also an eigenvector for the same eigenvalue if $c \neq 0$. Calculus shows that all solutions of the differential equation $Df = f$ are of the form $f(x) = ce^x$ for c in $\mathbb{R}$. This means that $\mathrm{sp}(e^x) = \{ce^x \mid c \in \mathbb{R}\}$ is the eigenspace E_1 of D.

A study of differential equations shows that for each integer $n > 0$, there exist n linearly independent functions $f_1, f_2, \ldots, f_n$ in F_∞ such that $D^n(f_i) = f_i$, and that these n functions form a basis for the eigenspace E_1 of D^n in F_∞. That is, this eigenspace has dimension n. The same is true for the eigenspace E_a for any a in $\mathbb{R}$. We give an illustration and, in the process, illustrate a technique for demonstrating independence of functions in F_∞.

Example 5   Drawing on experience with differentiation formulas in calculus, find four independent functions that are eigenvectors of $D^4: F_\infty \to F_\infty$ corresponding to eigenvalue 1, and prove that the functions are independent.

Solution   For those who have differentiated hundreds of functions when studying calculus, it is not difficult to come up with e^x, e^{-x}, $\sin x$, and $\cos x$ as four functions that are eigenvectors for D^4 with corresponding eigenvalue 1.

Let us show that these four functions are independent. Suppose there are constants $c_1, c_2, c_3,$ and c_4 such that

$$c_1 e^x + c_2 e^{-x} + c_3 \sin x + c_4 \cos x = 0 \quad \text{for all } x. \tag{3}$$

Differentiating this equation three times, we obtain

$$c_1 e^x - c_2 e^{-x} + c_3 \cos x - c_4 \sin x = 0$$
$$c_1 e^x + c_2 e^{-x} - c_3 \sin x - c_4 \cos x = 0 \tag{4}$$
$$c_1 e^x - c_2 e^{-x} - c_3 \cos x + c_4 \sin x = 0.$$

Setting $x = 0$ in Eq. (3) and the Eqs. (4), we obtain

$$
\begin{aligned}
c_1 + c_2 \quad\;\; + c_4 &= 0 \\
c_1 - c_2 + c_3 \quad\;\; &= 0 \\
c_1 + c_2 \quad\;\; - c_4 &= 0 \\
c_1 - c_2 - c_3 \quad\;\; &= 0,
\end{aligned}
$$

which is a square linear system easily seen to have only the trivial solution. Thus these functions are independent. ◁

INTEGRATION AS A LINEAR TRANSFORMATION

Let C consist of all continuous functions with domain $\mathbb{R}$ and let $C(a, b)$ consist of all continuous functions mapping the closed interval $[a, b]$ into $\mathbb{R}$. Since sums and scalar multiples of continuous functions are again continuous, it is easy to see that C is a subspace of F. Similarly, $C(a, b)$ is a vector space. We can regard $\int_a^b (\;) \, dx$ as a map of $C(a, b)$ into $\mathbb{R}$, for $\int_a^b f(x) \, dx$ is a real number. The familiar formulas

$$
\int_a^b (f + g)(x) \, dx = \int_a^b f(x) \, dx + \int_a^b g(x) \, dx
$$

and

$$
\int_a^b (cf)(x) \, dx = c \int_a^b f(x) \, dx
$$

show that $\int_a^b (\;) \, dx$ is a linear transformation of $C(a, b)$ into $\mathbb{R}$. The kernel of $\int_a^b (\;) \, dx$ consists of those functions f in $C(a, b)$ such that $\int_a^b f(x) \, dx = 0$. These are the functions in $C(a, b)$ of mean value (or average value) 0.

For each fixed a in $\mathbb{R}$ and each function f in C, the integral $\int_a^x f(t) \, dt$ defines another function in C. Thus we can view $\int_a^x (\;) \, dt$ as a map of C into itself, and it is easy to see that this map is also a linear transformation. (See Exercise 29.)

Example 6   Find the matrix representation A of

$$
\int_1^x (\;) \, dt: P_2 \to P_3
$$

relative to B, B' where

$$
B = (x^2, x, 1) \quad \text{and} \quad B' = (x^3, x^2, x, 1).
$$

Solution   Since

$$
\int_1^x t^n \, dt = \left[t^{n+1}/(n+1) \right]\Big|_1^x = x^{n+1}/(n+1) - 1/(n+1),
$$

we find that

$$
A = \begin{pmatrix} 1/3 & 0 & 0 \\ 0 & 1/2 & 0 \\ 0 & 0 & 1 \\ -1/3 & -1/2 & -1 \end{pmatrix}. \quad ◁
$$

SUMMARY

Let F_∞ consist of all real-valued functions of one real variable having domain $\mathbb{R}$ and having derivatives of all orders. Let C consist of all continuous functions with domain $\mathbb{R}$ and let $C(a, b)$ consist of all continuous real-valued functions with domain the closed interval $[a, b]$.

1. F_∞, C, and $C(a, b)$ are vector spaces with the usual operations of addition of functions and multiplication of a function by a scalar. The set P_n of all polynomial functions of degree $\leq n$ is a subspace of F_∞ and of C.

2. For a function f in F_∞, let $D^n f$ denote the nth derivative of f for each integer $n > 0$. Then $D^n : F_\infty \to F_\infty$ is a linear transformation.

3. For each positive integer n and constant a in $\mathbb{R}$, the eigenspace E_a of the linear transformation $D^n : F_\infty \to F_\infty$ has dimension n.

4. The operation $\int_a^b (\ \ ) \, dx$ can be viewed as a map of $C(a, b)$ into $\mathbb{R}$, and this map is a linear transformation.

5. For each a in $\mathbb{R}$, we can view $\int_a^x (\ \ ) \, dt$ as a map of C into itself, and this map is a linear transformation.

EXERCISES

In Exercises 1–9, determine whether the given map T is a linear transformation.

1. $T: F_\infty \to F_\infty$ defined by $T(f) = Df + 3$
2. $T: F_\infty \to F_\infty$ defined by $T(f) = D^2 f + 3Df$
3. $T: C \to C$ defined by $T(f) = \int_0^x [f(t)]^2 \, dt$
4. $T: C \to C$ defined by $T(f) = \int_0^x |f(t)| \, dt$
5. $T: C \to C$ defined by $T(f) = \int_0^x 4[f(t)] \, dt$
6. $T: C \to C$ defined by $T(f) = \int_0^x [3 + f(t)] \, dt$
7. $T: C \to C$ defined by $T(f) = 3f - 2\int_0^x f(t) \, dt$
8. $T: F_\infty \to F_\infty$ defined by $T(f) = D^2(f + 4) + \int_{-2}^x 3[f(t)] \, dt$
9. $T: F_\infty \to F_\infty$ defined by $T(f) = \int_3^x D^2[f(t) + t] \, dt$

10. Let $T: P_3 \to P_3$ be the linear transformation defined by $T(f) = D^2 f - 4Df + f$. Find the matrix representation A_T of T relative to B,B where $B = (x^3, x^2, x, 1)$. Use the result to compute $T(4x^3 - 5x^2 + 7x + 3)$.

11. Repeat Exercise 10 if $T(f) = 2D^3 f - 3D^2 f + 4f$.

12. Consider the subspace $S = \text{sp}(e^x, e^{3x}, \sin 2x, \cos 2x)$ of F_∞. Find the matrix representation A_D of $D: S \to S$ relative to B,B where $B = (e^x, e^{3x}, \sin 2x, \cos 2x)$.

13. For S and B in Exercise 12, find the matrix representation A_T relative to B,B if $T: S \to S$ is defined by $T(f) = D^2 f + 3Df - 4f$. Use the result to compute $T(3e^x - 4e^{3x} + 5 \sin 2x - \cos 2x)$.

14. Find the kernel of the linear transformation $D^n : F_\infty \to F_\infty$ for each integer $n \geq 1$.

15. Recall that for a linear transformation $T: V \to W$ where V is finite-dimensional, we have $\dim(V) = \text{rank}(T) + \text{null}(T)$. Illustrate this equation for each of the following.
    a) $D^2 : P_5 \to P_5$ b) $D^6 : P_5 \to P_5$

In Exercises 16–21, use the technique of Example 5 to show that the given set of functions is independent.

16. $\{e^x, e^{2x}\}$

17. $\{\sin x, \cos x\}$

18. $\{x, \sin x, \sin 2x\}$

19. $\{e^x, \sin 2x, \cos 2x\}$

20. $\{x, x^2, e^x, e^{-x}\}$

21. $\{x, x^2, \cos x, \cos x^2)$

22. Let $f_1, f_2, \ldots, f_n$ be n functions in F_∞. The **Wronskian** of these functions is the determinant

$$\begin{vmatrix} f_1 & f_2 & \cdots & f_n \\ f_1' & f_2' & \cdots & f_n' \\ & & \vdots & \\ f_1^{(n-1)} & f_2^{(n-1)} & \cdots & f_n^{(n-1)} \end{vmatrix}.$$

Show that if there is some real number x_0 such that this Wronskian evaluated at $x = x_0$ is nonzero, then the functions are linearly independent.

23. It can be shown that if $a_1, a_2, \ldots, a_n$ are n distinct constants, then the linear system

$$t_1 + t_2 + \cdots + t_n = 0$$
$$a_1 t_1 + a_2 t_2 + \cdots + a_n t_n = 0$$
$$a_1^2 t_1 + a_2^2 t_2 + \cdots + a_n^2 t_n = 0$$
$$\vdots$$
$$a_1^{n-1} t_1 + a_2^{n-1} t_2 + \cdots + a_n^{n-1} t_n = 0$$

has only the trivial solution. Use this fact and the technique of Example 5 to show that for distinct constants a_i, the set $\{e^{a_1 x}, e^{a_2 x}, \ldots, e^{a_n x}\}$ is independent.

In Exercises 24–28, use differentiation formulas of calculus and number (3) of the Summary to find a basis for the indicated eigenspace.

24. E_1 of $D^2 \colon F_\infty \to F_\infty$

25. E_{-1} of $D^2 \colon F_\infty \to F_\infty$

26. E_0 of $D^5 \colon F_\infty \to F_\infty$

27. E_{16} of $D^4 \colon F_\infty \to F_\infty$

28. E_{-3} of $D^2 \colon F_\infty \to F_\infty$

29. Show that for each a in $\mathbb{R}$, the operation $\int_a^x (\ ) \, dt$ is a linear transformation of C into itself.

In Exercises 30–34, find the matrix representation A of the linear transformation relative to B,B'. Use the matrix to compute the value of the transformation at the given polynomial.

30. $\int_0^x (\ ) \, dt \colon P_3 \to P_4$,
$B = (x^3, x^2, x, 1)$, $B' = (x^4, x^3, x^2, x, 1)$
Compute at $2x^3 - 5x + 7$.

31. $\int_5^x (\ ) \, dt \colon P_0 \to P_1$, $B = (1)$, $B' = (x, 1)$
Compute at -4.

32. $\int_{-1}^x (\ ) \, dt \colon P_2 \to P_3$, $B = (x^2, x, 1)$,
$B' = (x^3, x^2, x, 1)$
Compute at $4x^2 - 3x + 5$.

33. $\int_2^x 4(\ ) \, dt \colon P_2 \to P_3$, $B = (x^2, x, 1)$,
$B' = (x^3, x^2, x, 1)$
Compute at $-3x^2 + 7x$.

34. $\int_{-2}^x -2(\ ) \, dt \colon P_3 \to P_4$,
$B = (x^3, x^2, x, 1)$, $B' = (x^4, x^3, x^2, x, 1)$
Compute at $-3x^3 + x^2 + 5x - 3$.

8.2
Linear Algebra and Single-Variable Calculus

A central technique in calculus is to approximate functions by linear functions. Difficult nonlinear problems can sometimes be converted into more easily solved linear problems by this means. We introduced this idea in Section 1.1.

In this section we make no attempt to give a rigorous treatment of calculus. We use linear algebra to develop an intuitive understanding of some of the concepts and formulas in single-variable calculus.

LOCAL COORDINATES

Part of the graph of a differentiable function $f \colon \mathbb{R} \to \mathbb{R}$ is shown in Fig. 8.1. The figure shows the tangent line to the graph at a point (x_0, y_0) on the graph. Now this line is not the graph of a linear transformation mapping the x-axis $\mathbb{R}$ into the y-axis $\mathbb{R}$ because it does not go through the x,y-origin. Recall that a linear transformation must map $\mathbf{0}$ to $\mathbf{0}$, and consequently its graph goes through the origin.

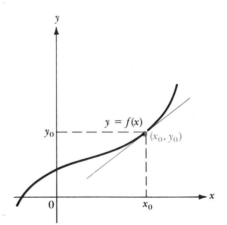

Figure 8.1
The tangent line to the graph of $f(x)$ at (x_0, y_0).

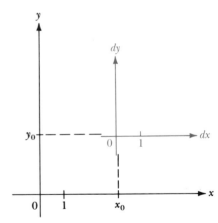

Figure 8.2
Local dx, dy-coordinate axes at (x_0, y_0).

We remedy this situation by taking new coordinates for the plane, translating our origin to (x_0, y_0). This is shown in Fig. 8.2. We are calling our new axes the dx-axis and the dy-axis. A point in the plane has both x,y-coordinates and dx,dy-coordinates. We easily see from Fig. 8.2 that

$$dx = x - x_0$$
$$dy = y - y_0.$$

Translation equations   (1)

Example 1   Find dx,dy-coordinates of the point $(x, y) = (-1.3, 2.2)$ if axes are translated with new origin at $(-1, 2)$.

Solution   From the translation equations (1), we obtain

$$dx = -1.3 - (-1) = -.3 \quad \text{and} \quad dy = 2.2 - 2 = .2$$

so the point has dx,dy-coordinates $(-.3, .2)$. ◁

The approximation of the graph of $y = f(x)$ by the tangent line in Fig. 8.1 is a good one only near (x_0, y_0). Thus we are interested in dx,dy-coordinates primarily for dx and dy of small magnitude. To approximate the entire graph linearly, we have to keep taking new tangent-line approximations at different points as we move along the graph. Consequently, we will call dx,dy-coordinates **local coordinates**.

We know from calculus that the derivative $f'(x_0)$ of $f(x)$ is the slope of the tangent line to the graph $y = f(x)$ at $x = x_0$. With respect to local dx,dy-coordinates at (x_0, y_0), this tangent line has equation

$$dy = f'(x_0)\, dx, \tag{2}$$

as shown in Fig. 8.3. We recognize Eq. (2) from calculus as the formula for the **differential of** f at x_0. This differential should be regarded as a linear transformation $T: \mathbb{R} \to \mathbb{R}$, where for numbers dx in $\mathbb{R}$,

$$T(dx) = f'(x_0)\, dx. \qquad \textbf{Differential at } x_0 \tag{3}$$

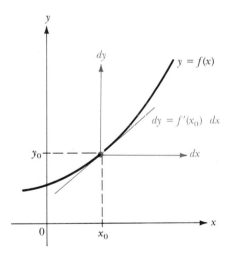

Figure 8.3
The local equation of the tangent line at (x_0, y_0).

We emphasize that Eq. (3) gives the differential of f at x_0. For a function having a derivative at all points of its domain, one should think of the **differential** df **of** f as a *map* assigning to each point x_0 in the domain of f the linear transformation (3). The differential of f provides the complete set of local linear approximations needed to approximate the entire graph.

> The differential of a function f provides a large collection of local linear approximations to f, one for each point in the domain of f.

ANOTHER VIEW OF NEWTON'S METHOD

We developed Newton's method in Section 1.1. Recall that the method is designed to find a solution of an equation $f(x) = 0$. However, in Section 1.1 we were not able to find the exact slope of the tangent line to the graph $y = f(x)$ at (x_0, y_0), for we did not assume knowledge of calculus. We came up with an approximation m for the slope, found the x,y-equation of the line through (x_0, y_0) of slope m, and used this as an approximating linear equation for $y = f(x)$ at x_0. We develop this again in the language of differentials, using the true tangent line to the graph.

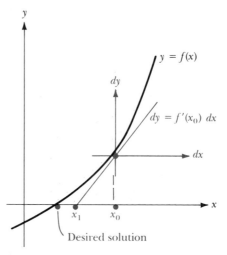

Figure 8.4
Solving $f(x) = 0$ by Newton's method.

Suppose a point x_0 has been found such that $f(x_0)$ is of fairly small magnitude, as shown in Fig. 8.4. We take our local linear approximation

$$dy = f'(x_0)\, dx \tag{4}$$

to $y = f(x)$ at x_0, and find where the line $dy = f'(x_0)\, dx$ crosses the original x-axis. Setting $y = 0$ in Eqs. (1), we obtain $dy = -y_0 = -f(x_0)$, which we substitute in

Eq. (4). Our line crosses the x-axis where

$$dx = -f(x_0)/f'(x_0),$$

assuming $f'(x_0) \neq 0$. From the translation equation $dx = x - x_0$ we see that the x-value corresponding to this dx is $x_1 = x_0 + dx$ or

$$x_1 = x_0 - \frac{f(x_0)}{f'(x_0)}. \tag{5}$$

From Fig. 8.4, we expect that x_1 might be a better approximation than x_0 to a solution of $f(x) = 0$. We then start with x_1 and repeat the process to obtain the next approximation x_2, and so on. The generalization of formula (5) used to compute x_{i+1} is

$$x_{i+1} = x_i - \frac{f(x_i)}{f'(x_i)}. \qquad \textbf{Newton's recursion formula} \tag{6}$$

Example 2 Using a calculator, find a solution of $x^5 + 7x^3 - 20 = 0$.

Solution Let $f(x) = x^5 + 7x^3 - 20$. Clearly $f(x) \leq 0$ for $x \leq 1$, while $f(x) \geq 0$ for $x \geq 2$. Since $f(1) = -12$ is closer to zero than $f(2) = 68$, we start with $x_1 = 1$. The recursion formula (6) becomes

$$x_{i+1} = x_i - \frac{x_i^5 + 7x_i^3 - 20}{5x_i^4 + 21x_i^2}.$$

Our calculator yields results given in Table 8.1. We always compute $f(x_i)$ also to be sure that we are actually approaching a solution of $f(x) = 0$. ◁

Table 8.1

i	x_i	$f(x_i)$
1	1	-12
2	1.461538462	8.522746187
3	1.335597387	0.9271519711
4	1.318225329	0.0155248157
5	1.317924405	0.0000045805
6	1.317924316	1×10^{-11}
7	1.317924316	1×10^{-11}

The differential $dy = f'(x_0)\,dx$ gives the *best* local linear approximation to f at x_0 because it corresponds to the tangent line to the graph. (The notion of *best local linear approximation* to $f(x)$ at x_0 is made precise in a course in advanced calculus. We rely here on your geometric intuition.) In Section 1.1, we used only an approximation of this tangent line. Iterates using the recursion formula (6) can be expected to approach a solution of $f(x) = 0$ a bit faster than the iterates computed in Section 1.1. However, the result is generally the same. If we use a program like NEWT123, which employs the method of Section 1.1, we do not have to supply the computer with the formula for the derivative of f, which is an advantage. With a computer, the method of Section 1.1 is generally a timesaver over using Eq. (6). We emphasize that the basic idea is the same for both methods: To obtain each iterate, one approximates the function locally by a linear one, and solves the linear problem instead. It might be an interesting exercise to solve the problem in Example 2 using NEWT123 and compare the data generated with those in Table 8.1.

MATRICES, DETERMINANTS, AND LENGTH-CHANGE FACTORS

The standard matrix representation of the transformation $T: \mathbb{R} \to \mathbb{R}$, where

$$T(dx) = f'(x_0)\, dx,$$

is the 1×1 matrix

$$A_T = \left(f'(x_0) \right).$$

Of course $\det(A_T) = f'(x_0)$. Recall from Section 6.6 that $|\det(A_T)|$ is the length-change factor for T. That is, a line segment of $\mathbb{R}$ of length s is carried by T into one of length $|\det(A_T)|s$. Since T approximates f in terms of local coordinates at (x_0, y_0), we regard $|f'(x_0)|$ as the **local length-change factor** at x_0 for f. Figure 8.5 gives illustrations of this idea. In Fig. 8.5(a), where $f'(x_0) = 1/2$, a small line segment of length s is carried by T into one of length $s/2$ and by f into one of length close to $s/2$. In Fig. 8.5(b), where $f'(x_0) = -3$, a small line segment of length s is carried by T into one of length $3s$ and by f into one of length close to $3s$. Note that in Fig. 8.5(a), the positively directed line segment is carried into a positively directed one. In Fig. 8.5(b), the positively directed segment is carried into a negatively directed one. That is, the **orientation** of the line segment in Fig. 8.5(b) is reversed by the map. We call a linear transformation $T: \mathbb{R}^n \to \mathbb{R}^n$ **orientation preserving** if $\det(A_T) > 0$ and **orientation reversing** if $\det(A_T) < 0$. Thus it is natural to consider a differentiable function $f: \mathbb{R} \to \mathbb{R}$ to be **orientation preserving** at x_0 if $f'(x_0) > 0$ and to be **orientation reversing** at x_0 if $f'(x_0) < 0$. Figure 8.5 illustrates these ideas for $T: \mathbb{R} \to \mathbb{R}$.

APPLICATION TO CHANGE OF VARIABLE IN AN INTEGRAL

In the integral notation

$$\int_a^b f(x)\, dx, \tag{7}$$

we think of dx as the length of a subinterval of $[a, b]$, and we think of $f(x)\, dx$ as the area of the rectangular strip shown in Fig. 8.6. The integral (7) should be regarded as summing the contributions $f(x)\, dx$ over a partition of $[a, b]$ and taking the limit of such sums as the lengths dx of the bases of the rectangles all approach zero.

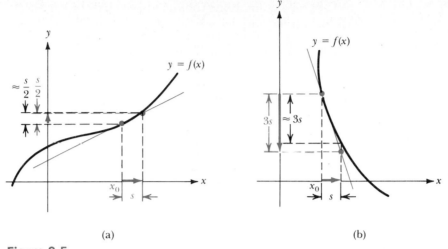

(a)   (b)

Figure 8.5
(a) $f'(x_0) = 1/2$ and f is orientation preserving at x_0;
(b) $f'(x_0) = -3$ and f is orientation reversing at x_0.

Suppose we now make a substitution $x = g(t)$ in the integral (7) using a differentiable function g as shown in Fig. 8.7. We know from calculus that we should replace dx by $g'(t)\,dt$. Let us use linear algebra to see geometrically the reason for this. Note that a partition of the t-interval $[c, d]$ into subintervals gives a partition of the x-interval $[a, b]$ into subintervals as illustrated in Fig. 8.7.

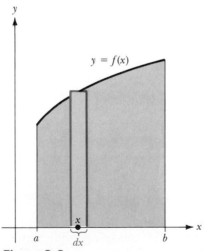

Figure 8.6
A strip of area $f(x)\,dx$.

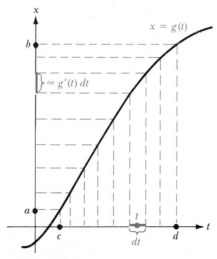

Figure 8.7
The graph of a substitution $x = g(t)$.

Let t be a point in the subinterval of $[c, d]$ of length dt shown in Fig. 8.7. As we have seen, the length of the corresponding subinterval of $[a, b]$ on the x-axis is approximately

$$\text{(Local length-change factor for } g \text{ at } t) \; dt = |g'(t)| \; dt.$$

Thus the numerical quantity

$$f(x) \; dx = f(x)(\text{length of } x\text{-subinterval})$$

is approximately equal to

$$f(x) \; |g'(t)| \; dt. \tag{8}$$

For $g(t)$ which is increasing, as in Fig. 8.7, we see that $g'(t) > 0$ and we can write formula (8) as

$$f(g(t))g'(t) \; dt. \tag{9}$$

Thus it seems geometrically reasonable that

$$\int_a^b f(x) \; dx = \int_c^d f(g(t))g'(t) \; dt \tag{10}$$

for an increasing substitution function g. Exercises 21 and 22 consider the cases where g is decreasing or sometimes increasing and sometimes decreasing, as shown in Fig. 8.8(a).

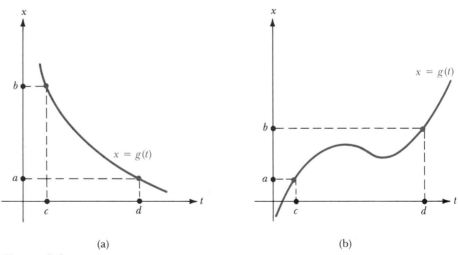

(a)   (b)

Figure 8.8
(a) A decreasing substitution $x = g(t)$; (b) a general substitution $x = g(t)$.

The preceding discussion should provide a good intuitive understanding, in terms of local linear approximation, of the necessity to replace dx by $g'(t) \; dt$ rather than by just dt when making a substitution $x = g(t)$ in an integral.

APPLICATIONS TO ARC LENGTH

Consider the problem of finding the length of the arc of a smooth graph $y = f(x)$ lying over an interval $[a, b]$, as shown in Fig. 8.9. (Recall that *smooth* means that $f'(x)$ exists and is continuous in $[a, b]$.) It is useful to think of the arc as the image in $\mathbb{R}^2$ of the interval $[a, b]$ in $\mathbb{R}$ if $[a, b]$ is picked up and put down by a map carrying x into $(x, f(x))$. That is, consider the map $h: \mathbb{R} \to \mathbb{R}^2$ where, using column vectors,

$$h(x) = \begin{pmatrix} x \\ f(x) \end{pmatrix}.$$

The colored arc in Fig. 8.9 is the image of $[a, b]$ under this map h.

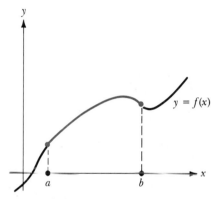

Figure 8.9
The arc of $y = f(x)$ over $[a, b]$.

We take local coordinates at x_0 in an x-axis $\mathbb{R}$ and at (x_0, y_0) in $\mathbb{R}^2$. Since $y = f(x)$ is approximated locally at x_0 by the linear map $dy = f'(x_0)\, dx$, we expect that the linear transformation $T: \mathbb{R} \to \mathbb{R}^2$, where

$$T(dx) = \begin{pmatrix} dx \\ f'(x_0)\, dx \end{pmatrix},$$

approximates h locally near x_0 in $\mathbb{R}$ using these local coordinates. The standard matrix representation A of T is

$$A = \begin{pmatrix} 1 \\ f'(x_0) \end{pmatrix}.$$

We find that

$$A^T A = (1, f'(x_0)) \begin{pmatrix} 1 \\ f'(x_0) \end{pmatrix} = (1 + f'(x_0)^2).$$

Thus by Theorem 6.19 on page 360, the local length-change factor for T is $\sqrt{1 + f'(x_0)^2}$, so the interval of length dx in Fig. 8.10 is carried by T into a

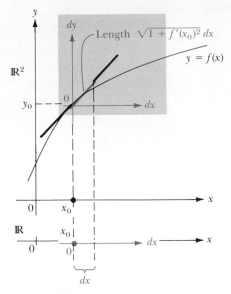

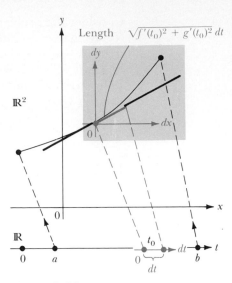

Figure 8.10
Approximating the mapping f of $\mathbb{R}$ into $\mathbb{R}^2$ using local coordinates.

Figure 8.11
Approximation in local coordinates of a parametrically defined curve.

line segment (actually tangent to the graph of f) of length $\sqrt{1 + f'(x_0)^2}\, dx$. This gives rise to the formula $\int_a^b \sqrt{1 + f'(x)^2}\, dx$ for the length of the arc in Fig. 8.9. We consider $\sqrt{1 + f'(x_0)^2}$ to be the **local length-change factor** for h at x_0.

Precisely the same analysis holds for the length of an arc of a smooth plane curve defined parametrically by

$$x = f(t), \qquad y = g(t) \quad \text{for } a \leq t \leq b,$$

shown in Fig. 8.11. We regard the curve as defined by

$$h(t) = \begin{pmatrix} f(t) \\ g(t) \end{pmatrix} \quad \text{for } a \leq t \leq b.$$

Let t_0 in $[a, b]$ be carried into $\begin{pmatrix} x_0 \\ y_0 \end{pmatrix}$ in $\mathbb{R}^2$ by h. In terms of local coordinates at t_0 in $\mathbb{R}$ and at (x_0, y_0) in $\mathbb{R}^2$, the function h is approximated linearly at t_0 by $T: \mathbb{R} \to \mathbb{R}^2$, where

$$T(dt) = \begin{pmatrix} f'(t_0)\, dt \\ g'(t_0)\, dt \end{pmatrix}.$$

The standard matrix representation A of T is

$$A = \begin{pmatrix} f'(t_0) \\ g'(t_0) \end{pmatrix}.$$

We find that

$$A^T A = (f'(t_0), g'(t_0)) \begin{pmatrix} f'(t_0) \\ g'(t_0) \end{pmatrix} = (f'(t_0)^2 + g'(t_0)^2).$$

Thus by Theorem 6.19 on page 360, the local length-change factor for T is

$$\sqrt{f'(t_0)^2 + g'(t_0)^2}. \tag{11}$$

The segment of length dt in Fig. 8.11 is carried by T into one of length $\sqrt{f'(t_0)^2 + g'(t_0)^2}\, dt$. We thus arrive at the integral

$$\int_a^b \sqrt{f'(t)^2 + g'(t)^2}\, dt$$

giving the length of the parametric curve. We consider (11) to be the **local length-change factor** for h at t_0.

Example 3   Find the local length-change factor for $h \colon \mathbb{R} \to \mathbb{R}^2$, where $h(t) = (t^2, t^3)$ at $t_0 = -1$.

Solution   Taking $f(t) = t^2$ and $g(t) = t^3$, we obtain from (11) the local length-change factor

$$\sqrt{(2t)^2 + (3t^2)^2} = \sqrt{4t^2 + 9t^4}$$

at any point t. Taking $t = -1$, we find that $\sqrt{13}$ is the local length-change factor there. ◁

SUMMARY

Let f be a differentiable function and x_0 a point in its domain.

1. Local dx,dy-coordinates with local origin at $(x, y) = (x_0, y_0)$ can be found using the translation equations

   $$dx = x - x_0, \qquad dy = y - y_0.$$

2. The linear transformation $T \colon \mathbb{R} \to \mathbb{R}$, where $T(dx) = f'(x_0)\, dx$, is the differential of f at x_0, and is the best local linear approximation to f at x_0.

3. The recursion formula for Newton's method to approximate a solution of $f(x) = 0$ is

   $$x_{i+1} = x_i - \frac{f(x_i)}{f'(x_i)}.$$

4. The function f is orientation preserving at x_0 if $f'(x_0) > 0$, and is orientation reserving there if $f'(x_0) < 0$.

5. The local length-change factor for f at x_0 is $|f'(x_0)|$.

6. Let $h \colon \mathbb{R} \to \mathbb{R}^2$ be defined by $h(t) = (f(t), g(t))$, where f and g are both continuously differentiable functions. The local length-change factor for h at t_0 is $\sqrt{f'(t_0)^2 + g'(t_0)^2}$.

EXERCISES

In Exercises 1–4, find local dx,dy-coordinates of the point (x, y) when the local origin is taken at the point with the given x,y-coordinates.

1. $(x, y) = (2.4, -1.3)$, origin at $(2, -1)$

2. $(x, y) = (-10.4, 9)$, origin at $(-10, 10)$

3. $(x, y) = (8.3, -7.2)$, origin at $(9, -8)$

4. $(x, y) = (-5.6, -8.4)$, origin at $(-6, -7)$

In Exercises 5–8, find $T(dx)$ if T is the best local linear approximation to f at the point x_0.

5. $f(x) = x^3 - 3x^2$, $x_0 = 1$

6. $f(x) = (x^2 - 1)/(x^2 + 1)$, $x_0 = 2$

7. $f(x) = \sin^2 3x$, $x_0 = 0$

8. $f(x) = e^{2x+3}$, $x_0 = 1$

In Exercises 9–12, use Newton's method to estimate the given quantity, starting with the given x_0 and finding x_2.

9. $\sqrt{3}$, that is, a solution of $x^2 - 3 = 0$, starting at $x_0 = 2$

10. $6^{1/3}$, that is, a solution of $x^3 - 6 = 0$, starting at $x_0 = 2$

11. A solution of $x^3 + x - 1 = 0$, starting at $x_0 = 1$

12. A solution of $x^3 + x^2 - 1 = 0$, starting at $x_0 = 1$

In Exercises 13–16, classify the given function as orientation preserving or orientation reversing at the given point.

13. $f(x) = x^3 - 3x^2 + 2x$ at $x_0 = 1$

14. $f(x) = \sin 3x$ at $x_0 = 0$

15. $f(x) = e^{-x}$ at $x_0 = 0$

16. $f(x) = \ln x$ at $x_0 = 1$

In Exercises 17–20, find the local length-change factor for the function at the given point.

17. $f: \mathbb{R} \to \mathbb{R}$, where $f(x) = (x^3 + 1)/(x^2 - 2)$ at $x_0 = 1$

18. $f: \mathbb{R} \to \mathbb{R}$, where $f(x) = \sqrt{x^2 + 3}$ at $x_0 = -1$

19. $h: \mathbb{R} \to \mathbb{R}^2$, where $h(t) = (t^2 - 1, 3t + 1)$ at $t_0 = 1$

20. $h: \mathbb{R} \to \mathbb{R}^2$, where $h(t) = (\sin t, \cos t)$ at $t_0 = \pi/4$

21. Let $f(x)$ be continuous for $a \leq x \leq b$, and let $g(t)$ be differentiable and decreasing for $c \leq t \leq d$ with $g(c) = b$ and $g(d) = a$.
  a) Using an intuitive argument like the ones in the text, explain that $\int_a^b f(x) \, dx = -\int_c^d f(g(t))g'(t) \, dt$.
  b) Explain why part (a) can be written as $\int_a^b f(x) \, dx = \int_d^c f(g(t))g'(t) \, dt$.

22. Let $f(x)$ be continuous for $a \leq x \leq b$ and let $g(t)$ be defined for $c \leq t \leq d$ with $g(c) = a$ and $g(d) = b$. Suppose that g is increasing over a finite number of subintervals of $[c, d]$ and decreasing over other subintervals. Give an intuitive argument that we still have $\int_a^b f(x) \, dx = \int_c^d f(g(t))g'(t) \, dt$. [*Hint:* Use an argument in the text and Exercise 21.]

23. Let $h: \mathbb{R} \to \mathbb{R}^n$ be defined by

$$h(t) = \begin{pmatrix} f_1(t) \\ f_2(t) \\ \cdot \\ \cdot \\ \cdot \\ f_n(t) \end{pmatrix} \quad \text{and let} \quad h(t_0) = \begin{pmatrix} c_1 \\ c_2 \\ \cdot \\ \cdot \\ \cdot \\ c_n \end{pmatrix} = \mathbf{c},$$

where each f_i is differentiable. Let local dt-coordinates be chosen at t_0 in $\mathbb{R}$ and local dx_1, dx_2, . . . , dx_n-coordinates be chosen at $\mathbf{c}$ in $\mathbb{R}^n$. Let $T: \mathbb{R} \to \mathbb{R}^n$ be the best local linear approximation to h at t_0. Use intuition to answer each of the following.

a) Find a formula for $T(dt)$.

b) Find the matrix representation for T.

c) Find the length-change factor for T.

d) Find the local length-change factor for h at t_0.

e) Express as an integral the length of the parametric curve in $\mathbb{R}$ defined by h for $a \leq t \leq b$.

f) Find an approximation for the length of the curve given by h and corresponding to a short length from t_0 to $t_0 + \Delta t$.

In Exercises 24–27, use the answer to Exercise 23 to approximate the length of the short parametrically defined curve. Although we write the vectors h(t) in row form to save space, they should be regarded as column vectors.

24. $h \colon \mathbb{R} \to \mathbb{R}^2$, where $h(t) = (t^2 - 1, t^2 + 3)$ for $1 \leq t \leq 1.01$

25. $h \colon \mathbb{R} \to \mathbb{R}^2$, where $h(t) = (\sin t, \cos t)$ for $\pi/4 \leq t \leq \pi/4 + .02$

26. $h \colon \mathbb{R} \to \mathbb{R}^3$, where $h(t) = (t^2, t^3, -t^4)$ for $-1.03 \leq t \leq -1$

27. $h \colon \mathbb{R} \to \mathbb{R}^4$, where $h(t) = (\sin t, 2t + 3, 3t - 1, \sin 4t)$ for $0 \leq t \leq .05$

28. Let f and g be differentiable functions with domain $\mathbb{R}$ and with $g(t_0) = x_0$. Use local length-change factor and orientation-preservation arguments to give an intuitive explanation of the chain rule $(f \circ g)'(t_0) = f'(x_0)g'(t_0)$.

8.3
Linear Algebra and Multivariable Calculus

In this section we assume that the reader has studied multivariable calculus. We extend the ideas of Section 8.2 to differentiable functions mapping a subset of $\mathbb{R}^n$ into $\mathbb{R}^m$, starting with the most familiar case from calculus.

LOCAL LINEAR APPROXIMATION OF FUNCTIONS OF TWO VARIABLES

Suppose $z = f(x, y)$ so that f maps a subset of $\mathbb{R}^2$ into $\mathbb{R}$. We will always assume that the domain of f is an **open set** in $\mathbb{R}^2$, so that it includes all points in some neighborhood of each of its points. We also assume that f has continuous partial derivatives $f_x = \partial z/\partial x$ and $f_y = \partial z/\partial y$. These are conditions that guarantee that f has a good local linear approximation at each point in its domain.

The graph $z = f(x, y)$ is a surface in $\mathbb{R}^2$, as shown in Fig. 8.12. We wish to approximate the surface at the point (x_0, y_0, z_0) by a graph having a linear equation. The graph of a linear equation $z = ax + by + c$ is a plane in $\mathbb{R}^3$. The plane that best approximates the surface near (x_0, y_0, z_0) is the tangent plane to the surface at that point, as shown in Fig. 8.12. This tangent plane is the analogue of the tangent line to the graph $y = f(x)$ discussed in Section 8.2. Just as in Section 8.2, a plane is the graph of a linear transformation only if it passes through the origin, so we take local dx,dy,dz-coordinates at (x_0, y_0, z_0), as indicated by the dx,dy,dz-axes in Fig. 8.12. We have the *translation equations*

$$dx = x - x_0, \qquad dy = y - y_0, \qquad dz = z - z_0. \qquad (1)$$

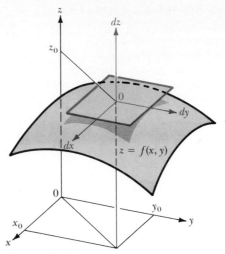

Figure 8.12
The tangent plane to the graph of f and local coordinates at the point (x_0, y_0, z_0).

From calculus, we know that the equation of the tangent plane in Fig. 8.12 is

$$z - z_0 = f_x(x_0, y_0)(x - x_0) + f_y(x_0, y_0)(y - y_0). \tag{2}$$

In terms of local dx,dy,dz-coordinates, Eq. (2) becomes

$$dz = f_x(x_0, y_0)dx + f_y(x_0, y_0)dy. \tag{3}$$

We recognize Eq. (3) from calculus as the expression for the **differential** of f at (x_0, y_0). This differential gives the best local linear approximation to f at (x_0, y_0). The corresponding linear transformation is $T: \mathbb{R}^2 \to \mathbb{R}$, where

$$T\begin{pmatrix} dx \\ dy \end{pmatrix} = f_x(x_0, y_0)dx + f_y(x_0, y_0)dy.$$

The standard matrix representation of T is

$$A_T = \big(f_x(x_0, y_0), f_y(x_0, y_0)\big). \tag{4}$$

The row vector (4) is known as the **gradient vector** $\nabla f(x_0, y_0)$ of f at (x_0, y_0).

Just as for the case $y = f(x)$, the **differential** df should be regarded as the collection of all these approximating local linear maps, one for each point in the domain of f.

Example 1   Let $z = f(x, y) = x^3 + 3xy^2$. Find the differential of f at $(1, -2)$ and the gradient vector there.

Solution   We have

$$f_x = \partial z/\partial x = 3x^2 + 3y^2 \quad \text{and} \quad f_y = \partial z/\partial y = 6xy.$$

Thus

$$f_x(1, -2) = 3 + 12 = 15 \quad \text{and} \quad f_y(1, -2) = -12.$$

The differential is $dz = 15dx - 12dy$ and the gradient vector is $\nabla f(1, -2) = (15, -12)$. ◁

SUBSCRIPTED NOTATION

When we consider a function f of n variables, we do not want to use different letters for the components of vectors in $\mathbb{R}^n$, so we turn to subscripted variables. The structure of calculus actually becomes clearer when we use x_1 for x, x_2 for y, and so on. Rather than work with $z = f(x, y)$ at (x_0, y_0), we will consider $y = f(\mathbf{x}) = f(x_1, x_2)$ at the point $\mathbf{a} = (a_1, a_2)$. The notation $y = f(\mathbf{x})$ at $\mathbf{a}$ is very similar to the simple notation $y = f(x)$ at x_0, which we used in Section 8.2. We will see that if we use vector notations from linear algebra, many formulas from multivariable calculus appear the same as formulas in single-variable calculus.

Let us denote the gradient vector $\nabla f(\mathbf{a})$ by $f'(\mathbf{a})$ as well, so that

$$f'(\mathbf{a}) = \nabla f(\mathbf{a}) = (f_{x_1}(\mathbf{a}), f_{x_2}(\mathbf{a})). \tag{5}$$

Thus $f'(\mathbf{a})$ is the standard matrix representation of the local linear transformation T that best approximates f at $\mathbf{a}$, and which is given by

$$dy = T\begin{pmatrix} dx_1 \\ dx_2 \end{pmatrix} = f_{x_1}(\mathbf{a})dx_1 + f_{x_2}(\mathbf{a})dx_2. \tag{6}$$

If we let

$$\mathbf{dx} = \begin{pmatrix} dx_1 \\ dx_2 \end{pmatrix},$$

then we can write Eq. (6) as

$$dy = T(\mathbf{dx}) = (f_{x_1}(\mathbf{a}), f_{x_2}(\mathbf{a}))\begin{pmatrix} dx_1 \\ dx_2 \end{pmatrix} = f'(\mathbf{a})\mathbf{dx}. \tag{7}$$

Using vectors, we have recovered the simple notation $dy = f'(\mathbf{a})\mathbf{dx}$ that we had in the single-variable case in Section 8.2. This is why we call the matrix representation $\nabla f(\mathbf{a})$ of T the **derivative** $f'(\mathbf{a})$ of f at the point $\mathbf{a}$. In other words, the role of the derivative for a function of a single variable is played here by the vector of partial derivatives.

Everything we have done so far in this section can be done just as well for a function f of n variables, written as

$$y = f(\mathbf{x}) = f(x_1, x_2, \ldots, x_n).$$

We started with $z = f(x, y)$ because we can draw a 3-space picture (Fig. 8.12). The equation $dy = f'(\mathbf{a})\mathbf{dx}$ is valid for $y = f(x_1, x_2, \ldots, x_n)$, where of course

$$f'(\mathbf{a}) = \nabla f(\mathbf{a}) = (f_{x_1}(\mathbf{a}), f_{x_2}(\mathbf{a}), \ldots, f_{x_n}(\mathbf{a}))$$

and

$$\mathbf{dx} = \begin{pmatrix} dx_1 \\ dx_2 \\ \vdots \\ dx_n \end{pmatrix}.$$

In local coordinates, $dy = f'(\mathbf{a})\mathbf{dx}$ becomes the equation of the **tangent hyperplane** to the graph of f in $\mathbb{R}^{n+1}$.

LOCAL LINEAR APPROXIMATION OF VECTOR-VALUED FUNCTIONS

We now turn to the most general case of a function given in vector notation by $\mathbf{y} = f(\mathbf{x})$, or written out,

$$(y_1, y_2, \ldots, y_m) = f(x_1, x_2, \ldots, x_n).$$

Each of the components $y_1, y_2, \ldots, y_m$ of $\mathbf{y}$ appears itself as a function $y_i = f_i(x_1, x_2, \ldots, x_n)$. These functions are known as the **coordinate functions** $f_1, f_2, \ldots, f_m$ of f.

Example 2   Let $f: \mathbb{R}^3 \to \mathbb{R}^4$ be defined by $\mathbf{y} = f(\mathbf{x}) = (x_1^2 x_2, -x_3^4, x_1 x_3, x_2 - x_3)$. Find the coordinate functions of f.

Solution   The four coordinate functions are

$$\begin{aligned} y_1 &= f_1(\mathbf{x}) = x_1^2 x_2, & y_3 &= f_3(\mathbf{x}) = x_1 x_3, \\ y_2 &= f_2(\mathbf{x}) = -x_3^4, & y_4 &= f_4(\mathbf{x}) = x_2 - x_3. \end{aligned} \quad \triangleleft$$

Let $\mathbf{y} = f(\mathbf{x})$ so that f maps a subset of $\mathbb{R}^n$ into $\mathbb{R}^m$. We will assume that the domain of f is an open subset of $\mathbb{R}^n$, and that the partial derivatives $\partial y_i / \partial x_j$ are all continuous. Under these conditions, it can be shown that f has a good local linear approximation at each point $\mathbf{a}$ in its domain. This local linear approximation is a linear transformation $T: \mathbb{R}^n \to \mathbb{R}^m$, and T itself has coordinate functions $T_i: \mathbb{R}^n \to \mathbb{R}$. Now T_i is a linear transformation of the type we just considered, and should be the local linear approximation to the coordinate function f_i of f. Thus we expect to have

$$dy_i = T_i(d\mathbf{x}) = \left(\frac{\partial y_i}{\partial x_1}, \frac{\partial y_i}{\partial x_2}, \ldots, \frac{\partial y_i}{\partial x_n} \right)\Bigg|_{\mathbf{x}=\mathbf{a}} \begin{pmatrix} dx_1 \\ dx_2 \\ \vdots \\ dx_n \end{pmatrix}. \tag{8}$$

(The notation $|_{\mathbf{x}=\mathbf{a}}$ means *evaluated at* $\mathbf{x} = \mathbf{a}$.) We can express all the Eqs. (8) for $i = 1, 2, \ldots, m$ as the single matrix equation

$$\begin{pmatrix} dy_1 \\ dy_2 \\ \vdots \\ dy_m \end{pmatrix} = \begin{pmatrix} \dfrac{\partial y_1}{\partial x_1} & \dfrac{\partial y_1}{\partial x_2} & \cdots & \dfrac{\partial y_1}{\partial x_n} \\[2mm] \dfrac{\partial y_2}{\partial x_1} & \dfrac{\partial y_2}{\partial x_2} & \cdots & \dfrac{\partial y_2}{\partial x_n} \\[2mm] \vdots & \vdots & & \vdots \\[2mm] \dfrac{\partial y_m}{\partial x_1} & \dfrac{\partial y_m}{\partial x_2} & \cdots & \dfrac{\partial y_m}{\partial x_n} \end{pmatrix}_{\mathbf{x=a}} \begin{pmatrix} dx_1 \\ dx_2 \\ \vdots \\ dx_n \end{pmatrix},$$

which we write as

$$\mathbf{dy} = f'(\mathbf{a})\mathbf{dx}. \tag{9}$$

Note that $f'(\mathbf{a})$ in Eq. (9) stands for the $m \times n$ matrix in the preceding equation. We have recovered the simple equation $dy = f'(a)dx$ of single-variable calculus, but in vector form. Thus the $m \times n$ matrix $f'(\mathbf{a}) = (\partial y_i/\partial x_j)$ is the **derivative** of the function f mapping a subset of $\mathbb{R}^n$ into $\mathbb{R}^m$. This derivative at $\mathbf{x = a}$ is the standard matrix representation of the local linear transformation $T: \mathbb{R}^n \to \mathbb{R}^m$ best approximating f at the point $\mathbf{a}$. Calculus texts generally refer to this matrix as the **Jacobian matrix** of f, and denote it by

$$f'(\mathbf{a}) = \left(\frac{\partial y_i}{\partial x_j} \Big|_{\mathbf{x=a}} \right). \quad \textbf{Jacobian matrix} \tag{10}$$

Example 3 Find the derivative (Jacobian matrix) $f'(1, -2)$ if

$$\mathbf{y} = (y_1, y_2, y_3) = f(x_1, x_2) = (x_1{}^2 - 2x_1x_2, \ x_2{}^3, \ -3x_1{}^2x_2).$$

Solution We have the coordinate functions

$$y_1 = f_1(x_1, x_2) = x_1{}^2 - 2x_1x_2,$$
$$y_2 = f_2(x_1, x_2) = x_2{}^3,$$
$$y_3 = f_3(x_1, x_2) = -3x_1{}^2x_2.$$

We find that

$$\frac{\partial y_1}{\partial x_1}\Big|_{(1, -2)} = (2x_1 - 2x_2)\big|_{(1, -2)} = 6,$$

$$\frac{\partial y_1}{\partial x_2}\Big|_{(1, -2)} = -2x_1\big|_{(1, -2)} = -2,$$

$$\frac{\partial y_2}{\partial x_1}\Big|_{(1, -2)} = 0,$$

$$\frac{\partial y_2}{\partial x_2}\Big|_{(1, -2)} = 3x_2{}^2\big|_{(1, -2)} = 12,$$

$$\frac{\partial y_3}{\partial x_1}\Big|_{(1, -2)} = -6x_1x_2\big|_{(1, -2)} = 12,$$

$$\frac{\partial y_3}{\partial x_2}\Big|_{(1, -2)} = -3x_1{}^2\big|_{(1, -2)} = -3.$$

Thus we have

$$f'(1, -2) = \begin{pmatrix} 6 & -2 \\ 0 & 12 \\ 12 & -3 \end{pmatrix}. \quad \triangleleft$$

APPLICATION: SOLVING NONLINEAR SYSTEMS

We discussed solving nonlinear systems by Newton's method in Section 1.1. We treat this again from the calculus point of view.

Consider a nonlinear system

$$f(x_1, x_2) = 0 \qquad (11)$$
$$g(x_1, x_2) = 0$$

where we wish to solve simultaneously for x_1 and x_2. Let h be the map of a subset of $\mathbb{R}^2$ into $\mathbb{R}^2$ having coordinate functions f and g, so that

$$(y_1, y_2) = h(x_1, x_2) = (f(x_1, x_2), g(x_1, x_2)).$$

The system (11) can be expressed as the single linear equation

$$\mathbf{y} = h(\mathbf{x}) = (0, 0) = \mathbf{0}. \qquad (12)$$

Suppose that $\mathbf{a}_0 = (a_{01}, a_{02})$ is a first approximation for a solution of (12). We attempt to refine $\mathbf{a}_0$ to a better approximation $\mathbf{a}_1$ using Newton's method. Taking local coordinates at $\mathbf{a}_0$, we have the local linear approximation

$$\mathbf{dy} = h'(\mathbf{a}_0)\mathbf{dx} \qquad (13)$$

to h at $\mathbf{a}_0$. In Eq. (13), we set $\mathbf{dy} = -h(\mathbf{a}_0)$, which represents the desired change in $\mathbf{y}$ to make $h(\mathbf{x}) = \mathbf{0}$. We obtain

$$h'(\mathbf{a}_0)\mathbf{dx} = -h(\mathbf{a}_0), \qquad (14)$$

and now we find the vector $\mathbf{dx}$ that is the solution of this linear system. This vector $\mathbf{dx}$ is the adjustment we should make to $\mathbf{a}_0$ to obtain the next approximation $\mathbf{a}_1$. Now $h'(\mathbf{a}_0)$ is a 2×2 matrix while $\mathbf{dx}$ and $-h(\mathbf{a}_0)$ are both column vectors of length 2. Assuming that the linear system (14) has a unique solution, this solution is given in matrix form by

$$\mathbf{dx} = h'(\mathbf{a}_0)^{-1}(-h(\mathbf{a}_0)), \qquad (15)$$

where $h'(\mathbf{a}_0)^{-1}$ is of course the inverse of the matrix $h'(\mathbf{a}_0)$. Adjusting $\mathbf{a}_0$ by $\mathbf{dx}$, we take for $\mathbf{a}_1$

$$\mathbf{a}_1 = \mathbf{a}_0 - h'(\mathbf{a}_0)^{-1}h(\mathbf{a}_0).$$

Turning to a general iterate, we see that the recursion relation for Newton's method can be expressed in matrix form as

$$\mathbf{a}_{i+1} = \mathbf{a}_i - h'(\mathbf{a}_i)^{-1}h(\mathbf{a}_i). \qquad (16)$$

Recall that for a single equation $f(x) = 0$ in one variable, the Newton's method recursion relation (6) from Section 8.2 is

$$x_{i+1} = x_i - \frac{f(x_i)}{f'(x_i)}. \tag{17}$$

Since the inverse of the 1×1 matrix $(f'(x_i))$ is the 1×1 matrix $(1/f'(x_i))$ containing the reciprocal entry, we see that Eq. (17) is indeed a special case of Eq. (16). Of course the matrix form (16) of the recursion relation is equally valid for the general case of n equations in n unknowns.

Pencil-and-paper illustrations of Eq. (16), even for $n = 2$, can be very tedious, hence we will not give them. The available program NEWT123 essentially uses Eq. (16), although the partial derivatives appearing in the matrix $h'(\mathbf{a}_i)$ are approximated numerically so that the user need not take the time to differentiate and enter into the program the formulas for the partial derivatives. From personal experience, we can say that when using the partial derivatives with a computer, there is a good chance of making an error either in differentiating or in entering the results into the computer. Having the computer approximate the partial derivatives numerically is apt to be less frustrating, as well as quicker.

LOCAL VOLUME-CHANGE FACTORS

Let f be a differentiable map of a subset of $\mathbb{R}^n$ into $\mathbb{R}^m$. For each point $\mathbf{a}$ in the domain of f, the $m \times n$ Jacobian matrix $f'(\mathbf{a})$ is the matrix representation A of the local linear approximation T of f at $\mathbf{a}$. We know from Section 6.6 that

$$\sqrt{\det(A^T A)} = \sqrt{\det(f'(\mathbf{a})^T f'(\mathbf{a}))}$$

is the volume-change factor for the linear transformation T. Since T approximates f at $\mathbf{a}$, we consider

$$\sqrt{\det(f'(\mathbf{a})^T f'(\mathbf{a}))} \tag{18}$$

to be the **local volume-change factor** for f at the point $\mathbf{a}$.

Example 4 Find the local volume-change factor for f in Example 3 at $(1, -2)$.

Solution In Example 3, we found that

$$f'(1, -2) = \begin{pmatrix} 6 & -2 \\ 0 & 12 \\ 12 & -3 \end{pmatrix}.$$

We now compute

$$f'(-1, 2)^T f'(-1, 2) = \begin{pmatrix} 6 & 0 & 12 \\ -2 & 12 & -3 \end{pmatrix} \begin{pmatrix} 6 & -2 \\ 0 & 12 \\ 12 & -3 \end{pmatrix} = \begin{pmatrix} 180 & -48 \\ -48 & 157 \end{pmatrix}.$$

The volume-change factor at $(-1, 2)$ for f is

$$\sqrt{\det(f'(-1, 2)^T f'(-1, 2))} = \sqrt{(180)(157) - (48)(48)}$$
$$= \sqrt{25956} \approx 161.11. \quad \triangleleft$$

If f maps a subset of $\mathbb{R}^n$ into $\mathbb{R}^n$, then $f'(\mathbf{a})$ is a square matrix and formula (18) reduces to $|\det(f'(\mathbf{a}))|$. Again, f is said to be **orientation preserving** at $\mathbf{a}$ if $\det(f'(\mathbf{a})) > 0$ and is **orientation reversing** if $\det(f'(\mathbf{a})) < 0$.

APPLICATION TO INTEGRALS

Let $\mathbf{x} = f(\mathbf{t})$ be a map of a subset S of $\mathbb{R}^n$ into $\mathbb{R}^m$, where $\mathbf{x} = (x_1, x_2, \ldots, x_m)$ and $\mathbf{t} = (t_1, t_2, \ldots, t_n)$. We consider such a map to be a **parametric representation** of the image $f[S]$ in $\mathbb{R}^m$. Suppose f is a one-to-one map, so that no point of the image $f[S]$ is covered by more than one point in S. We can find the volume of $f[S]$ in $\mathbb{R}^m$ by integration, just as we found arc length in Section 8.2.

Let $\mathbf{t}$ be a point of S, as shown in Fig. 8.13, for the case $n = 2$, $m = 3$. A small box with edges parallel to the coordinate axes and containing $\mathbf{t}$ has volume $dt_1 dt_2 \ldots dt_n$. Under the map f, this volume is multiplied by approximately the local volume-change factor $\sqrt{\det(f'(\mathbf{t})^T f'(\mathbf{t}))}$. Thus the volume of $f[S]$ should be given by the integral

$$\int_S \sqrt{\det(f'(\mathbf{t})^T f'(\mathbf{t}))} \, dt_1 dt_2 \ldots dt_n. \tag{19}$$

We use just one integral sign in (19) for simplicity; for this multiple integral, n integral signs would also be appropriate.

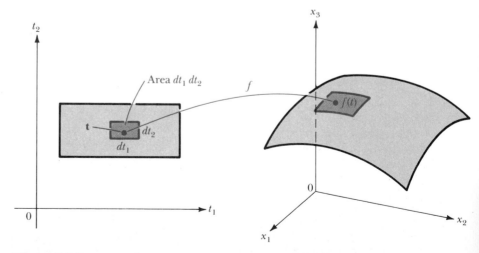

Figure 8.13
Finding the area of a parametrically defined surface in $\mathbb{R}^3$.

Example 5 Let S be the rectangular region $1 \leq u \leq 2$, $0 \leq v \leq 2\pi$ in $\mathbb{R}^2$. The map $f: \mathbb{R}^2 \to \mathbb{R}^3$ given by

$$(x, y, z) = f(u, v) = (u \cos v, u \sin v, v)$$

has as image the single-turn spiral ramp shown in Fig. 8.14. Express the area of this ramp as an integral.

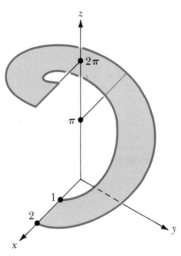

Figure 8.14
A spiral ramp surface.

Solution Computing, we find

$$f'(u, v) = \begin{pmatrix} \cos v & -u \sin v \\ \sin v & u \cos v \\ 0 & 1 \end{pmatrix}$$

so

$$f'(u, v)^T f'(u, v) = \begin{pmatrix} \cos v & \sin v & 0 \\ -u \sin v & u \cos v & 1 \end{pmatrix} \begin{pmatrix} \cos v & -u \sin v \\ \sin v & u \cos v \\ 0 & 1 \end{pmatrix}$$

$$= \begin{pmatrix} 1 & 0 \\ 0 & u^2 + 1 \end{pmatrix}.$$

Thus $\det(f'(u, v)^T f'(u, v)) = u^2 + 1$, and the area of the spiral ramp is given by the integral

$$\int_S \sqrt{u^2 + 1} \, du \, dv = \int_0^{2\pi} \int_1^2 \sqrt{u^2 + 1} \, du \, dv. \quad \triangleleft$$

CHANGE OF VARIABLE IN MULTIPLE INTEGRALS

Let $y = f(\mathbf{x}) = f(x_1, x_2, \ldots, x_n)$ be a continuous function of n variables, mapping a subset of $\mathbb{R}^n$ into $\mathbb{R}$. Consider an integral

$$\int_Q f(\mathbf{x}) \, dx_1 dx_2 \cdots dx_n, \tag{20}$$

where Q is a region of $\mathbb{R}^n$. Again we use one integral sign rather than trying to write n of them. Suppose we wish to make a change of variable, using a function $\mathbf{x} = g(\mathbf{t})$, which carries a region S of $\mathbb{R}^n$ one to one onto the region Q. A partition of the region S into cells of volume $dt_1 dt_2 \cdots dt_n$ gives rise to a partition of Q. If $\mathbf{t}$ is a point in a cell of the partition of S, then the image of this cell under g has approximate volume

$$|\det(g'(\mathbf{t}))| \, dt_1 dt_2 \cdots dt_n. \tag{21}$$

Thus, in changing to variables t_i, the $dx_1 dx_2 \cdots dx_n$ in the integral (20) should be replaced by formula (21), and we find that

$$\int_Q f(\mathbf{x}) \, dx_1 dx_2 \cdots dx_n = \int_S f(g(\mathbf{t})) |\det(g'(\mathbf{t}))| \, dt_1 dt_2 \cdots dt_n. \tag{22}$$

The notation used for $\det(g'(\mathbf{t}))$ in classical texts on advanced calculus is

$$\frac{\partial(x_1, x_2, \ldots, x_n)}{\partial(t_1, t_2, \ldots, t_n)}. \tag{23}$$

It is often called the **Jacobian** of the function g; it is a *scalar* and must not be confused with the Jacobian *matrix*.

Example 6 Find the replacement formula for an integral

$$\iint_Q f(x, y) \, dx \, dy$$

using the polar coordinate substitution

$$(x, y) = g(r, \theta) = (r \cos \theta, r \sin \theta).$$

THE JACOBIAN DETERMINANT J was first introduced by Euler in 1769 when he developed the notion of a double integral. He used formal methods to calculate J by changing one variable at a time. Four years later, Lagrange used the Jacobian in three dimensions when he was dealing with the attraction that an elliptical spheroid exercised on any point in its interior. He had to integrate a function over the entire body; to perform this integration it was necessary to change coordinates.

Mikhail Ostrogradskii generalized Lagrange's result and his method of proof when he dealt with multiple integrals in n-dimensional space in 1836. His proofs, however, like those of Euler and Lagrange, do not meet modern standards.

In 1841 Carl Gustav Jacob Jacobi (1804–1851) gave a detailed treatment of what he called the functional determinant. His proofs were generally recognized as valid, and the determinant was soon named after him. Jacobi was born of Jewish parents, but converted to Christianity before the age of 20. He was therefore able to begin a university career at the University of Berlin. In a relatively short professional life, he made major contributions to the theory of elliptic functions, number theory, and the theory of determinants.

Solution Let g map region S in $\mathbb{R}^2$ one to one onto Q. The coordinate functions of g are

$$x = r \cos \theta \quad \text{and} \quad y = r \sin \theta.$$

Using the notation in (23), we find that

$$\frac{\partial(x, y)}{\partial(r, \theta)} = \begin{vmatrix} \partial x/\partial r & \partial x/\partial \theta \\ \partial y/\partial r & \partial y/\partial \theta \end{vmatrix} = \begin{vmatrix} \cos \theta & -r \sin \theta \\ \sin \theta & r \cos \theta \end{vmatrix}$$
$$= r \cos^2\theta + r \sin^2\theta = r.$$

Thus

$$\iint_Q f(x, y) \, dx \, dy = \iint_S f(r \cos \theta, r \sin \theta) \, |r| \, dr \, d\theta.$$

We have obtained the familiar polar coordinate integral. ◁

Exercise 20 involves a derivation of the usual spherical coordinate substitution in an integral, following the model of Example 6.

Example 7 Make a variable substitution that will simplify the computation of the integral

$$\int_{-1}^0 \int_y^{-y} (x - y)(x + y)^4 \, dx \, dy.$$

Solution The form of the integrand suggests that we let

$$u = x - y \quad \text{and} \quad v = x + y. \tag{24}$$

Solving for x and y, we obtain

$$x = (1/2)(u + v) \quad \text{and} \quad y = (1/2)(v - u). \tag{25}$$

Equation (25) exhibits the coordinate functions of the substitution $g: \mathbb{R}^2 \to \mathbb{R}^2$ given by

$$(x, y) = g(u, v) = ((u + v)/2, (v - u)/2).$$

The region of integration Q in the x,y-plane is bounded by the lines

$$y = x, \quad y = -x, \quad \text{and} \quad y = -1,$$

and is shown in Fig. 8.15(b). From Eqs. (24) and (25), we see that this region Q is the image of the region S in the u,v-plane bounded by the lines

$$u = 0, \quad v = 0, \quad \text{and} \quad u - v = 2,$$

which is shown in Fig. 8.15(a). From Eqs. (25), we find that

$$\frac{\partial(x, y)}{\partial(u, v)} = \begin{vmatrix} 1/2 & 1/2 \\ -1/2 & 1/2 \end{vmatrix} = \frac{1}{4} + \frac{1}{4} = \frac{1}{2}.$$

Referring to Fig. 8.15(a) for the appropriate limits for the integral over S in the u,v-plane, we find that

$$\int_{-1}^0 \int_y^{-y} (x - y)(x + y)^4 \, dx \, dy = \int_{-2}^0 \int_0^{v+2} uv^4(1/2) \, du \, dv$$

and the latter integral is not difficult to compute. ◁

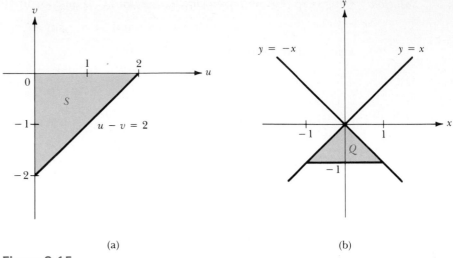

(a)   (b)

Figure 8.15
(a) The region *S* in the *u,v*-plane; (b) the region *Q* in the *x,y*-plane.

SUMMARY

Let $\mathbf{y} = f(\mathbf{x})$, written out as $(y_1, y_2, \ldots, y_m) = f(x_1, x_2, \ldots, x_n)$, be a differentiable function mapping a subset of $\mathbb{R}^n$ into $\mathbb{R}^m$, and let $\mathbf{a} = (a_1, a_2, \ldots, a_n)$ be a point in the domain of f.

1. The derivative $f'(\mathbf{a})$ is the $m \times n$ matrix

$$\left(\frac{\partial y_i}{\partial x_j} \right)\Bigg|_{\mathbf{x}=\mathbf{a}}.$$

It is also called the Jacobian matrix of f at $\mathbf{a}$. If $m = 1$, it is also called the gradient vector $\nabla f(\mathbf{a})$.

2. The linear transformation $T: \mathbb{R}^n \to \mathbb{R}^m$, where $\mathbf{dy} = T(\mathbf{dx}) = f'(\mathbf{a})\mathbf{dx}$, is the differential of f at $\mathbf{a}$. In terms of local coordinate vectors $\mathbf{dx}$ and $\mathbf{dy}$, it gives the best linear approximation to f near $\mathbf{a}$.

3. A nonlinear system of n equations in n unknowns can be written in the form $h(\mathbf{x}) = \mathbf{0}$, where h maps a subset of $\mathbb{R}^n$ into $\mathbb{R}^n$. The Newton recursion formula for approximating a solution of the system is $\mathbf{a}_{i+1} = \mathbf{a}_i - h'(\mathbf{a}_i)^{-1}h(\mathbf{a}_i)$.

4. If $m = n$, the function f is orientation preserving at $\mathbf{a}$ if $\det(f'(\mathbf{a})) > 0$ and is orientation reversing if $\det(f'(\mathbf{a})) < 0$.

5. The local volume-change factor for f at $\mathbf{a}$ is

$$\sqrt{\det(f'(\mathbf{a})^T f'(\mathbf{a}))}.$$

6. If $n = m$, a notation for $\det(f'(\mathbf{a}))$ is

$$\frac{\partial(y_1, y_2, \ldots, y_n)}{\partial(x_1, x_2, \ldots, x_n)}.$$

7. If $\mathbf{x} = g(\mathbf{t})$ for a differentiable function g mapping a region S of $\mathbb{R}^n$ one to one onto a region Q of $\mathbb{R}^n$, and if f is a continuous map of Q into $\mathbb{R}$, then

$$\int_Q f(\mathbf{x}) \, dx_1 dx_2 \cdots dx_n = \int_S f(g(\mathbf{t})) \, |\det(g'(\mathbf{t}))| \, dt_1 dt_2 \cdots dt_n.$$

EXERCISES

In Exercises 1–8, find the derivative (Jacobian matrix) of the given function at the indicated point.

1. $f: \mathbb{R}^2 \to \mathbb{R}$, where $f(x_1, x_2) = x_1^2 - 2x_1 x_2^3$ at $(2, -1)$

2. $f: \mathbb{R}^3 \to \mathbb{R}$, where $f(x_1, x_2, x_3) = x_1^2 + 2x_2 x_3$ at $(4, -1, 2)$

3. $f: \mathbb{R}^4 \to \mathbb{R}$, where $f(x_1, x_2, x_3, x_4) = 2x_1 x_2 x_3^2 x_4$ at $(2, 3, -2, -1)$

4. $f: \mathbb{R}^2 \to \mathbb{R}^2$, where $f(x_1, x_2) = (x_1 - 2x_2, 3x_2^2)$ at $(2, 3)$

5. $f: \mathbb{R}^3 \to \mathbb{R}^3$, where $f(x_1, x_2, x_3) = (\ln x_1 x_2, e^{x_2 x_3}, \sin x_1 x_3)$ at $(2, 4, \pi)$

6. $f: \mathbb{R}^2 \to \mathbb{R}^3$, where $f(x_1, x_2) = (x_1^2 - 2x_2^2, \tan x_1 x_2, 1/x_1)$ at $(-2, 0)$

7. $f: \mathbb{R} \to \mathbb{R}^3$, where $f(x) = (3x^2, \sec x, \ln x)$ at π

8. $f: \mathbb{R}^2 \to \mathbb{R}^4$, where $f(x_1, x_2) = (x_1 - x_2, x_1 + x_2, x_1 \cos x_2, x_1 \sin x_2)$ at $(3, \pi)$

In Exercises 9–14, find the local volume-change factor for the given function at the indicated point. If the derivative matrix is square, classify the function as orientation preserving or orientation reversing at the point.

9. $f: \mathbb{R}^2 \to \mathbb{R}^2$, where $f(x, y) = (xy, x + \sin y)$ at $(1, 0)$

10. $f: \mathbb{R}^2 \to \mathbb{R}^2$, where $f(x, y) = (x^2 - y^2, 2xy)$ at $(1, -2)$

11. $f: \mathbb{R}^3 \to \mathbb{R}^3$, where $f(x, y, z) = (2x + y, xyz, y^2 z)$ at $(1, -1, 1)$

12. $f: \mathbb{R} \to \mathbb{R}^2$, where $f(x) = (x, 1 + x^2)$ at 3

13. $f: \mathbb{R}^2 \to \mathbb{R}^3$, where $f(x, y) = (xy, x + y, y^2)$ at $(2, 1)$

14. $f: \mathbb{R}^2 \to \mathbb{R}^4$, where $f(x, y) = (e^x, y \sin x, x + y, xy)$ at $(0, 1)$

15. For f in Exercise 9, estimate the area of the image under f of the disk
$$(x - 1)^2 + y^2 \le .09.$$

16. For f in Exercise 11, estimate the volume of the image under f of the ball
$$(x - 1)^2 + (y + 1)^2 + (z - 1)^2 \le .04.$$

17. For f in Exercise 13, estimate the volume of the image under f of the rectangle
$$1.8 \le x \le 2.2, \, 0.9 \le y \le 1.1.$$

18. Estimate the area of that part of the surface $z = x^2 + 2y^2$ that lies over the disk
$$(x - 1)^2 + (y + 1)^2 \le 0.16$$
in the x,y-plane. [*Hint:* Consider the function $f: \mathbb{R}^2 \to \mathbb{R}^3$ defined by $f(u, v) = (u, v, u^2 + v^2)$.]

19. Repeat Exercise 18 for the surface $z = xy - 3x + 4y$ and the disk $x^2 + y^2 \leq .09$.

20. With Example 6 as a model, find the replacement formula when changing $\iiint_Q f(x, y, z)\, dx\, dy\, dz$ to spherical coordinates. Recall that spherical coordinates are related to x, y, z-coordinates by the equations $x = \rho \sin \phi \cos \theta$, $y = \rho \sin \phi \sin \theta$, $z = \rho \cos \phi$.

21. Let $f: \mathbb{R}^n \to \mathbb{R}^m$ and $g: \mathbb{R}^m \to \mathbb{R}^s$ be differentiable functions. Let $f(\mathbf{a}) = \mathbf{b}$ lie in the domain of g. Use the chain rule of multivariable calculus to derive the general matrix chain rule $(g \circ f)'(\mathbf{a}) = g'(\mathbf{b})f'(\mathbf{a})$. Note that if we use notations $\mathbf{y} = g(\mathbf{x})$, $\mathbf{x} = f(\mathbf{t})$, $d\mathbf{y}/d\mathbf{x} = g'(\mathbf{x})$, and $d\mathbf{x}/d\mathbf{t} = f'(\mathbf{t})$, then we obtain the familiar formula

$$\frac{d\mathbf{y}}{d\mathbf{t}} = \frac{d\mathbf{y}}{d\mathbf{x}} \frac{d\mathbf{x}}{d\mathbf{t}}.$$

22. Let $f: \mathbb{R}^n \to \mathbb{R}^n$ be a differentiable function and let $\mathbf{a}$ be a point in its domain. Thinking in terms of the best local linear approximation at $\mathbf{a}$, conjecture a condition for f to be a one-to-one map in some neighborhood of $\mathbf{a}$.

23. In Example 7, we solved the equations $u = x - y$, $v = x + y$ for x and y in terms of u and v to compute $\partial(x, y)/\partial(u, v)$. Find an easier way to compute this Jacobian. [*Hint:* If $T: \mathbb{R}^n \to \mathbb{R}^n$ is a one-to-one linear transformation, find a relation between the volume-change factors for T and for T^{-1}.]

24. Use the result of Exercise 23 to find $\partial(x, y, z)/\partial(u, v, w)$ if $u = x - y + 3z$, $v = 2x + y + z - 4$, and $w = y + z + 6$.

In Exercises 25 and 26, proceed as in Example 7 to make a variable substitution that will simplify the computation of the integral. Do not bother to evaluate the integral.

25. $\iint_Q (x - y)^2(x + y)^3\, dx\, dy$, where Q is the region bounded by $x - y = -1$, $x - y = 1$, $x + y = -2$, and $x + y = 1$

26. $\iint_Q (x - 2y + 1)^3/(2x + y - 1)^2\, dx\, dy$, where Q is the region bounded by $x - 2y = -1$, $x - 2y = 1$, $2x + y = 2$, and $2x + y = 4$

■ *For Exercises 27–31, do Exercises 25–29 of Section 1.1, respectively, using the available program NEWT123 or similar software, if you did not do them before.*

8.4
Systems of Linear Differential Equations

We consider the problem of solving the system of linear differential equations

$$x_1' = a_{11}x_1 + a_{12}x_2 + \cdots + a_{1n}x_n$$
$$x_2' = a_{21}x_1 + a_{22}x_2 + \cdots + a_{2n}x_n$$
$$\vdots$$
$$x_n' = a_{n1}x_1 + a_{n2}x_2 + \cdots + a_{nn}x_n \tag{1}$$

where each x_i is a differentiable function of the real variable t and each a_{ij} is a scalar. The simplest such system is the single differential equation

$$x' = ax. \tag{2}$$

If we divide Eq. (2) by x and integrate both sides of the resulting equation, we see that every solution $x(t)$ of Eq. (2) must satisfy $\ln|x(t)| = at + c$, so $|x(t)| = e^c e^{at}$. This shows that every solution $x(t)$ of Eq. (2) must be of the form

$$x(t) = ke^{at}, \tag{3}$$

where k is a scalar. Direct computation easily verifies that the function (3) is the general solution of Eq. (2).

Turning to the solution of the system (1), we write it in matrix form as

$$\mathbf{x}' = A\mathbf{x}, \tag{4}$$

where

$$\mathbf{x} = \begin{pmatrix} x_1(t) \\ x_2(t) \\ \vdots \\ x_n(t) \end{pmatrix}, \qquad \mathbf{x}' = \begin{pmatrix} x_1'(t) \\ x_2'(t) \\ \vdots \\ x_n'(t) \end{pmatrix}, \quad \text{and} \quad A = (a_{ij}).$$

If the matrix A is a diagonal matrix so that $a_{ij} = 0$ for $i \neq j$, then the system (1) reduces to a system of n equations, each like Eq. (2), namely:

$$
\begin{aligned}
x_1' &= a_{11}x_1 \\
x_2' &= a_{22}x_2 \\
&\ \ \vdots \\
x_n' &= a_{nn}x_n .
\end{aligned}
\tag{5}
$$

The general solution is given by

$$\mathbf{x} = \begin{pmatrix} x_1 \\ x_2 \\ \vdots \\ x_n \end{pmatrix} = \begin{pmatrix} k_1 e^{a_{11}t} \\ k_2 e^{a_{22}t} \\ \vdots \\ k_n e^{a_{nn}t} \end{pmatrix}$$

In the general case, we try to diagonalize A and reduce the system (4) to a system like (5). If A is diagonalizable, we have

$$D = C^{-1}AC = \begin{pmatrix} \lambda_1 & & & \\ & \lambda_2 & & \bigcirc \\ & & \ddots & \\ \bigcirc & & & \lambda_n \end{pmatrix}$$

for some invertible $n \times n$ matrix C. If we substitute $A = CDC^{-1}$, Eq. (4) takes the form $C^{-1}\mathbf{x}' = D(C^{-1}\mathbf{x})$, or

$$\mathbf{y}' = D\mathbf{y}, \tag{6}$$

where

$$\mathbf{x} = C\mathbf{y}. \tag{7}$$

(It is easily checked that if we let $\mathbf{x} = C\mathbf{y}$, then indeed $\mathbf{x}' = C\mathbf{y}'$.) The general solution of the system (6) is

$$
\mathbf{y} = \begin{pmatrix} y_1 \\ y_2 \\ \vdots \\ y_n \end{pmatrix} = \begin{pmatrix} k_1 e^{\lambda_1 t} \\ k_2 e^{\lambda_2 t} \\ \vdots \\ k_n e^{\lambda_n t} \end{pmatrix},
$$

where the λ_i are eigenvalues of the diagonalizable matrix A. The general solution $\mathbf{x}$ of the system (4) is then obtained from Eq. (7) using the corresponding eigenvectors $\mathbf{v}_i$ of A appearing as columns of the matrix C.

We emphasize that we have described the general solution (7) only in the case where the matrix A is diagonalizable. The algebraic multiplicity of each eigenvalue of A must equal its geometric multiplicity. The following example illustrates that the eigenvalues need not be distinct, so long as their algebraic and geometric multiplicities are the same.

Example 1 Solve the linear differential system

$$
\begin{aligned}
x_1' &= x_1 - x_2 - x_3 \\
x_2' &= -x_1 + x_2 - x_3 \\
x_3' &= -x_1 - x_2 + x_3 .
\end{aligned}
$$

Solution The first step is to diagonalize the matrix

$$
A = \begin{pmatrix} 1 & -1 & -1 \\ -1 & 1 & -1 \\ -1 & -1 & 1 \end{pmatrix}.
$$

From Example 2 of Section 6.5, we see that if the matrix C there is replaced by

$$
C = \begin{pmatrix} 1 & -1 & -1 \\ 1 & 1 & -1 \\ 1 & 0 & 2 \end{pmatrix},
$$

then

$$
D = C^{-1}AC = \begin{pmatrix} -1 & 0 & 0 \\ 0 & 2 & 0 \\ 0 & 0 & 2 \end{pmatrix}.
$$

Putting $\mathbf{x} = C\mathbf{y}$, the system becomes

$$
\begin{aligned}
y_1' &= -y_1 \\
y_2' &= 2y_2 \\
y_3' &= 2y_3
\end{aligned}
$$

whose solution is given by

$$\begin{pmatrix} y_1 \\ y_2 \\ y_3 \end{pmatrix} = \begin{pmatrix} k_1 e^{-t} \\ k_2 e^{2t} \\ k_3 e^{2t} \end{pmatrix}.$$

Therefore, the solution to the original system is

$$\begin{pmatrix} x_1 \\ x_2 \\ x_3 \end{pmatrix} = \begin{pmatrix} 1 & -1 & -1 \\ 1 & 1 & -1 \\ 1 & 0 & 2 \end{pmatrix} \begin{pmatrix} y_1 \\ y_2 \\ y_3 \end{pmatrix} = \begin{pmatrix} k_1 e^{-t} - k_2 e^{2t} - k_3 e^{2t} \\ k_1 e^{-t} + k_2 e^{2t} - k_3 e^{2t} \\ k_1 e^{-t} \qquad + 2k_3 e^{2t} \end{pmatrix}. \quad \triangleleft$$

In a more advanced text in linear algebra, it is shown that every square matrix A is similar to a matrix J in *Jordan canonical form*, so that $J = C^{-1}AC$. The Jordan canonical form J and the invertible matrix C may have complex entries. In the Jordan canonical form J, all nondiagonal entries are zero except for some entries 1 immediately above the diagonal entries. If the Jordan form $J = C^{-1}AC$ is known, then the substitution $\mathbf{x} = C\mathbf{y}$ again reduces a system $\mathbf{x}' = A\mathbf{x}$ of linear differential equations to one that can be solved easily for $\mathbf{y}$. The solution of the original system is computed as $\mathbf{x} = C\mathbf{y}$, just as we did in Example 1. This is an extremely important technique in the study of differential equations.

SUMMARY

The system $\mathbf{x}' = A\mathbf{x}$ of linear differential equations can be solved, if A is diagonalizable, using these steps.

Step 1 Find a matrix C so that $D = C^{-1}AC$ is a diagonal matrix.

Step 2 Solve the simpler diagonal system $\mathbf{y}' = D\mathbf{y}$.

Step 3 The solution of the original system is $\mathbf{x} = C\mathbf{y}$.

EXERCISES

In Exercises 1–8, solve the given system of linear differential equations as outlined in the summary.

1. $x_1' = 3x_1 - 5x_2$
$\quad x_2' = \qquad 2x_2$

2. $x_1' = \quad x_1 + 4x_2$
$\quad x_2' = 3x_1$

3. $x_1' = \quad x_1 + 2x_2$
$\quad x_2' = 2x_2 + \quad x_1$

4. $x_1' = 2x_1 + 2x_2$
$\quad x_2' = 5x_1 + 3x_2$

5. $x_1' = \quad 6x_1 + 3x_2 - 3x_3$
$\quad x_2' = -2x_1 - \quad x_2 + 2x_3$
$\quad x_3' = 16x_1 + 8x_2 - 7x_3$

6. $x_1' = -3x_1 + 10x_2 - 6x_3$
$\quad x_2' = \qquad 7x_2 - 6x_3$
$\quad x_3' = \qquad\qquad x_3$

7. $x_1' = -3x_1 + 5x_2 - 20x_3$
$\quad x_2' = \quad 2x_1 \qquad + \quad 8x_3$
$\quad x_3' = \quad 2x_1 + \quad x_2 + \quad 7x_3$

8. $x_1' = -2x_1 \qquad - \quad x_3$
$\quad x_2' = \qquad 2x_2$
$\quad x_3' = \quad 3x_1 \qquad + 2x_3$

Linear Programming

Linear programming was introduced in Section 1.2 and should be reviewed at this time. Linear programming deals with the problem of maximizing or minimizing a linear function

$$c_1 x_1 + c_2 x_2 + \cdots + c_n x_n$$

subject to constraints that are linear equalities or linear inequalities in the variables x_i. In Section 1.2 we gave a geometric approach to the solution of such problems for the case $n = 2$. In this chapter we will see how solutions can be obtained algebraically. In Section 9.1 we describe Dantzig's simplex method for constraints of one type. The simplex method for general constraints is treated in Section 9.2, and duality is discussed in Section 9.3. Duality shows that every maximizing linear programming problem has an associated dual minimizing problem, and vice versa. Moreover, both problems can be solved at the same time.

We make no attempt to give a complete or rigorous treatment of the simplex method. Whole books have been written on the subject. We merely explain the technique while trying as best we can to show in an intuitive way why it works.

9.1
The Simplex Method for Constraints $Ax \le b$

THE STANDARD MAXIMUM PROBLEM

In practice, linear programming problems that arise deal with maximizing or minimizing a linear function in which the variables can assume only nonnegative values. We start by presenting the simplex method for the *standard maximum problem,* which may be stated as follows.

Standard Maximum Problem

Maximize the linear function

$$P = c_1 x_1 + c_2 x_2 + \cdots + c_n x_n \tag{1}$$

subject to the constraints

$$
\begin{aligned}
a_{11} x_1 + a_{12} x_2 + \cdots + a_{1n} x_n &\le b_1 \\
a_{21} x_1 + a_{22} x_2 + \cdots + a_{2n} x_n &\le b_2 \\
&\vdots \\
a_{m1} x_1 + a_{m2} x_2 + \cdots + a_{mn} x_n &\le b_m,
\end{aligned}
\tag{2}
$$

where all b_i are nonnegative, and where

$$x_1 \ge 0, \, x_2 \ge 0, \cdots, x_n \ge 0. \tag{3}$$

The linear function in Eq. (1) is the **objective function;** in this maximizing situation it is called the **profit function.** For a minimum problem, the objective function is referred to as the **cost function.** The solution set of the constraints (2) and (3) is called the **feasible set** and any point in the feasible set is a **feasible solution.**

In Section 1.2 we saw that the feasible set for a linear programming problem in two variables is a convex set obtained as an intersection of half planes in $\mathbb{R}^2$. In the present situation the feasible set is an intersection of half spaces in $\mathbb{R}^n$, each given by an inequality (2) or (3). In Section 1.2 with $n = 2$, we saw that if the maximum value of the objective function over the feasible set is attained, then it is attained at a *corner point* of the feasible set in $\mathbb{R}^2$; the corner point appears as an intersection of two lines. Generalizing, a **corner point** is a *feasible solution* in $\mathbb{R}^n$ that can be obtained by turning n of the $n + m$ inequalities (2) and (3) into equations, replacing $\le$ by $=$, and solving the resulting $n \times n$ linear system. If $n = 2$, a corner point becomes the intersection of two lines; if $n = 3$, it becomes the intersection of three planes; and so on. We state an important theorem without proof. An intuitive geometric argument in terms of convex sets and hyperplanes can be made just as for the case $n = 2$ in Section 1.2.

Theorem 9.1   Achieving Optimal Values

If a linear programming problem has an optimal solution, then the optimal value of the objective function is achieved at some corner point of the feasible set.

Feasible solutions that are not corner points may also give the same optimal value. Any feasible solution giving the optimal value is called an **optimal solution.**

The standard maximum problem can be stated more concisely in matrix form. An inequality $A \leq B$ for two matrices of the same size means that $a_{ij} \leq b_{ij}$ for each index pair i, j. In a similar way, we can summarize our requirement $b_i \geq 0$ for all i by $\mathbf{b} \geq \mathbf{0}$.

Standard Maximum Problem

(Matrix form)

Maximize

$$\mathbf{c} \cdot \mathbf{x}$$

subject to

$$A\mathbf{x} \leq \mathbf{b}$$

where

$$\mathbf{b} \geq \mathbf{0} \text{ and } \mathbf{x} \geq \mathbf{0}.$$

In this matrix form $A = (a_{ij})$ is an $m \times n$ matrix, $\mathbf{c}$ and $\mathbf{x}$ are n-component column vectors, and $\mathbf{b}$ is an m-component column vector. Note that the objective function (1) is written as a dot product.

INTRODUCING SLACK VARIABLES

The first step in solving a linear programming problem by the simplex method is to convert the inequalities (2) to a linear system of equations. For the standard maximum problem, this is done by introducing the **slack variables** y_i defined by

$$a_{i1}x_1 + a_{i2}x_2 + \cdots + a_{in}x_n + y_i = b_i \tag{4}$$

for each $i = 1, 2, \ldots, m$. The inequalities (2) correspond to the requirements $y_i \geq 0$ for each i, or more briefly, the requirement $\mathbf{y} \geq \mathbf{0}$. Note that setting $y_i = 0$ in Eq. (4) gives the equation $a_{i1}x_1 + a_{i2}x_2 + \cdots + a_{in}x_n = b_i$, which is usually a bounding hyperplane of the feasible set. This is illustrated in Fig. 9.1.

The linear system (4) of equations can be written in matrix form as

$$(A \quad I) \begin{pmatrix} \mathbf{x} \\ \mathbf{y} \end{pmatrix} = \mathbf{b}, \tag{5}$$

where the $m \times m$ identity matrix I has been appended to the matrix A to form the larger $m \times (n + m)$ coefficient matrix in (5). A solution to (5) with non-

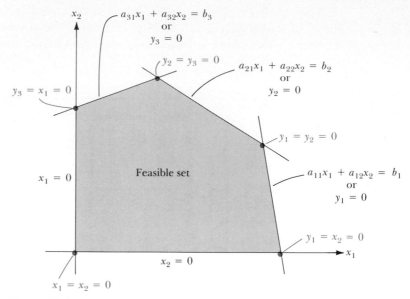

Figure 9.1
Bounding lines and corner points.

negative components is an $(n + m)$-component vector

$$\begin{pmatrix} \mathbf{x} \\ \mathbf{y} \end{pmatrix} = \begin{pmatrix} x_1 \\ \vdots \\ x_n \\ y_1 \\ \vdots \\ y_m \end{pmatrix}$$

and will also be called a **feasible solution** for the linear programming problem.

The corner points of the feasible set are *represented by* those feasible solutions of Eq. (5) in which n of the $n + m$ components are zero, as illustrated in Fig. 9.1 for the case $n = 2$, $m = 3$. The variables that are set equal to zero to produce a corner point are called **nonbasic variables** for that corner point. The remaining variables are **basic variables** there. Such a feasible solution of Eq. (5) representing a corner point is called a **basic feasible solution.**

In theory, the method of Section 1.2 could be used to solve linear programming problems: Find all corner points and substitute them in the objective function. In practical situations, there are too many corner points to make such a computation feasible (see Exercise 1).

We are interested in finding a corner point (basic feasible solution) at which the maximum of the objective function is attained. The simplex method achieves this by beginning with any corner point and moving from it to an adjacent corner point in a way that increases the objective function as quickly as possible. We continue until an optimal solution is found.

> Note that for the standard maximum problem, the origin $(0, 0, \ldots, 0)$ is always a corner point.

Figure 9.2 illustrates the simplex method for the case $n = 2$; start at the origin and follow the arrows along edges to the corner point that is an optimal solution. We will explain the simplex method, developed by George Dantzig, in the context of a simple problem.

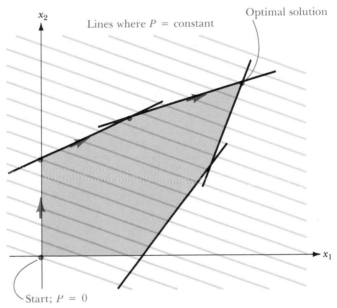

Figure 9.2
Going from (0, 0) to an optimal solution.

Problem   A boatyard produces three kinds of boats: cabin cruisers, racing sailboats, and cruising sailboats. Beginning with fiberglass hulls and decks only, the boatyard finishes the boats taking 2, 1, and 3 weeks, respectively, to do each job. The boatyard is closed during the last week of December. The net unit profit on each kind of boat is $5,000, $4,000, and $6,000, respectively. Because of space considerations, the yard can finish no more than 25 boats in one year (51 weeks). Find the number of boats of each kind that will maximize the annual profit.

The data is summarized in Table 9.1. We wish to maximize the annual profit

$$P = 5x_1 + 4x_2 + 6x_3 \tag{6}$$

Table 9.1

	Cabin Cruiser	Racing Sailboat	Cruising Sailboat	Total for One Year
Number of boats per year	x_1	x_2	x_3	25
Unit time required (weeks)	2	1	3	51
Unit profit (thousands of dollars)	5	4	6	$5x_1 + 4x_2 + 6x_3$

subject to the constraints

$$x_1 + x_2 + x_3 \le 25$$
$$2x_1 + x_2 + 3x_3 \le 51 \tag{7}$$

and

$$x_1 \ge 0, \quad x_2 \ge 0, \quad x_3 \ge 0.$$

The two inequalities (7) are turned into the equations

$$x_1 + x_2 + x_3 + y_1 \qquad = 25$$
$$2x_1 + x_2 + 3x_3 \qquad + y_2 = 51 \tag{8}$$

by introducing the slack variables $y_1 \ge 0$ and $y_2 \ge 0$. The matrix equation (5) for the system (8) is

$$\begin{pmatrix} 1 & 1 & 1 & 1 & 0 \\ 2 & 1 & 3 & 0 & 1 \end{pmatrix} \begin{pmatrix} x_1 \\ x_2 \\ x_3 \\ y_1 \\ y_2 \end{pmatrix} = \begin{pmatrix} 25 \\ 51 \end{pmatrix}. \tag{9}$$

THE INITIAL TABLEAU

To begin the simplex method, we represent our problem in this *initial tableau*

$$\begin{array}{c|ccccc|c} & x_1 & x_2 & x_3 & y_1 & y_2 & \\ \hline y_1 & 1 & 1 & 1 & 1 & 0 & 25 \\ y_2 & 2 & 1 & 3 & 0 & 1 & 51 \\ \hline P & -5 & -4 & -6 & 0 & 0 & 0 \end{array} \tag{10}$$

which contains the augmented matrix of the system (8). The last row in (10) represents the objective function and Eq. (6), but in the form $P - 5x_1 - 4x_2 - 6x_3 - 0y_1 - 0y_2 = 0$. This bottom row is called the **objective row** and is always formed as illustrated.

At a corner point in the feasible set, $n = 3$ of the variables x_i, y_j are zero, and these are the nonbasic variables at that corner point. In the initial tableau, the corner point is the origin where we have $x_1 = x_2 = x_3 = 0$. If we set $x_1 = x_2 = x_3 = 0$ in the system (9) and solve for y_1 and y_2, we obtain the basic feasible solution

$$(x_1, x_2, x_3, y_1, y_2) = (0, 0, 0, 25, 51) \tag{11}$$

representing the corner point $(0, 0, 0)$. For this corner point the nonbasic variables are x_1, x_2, x_3 and the basic variables are y_1, y_2. Note that the column vectors under y_1 and y_2 above the objective row in the initial tableau (10) comprise the standard basis for $\mathbb{R}^m = \mathbb{R}^2$. In the initial tableau (10) the first two rows are labeled at the left with the basic variables; their respective values, shown in Eq. (11), appear in the last column. The value of the objective function

$$P = 5x_1 + 4x_2 + 6x_3 + 0y_1 + 0y_2 \tag{12}$$

at the corner point represented by $(0, 0, 0, 25, 51)$ is 0, and appears at the bottom of the last column in the tableau. In summary, this initial tableau (10) represents the situation at the corner point $(0, 0, 0)$, that is, at the origin.

In a similar manner, each subsequent simplex tableau exhibits data at a single corner point as explained in the following box.

Interpreting a Simplex Tableau

1. The basic variables are those that are used both as row labels and column labels. They can also be described as those labeling columns containing standard basis vectors, that is, columns with all zero entries except for the entry 1 in the row and column having a common label.

2. The value of each basic variable and of the objective function at the corner point appears in the final column of the tableau. Nonbasic variables have value zero.

OBTAINING THE NEXT TABLEAU

We wish to move to an adjacent corner point where the value P of the objective function will be increased, and continue until the maximum value is found. In order to move from $(0, 0, 0)$ to an *adjacent* corner point, *just one* of the basic variables in the variable list x_1, x_2, x_3, y_1, y_2 must become nonbasic, and vice versa. First we decide which nonbasic variable will become basic, that is, which variable will be increased from 0 to a positive quantity. Since we want to increase P as rapidly as possible, we choose a variable in (12), in this case x_3, corresponding to the largest coefficient. In the initial tableau (10), this corresponds to a *negative entry of greatest magnitude* in the objective row. Thus x_3 has been selected as the *entering variable* to become basic. The column labeled x_3 in the initial tableau is the **pivotal column.**

Now we must decide which basic variable, y_1 or y_2, to choose as *exiting variable*, that is, to become nonbasic with value 0. This is done by computing the ratios of entries in the last column of the tableau (10) by corresponding entries in the pivotal column. We choose the variable labeling the row corresponding to the *smallest nonnegative ratio*.† We can see the reason for this as follows. Since we are moving from the corner point (11) by allowing x_3 to become positive, we must make one of the basic variables y_1, y_2 zero. Equations (8) with $x_1 = x_2 = 0$ can be written

$$y_1 = 25 - x_3 = 25[1 - (1/25)x_3]$$
$$y_2 = 51 - 3x_3 = 51[1 - (1/17)x_3].$$

If y_1 is allowed to decrease to zero, then $x_3 = 25$, which makes y_2 negative and thus does not result in a feasible solution. Therefore, we choose y_2 to be zero corresponding to $x_3 = 17$; this accommodates the requirements $y_1, y_2 \geq 0$. The largest *allowable* increase in x_3 corresponds to the smallest of the aforementioned ratios. The **pivotal row** for our initial tableau is then the row labeled y_2. The entry 3 in the pivotal row and pivotal column is the **pivot** for this step of the simplex method; see the first tableau on page 451.

In choosing the pivotal row, note that only nonnegative ratios are considered because negative entries in the pivotal column would cause the objective function to decrease as the entering basic variable increases. In fact, if all the entries in the pivotal column should be zero or negative, then x_3 could be increased as much as we like without leaving the feasible set. The values of the objective function on the feasible set are then **unbounded** and the linear programming problem has no optimal solution.

Using the pivot, we transform the initial tableau to a simplex tableau in which the nonbasic variables are x_1, x_2, y_2 and the basic variables are y_1, x_3. This is accomplished by using a Gauss–Jordan type reduction to convert the pivot to 1 and all the numerical entries above and below it to zero. Thus the column vector under x_3 becomes a standard basis vector. Such a reduction of an augmented matrix does not alter the solutions of the corresponding linear system (9). On the objective row, the reduction amounts to adding a multiple of a valid equation to the objective equation

$$P - 5x_1 - 4x_2 - 6x_3 + 0y_1 + 0y_2 = 0 \qquad (13)$$

---

† If a ratio is zero, then one of the basic variables has value zero. The corresponding basic feasible solution

is said to be degenerate. It is possible under these circumstances that the simplex method may enter an infinite cycle in which no progress is made. Such cyclic behavior is extremely rare in practice, and shall not concern us further.

and the resulting equation is still a valid expression for our objective function. We obtain

	x_1	x_2	x_3	y_1	y_2	
y_1	1	1	1	1	0	25
y_2	2	1	3	0	1	51
P	-5	-4	-6	0	0	0

$\sim$

	x_1	x_2	x_3	y_1	y_2	
y_1	$\frac{1}{3}$	$\frac{2}{3}$	0	1	$-\frac{1}{3}$	8
x_3	$\frac{2}{3}$	$\frac{1}{3}$	1	0	$\frac{1}{3}$	17
P	-1	-2	0	0	2	102.

This new tableau exhibits the basic variables y_1 and x_3 as row labels in the left-most column, and their values 8 and 17 at the corner point where $x_1 = x_2 = y_2 = 0$ appear in the right-most column. Our choice of negative entry in the objective row to determine the pivotal column and the test using ratios to find the pivotal row ensures that the entries in the last column of a tableau will always be nonnegative. Our new tableau shows the current corner point to be $(x_1, x_2, x_3) = (0, 0, 17)$, which is represented by the basic feasible solution

$$(x_1, x_2, x_3, y_1, y_2) = (0, 0, 17, 8, 0).$$

Our original objective equation (13) has been transformed into

$$P - x_1 - 2x_2 + 0x_3 + 0y_1 + 2y_2 = 102$$

so $P = 102$ at the basic feasible solution where $x_1 = x_2 = y_2 = 0$. Of course, we could also evaluate the original objective function $P = 5x_1 + 4x_2 + 6x_3$ at the corner point $(0, 0, 17)$, obtaining again 102.

The process of finding a pivot corresponding to swapping a basic for a nonbasic variable is repeated until the objective row contains no negative entry. When this occurs, we have found an optimal solution of the linear programming problem. We summarize the steps involved in the following box.

Steps in the Simplex Method for the Standard Maximum Problem

1. Set up the initial tableau, exhibiting data at the origin.

2. *Test for an optimal solution:* If the entries in the objective row are nonnegative, an optimal solution has been found. Otherwise, continue.

3. *Find the pivotal column:* Choose the column in the tableau corresponding to a negative entry of largest magnitude in the bottom row.

4. *Test for unboundedness:* If all the entries in the pivotal column are zero or negative, then the objective function is unbounded on the feasible set and the problem has no optimal solution.

5. *Find the pivotal row:* Compute the ratios of entries in the last column by corresponding entries in the pivotal column that are positive. Choose the row corresponding

to the smallest nonnegative ratio. (In case of a tie, either row may be chosen.)

6. The entry in the row and column described in steps (2) and (4) is the pivot. Use a Gauss–Jordan type reduction on the tableau to convert this pivot to 1 with zeros above and below it. Relabel the rows appropriately. Go to step (2).

Following the rules in the box, we take our current simplex tableau above and find the pivot. The pivot column is the column containing -2 in the objective row and labeled x_2, the nonbasic variable to enter the basic category. The ratios of entries in the last column by entries in the pivot column are $8/(2/3) = 12$ and $17/(1/3) = 51$; thus the pivot row is the one labeled y_1, the variable to exit the basic category and become nonbasic. We use a Gauss–Jordan reduction with the indicated pivot as follows.

	x_1	x_2	x_3	y_1	y_2	
y_1	$\frac{1}{3}$	$\frac{2}{3}$	0	1	$-\frac{1}{3}$	8
x_3	$\frac{2}{3}$	$\frac{1}{3}$	1	0	$\frac{1}{3}$	17
P	-1	-2	0	0	2	102

$\sim$

	x_1	x_2	x_3	y_1	y_2	
x_2	$\frac{1}{2}$	1	0	$\frac{3}{2}$	$-\frac{1}{2}$	12
x_3	$\frac{1}{2}$	0	1	$-\frac{1}{2}$	$\frac{1}{2}$	13
P	0	0	0	3	1	126

Since all entries in the objective row of the new tableau are nonnegative, we have found the optimal solution. The maximum profit is $P = 126$ and it occurs at the corner point $(0, 12, 13)$ represented by

$$(x_1, x_2, x_3, y_1, y_2) = (0, 12, 13, 0, 0).$$

Since this result was obtained, the boatyard no longer produces cabin cruisers because $x_1 = 0$. It now produces $x_2 = 12$ racing sailboats and $x_3 = 13$ cruising sailboats per year, earning a maximum annual profit of $126,000.

MINIMIZING SUBJECT TO $A\mathbf{x} \le \mathbf{b}$, $\mathbf{x} \ge \mathbf{0}$

Suppose that we wish to *minimize* a linear function $f(\mathbf{x})$ subject to constraints $A\mathbf{x} \le \mathbf{b}$, $\mathbf{x} \ge \mathbf{0}$, where $\mathbf{b} \ge \mathbf{0}$. A moment of thought shows that if the minimum of $f(\mathbf{x})$ subject to these constraints exists, then

$$[\text{minimum of } f(\mathbf{x})] = -[\text{maximum of } -f(\mathbf{x})],$$

where the maximum of $-f(\mathbf{x})$ is subject to these same constraints. For example, in our boatyard problem we found that the maximum of the objective function $P = 5x_1 + 4x_2 + 6x_3$ subject to the constraints (7) is 126, attained when $x_1 = 0$, $x_2 = 12$, and $x_3 = 13$. The minimum of $f(x) = -5x_1 - 4x_2 - 6x_3$ subject to these same constraints is -126, and is attained for the same values of x_1, x_2, and x_3. Thus when we encounter a minimizing linear programming problem with constraints $A\mathbf{x} \le \mathbf{b}$, $\mathbf{x} \ge \mathbf{0}$, where $\mathbf{b} \ge \mathbf{0}$, we may simply maximize the negative of our objective function, and then change the sign of the maximum value to obtain the desired minimum value.

We give two examples. In Example 1 we maximize an objective function to emphasize the simplex method described above. Then in Example 2, we minimize the *same* objective function, using the technique just described.

Example 1   Maximize $-2x_1 + x_2$ subject to the constraints

$$x_1 - 2x_2 \leq 4, \qquad -3x_1 + x_2 \leq 3, \qquad 4x_1 + 7x_2 \leq 46.$$

Solution   We form the initial tableau and proceed to reduce it, following the boxed outline.

	x_1	x_2	y_1	y_2	y_3	
y_1	1	-2	1	0	0	4
y_2	-3	1	0	1	0	3
y_3	4	7	0	0	1	46
P	2	-1	0	0	0	0

$\sim$

	x_1	x_2	y_1	y_2	y_3	
y_1	-5	0	1	2	0	10
x_2	-3	1	0	1	0	3
y_3	25	0	0	-7	1	25
P	-1	0	0	1	0	3

$\sim$

	x_1	x_2	y_1	y_2	y_3	
y_1	0	0	1	$\frac{3}{5}$	$\frac{1}{5}$	15
x_2	0	1	0	$\frac{4}{25}$	$\frac{3}{25}$	6
x_1	1	0	0	$-\frac{7}{25}$	$\frac{1}{25}$	1
P	0	0	0	$\frac{18}{25}$	$\frac{1}{25}$	4

Thus the maximum value of the objective function is 4 and is attained at $(x_1, x_2) = (1, 6)$. ◁

Example 2   Minimize the same objective function $-2x_1 + x_2$ as in Example 1, subject to the same constraints.

Solution   We maximize the function $2x_1 - x_2$, following again the steps in the boxed outline.

	x_1	x_2	y_1	y_2	y_3	
y_1	1	-2	1	0	0	4
y_2	-3	1	0	1	0	3
y_3	4	7	0	0	1	46
P	-2	1	0	0	0	0

$\sim$

	x_1	x_2	y_1	y_2	y_3	
x_1	1	-2	1	0	0	4
y_2	0	-5	3	1	0	15
y_3	0	15	-4	0	1	30
P	0	-3	2	0	0	8

$\sim$

	x_1	x_2	y_1	y_2	y_3	
x_1	1	0	$\frac{7}{15}$	0	$\frac{2}{15}$	8
y_2	0	0	$\frac{5}{3}$	1	$\frac{1}{3}$	25
x_2	0	1	$-\frac{4}{15}$	0	$\frac{1}{15}$	2
P	0	0	$\frac{6}{5}$	0	$\frac{1}{5}$	14

Thus the maximum of $2x_1 - x_2$ is 14, attained at the point $(x_1, x_2) = (8, 2)$. Consequently, the minimum of our given objective function is -14, attained at this same point $(8, 2)$. ◁

We close with an example illustrating the case in which the objective function can assume arbitrarily large values on the feasible set, so that no maximum value exists.

Example 3   Maximize $-x_1 + 3x_2$ subject to the constraints

$$-2x_1 + x_2 \le 1, \qquad -x_1 + x_2 \le 4, \qquad x_1 \ge 0, \qquad x_2 \ge 0.$$

Solution   We form the initial tableau and reduce it as usual.

	x_1	x_2	y_1	y_2	
y_1	-2	1	1	0	1
y_2	-1	1	0	1	4
P	1	-3	0	0	0

$\sim$

	x_1	x_2	y_1	y_2	
x_2	-2	1	1	0	1
y_2	1	0	-1	1	3
P	-5	0	3	0	3

$\sim$

	x_1	x_2	y_1	y_2	
x_2	0	1	-1	2	7
x_1	1	0	-1	1	4
P	0	0	-2	5	18

The pivotal column labeled y_1 in this final tableau contains only negative entries, so the objective function is unbounded on the feasible set and the problem has no optimal solution. (See number (4) of the boxed outline.)   ◁

SUMMARY

1. A linear programming problem consists of maximizing or minimizing a linear function, called the objective function, in nonnegative variables $x_1, x_2, \ldots, x_n$ that satisfy additional constraints in the form of linear equalities or inequalities.

2. The standard maximum problem is to maximize an objective function $c_1x_1 + c_2x_2 + \cdots + c_nx_n$ in which the variables x_i are required to be nonnegative and satisfy other linear inequality constraints of the form (linear expression) $\le b_i$, where $b_i \ge 0$. In matrix notation the standard maximum problem is to maximize $\mathbf{c} \cdot \mathbf{x}$ subject to $A\mathbf{x} \le \mathbf{b}$ and $\mathbf{x} \ge \mathbf{0}$, where $\mathbf{x}$ is the column vector of n variables, and A, $\mathbf{b}$, and $\mathbf{c}$ have appropriate size with $\mathbf{b} \ge \mathbf{0}$. A vector $\mathbf{x}$ that satisfies all these conditions is a feasible solution for the problem, and a feasible solution that maximizes the linear function $\mathbf{c} \cdot \mathbf{x}$ is an optimal solution. The set of all

feasible solutions is the feasible set. If an optimal solution exists, it occurs at a corner point of the feasible set.

3. For $i = 1, 2, \ldots, m$, the ith inequality represented by $A\mathbf{x} \leq \mathbf{b}$, namely

$$a_{i1}x_1 + a_{i2}x_2 + \cdots + a_{in}x_n \leq b_i,$$

can be turned into an equation

$$a_{i1}x_1 + a_{i2}x_2 + \cdots + a_{in}x_n + y_i = b_i$$

by inserting the slack variable y_i. The standard maximum problem then takes the following form: Maximize $\mathbf{c} \cdot \mathbf{x}$ subject to

$$(A \quad I) \begin{pmatrix} \mathbf{x} \\ \mathbf{y} \end{pmatrix} = \mathbf{b}, \quad \mathbf{x} \geq \mathbf{0}, \quad \mathbf{y} \geq \mathbf{0}.$$

4. A corner point of the feasible set is a feasible solution $\mathbf{x}$ obtained by setting n of the $n + m$ variables x_i, y_j equal to zero and solving the equations in number (3) of the Summary for the other variables. The variables set equal to zero are called nonbasic at the corner point; the others are basic variables there.

5. A simplex tableau summarizes the status of the linear programming problem at a particular corner point of the feasible set. The simplex method consists of a Gauss–Jordan type of reduction, which transforms one simplex tableau into another in which the objective function has been increased. The steps for the simplex method are summarized in the box on page 451.

EXERCISES

1. Consider a standard maximum problem in 7 variables with 10 constraints in addition to the constraints $x_i > 0$. How many *corner points* must be found and checked to be sure they are feasible to solve such a problem by the method in Section 1.2? (Use a counting argument involving the number of combinations of k things taken r at a time.)

2. Solve the boatyard problem stated in this section if the units of profit in Table 9.1 are changed to 8 units for a cabin cruiser, 3 units for a racing sailboat, and 6 units for a cruising sailboat.

3. Solve Exercise 15 of Section 1.2 (the coffee problem) using the simplex method.

4. At first, one might think that if an objective function has some positive and some negative coefficients, then a maximum would have to be achieved at a point where all variables with negative coefficients are zero, and a minimum at a point where all variables with positive coefficients are zero. Of course, Examples 1 and 2 show that this is not the case. Draw a sketch of the feasible set given by the constraints in Example 1, as in Section 1.2. Find the corner points, draw some lines where the objective function has constant values, and see graphically why one obtains the answers in Examples 1 and 2. Trace the path from the origin to the optimal solution in each example, and note how the intermediate corner point appears in the second tableau in each case.

In Exercises 5–8, write the given linear programming problem as a standard maximum problem using the matrix form boxed on page 444.

5. Maximize $x_1 + 2x_2 + 4x_3$
subject to the constraints

$$12 - x_3 \geq 3x_1 + 4x_2,$$
$$2x_1 + 2x_2 - x_3 \geq -4,$$
$$x_2 \leq x_1 + 1,$$
$$\text{all } x_i \geq 0.$$

6. Minimize $x_1 - 3x_2 - 8x_3$
subject to the constraints

$$2x_2 - 20 \leq 2x_1 - x_3,$$
$$-x_1 + 2x_2 - 2x_3 \leq 4,$$
$$16 - 4x_1 - 8x_2 + x_3 \geq 0,$$
$$\text{all } x_i \geq 0.$$

7. Minimize $2x_1 + x_2 + x_3 - x_4$
subject to the constraints

$$3x_2 + 5x_3 \leq 10 - x_1,$$
$$-2x_2 - 5x_3 + x_4 \leq 5 - x_1,$$
$$\text{all } x_i \geq 0.$$

8. Maximize $5x_1 - x_2 - x_3 - 5x_4$
subject to the constraints

$$-x_1 + x_2 - x_3 + x_4 \geq -2,$$
$$4x_1 + x_2 - 5x_3 - 4x_4 \leq 3,$$
$$x_1 + 3x_2 \leq 5 - 3x_1 - x_2,$$
$$\text{all } x_i \geq 0.$$

In Exercises 9–16, solve the given linear programming problem using the simplex method. All variables x_i given are assumed to be nonnegative.

9. Maximize $3x_1 + 2x_2$
subject to the constraints

$$x_1 + x_2 \leq 3,$$
$$2x_1 + x_2 \leq 5.$$

10. Minimize $-2x_1 + x_2$
subject to the constraints

$$2x_1 - x_2 \leq 8,$$
$$x_1 + x_2 \leq 10,$$
$$x_1 + 2x_2 \leq 14.$$

11. Maximize $4x_1 + 8x_2 - 4x_3$
subject to the constraints

$$x_1 + 2x_2 + x_3 \leq 2,$$
$$x_1 + x_2 + 2x_3 \leq 3.$$

12. Maximize $3x_1 + x_2 - 4x_3$
subject to the constraints

$$3x_1 + 2x_2 + x_3 \leq 5,$$
$$x_1 + x_2 - x_3 \leq 8,$$
$$2x_1 + x_2 + 3x_3 \leq 15.$$

13. Minimize $2x_1 - 4x_2$
subject to the constraints

$$x_1 + x_2 \leq 6,$$
$$-x_1 + 3x_2 \leq 18,$$
$$x_1 - x_2 \leq 8.$$

14. Minimize $-x_1 + 3x_2$
subject to the constraints

$$x_1 + 2x_2 \leq 10,$$
$$-x_1 + x_2 \leq 10.$$

15. Maximize $3x_1 + 2x_2 + x_3$
subject to the constraints

$$x_1 + x_2 - x_3 \leq 10,$$
$$2x_1 + x_2 + x_3 \leq 6.$$

16. Maximize $2x_1 + 2x_2 - x_3 + 3x_4$
subject to the constraints

$$4x_1 + 6x_2 + 2x_3 - 2x_4 \leq 7,$$
$$3x_1 + x_2 - 4x_3 + 5x_4 \leq 8.$$

17. Working graphically, construct a standard maximum problem in two variables with three constraints other than $x_i \geq 0$ having the following property:

Solving the problem using the simplex method and choosing the pivotal column in the initial tableau with negative objective row entry of maximum magnitude, as described in the text, requires 3 iterations, while another choice of initial pivotal column requires only 2 iterations. Check your problem by solving it both ways.

18. Continuing the idea of Exercise 17, draw a graph indicating that one can construct a standard maximum problem in two variables such that with initial choice of pivotal column as described in the text, 10 iterations will be required to solve the problem, but a different choice of pivotal column will solve the problem in a single iteration. (A great deal of effort has been expended in attempting to maximize the efficiency of the simplex method.)

19. Solve the problem given as text answer to Exercise 17 using the graphic program LINPROG, and observe graphically whether the text answer satisfies the requirements of the problem.

20. Show graphically that the problem in Example 3 has no optimal solution.

 *The available program SIMPLEX has three options:*

1. an option allowing the user to specify the next step in a reduction of a tableau,
2. a quiz option on the simplex method as described in the text, and
3. an option to have the computer execute the simplex method and print the results from supplied data, with tableaux shown if desired.

In Exercises 21–24, use the first two options of SIMPLEX or similar software to find the indicated optimal value. Work at least the odd-numbered problems in the quiz format. It may be interesting to experiment with different choices for pivotal columns, and see how the numbers of iterations needed to obtain a solution compare.

21. The problem in Exercise 5   **22.** The problem in Exercise 6
23. The problem in Exercise 7   **24.** The problem in Exercise 8

9.2
The Simplex Method for General Constraints

A constraint in a general linear programming problem has one of the forms

$$d_1x_1 + d_2x_2 + \cdots + d_nx_n \begin{cases} \leq b \\ = b \\ \geq b \end{cases} \tag{1}$$

where we require that $b \geq 0$.

Recall that we always have the constraints $x_i \geq 0$. In the preceding section, we treated the case in which all other constraints have the ($\leq b$)-form in (1). We now discuss the situation in which any or all of the types in (1) may occur.

In the literature, one finds the simplex method divided geometrically into two parts:

Phase 1 Find a corner point of the feasible set.

Phase 2 Move in turn to adjacent corner points until the optimal value (if it exists) of the objective function is obtained.

For the standard maximum problem in Section 9.1, phase 1 is easy, for the origin is *always* a corner point. This need not be true if we have constraints of the other types. For example, the origin $(0, 0)$ in $\mathbb{R}^2$ does not satisfy the constraints

$$x_1 + x_2 \geq 1, \qquad x_1 \geq 0, \qquad x_2 \geq 0.$$

We proceed to show how to execute phase 1 by passing to an augmented problem formed by introducing *surplus* and *artificial* variables as well as the *slack* variables of the preceding section. We then show how to form an initial tableau that can be reduced just as in the preceding section. In the process we make a smooth transition from phase 1 to phase 2.

FINDING A BASIC FEASIBLE SOLUTION

Recall these two features of the simplex method for the standard maximum problem:

1. *Converting to equations:* The (≤ 1)-type of inequalities are converted to equalities by inserting slack variables.

2. *Property of the initial tableau:* If m is the number of constraints other than the $x_i \geq 0$ constraints, then the m columns above the objective row in the initial tableau labeled by the slack variables give the m column vectors in the standard basis for $\mathbb{R}^m$.

We show how to achieve these two features for a general linear programming problem with m constraints other than $x_i \geq 0$.

We introduce a slack variable y_i if the ith constraint is of the ($\leq b$)-type just as in the preceding section. Such a slack variable converts the ith inequality to an equality, and contributes one of the standard basis vectors of $\mathbb{R}^m$ in a column of the initial tableau.

Consider the first constraint listed of ($\geq b$)-type. To use a specific illustration, suppose the kth constraint is $2x_1 - x_2 + 4x_3 \geq 7$. We introduce a **surplus variable** term $-y_k$ to convert this constraint to an inequality, requiring all four variables to be positive. That is, we form the equality

$$2x_1 - x_2 + 4x_3 - y_k = 7$$

in which $y_k \geq 0$. We denote slack and surplus variables consecutively as y_1, y_2, y_3, etc., numbered according to the order in which we list the constraints. We have converted the constraint to an equality with all variables positive, but in the initial tableau, the surplus variable column would be the *negative* of a standard basis vector of $\mathbb{R}^m$. Thus we also introduce an **artificial variable** q_1, which we recognize should eventually take on the value zero in order to provide a feasible solution of the original problem. Our constraint then becomes

$$2x_1 - x_2 + 4x_3 - y_k + q_1 = 7$$

and the column of the artificial variable q_1 in the initial tableau will provide the desired standard basis vector of $\mathbb{R}^m$.

A constraint that is originally stated as an equality is altered to contribute a standard basis vector to the columns of the initial tableau by inserting an artificial variable. For example, the constraint $x_1 - x_2 + 5x_3 = 4$ might become

$x_1 - x_2 + 5x_3 + q_2 = 4$, and we must again eventually require that $q_2 = 0$ to provide a feasible solution of our original problem.

Each constraint now contains either a slack variable or an artificial variable. These variables are initially assigned the nonnegative values b_i on the right-hand side of their constraints and all other variables (problem and surplus variables) are assigned the value zero. Thus the slack and artificial variables are initially taken as basic variables, and correspond to a basic feasible solution (corner point) of the *augmented problem*. We still may not have a basic feasible solution for the original problem. That is, phase 1 continues until all artificial variables become zero.

FORMING THE INITIAL TABLEAU

The question now arises, after augmenting the problem, "how do we form an initial tableau and reduce it in such a way that the artificial variables are driven to zero?" An ingenious technique is to modify the objective function so that it could not be optimized without the artificial variables being zero. For example, suppose we wish to maximize

$$P = 3x_1 - x_2 + 7x_3$$

and have introduced two artificial variables q_1 and q_2 for the reasons described above. We modify our objective function to become

$$P = 3x_1 - x_2 + 7x_3 - Mq_1 - Mq_2,$$

where we think of M as a positive number that is so huge compared with the magnitude of other data in our problem that this new objective function could not achieve a maximum without having $q_1 = q_2 = 0$. In pencil-and-paper computations, which can be terribly tedious, we generally carry M along as a particular but unspecified huge number—so huge that for any nonzero k that we compute, the number kM will be of greater magnitude than any computed magnitude that doesn't involve M. Computer programs sometimes set M equal to something like 10,000 times the largest magnitude of the given data. Our program in SIMPLEX does not do this; it carries the designation M throughout its computations, just as we will do in pencil-and-paper work below.

The initial tableau for this general situation is formed just as for the standard maximum problem except for the objective row. We will illustrate using the plywood mill problem that we discussed in Section 1.2, and which we restate here for easy reference.

Problem A lumber company owns two mills that produce hardwood veneer sheets of plywood. Each mill produces the same three types of plywood. Table 9.2 shows the daily production and the daily cost of operation of each mill. The rightmost column shows the plywood required by the lumber company for a period of six months. Find the number of days each mill should operate during the six months in order to supply the required sheets most economically.

Table 9.2

Plywood type	Mill 1 per day	Mill 2 per day	6-month demand
A	100 sheets	20 sheets	2000 sheets
B	40 sheets	80 sheets	3200 sheets
C	60 sheets	60 sheets	3600 sheets
Daily costs	$3000	$2000	

If mill 1 operates x_1 days and mill 2 operates x_2 days, then the cost is

$$C = 3000x_1 + 2000x_2.$$

To fulfill the company's requirements, these constraints must be satisfied:

$$100x_1 + 20x_2 \geq 2000$$
$$40x_1 + 80x_2 \geq 3200 \tag{2}$$
$$60x_1 + 60x_2 \geq 3600,$$

where $x_1 \geq 0$ and $x_2 \geq 0$. We wish to minimize C subject to these constraints.

We minimize C by maximizing $P = -C$ as explained in the preceding section. Converting our constraints to equalities with surplus and artificial variables as described above, we obtain

$$100x_1 + 20x_2 - y_1 \qquad\qquad + q_1 \qquad\qquad = 2000$$
$$40x_1 + 50x_2 \qquad - y_2 \qquad\qquad + q_2 \qquad = 3200 \tag{3}$$
$$60x_1 + 60x_2 \qquad\qquad - y_3 \qquad\qquad + q_3 = 3600,$$

where all variables are to be nonnegative, and the q_i must eventually become zero. We take as objective function

$$P = -3000x_1 - 2000x_2 - Mq_1 - Mq_2 - Mq_3 \tag{4}$$

as explained above. To form the objective row of the initial tableau, we first write Eq. (4) as

$$P + 3000x_1 + 2000x_2 + Mq_1 + Mq_2 + Mq_3 = 0 \tag{5}$$

just as we did for the standard maximum problem. However, formation of the objective row here is a bit more complicated than for the standard maximum

---

THE BASIC IDEAS OF LINEAR PROGRAMMING grew out of a World War II Air Force study of resource allocation problems, although there were many results over the years dealing with maximizing functions under certain constraints. George Dantzig, a member of this Air Force Project SCOOP (Scientific Computation of Optimum Programs), formulated the general linear programming problem presented here and devised the simplex method of solution in 1947. Since that time, it has been successfully applied to numerous problems, especially military and economic ones. The first successful solution of such a linear programming problem via the simplex method on an electronic computer took place in 1952 at the National Bureau of Standards. In recent years, faster methods of solving such problems have been developed.

Born in 1914, George Dantzig worked for the Rand Corporation after his Air Force stint. He is currently a Professor of Operations Research at Stanford University.

problem. We give the initial tableau, and then explain how the objective row is obtained.

	x_1	x_2	y_1	y_2	y_3	q_1	q_2	q_3	
q_1	100	20	-1	0	0	1	0	0	2000
q_2	40	80	0	-1	0	0	1	0	3200
q_3	60	60	0	0	-1	0	0	1	3600
P	3000	2000				0	0	0	
	$-200M$	$-160M$	M	M	M				$-8800M$

Recall that the objective row should contain data that tell us which nonbasic variable should be increased to become basic. In Section 1.2 where we wished to maximize a function such as $P = 4x_1 + 3x_2$, a unit increase in x_1 provides 4 units increase in P whereas a unit increase in x_2 provides only 3 units increase in P. Thus the rate of increase of P is measured by the coefficients of x_1 and x_2. But in our present situation, our objective function contains artificial variables. If we select x_1 as the entering basic variable, then the constraints (3) show that one unit increase in x_1 must cause q_1 to decrease by 100 units, 40 units decrease in q_2, and 60 units decrease in q_3. We find that this unit increase in x_1 and corresponding decreases in the q_i would produce a change of

$$3000 - 100M - 40M - 60M = 3000 - 200M$$

in the expression following P on the left-hand side of Eq. (5). This is the reason for the entry $3000 - 200M$ in the objective row in the column labeled x_1; it measures the change in P corresponding to a unit increase in x_1.

The entries in the other positions of the objective row are found in a similar way. In particular, Eqs. (3) show that an increase of one unit in a surplus variable y_i must be matched by an increase of one unit in the corresponding q_i, and that produces a change of M in the expression following P in the left-hand side of Eq. (5). Of course, the q_i are already basic variables in the initial tableau and are not candidates to enter the basic set. Our illustration using the plywood mill problem has no slack variables. Slack variables do not appear in those modified constraints that contain artificial variables. However, when slack variables do appear in a problem, then the objective row entry in a column labeled by a slack variable will be 0 in the initial tableau. Finally, the last entry in the objective row is the negative of the value of P in Eq. (4) at our initial basic solution (corner point). This last entry is thus vM, where v is the negative of the sum of the values of the *artificial* variables in the initial tableau. We summarize in the following box.

Objective Row of the Initial Tableau to Maximize P

1. A column labeled with a problem variable x_i has, in the objective row, the negative of the x_i-coefficient in the objective function minus M times the sum of the x_i-coefficients in those constraints that involve artificial variables.

2. A column labeled with a surplus variable has M in the objective row.

3. A column labeled with a slack or an artificial variable has 0 in the objective row.

4. The final entry in the objective row is the negative of the sum of the values of the artificial variables multiplied by M.

This initial tableau may be reduced using the same criteria for entering and exiting variables as in Section 1.2, for the same arguments go through. With a little thought, one can see that the reduction continues to generate the data in the objective row needed to decide what entering variable corresponds to the fastest increase in the objective function. We will not repeat the arguments here.

Returning to our illustration, we give again the initial tableau and proceed to give the successive reductions with each pivot shown in color. The arithmetic is nothing people care to do in their heads.

	x_1	x_2	y_1	y_2	y_3	q_1	q_2	q_3	
q_1	100	20	-1	0	0	1	0	0	2000
q_2	40	80	0	-1	0	0	1	0	3200
q_3	60	60	0	0	-1	0	0	1	3600
	3000	2000				0	0	0	
	$-200M$	$-160M$	M	M	M				$-8800M$

		x_1	x_2	y_1	y_2	y_3	q_1	q_2	q_3	
	x_1	1	.2	$-.01$	0	0	.01	0	0	20
~	q_2	0	72	.4	-1	0	$-.4$	1	0	2400
	q_3	0	48	.6	0	-1	$-.6$	0	1	2400
	P	0	1400	30			-30	0	0	-60000
			$-120M$	$-M$	M	M	$+2M$			$-4800M$

		x_1	x_2	y_1	y_2	y_3	q_1	q_2	q_3	
	x_1	1	0	$-.0111$	.0028	0	.0111	$-.0028$	0	13.33
~	x_2	0	1	.0056	$-.0139$	0	$-.0056$	.0139	0	33.33
	q_3	0	0	.3333	.6667	-1	$-.3333$	$-.6667$	1	800
	P	0	0	22.222	19.44		-22.22	-19.44	0	-106667
				$-.333M$	$-.667M$	M	$+1.333M$	$+1.667M$		$-800M$

	x_1	x_2	y_1	y_2	y_3	q_1	q_2	q_3	
x_1	1	0	−.0125	0	.0042	.0125	0	−.0042	10
x_2	0	1	.0125	0	−.0208	−.0125	0	.0208	50
y_2	0	0	.5	1	−1.5	−.5	−1	1.5	1200
P	0	0	12.5	0	29.17	−12.5 +M	M	−29.17 +M	−130000

The objective row contains no more negative entries (remember that M is positive and huge), so we are done. The maximum of P is $-130,000$ when $x_1 = 10$ and $x_2 = 50$. Thus the minimum cost C is \$130,000 when $x_1 = 10$ and $x_2 = 50$. Note that in this problem, the artificial variables were not driven to zero until the final tableau was computed.

The preceding tableaux indicate that paper-and-pencil execution of the simplex method can be a formidable task. However, a computer has no difficulty. The available program SIMPLEX has an option to allow the user to control the execution of the simplex method without having to do any actual computations.

We need only modify step 2 of the boxed outline on page 451 in Section 9.1 to obtain an outline for the simplex method for a general linear programming problem. Note that each linear programming problem is described as a maximizing one before the initial tableau is formed.

**Modified Step 2 in the Boxed Procedure
on Page 451**

Test for an optimal solution: If all the entries in the objective row are nonnegative and if either no artificial variable is basic or all basic artificial variables have value zero, then an optimal solution has been found. If the entries in the objective row are nonnegative and some artificial variable has positive value, then the original linear programming problem has no optimal solution.

We now give an example involving three constraints: an equality and one inequality of each type.

Example 1   Maximize $P = x_1 + x_2$ subject to

$$-2x_1 + x_2 \leq 2$$
$$2x_1 + x_2 = 9$$
$$3x_1 + x_2 \geq 11,$$

where $x_1 \geq 0$ and $x_2 \geq 0$.

Solution   If we supply slack, surplus, and artificial variables, the constraints become

$$-2x_1 + x_2 + y_1 \qquad\qquad\quad = 2$$
$$2x_1 + x_2 \qquad\quad + q_1 \qquad = 9$$
$$3x_1 + x_2 \quad - y_2 \quad + q_2 = 11.$$

Our profit function becomes

$$P = x_1 + x_2 - Mq_1 - Mq_2.$$

We now form the initial tableau and reduce it.

	x_1	x_2	y_1	y_2	q_1	q_2	
y_1	-2	1	1	0	0	0	2
q_1	2	1	0	0	1	0	9
q_2	3	1	0	-1	0	1	11
P	-1	-1	0		0	0	
	$-5M$	$-2M$		M			$-20M$

	x_1	x_2	y_1	y_2	q_1	q_2	
y_1	0	5/3	1	$-2/3$	0	2/3	28/3
q_1	0	1/3	0	2/3	1	$-2/3$	5/3
x_1	1	1/3	0	$-1/3$	0	1/3	11/3
P	0	$-2/3$	0	$-1/3$	0	1/3	11/3
		$-M/3$		$-2M/3$		$+5M/3$	$-5M/3$

	x_1	x_2	y_1	y_2	q_1	q_2	
y_1	0	2	1	0	1	0	11
y_2	0	1/2	0	1	3/2	-1	5/2
x_1	1	1/2	0	0	1/2	0	9/2
P	0	$-1/2$	0	0	1/2		9/2
					$+M$	M	

	x_1	x_2	y_1	y_2	q_1	q_2	
y_1	0	0	1	-4	-5	4	1
x_2	0	1	0	2	3	-2	5
x_1	1	0	0	-1	-1	1	2
P	0	0	0	1	2	-1	7
					$+M$	$+M$	

The maximum value of 7 is attained at the point $(x_1, x_2) = (2, 5)$. Note that all artificial variables have been driven to zero in the third tableau. This tableau marks the end of phase 1 since we have attained a basic feasible solution to our *original* problem. ◁

We conclude with an illustration of the case in which there is no solution because the feasible set is empty, as shown by Exercise 8.

Example 2   Maximize $P = 2x_1 + 3x_2$ subject to the constraint

$$x_1 + x_2 \leq -7,$$

where $x_1 \geq 0$ and $x_2 \geq 0$.

Solution   We must write the constraint so that the right-hand side is positive. Multiplying it by -1, we obtain the constraint

$$-x_1 - x_2 \geq 7.$$

If we supply a surplus and an artificial variable, the constraint becomes

$$-x_1 - x_2 - y_1 + q_1 = 7.$$

We form the initial tableau and try to reduce it.

	x_1	x_2	y_1	q_1	
q_1	-1	-1	-1	1	7
P	-2 $+M$	-3 $+M$	M	0	$-7M$

Since the entries in the objective row are all nonnegative (recall that M is huge and positive), no reduction is necessary. Since the artificial variable has value $7 > 0$, our boxed test for a solution given above indicates that there is no solution to the problem. ◁

In the literature, one often finds all variables (problem variables, slack variables, surplus variables, and artificial variables) denoted by x_i. For example, x_1 through x_4 might be problem variables, x_5 and x_6 slack variables, x_7 a surplus variable, and x_8 and x_9 artificial variables. The simplex method treats all these variables in the same way, except that one must keep track of which are surplus and artificial variables in forming the objective row. In more advanced treatments where the simplex method is carefully developed and proved, calling all variables x_i makes the notation cleaner. The development is usually in terms of matrices and generating vectors for their column spaces. Often tableaux are mentioned only in passing.

SUMMARY

1. Constraints other than $x_i \geq 0$ in a linear programming problem are modified as follows to obtain equalities and create, in the columns of the initial tableau, a standard basis for the Euclidean space $\mathbb{R}^m$.

   a) A slack variable y_i is inserted in each ($\leq b$)-type constraint.

   b) An artificial variable q_i is inserted in each ($= b$)-type constraint.

c) A surplus variable term $-y_i$ and an artificial variable q_j are inserted in each ($\geq b$)-type constraint.

In each of the cases, we require that the constraint first be written so that $b_i \geq 0$. For a feasible solution of the original problem, we require that all variables be nonnegative and the artificial variables be zero.

2. Formation of the objective row in the initial tableau is described in the box on page 461. Otherwise, the initial tableau is formed just as in the standard maximum problem.

3. The tableaux are reduced as described in the box on page 451 of Section 9.1 except for the test for an optimal solution. This test is described in the box on page 463.

EXERCISES

In Exercises 1–4, find the initial tableau for the given linear programming problem. All variables x_i given are assumed to be nonnegative.

1. Maximize $3x_1 - 2x_2 + 5x_3$
   subject to the constraints

$$\begin{aligned} x_1 + 2x_2 + x_3 &\leq 5, \\ 4x_1 + 2x_2 - 3x_3 &\geq 3, \\ 5x_1 \quad\quad - x_3 &\geq 4. \end{aligned}$$

2. Maximize $x_1 + 2x_2 - 3x_3$
   subject to the constraints

$$\begin{aligned} x_1 - x_2 \quad\quad &= 1, \\ 3x_1 \quad\quad + x_3 &\geq 2, \\ 4x_2 + x_3 &\leq 10. \end{aligned}$$

3. Minimize $-x_1 + 2x_2 - 5x_3$
   subject to the constraints

$$\begin{aligned} 3x_1 - x_2 \quad\quad &= 5, \\ 2x_2 + x_3 &= 2, \\ x_1 + 4x_2 + 5x_3 &\leq 20, \\ 4x_1 - 2x_2 - x_3 &\geq 3. \end{aligned}$$

4. Minimize $20x_1 + 40x_2 + 10x_3$
   subject to the constraints

$$\begin{aligned} x_1 + 3x_2 \quad\quad &\geq 2, \\ 5x_1 \quad\quad + 2x_3 &\geq 5, \\ 2x_2 + 4x_3 &\geq 3, \\ x_3 &\leq 12. \end{aligned}$$

5. Use the simplex method to solve the vitamin problem in Exercise 17 of Section 1.2. Use a calculator.

6. Use the simplex method to solve the fertilizer problem in Exercise 19 of Section 1.2.

7. Use the simplex method to find the *minimum* value of the objective function in Example 1, subject to the constraints given there. Use a calculator.

8. Draw a sketch to show that there is no solution to the linear programming problem stated in Example 2.

9. *(A transportation problem).* A manufacturer has warehouses in towns W_1 and W_2 with 30 units of a product available at the warehouse in W_1 and 20 units of the product available at W_2. Buyer B_1 needs 10 units of this product and buyer B_2 wants 40 units of it. The cost per unit for shipment from the warehouses to the buyers is as follows:

from W_1 to B_1: $10 per unit,
from W_1 to B_2: $ 5 per unit,
from W_2 to B_1: $ 4 per unit,
from W_2 to B_2: $ 8 per unit.

We wish to find how much of the product should be shipped from each warehouse to each buyer in order to minimize the cost of shipment.

a) Formulate this problem as a linear programming problem.
b) Set up the initial tableau for the problem.
c) Solve the problem by the simplex method.

In Exercises 10–18, use the user's choice option of the available program SIMPLEX to solve the indicated linear programming problem.

10. The mill problem with the data in Table 9.2, which was solved in the text

11. The problem in Exercise 7

12. The problem in Exercise 5

13. The problem in Exercise 6

14. The problem in Exercise 9

15. The problem in Exercise 1

16. The problem in Exercise 2

17. The problem in Exercise 3

18. The problem in Exercise 4

9.3
Duality

THE DUAL OF A STANDARD MAXIMUM PROBLEM

Every linear programming problem has an associated dual problem. We begin our presentation with an economic motivation for the dual of a standard maximum problem.

Let us consider a modification of our coffee problem in Exercise 15 of Section 1.2. Suppose a food packaging house has m different types of coffee beans in stock, with b_i lb of type i beans for $i = 1, 2, \ldots , m$. The house can create n different blends of coffee to sell to retailers. Suppose each pound of blend j requires a_{ij} lb of type i coffee, and can be sold to the retailers for c_j dollars. The house wishes to maximize its revenue R from the retailers for this coffee. Let x_j be the number of pounds of blend j created. The house wants to maximize the revenue

$$R = \mathbf{c} \cdot \mathbf{x} = c_1 x_1 + c_2 x_2 + \cdots + c_n x_n$$

subject to the constraints

$$a_{11} x_1 + a_{12} x_2 + \cdots + a_{1n} x_n \leq b_1$$
$$a_{21} x_1 + a_{22} x_2 + \cdots + a_{2n} x_n \leq b_2$$
$$\vdots$$
$$a_{m1} x_1 + a_{m2} x_2 + \cdots + a_{mn} x_n \leq b_m,$$

that is, subject to $A\mathbf{x} \leq \mathbf{b}$, where $A = (a_{ij})$ is an $m \times n$ matrix and $\mathbf{x} \geq \mathbf{0}$. We take this linear programming problem as our *primal* problem:

$$\text{maximize } \mathbf{c} \cdot \mathbf{x} \text{ subject to } A\mathbf{x} \leq \mathbf{b}, \text{ where } \mathbf{x} \geq \mathbf{0}. \tag{1}$$

The *dual* problem arises if we consider the value $y_i \geq 0$ to the house of each pound of type i coffee beans. The total value of the stock of coffee beans becomes

$$V = \mathbf{b} \cdot \mathbf{y} = b_1 y_1 + b_2 y_2 + \cdots + b_m y_m. \tag{2}$$

Since each pound of blend j coffee that the house makes requires a_{ij} lb of type i beans for $i = 1, 2, \ldots, n$, the value of the beans in each pound of blend j is

$$a_{1j} y_1 + a_{2j} y_2 + \cdots + a_{mj} y_m. \tag{3}$$

Now we know that a pound of blend j can be sold to the retailers for c_j dollars, so we must have

$$a_{1j} y_1 + a_{2j} y_2 + \cdots + a_{mj} y_m \geq c_j \tag{4}$$

for $j = 1, 2, \ldots, m$. The inequalities (4) can be written as

$$A^T \mathbf{y} \geq \mathbf{c}. \tag{5}$$

The house would like to find (for tax purposes) the minimum value for this existing stock of coffee beans. That is, the house wishes to

$$\text{minimize } \mathbf{b} \cdot \mathbf{y} \text{ subject to } A^T \mathbf{y} \geq \mathbf{c}, \text{ where } \mathbf{y} \geq \mathbf{0}. \tag{6}$$

This linear programming problem (6) is called the *dual* of the *primal* linear programming problem (1).

Let $R_{\max}$ be the optimal value for the problem (1) and $V_{\min}$ be the optimal value for the problem (6). Since $R_{\max}$ will be the revenue obtained if the beans are retailed in blends in accordance with the solution of problem (1), it is clear that the total value V of the beans is at least as large as $R_{\max}$. In particular, $V_{\min} \geq R_{\max}$. Exercise 10 requests an algebraic proof of this. The duality theorem, which we state later, asserts that actually $V_{\min} = R_{\max}$. This is not surprising since the sale prices c_j that appear in the constraints in (6) are the only price information we have available to use in computing the values y_i. The values y_i that produce the optimal value $V_{\min}$ are called the **shadow prices** or **accounting prices.**

Here is an illustration showing how the dual of a standard maximum problem is formed.

Example 1   Find the dual of the boatyard problem in Section 9.1, which is to maximize the profit function $P = 5x_1 + 4x_2 + 6x_3$ subject to the constraints

$$\begin{aligned}
x_1 + x_2 + x_3 &\leq 25 \\
2x_1 + x_2 + 3x_3 &\leq 51 \\
x_1, x_2, x_3 &\geq 0.
\end{aligned}$$

Solution   The given primal problem has the matrix form (1) with

$$\mathbf{c} = \begin{pmatrix} 5 \\ 4 \\ 6 \end{pmatrix}, \quad \mathbf{x} = \begin{pmatrix} x_1 \\ x_2 \\ x_3 \end{pmatrix}, \quad \mathbf{b} = \begin{pmatrix} 25 \\ 51 \end{pmatrix}, \quad \text{and} \quad A = \begin{pmatrix} 1 & 1 & 1 \\ 2 & 1 & 3 \end{pmatrix}.$$

Its dual has the form (6), that is,

$$\text{minimize} \quad \mathbf{b} \cdot \mathbf{y} = \begin{pmatrix} 25 \\ 51 \end{pmatrix} \cdot \begin{pmatrix} y_1 \\ y_2 \end{pmatrix} = 25y_1 + 51y_2$$

$$\text{subject to} \quad A^T \mathbf{y} = \begin{pmatrix} 1 & 2 \\ 1 & 1 \\ 1 & 3 \end{pmatrix} \begin{pmatrix} y_1 \\ y_2 \end{pmatrix} \geq \begin{pmatrix} 5 \\ 4 \\ 6 \end{pmatrix} \quad \text{and} \quad \mathbf{y} \geq \mathbf{0}.$$

The constraints of the dual are then

$$y_1 + 2y_2 \geq 5$$
$$y_1 + y_2 \geq 4$$
$$y_1 + 3y_2 \geq 6$$
$$y_1, y_2 \geq 0. \quad \triangleleft$$

The initial tableau for the primal problem in Example 1 is

	x_1	x_2	x_3	y_1	y_2	C
y_1	1	1	1	1	0	25
y_2	2	1	3	0	1	51
P	-5	-4	-6	0	0	0 .

If we consider just the entries in color, ignoring the columns labeled with the slack variables, we obtain the **condensed tableau**

	x_1	x_2	x_3	C
y_1	1	1	1	25
y_2	2	1	3	51
P	-5	-4	-6	0 .

Reading from left to right across the rows of the condensed tableau, we obtain the data for the primal problem. The final row represents the objective function $P = 5x_1 + 4x_2 + 6x_3$. Recall that the negative entries are present because we want to view the entire last row as the equation $P - 5x_1 - 4x_2 - 6x_3 = 0$ so that the final entry gives the value of the objective function P. Mentally changing signs of elements in the last row and last column of either tableau and then reading from top to bottom down the colored columns gives the data for the dual problem. We place C in the upper right corner of the tableau, analogous to our placement of P in the lower left corner. (Recall that for a maximizing problem, the objective function is generally referred to as the *profit* while for a minimizing problem, it is usually called the *cost*.) This condensed tableau exhibits the symmetry of a standard maximum primal problem with its dual. Note how the basic variables of the primal problem are nonbasic variables of the dual and vice versa. In implementing the simplex method, we choose to work with the original tableau.

THE DUAL OF A GENERAL LINEAR PROGRAMMING PROBLEM

We now describe formation of the dual of any linear programming problem. To do this, we relax the condition that the constants on the right-hand side of our constraints be nonnegative, and write all constraints other than $x_i \geq 0$ in the ($\leq$ type)-form. For example,

$$x_1 - 3x_2 \geq 5 \quad \text{should be written} \quad -x_1 + 3x_2 \leq -5.$$

Also, any equality constraint should be expressed as *two* inequality constraints. To illustrate, $2x_1 - 3x_2 = 5$ can be expressed as $2x_1 - 3x_2 \leq 5$ and $2x_1 - 3x_2 \geq 5$, or using only ($\leq$ type)-inequalities,

$$2x_1 - 3x_2 = 5 \quad \text{becomes} \quad 2x_1 - 3x_2 \leq 5 \quad \text{and} \quad -2x_1 + 3x_2 \leq -5.$$

Thus any maximizing problem can be written in the form

$$\text{maximize} \quad \mathbf{c} \cdot \mathbf{x} \quad \text{subject to} \quad A\mathbf{x} \leq \mathbf{b}, \text{ where } \mathbf{x} \geq \mathbf{0} \tag{7}$$

and where entries in $\mathbf{b}$ may be positive, zero, or negative. In a similar manner, any minimizing problem can be written as

$$\text{minimize} \quad \mathbf{s} \cdot \mathbf{y} \quad \text{subject to} \quad B\mathbf{y} \geq \mathbf{r}, \text{ where } \mathbf{y} \geq \mathbf{0}. \tag{8}$$

Having written a linear programming problem in the form (7) or (8), its *dual* may be defined as follows.

Definition 9.1 Dual Problems

The linear programming problems

$$\text{maximize} \quad \mathbf{c} \cdot \mathbf{x} \quad \text{subject to} \quad A\mathbf{x} \leq \mathbf{b}, \text{ where } \mathbf{x} \geq \mathbf{0} \tag{9}$$

and

$$\text{minimize} \quad \mathbf{b} \cdot \mathbf{y} \quad \text{subject to} \quad A^T\mathbf{y} \geq \mathbf{c}, \text{ where } \mathbf{y} \geq \mathbf{0} \tag{10}$$

are **dual problems.** If problem (9) is given first, then it is the **primal** problem and problem (10) is its **dual.** Similarly, if (10) is given first, it is the **primal** problem and problem (9) is its **dual.**

Example 2 Find the dual of the following minimizing linear programming problem:

minimize the cost function $C = 3y_1 + 2y_2 + 5y_3$

subject to

$$\begin{aligned}
y_1 + 4y_2 + y_3 &\geq 2 \\
2y_2 + 10y_3 &\geq 4 \\
y_1 + y_3 &\geq 5 \\
y_1, y_2, y_3 &\geq 0.
\end{aligned}$$

Solution   Our primal problem has the form (10). Thus its dual has the form (9) and is:

maximize the profit function $P = 2x_1 + 4x_2 + 5x_3$

subject to

$$
\begin{aligned}
x_1 \qquad\quad + x_3 &\le 3 \\
4x_1 + \ 2x_2 \qquad &\le 2 \\
x_1 + 10x_2 + x_3 &\le 5
\end{aligned}
$$

where $x_1, x_2, x_3 \ge 0$.   ◁

We now state the Duality Theorem. Our economic example at the beginning of the section anticipates the result.

Theorem 9.2   **Duality Theorem**

> If a linear programming problem has an optimal solution, then its dual also has an optimal solution. Furthermore, the optimal values attained by the primal problem and by its dual are the same.

In a linear programming problem, one generally wants to know not only the optimal value attained but also a basic feasible solution (corner point) at which it is attained. The importance of duality lies partly in the fact that use of the simplex method to solve the primal problem solves at the same time the dual problem. Recall how a simplex tableau or condensed tableau for a primal problem exhibits the data for the dual problem also. We choose not to describe how to find the dual solution from the final primal tableau when some of the constraints are equalities. If equality constraints appear, we can always replace each of them by two inequality constraints and use the following boxed result.

> ### Solution of the Dual Problem from the Final Primal Tableau
>
> For a primal linear programming problem (9) not containing equality constraints, an optimal basic feasible solution of the dual problem (10) is provided by values in the bottom row of the final tableau for (9). The value of y_i in the dual solution appears in the objective row in the column labeled by the y_i at the top.

For illustration, consider the boatyard problem stated in Example 1 above and solved in Section 9.1. The final tableau for this problem was found in Section 9.1 to be

	x_1	x_2	x_3	y_1	y_2	C
x_2	$\frac{1}{2}$	1	0	$\frac{3}{2}$	$-\frac{1}{2}$	12
x_3	$\frac{1}{2}$	0	1	$-\frac{1}{2}$	$\frac{1}{2}$	13
P	0	0	0	3	1	126 .

From this tableau and the boxed statement above, we see that a solution of the dual problem is given by $y_1 = 3$, $y_2 = 1$. Looking back at Example 1, we see that the objective function to be minimized for the dual to the boatyard problem is $C = 25y_1 + 51y_2$, and we can easily check that $25(3) + 51(1) = 126$.

Remember that to solve the problem (9) by the simplex method, we must write all constraints so that the constants on the right-hand sides are nonnegative.

Example 3 Let the primal linear programming problem be:

maximize $3x_1 + 2x_2$ subject to

$$x_1 + 2x_2 \leq 10$$
$$5x_1 + x_2 \geq 10$$
$$x_1 + 10x_2 \geq 20$$

where $x_1 \geq 0$, $x_2 \geq 0$.

Formulate the dual problem, and solve it by solving the primal problem and using the boxed statement above.

Solution To form the dual problem, we rewrite the constraints of the primal problem as

$$x_1 + 2x_2 \leq 10$$
$$-5x_1 - x_2 \leq -10$$
$$-x_1 - 10x_2 \leq -20.$$

The dual problem is

minimize $10y_1 - 10y_2 - 20y_3$ subject to

$$y_1 - 5y_2 - y_3 \geq 3$$
$$2y_1 - y_2 - 10y_3 \geq 2$$

where $y_1 \geq 0$, $y_2 \geq 0$, and $y_3 \geq 0$.

We form the initial tableau as usual from the problem as originally stated and reduce it as usual.

	x_1	x_2	y_1	y_2	y_3	q_1	q_2	C
y_1	1	2	1	0	0	0	0	10
q_1	5	1	0	-1	0	1	0	10
q_2	1	10	0	0	-1	0	1	20
P	-3	-2	0			0	0	
	$-6M$	$-11M$		M	M			$-30M$

	x_1	x_2	y_1	y_2	y_3	q_1	q_2	C
y_1	.8	0	1	0	.2	0	−.2	6
q_1	4.9	0	0	−1	.1	1	−.1	8
x_2	.1	1	0	0	−.1	0	.1	2
P	−2.8	0	0		−.2	0	.2	4
	−4.9M			M	−.1M		+1.1M	−8M

	x_1	x_2	y_1	y_2	y_3	q_1	q_2	C
y_1	0	0	1	.1633	.1837	−.1633	−.1837	4.6939
x_1	1	0	0	−.2041	.0204	.2041	−.0204	1.6327
x_2	0	1	0	.0204	−.1020	−.0204	.1020	1.8367
P	0	0	0	−.5714	−.1429	.5714	.1429	8.5714
						+M	+M	

	x_1	x_2	y_1	y_2	y_3	q_1	q_2	C
y_2	0	0	6.125	1	1.125	−1	−1.125	28.75
x_1	1	0	1.25	0	.25	0	−.25	7.5
x_2	0	1	−.125	0	−.125	0	.125	1.25
P	0	0	3.5	0	.5		−.5	25
						M	+M	

From this final tableau, we see that the maximum value of the primal objective function $3x_1 + 2x_2$ is 25, attained at the basic feasible solution, $x_1 = 7.5$ and $x_2 = 1.25$. We also find that the minimum value of the dual objective function $10y_1 - 10y_2 - 20y_3$ is 25, attained at the basic feasible solution $y_1 = 3.5$, $y_2 = 0$, $y_3 = .5$. These are read off in the bottom row in the columns labeled y_1, y_2, and y_3, respectively. ◁

We are accustomed to working with variables x_j as our problem variables in a simplex tableau. Given a primal linear programming problem stated with problem variables x_j, we may attempt to solve the problem directly by the simplex method, or we may choose to form the dual problem and solve the dual using the simplex method. In the latter case, we suggest that the dual be restated using variables x_j as problem variables, and that the primal problem be restated using variables y_i. Then the labels on our tableaux will be in their accustomed positions.

Duality is useful when a linear programming problem has a large number m of constraints compared with the number n of variables. Inserting a slack or surplus variable in each constraint, we find that the matrix of data in the initial tableau has $m + 1$ rows and at least $n + m + 1$ columns. On the other hand, the initial tableau of the dual problem has only $n + 1$ rows and again at least

$n + m + 1$ columns. If m is much greater than n, it is advantageous to solve the dual problem instead. We close with an illustration of this.

Example 4   Minimize the function $C = 2y_1 + y_2$ subject to the constraints

$$
\begin{aligned}
10y_1 + y_2 &\geq 10 \\
2y_1 + y_2 &\geq 8 \\
y_1 + y_2 &\geq 6 \\
y_1 + 2y_2 &\geq 10 \\
y_1 + 12y_2 &\geq 12,
\end{aligned}
$$

where $\mathbf{y} \geq \mathbf{0}$.

Solution   In order to work with a smaller tableau we choose to solve the dual problem. The dual problem is:

maximize $10x_1 + 8x_2 + 6x_3 + 10x_4 + 12x_5$ subject to

$$
\begin{aligned}
10x_1 + 2x_2 + x_3 + x_4 + x_5 &\leq 2 \\
x_1 + x_2 + x_3 + 2x_4 + 12x_5 &\leq 1
\end{aligned}
$$

where $x \geq \mathbf{0}$.

We form the initial tableau and reduce it.

	x_1	x_2	x_3	x_4	x_5	y_1	y_2	C
y_1	10	2	1	1	1	1	0	2
y_2	1	1	1	2	12	0	1	1
P	-10	-8	-6	-10	-12	0	0	0

	x_1	x_2	x_3	x_4	x_5	y_1	y_2	C
y_1	9.9167	1.9167	.9167	.8333	0	1	$-.0833$	1.9167
x_5	.0833	.0833	.0833	.1667	1	0	.0833	.0833
P	-9	-7	-5	-8	0	0	1	1

	x_1	x_2	x_3	x_4	x_5	y_1	y_2	C
x_1	1	.1933	.0924	.0840	0	.1008	$-.0084$	.1933
x_5	0	.0672	.0756	.1597	1	$-.0084$	.0840	.0672
P	0	-5.2605	-4.1681	-7.2437	0	.9076	.9244	2.7395

	x_1	x_2	x_3	x_4	x_5	y_1	y_2	C
x_1	1	.1579	.0526	0	−.5263	.1053	−.0526	.1579
x_4	0	.4211	.4737	1	6.2632	−.0526	.5263	.4211
P	0	−2.2105	−.7368	0	45.3684	.5263	4.7368	5.7895

	x_1	x_2	x_3	x_4	x_5	y_1	y_2	C
x_1	1	0	−.125	−.375	−2.875	.125	−.25	0
x_2	0	1	1.125	2.375	14.875	−.125	1.25	1
P	0	0	1.75	5.25	78.25	.25	7.5	8

Thus the objective function $2y_1 + y_2$ in the original primal problem has minimum value 8 at the basic feasible solution $y_1 = .25$ and $y_2 = 7.5$.

In the next to the last tableau, we could have chosen the top row as pivot row, with .1579 as pivot entry. Note that both of the ratios .1579/.1579 and .4211/.4211 determining the choice of pivot row are 1. The program SIMPLEX always chooses the candidate row closest to the top of the tableau as pivot row in such a tie case. Using SIMPLEX or pencil and paper, we easily see that the other choice leads to the solution $y_1 = 2, y_2 = 4$. Of course, the same value 8 is obtained for the objective function. ◁

Solving the original primal problem in the preceding example using the program **SIMPLEX**, we would work with a tableau containing 13 columns and 6 rows of data; 6 iterations are required. With the dual in the example, we work with 8 columns and 3 rows of data, and 4 iterations were used. Thus it is advantageous for us to solve the dual in this case.

SUMMARY

1. To form the dual of a maximizing linear programming problem, write the problem as a maximizing one with all constraints in (≤ type)-form except for the nonnegativity-of-variables constraints. An equality should be expressed as two such constraints. The constants on the right-hand sides of the constraints may be negative, positive, or zero. If the resulting problem is to maximize $\mathbf{c} \cdot \mathbf{x}$ subject to $A\mathbf{x} \le \mathbf{b}$ and $\mathbf{x} \ge \mathbf{0}$, then the dual problem is to minimize $\mathbf{b} \cdot \mathbf{y}$ subject to $A^T\mathbf{y} \ge \mathbf{c}$ and $\mathbf{y} \ge \mathbf{0}$. Similarly, the dual of the latter problem is the former. The problem that is stated first is regarded as the primal problem; the other is its dual.

2. The Duality Theorem asserts that if a primal problem has an optimal solution, then so does its dual. Moreover, the optimal value satisfying the primal problem equals the optimal value satisfying the dual.

3. The box on page 471 explains how the basic feasible solution of the dual problem can be found from the final tableau for the primal problem, assuming that there are no equality constraints.

EXERCISES

In Exercises 1–8, formulate the dual of the given linear programming problem. Constraints asserting nonnegativity of variables are assumed, but are omitted in the statement of the problem.

1. Maximize $3x_1 - 2x_2$
subject to

$$x_1 + 2x_2 \leq 10,$$
$$3x_1 + x_2 \leq 15.$$

2. Maximize $x_1 - 2x_2 + 4x_3$
subject to

$$2x_1 + x_2 + 4x_3 \leq 40,$$
$$-x_1 + x_2 + x_3 \leq 10.$$

3. Minimize $-4y_1 + 3y_2$
subject to

$$2y_1 + 3y_2 \geq 6,$$
$$y_1 + 3y_2 \geq 4.$$

4. Minimize $2y_1 + 2y_2 + y_3$
subject to

$$y_1 + 3y_2 + 4y_3 \geq 24,$$
$$3y_1 + 2y_2 - 4y_3 \geq 4,$$
$$2y_1 - 5y_2 + y_3 \geq 8.$$

5. Maximize $3x_1 + 4x_2$
subject to

$$x_1 + 2x_2 \leq 8,$$
$$x_1 - x_2 \geq 1,$$
$$x_1 + x_2 \geq 2.$$

6. Maximize $x_1 - x_2 + 3x_3$
subject to

$$2x_1 + x_2 + 4x_3 \leq 12,$$
$$-2x_1 + x_2 + x_3 = 4.$$

7. Minimize $y_1 - 2y_3$
subject to

$$y_1 + 3y_2 + y_3 = 12,$$
$$y_2 + y_3 = 4.$$

8. Minimize $y_1 + 3y_2 + y_4$
subject to

$$y_1 + 3y_2 + 9y_3 + 2y_4 \geq 18,$$
$$y_1 - 4y_2 + 2y_3 \leq 8,$$
$$y_3 + 3y_4 \leq 6.$$

9. Consider a linear programming problem

$$\text{maximize} \quad \mathbf{c} \cdot \mathbf{x} \quad \text{subject to} \quad A\mathbf{x} \leq \mathbf{b}, \mathbf{x} \geq \mathbf{0},$$

where components of $\mathbf{b}$ may be positive, negative or zero.

a) State the vector form of the dual of this problem.
b) Restate the answer to (a) as a maximizing problem, as described in Section 9.1.
c) Form the dual of the maximizing problem obtained as answer to (b).
d) Show that the minimizing problem found in (c) is equivalent to the original given maximizing problem.

[Some authors define the dual of any linear programming problem as follows:

i) Write the problem as a maximizing problem of the form given at the start of this exercise.
ii) Form the dual minimizing problem as described in this section.

This exercise shows that if one repeats by this method steps (i) and (ii) to form the dual of the dual just obtained, one essentially obtains the original problem. That is, the dual of the dual is the original problem.]

10. Let a primal problem to maximize $P = \mathbf{c} \cdot \mathbf{x}$ subject to $A\mathbf{x} \leq \mathbf{b}$ as in (9) have optimal solution P_{max}. Let the dual problem (10) to minimize $C = \mathbf{b} \cdot \mathbf{y}$ subject to $A^T\mathbf{y} \geq \mathbf{c}$ have optimal solution C_{min}. Proceeding algebraically, without using the Duality Theorem, show that $P_{max} \leq C_{min}$. [*Hint:*

Let $\mathbf{x}$ and $\mathbf{y}$ be vectors with nonnegative components such that $A\mathbf{x} \leq \mathbf{b}$ and $A^T\mathbf{y} \geq \mathbf{c}$. Show from these inequalities that $\mathbf{c} \cdot \mathbf{x} \leq \mathbf{b} \cdot \mathbf{y}$.]

11. Solve the vitamin problem (Exercise 17 in Section 1.2) by using the simplex method to solve the dual problem.

12. Repeat Exercise 11 for the fertilizer problem (Exercise 19 in Section 1.2).

13. Repeat Exercise 11 for the coffee problem (Exercise 15 in Section 1.2).

14. Repeat Exercise 11 for the mill problem stated and solved in Section 1.2.

15. Suppose a linear programming problem has the same number of constraints as problem variables with some ($\leq$ type)-constraints and some ($\geq$ type)-constraints. (The constants on the right-hand sides of the constraints are nonnegative.) Under what circumstances might it be advantageous to solve the problem by using the simplex method on the dual problem, rather than on the given primal one?

In Exercises 16–19, solve the given linear programming problem by using the simplex method either on the given primal problem or on its dual, whichever appears to be easier. Constraints asserting nonnegativity of variables are assumed but are omitted in the statement of the problem.

16. Minimize $2x_1 + 3x_2$
subject to

$$2x_1 + x_2 \geq 6,$$
$$x_1 + x_2 \geq 5,$$
$$x_1 + 2x_2 \geq 8,$$
$$x_1 + 4x_2 \geq 12.$$

17. Minimize $x_1 + 3x_2$
subject to

$$x_1 + 2x_2 \geq 4,$$
$$3x_1 + 4x_2 \geq 10.$$

18. Maximize $2x_1 + 3x_2$
subject to

$$x_1 + 2x_2 \leq 8,$$
$$-x_1 + x_2 \leq 1,$$
$$x_1 \leq 4.$$

19. Maximize $3x_1 + 5x_2$
subject to

$$-x_1 + x_2 \leq 2,$$
$$x_2 \leq 4,$$
$$x_1 + 3x_2 \leq 15,$$
$$x_1 + 2x_2 \leq 12,$$
$$x_1 + x_2 \leq 10.$$

In Exercises 20–24, use the available program SIMPLEX to solve the given problem. Then form its dual, and use SIMPLEX again to solve the dual. Decide whether the primal or dual problem would be easier to solve if one were to use the simplex method with pencil and paper.

20. The problem in Exercise 1

21. The problem in Exercise 2

22. The problem in Exercise 3

23. The problem in Exercise 4

24. The problem in Exercise 5

25. The problem in Exercise 8

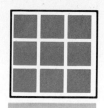

Appendix A

Mathematical Induction

Sometimes we want to prove that a statement about positive integers is true for all positive integers or perhaps for some finite or infinite sequence of consecutive integers. Such proofs are accomplished using *mathematical induction*. The validity of the method rests on this axiom of the positive integers. The set of all positive integers is denoted by Z^+.

Induction Axiom

Let S be a subset of Z^+ satisfying

1. $1 \in S$,

2. if $k \in S$, then $(k + 1) \in S$.

Then $S = Z^+$.

This axiom leads immediately to the method of mathematical induction.

Mathematical Induction

Let $P(n)$ be a statement concerning the positive integer n. Suppose that

1. $P(1)$ is true,

2. if $P(k)$ is true, then $P(k + 1)$ is true.

Then $P(n)$ is true for all $n \in Z^+$.

Most of the time, we want to show that $P(n)$ holds for all $n \in Z^+$. If we wish only to show that it holds for $r, r + 1, r + 2, \ldots, s - 1, s$, then we show that

$P(r)$ is true and that $P(k)$ implies $P(k + 1)$ for $r \leq k \leq s - 1$. Note that r may be any integer—positive, negative, or zero.

Example A.1 Prove the formula

$$1 + 2 + \cdots + n = \frac{n(n + 1)}{2} \tag{A.1}$$

for the sum of the arithmetic progression, using mathematical induction.

Solution We let $P(n)$ be the statement that Formula (A.1) is true. For $n = 1$, we obtain

$$\frac{n(n + 1)}{2} = \frac{1(2)}{2} = 1,$$

so $P(1)$ is true.

Suppose that $k \geq 1$ and $P(k)$ is true (our *induction hypothesis*), so

$$1 + 2 + \cdots + k = \frac{k(k + 1)}{2}.$$

To show that $P(k + 1)$ is true, we compute

$$1 + 2 + \cdots + (k + 1) = (1 + 2 + \cdots + k) + (k + 1)$$

$$= \frac{k(k + 1)}{2} + (k + 1) = \frac{k^2 + k + 2k + 2}{2}$$

$$= \frac{k^2 + 3k + 2}{2} = \frac{(k + 1)(k + 2)}{2}.$$

Thus $P(k + 1)$ holds, and Formula (A.1) is true for all $n \in Z^+$. ◁

Example A.2 Show that a set of n elements has exactly 2^n subsets for any nonnegative integer n.

Solution This time we start the induction with $n = 0$. Let S be a finite set having n elements. We wish to show

$$P(n)\text{: } S \text{ has } 2^n \text{ subsets.} \tag{A.2}$$

If $n = 0$, then S is the empty set and has only one subset, namely, the empty set itself. Since $2^0 = 1$, we see that $P(0)$ is true.

Suppose that $P(k)$ is true. Let S have $k + 1$ elements, and let one element of S be c. Then $S - \{c\}$ has k elements, hence 2^k subsets. Now every subset of S either contains c or does not contain c. Those not containing c are subsets of $S - \{c\}$, so there are 2^k of them by the induction hypothesis. Each subset containing c consists of one of the 2^k subsets not containing c with c adjoined. There are 2^k such subsets also. The total number of subsets of S is then

$$2^k + 2^k = 2^k(2) = 2^{k+1},$$

so $P(k + 1)$ is true. Thus $P(n)$ is true for all nonnegative integers n. ◁

Example A.3   Let $x \in \mathbb{R}$ with $x > -1$ and $x \neq 0$. Show that $(1 + x)^n > 1 + nx$ for every positive integer $n \geq 2$.

Solution   We let $P(n)$ be the statement

$$(1 + x)^n > 1 + nx. \tag{A.3}$$

(Note that $P(1)$ is false.) Then $P(2)$ is the statement $(1 + x)^2 > 1 + 2x$. Now $(1 + x)^2 = 1 + 2x + x^2$, and $x^2 > 0$ since $x \neq 0$. Thus $(1 + x)^2 > 1 + 2x$, so $P(2)$ is true.

Suppose that $P(k)$ is true, so

$$(1 + x)^k > 1 + kx. \tag{A.4}$$

Now $1 + x > 0$ since $x > -1$. Multiplying both sides of Ineq. (A.4) by $1 + x$, we obtain

$$(1 + x)^{k+1} > (1 + kx)(1 + x) = 1 + (k + 1)x + kx^2.$$

Since $kx^2 > 0$, we see that $P(k + 1)$ is true. Thus $P(n)$ is true for every positive integer $n \geq 2$.   ◁

In a frequently used form of induction known as *complete induction*, the statement

if $P(k)$ is true, then $P(k + 1)$ is true

in the box on page A-1 is replaced by the statement

if $P(m)$ is true for $1 \leq m \leq k$, then $P(k + 1)$ is true.

Again, we are trying to show that $P(k + 1)$ is true, knowing $P(k)$ is true. But if we have reached the stage of induction where $P(k)$ has been proved, then we know that $P(m)$ is true for $1 \leq m \leq k$, so the strengthened hypothesis in the second statement is permissible.

Example A.4   Recall that the set of all polynomials with real coefficients is denoted by P. Show that every polynomial in P of degree $n \in Z^+$ either is irreducible itself or is a product of irreducible polynomials in P. (An **irreducible polynomial** is one that cannot be factored into polynomials in P all of lower degree.)

Solution   We will use complete induction. Let $P(n)$ be the statement that is to be proved. Clearly $P(1)$ is true since a polynomial of degree 1 is already irreducible.

Let k be a positive integer. Our induction hypothesis is then: Every polynomial in P of degree less than $k + 1$ either is irreducible or can be factored into irreducible polynomials. Let $f(x)$ be a polynomial of degree $k + 1$. If $f(x)$ is irreducible, we have nothing more to do. Otherwise, we may factor $f(x)$ into polynomials $g(x)$ and $h(x)$ of lower degree than $k + 1$, obtaining $f(x) = g(x)h(x)$. The induction hypothesis indicates that each of $g(x)$ and $h(x)$ can be factored into irreducible polynomials, thus providing such a factorization of $f(x)$. This proves $P(k + 1)$. It follows that $P(n)$ is true for all $n \in Z^+$.   ◁

EXERCISES

1. Show that

$$1^2 + 2^2 + 3^2 + \cdots + n^2 = \frac{n(n+1)(2n+1)}{6}$$

for $n \in Z^+$.

2. Show that

$$1^3 + 2^3 + 3^3 + \cdots + n^3 = \frac{n^2(n+1)^2}{4}$$

for $n \in Z^+$.

3. Show that

$$1 + 3 + 5 + \cdots + (2n - 1) = n^2$$

for $n \in Z^+$.

4. Show that

$$\frac{1}{1 \cdot 2} + \frac{1}{2 \cdot 3} + \frac{1}{3 \cdot 4} + \cdots + \frac{1}{n(n+1)} = \frac{n}{n+1}$$

for $n \in Z^+$.

5. Prove by induction that if $a, r \in \mathbb{R}$ and $r \neq -1$, then

$$a + ar + ar^2 + \cdots + ar^n = a(1 - r^{n+1})/(1 - r) \quad \text{for } n \in Z^+.$$

6. Find the flaw in the argument.

We prove that any two integers i and j in Z^+ are equal. Let

$$\max(i, j) = \begin{cases} i & \text{if } i \geq j, \\ j & \text{if } j > i. \end{cases}$$

Let $P(n)$ be the statement

$$P(n): \text{Whenever } \max(i, j) = n, \text{ then } i = j.$$

Note that if $P(n)$ is true for all positive integers n, then any two positive integers i and j are equal. We proceed to prove $P(n)$ for positive integers n by induction.

Clearly $P(1)$ is true since if $i, j \in Z^+$ and $\max(i, j) = 1$, then $i = j = 1$.

Assume that $P(k)$ is true. Let i and j be such that $\max(i, j) = k + 1$. Then $\max(i - 1, j - 1) = k$ so that $i - 1 = j - 1$ by the induction hypothesis. Therefore, $i = j$ and $P(k + 1)$ is true. Consequently, $P(n)$ is true for all n.

7. Criticize this argument.

Let us show that every positive integer has some interesting property. Let $P(n)$ be the statement that n has an interesting property. We use complete induction.

Of course $P(1)$ is true, since 1 is the only positive integer that equals its own square, which is surely an interesting property of 1.

Suppose that $P(m)$ is true for $1 \leq m \leq k$. If $P(k + 1)$ were not true, then $k + 1$ would be the smallest integer without an interesting property, which would, in itself, be an interesting property of $k + 1$. So $P(k + 1)$ must be true. Thus $P(n)$ is true for all $n \in Z^+$.

8. We have never been able to see any flaw in (a). Try your luck with it and then answer (b).

a) A murderer is sentenced to be executed. He asks the judge not to let him know the day of the execution. The judge says, "I sentence you to be executed at 10 A.M. some day of this coming January, but I promise that you will not be aware that you are being executed that day until they come to get you at 8 A.M." The criminal goes to his cell and proceeds to prove, as follows, that he can't be executed in January.

Let $P(n)$ be the statement that I can't be executed on January $(31 - n)$. I want to prove $P(n)$ for $0 \leq n \leq 30$. Now I can't be executed on January 31, for since that is the last day of the month and I am to be executed that month, I would know that was the day before 8 A.M., contrary to the judge's sentence. Thus $P(0)$ is true. Suppose that $P(m)$ is true for $0 \leq m \leq k$, where $k \leq 29$. That is, suppose I can't be executed on January $(31 - k)$ through January 31. Then January

$(31 - k - 1)$ must be the last possible day for execution, and I would be aware that was the day before 8 A.M., contrary to the judge's sentence. Thus I can't be executed on January $(31 - (k + 1))$, so $P(k + 1)$ is true. Therefore, I can't be executed in January.

(Of course, the criminal was executed on January 17.)

b) An instructor teaches a class five days a week, Monday through Friday. She tells her class that she will give one more quiz on one day during the final week of classes, but that the students will not know for sure the quiz will be that day until they come to the classroom. What is the last day of the week on which she can give the quiz in order to satisfy these conditions?

Appendix B

Available Software

Following is a list of programs that make up the software available with this text, including a brief description of each.

NEWT123   A program showing the iterates in Newton's method to find solutions of n equations in n unknowns for $n = 1, 2,$ or 3. For use with Section 1.1

LINPROG   A graphic program allowing the user to estimate solutions of two-variable linear programming problems. For use with Section 1.2.

MATCOMP   Performs matrix computations, solves linear systems, and finds real eigenvalues and eigenvalues. For use throughout the course.

YUREDUCE   The user selects items from a menu for step-by-step row reduction of a matrix. For Sections 1.4–1.7 and Section 2.3.

TIMING   Times operations in interpretive BASIC or compiled BASIC. Also times various methods for solving linear systems. For Sections 2.1 and 2.2.

LUFACTOR   Gives the factorization $A = LU$, which can then be used to solve $A\mathbf{x} = \mathbf{b}$. For Section 2.2.

HILBERT   The user can experiment with ill-conditioned Hilbert matrices. For Section 2.3.

VECTGRPH   Gives graded quizzes on vector geometry based on displayed graphics. Useful from Section 3.1 on.

EBYMTIME   A specialized program giving a lower bound for the time required to find the determinant of a matrix using only repeated expansion by minors. For Section 4.3.

ALLROOTS   Step-by-step execution of Newton's method to find both real and complex roots of a polynomial with real or complex coefficients. May be used in conjunction with MATCOMP to find complex as well as real eigenvalues of a small matrix. For Section 4.5.

YOUFIT   The user can experiment with a graphic to try to find the least-squares linear, quadratic, or exponential fit of data. The computer can then be asked to find the best fit. For Section 5.2.

QRFACTOR   Executes the Gram–Schmidt process, gives the QR factorization of a suitable matrix A, and can be used to find least-squares solutions. For Section 5.4. Also step-by-step computation by the QR algorithm of both real and complex eigenvalues of a matrix. The user specifies shifts and when to decouple. For Section 7.4.

POWER   Executes, one step at a time, the power method (with deflation for symmetric matrices) for finding eigenvalues and eigenvectors. For Section 7.4.

JACOBI   Executes, one "rotation" at a time, the method of Jacobi for diagonalizing a symmetric matrix. For Section 7.4.

SIMPLEX   Allows the user to execute the simplex method by making choices from a menu of possible operations to be performed. A graded quiz option is included. Alternatively, the computer may be asked to simply execute the method and display the tableaux. For Chapter 9.

MATFILES   The user can create files of matrices that can be accessed from the programs MATCOMP, YUREDUCE, LUFACTOR, QRFACTOR, POWER, JACOBI, and SIMPLEX. Existing files can be viewed.

Answers to Odd-Numbered Exercises

Chapter 1

Section 1.1

1. (a) $2ax_0 + b$; **(b)** $(2ax_0 + b)x + c - ax_0^2$ **3.** $23x + 12$
5. $f(x) \approx -3x + 1$, $x_1 = \frac{1}{3}$; yes: $f(x_1) = \frac{1}{9} < 1 = f(0)$
7. $f(x) \approx 4.01x - 3.01$, $x \approx .751$; yes: $f(x_1) \approx .174 < f(1) = 1$
9. $f(x) \approx 8.1001x - 5.1001$, $x_1 \approx .630$; yes: $f(x_1) \approx .988 < f(1) = 3$
11. $(6x_0 - 4y_0)x + (10y_0 - 4x_0)y + (-3x_0^2 + 4x_0y_0 - 5y_0^2)$ **13.** $1.01x + 4y - 1.99$
15. $(x_1, y_1) = (3.25, -4.5)$; no: $f(x_1, y_1) = 3.875 > f(1, -2) = 1$; no: $g(x_1, y_1) = 4.375 > |g(1, -2)| = 1$.
   Continued iteration shows that a solution is approximately $(x_1, y_1) = (2.838, -4.371)$.
17. 1.895494267033981 **19.** 2.129372482760157 **21.** 4.555535705195128
23. 4.136687872473846, -2.13746608117813
25. $(1.860805853111703, \pm.7330767879460008)$, $(.2541016883650524, \pm1.983792411511353)$
27. $(3.077228234333831, 3.386640460128785)$, $(-4.957849742483486, 5.155606081640021)$
29. $(.8518838649739748, -2.173868928754215, -.9867431398353754)$,
   $(-.3303492803732686, 2.027099071837175, -.8530282645370337)$

Section 1.2

1.

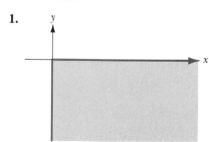

3.

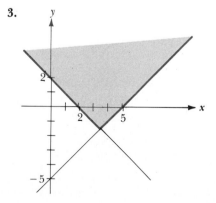

5.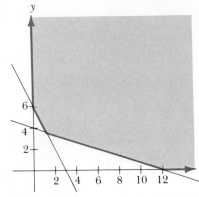

7. (a) No maximum, minimum 7;
  (b) no maximum, no minimum;
  (c) maximum -13, no minimum
9. (a) Maximum 0, no minimum;
  (b) no maximum, minimum 0;
  (c) no maximum, no minimum

11. (a) No maximum, minimum 14; **(b)** no maximum, minimum 6; **(c)** no maximum, minimum 3
13. (a) Ratio $< \frac{1}{2}$; **(b)** ratio > 5 **15.** 1000 lb of blend A, 2500 lb of blend B
17. 3.5 oz of food 1, .75 oz of food 2   **19.** 60 lb of bag 1, 10 lb of bag 2
21. (a) Maximum 28.9, minimum 10; **(b)** maximum 37, minimum 13.3
23. (a) Maximum 19, minimum 9.4; **(b)** maximum 14.7, minimum 8; **(c)** maximum 11, minimum 4;
  (d) maximum 0, minimum -12.4
25. (a) Maximum 72, minimum 2.1; **(b)** maximum 11.2, minimum -12; **(c)** maximum 48,
  minimum 2; **(d)** maximum 12, minimum -5

Section 1.3

1. $\begin{pmatrix} -6 & 3 & 9 \\ 12 & 0 & -3 \end{pmatrix}$ **3.** $\begin{pmatrix} 2 & 2 & 1 \\ 9 & -1 & 2 \end{pmatrix}$ **5.** $\begin{pmatrix} 6 & -3 \\ -3 & 1 \\ -2 & 5 \end{pmatrix}$ **7.** $\begin{pmatrix} -16 & 2 & 16 \\ 6 & 2 & -10 \end{pmatrix}$ **9.** Undefined

11. $\begin{pmatrix} -13 & 14 \\ 11 & -6 \end{pmatrix}$ **13.** Undefined   **15.** $\begin{pmatrix} 27 & -35 \\ -4 & 26 \end{pmatrix}$ **17.** $\begin{pmatrix} -8 & 16 & -40 \\ -5 & -26 & 63 \end{pmatrix}$

19. $\begin{pmatrix} 352 & -268 \\ 294 & -340 \end{pmatrix}$ **21.** $\begin{pmatrix} 14 & -11 \\ -11 & 17 \end{pmatrix}$ **23.** $\begin{pmatrix} -5 & 13 \\ 20 & 22 \end{pmatrix}$

25. $\mathbf{xy} = (-14)$, $\mathbf{yx} = \begin{pmatrix} -8 & 12 & -4 \\ 2 & -3 & 1 \\ -6 & 9 & -3 \end{pmatrix}$

27. Let A be $n \times m$ and $\mathbf{y}$ be $m \times 1$. Then $A\mathbf{y}$ is $n \times 1$, again a column vector.
29. Let $\mathbf{u}, \mathbf{v},$ and $\mathbf{w}$ have the same number n of components. Then $\mathbf{u} \cdot (\mathbf{v} + \mathbf{w}) = u_1(v_1 + w_1) +$
  $\cdots + u_n(v_n + w_n) = (u_1v_1 + \cdots + u_nv_n) + (u_1w_1 + \cdots + u_nw_n) = \mathbf{u} \cdot \mathbf{v} + \mathbf{u} \cdot \mathbf{w}$.
31. Let $A, B,$ and C be $m \times n$. Then $(A + B) + C = (a_{ij} + b_{ij}) + (c_{ij})$
  $= ((a_{ij} + b_{ij}) + c_{ij}) = (a_{ij} + (b_{ij} + c_{ij})) = A + (B + C)$.
33. Let A be $m \times n$. Then $(rs)(a_{ij}) = ((rs)a_{ij}) = (r(sa_{ij})) = r(sA)$.
35. Let A be $m \times n$. Then $A^T = (a_{ij})^T = B = (b_{ij})$, where $b_{rs} = a_{sr}$. Then $(A^T)^T = B^T = (c_{ij})$, where $c_{ij} = b_{ji} = a_{ij}$, so $(A^T)^T = (a_{ij}) = A$.
37. Let A be an $m \times n$ matrix, and let B be $n \times s$. Note that both $(AB)^T$ and B^TA^T are $s \times m$ matrices.
  Let $(AB)^T = (c_{ij})$ and $B^TA^T = (d_{ij})$. Then

$$c_{ij} = \text{(entry in } j\text{th row and } i\text{th column of } AB)$$
$$= (j\text{th row of } A) \cdot (i\text{th column of } B).$$

Also

$$d_{ij} = (i\text{th row of } B^T) \cdot (j\text{th column of } A^T)$$
$$= (i\text{th column of } B) \cdot (j\text{th row of } A).$$

Since the dot product is commutative, we see that $c_{ij} = d_{ij}$.

39. Let A be symmetric and $m \times n$. Then $A = A^T$ and A^T is $n \times m$. Thus $m = n$ and A is square.

41. $\begin{pmatrix} 1 & -1 & 2 & 5 \\ -1 & 4 & -7 & 8 \\ 2 & -7 & -1 & 6 \\ 5 & 8 & 6 & 3 \end{pmatrix}$ **43.** If A is square, then $(A^n)^T = (A^T)^n$.

45. $\begin{pmatrix} 588 & -398 & -266 & 494 \\ 646 & 1038 & -55 & 290 \\ -581 & 2175 & 953 & -502 \\ -676 & 2762 & 2150 & -7 \\ 248 & 1424 & -602 & 105 \end{pmatrix}$ **47.** Undefined

49. $\begin{pmatrix} 36499 & 30055 & -6890 & 10010 & 50800 \\ 30055 & 32653 & -11723 & 3622 & 44680 \\ -6890 & -11723 & 17179 & 14629 & -15569 \\ 10010 & 3622 & 14629 & 22237 & 10107 \\ 50800 & 44680 & -15569 & 10107 & 75753 \end{pmatrix}$

51. $\begin{pmatrix} 26918 & -3837 & -13284 & 14372 & -6772 \\ 41446 & 353741 & 209672 & 310468 & 194032 \\ 934 & 34811 & 44048 & 53254 & 27054 \\ 20416 & 82936 & 44566 & 82892 & 52440 \\ -2966 & -64951 & -27916 & -52777 & -9719 \end{pmatrix}$

Section 1.4

1. $x = 2, y = -4$ **3.** $x = -3, y = 2, z = 4$ **5.** $x_1 = -1, x_2 = 0, x_3 = 2$
7. $x_1 = 2, x_2 = -1, x_3 = 3, x_4 = -1$ **9.** $x_1 = -2, x_2 = 1$ **11.** $x_1 = 2, x_2 = 0, x_3 = -1$
13. $x_1 = 3, x_2 = 1, x_3 = -1, x_4 = 2$
15. System 1: $x = -1, y = 1$; system 2: $x = 2, y = -5$; system 3: $x = 3, y = 0$
17. System 1: $x_1 = -3, x_2 = -5, x_3 = 2$; system 2: $x_1 = 0, x_2 = 0, x_3 = 1$;
system 3: $x_1 = 2, x_2 = -1, x_3 = 3$
19. **(a)** The intersection of the solution sets of equations depends only on the sets, and not on any way in which they may be ordered. **(b)** The solution set of each equation is unchanged, so the intersection of all their solution sets is unchanged.

21. $\begin{pmatrix} 0 & 1 & 0 & 0 \\ 1 & 0 & 0 & 0 \\ 0 & 0 & 1 & 0 \\ 0 & 0 & 0 & 1 \end{pmatrix}$ **23.** $\begin{pmatrix} 5 & 0 & 0 & 0 \\ 0 & 0 & 1 & 0 \\ 0 & 1 & 0 & 0 \\ 0 & 2 & 0 & 1 \end{pmatrix}$ **25.** $\begin{pmatrix} 0 & 6 & 0 & 1 \\ -2 & 1 & 0 & 0 \\ 0 & -18 & 1 & -3 \\ 1 & 0 & 0 & 0 \end{pmatrix}$

27. $x_1 = 1, x_2 = 0, x_3 = -2, x_4 = 4$ **29.** $x_1 = -43, x_2 = -12, x_3 = 7, x_4 = 1$
31. System 1: $x_1 = 2, x_2 = 1, x_3 = 3, x_4 = -1$; system 2: $x_1 = 4, x_2 = -1, x_3 = 2, x_4 = -3$
33. $x_1 = -1, x_2 = 0, x_3 = 2$ **35.** $x_1 = -3, x_2 = 0, x_3 = -2, x_4 = -6$
37. System 1: $x_1 = 1, x_2 = 1, x_3 = -1, x_4 = 1, x_5 = 2$;
system 2: $x_1 \approx .0345, x_2 \approx -.5370, x_3 \approx -1.6240, x_4 \approx .1222, x_5 \approx .8188$

Section 1.5

1. Inverse $\begin{pmatrix} 1 & -1 \\ 0 & 1 \end{pmatrix}$, $\begin{pmatrix} 1 & 1 \\ 0 & 1 \end{pmatrix}$ is an elementary matrix.   **3.** Not invertible

5. $\begin{pmatrix} 1 & 0 & 1 \\ 0 & 1 & 1 \\ 0 & 0 & -1 \end{pmatrix}$, $\begin{pmatrix} 1 & 0 & 1 \\ 0 & 1 & 1 \\ 0 & 0 & -1 \end{pmatrix} = \begin{pmatrix} 1 & 0 & 0 \\ 0 & 1 & 0 \\ 0 & 0 & -1 \end{pmatrix} \begin{pmatrix} 1 & 0 & 1 \\ 0 & 1 & 0 \\ 0 & 0 & 1 \end{pmatrix} \begin{pmatrix} 1 & 0 & 0 \\ 0 & 1 & 1 \\ 0 & 0 & 1 \end{pmatrix}$

7. $\begin{pmatrix} -7 & 5 & 3 \\ 3 & -2 & -2 \\ 3 & -2 & -1 \end{pmatrix}$, $\begin{pmatrix} 2 & 1 & 4 \\ 3 & 2 & 5 \\ 0 & -1 & 1 \end{pmatrix} = \begin{pmatrix} 2 & 0 & 0 \\ 0 & 1 & 0 \\ 0 & 0 & 1 \end{pmatrix} \begin{pmatrix} 1 & 0 & 0 \\ 3 & 1 & 0 \\ 0 & 0 & 1 \end{pmatrix} \begin{pmatrix} 1 & 0 & 0 \\ 0 & .5 & 0 \\ 0 & 0 & 1 \end{pmatrix} \begin{pmatrix} 1 & 0 & 0 \\ 0 & 1 & 0 \\ 0 & -1 & 1 \end{pmatrix} \begin{pmatrix} 1 & .5 & 0 \\ 0 & 1 & 0 \\ 0 & 0 & 1 \end{pmatrix}$

$\begin{pmatrix} 1 & 0 & 0 \\ 0 & 1 & 0 \\ 0 & 0 & -1 \end{pmatrix} \begin{pmatrix} 1 & 0 & 3 \\ 0 & 1 & 0 \\ 0 & 0 & 1 \end{pmatrix} \begin{pmatrix} 1 & 0 & 0 \\ 0 & 1 & -2 \\ 0 & 0 & 1 \end{pmatrix}$ **9.** $\begin{pmatrix} -1 & 0 & -1 & -1 \\ -3 & -1 & 0 & -1 \\ 5 & 0 & 4 & 3 \\ 3 & 0 & 3 & 2 \end{pmatrix}$ **11.** $\begin{pmatrix} 1 & 0 & 0 & 0 & 0 & 0 \\ 0 & -1 & 0 & 0 & 0 & 0 \\ 0 & 0 & \frac{1}{2} & 0 & 0 & 0 \\ 0 & 0 & 0 & \frac{1}{3} & 0 & 0 \\ 0 & 0 & 0 & 0 & \frac{1}{4} & 0 \\ 0 & 0 & 0 & 0 & 0 & \frac{1}{5} \end{pmatrix}$

13. (a) $A^{-1} = \begin{pmatrix} -7 & 3 \\ -5 & 2 \end{pmatrix}$; (b) $x_1 = -37$, $x_2 = -26$

15. $AA^{-1} = A^{-1}A = I$ shows that A^{-1} is invertible and that $(A^{-1})^{-1} = A$.

17. (a) $\begin{pmatrix} 0 & 1 \\ 0 & 0 \end{pmatrix}$; (b) If A is invertible, then $(A^{-1})^r A^r = I$ so A^r cannot be the zero matrix.

19. If E is obtained from I by interchanging rows i and j, then it can be obtained from I by interchanging columns i and j.

If E is obtained from I by multiplying row i by r, then it can be obtained from I by multiplying column i by r.

If E is obtained from I by adding r times row i to row j, then it can be obtained from I by adding r times column j to row i.

21. (a) If B is obtained from A by interchanging rows i and j, then B^{-1} is obtained from A^{-1} by interchanging columns i and j. (b) If B is obtained from A by multiplying row i by $r \neq 0$, then B^{-1} is obtained from A^{-1} by multiplying column i by $1/r$. (c) If B is obtained from A by adding r times row i to row j, then B^{-1} is obtained from A^{-1} by subtracting r times column j from column i.

23. $\begin{pmatrix} -.5879397 & .3115578 & -.08542714 \\ .4874372 & -.241206 & .1306533 \\ -.1658292 & .2160804 & -.07537688 \end{pmatrix}$ **25.** Not invertible   **27.** Not invertible

29. $\begin{pmatrix} .3548387 & -.06451613 & .1612903 \\ -.1290323 & .3870968 & .03225806 \\ -.09677419 & .2903226 & -.2258064 \end{pmatrix}$ **31.** Not invertible

Section 1.6

1. (a) $\begin{pmatrix} 1 & 3 & 2 \\ 0 & -5 & 0 \\ 0 & 0 & 0 \end{pmatrix}$; (b) $\begin{pmatrix} 1 & 0 & 2 \\ 0 & 1 & 0 \\ 0 & 0 & 0 \end{pmatrix}$ **3.** (a) $\begin{pmatrix} 1 & 1 & -3 & 3 \\ 0 & 2 & -1 & 3 \\ 0 & 0 & 10 & 0 \\ 0 & 0 & 0 & 0 \end{pmatrix}$; (b) $\begin{pmatrix} 1 & 0 & 0 & 1.5 \\ 0 & 1 & 0 & 1.5 \\ 0 & 0 & 1 & 0 \\ 0 & 0 & 0 & 0 \end{pmatrix}$

5. (a)
$$\begin{pmatrix} 1 & -3 & 0 & 0 & -1 \\ 0 & 0 & 1 & 3 & -4 \\ 0 & 0 & 0 & 1 & 3 \\ 0 & 0 & 0 & 0 & 16 \end{pmatrix}$$
; (b)
$$\begin{pmatrix} 1 & -3 & 0 & 0 & 0 \\ 0 & 0 & 1 & 0 & 0 \\ 0 & 0 & 0 & 1 & 0 \\ 0 & 0 & 0 & 0 & 1 \end{pmatrix}$$
7. $\mathbf{x} = \begin{pmatrix} 5 - 6r \\ 2 - 4r \\ r \end{pmatrix}, \begin{pmatrix} -7 \\ -6 \\ 2 \end{pmatrix}$

9. $\mathbf{x} = \begin{pmatrix} 1 - 2r \\ -2 - r - 3s \\ r \\ s \end{pmatrix}, \begin{pmatrix} -5 \\ 1 \\ 3 \\ -2 \end{pmatrix}$ **11.** $\begin{pmatrix} 0 \\ 2 \\ -5 \\ 2 \end{pmatrix}$ **13.** $\mathbf{x} = \begin{pmatrix} 5 \\ 1 \end{pmatrix}$ **15.** No solution

17. $\mathbf{x} = \begin{pmatrix} -8 \\ -23 - 5r/2 \\ -7 + r/2 \\ r \end{pmatrix}$ **19.** $(x_1, x_2) = (1, 9)$ **21.** $(x_1, x_2) = (2 + 3r, r)$

23. $\begin{pmatrix} x_1 \\ x_2 \end{pmatrix} = \begin{pmatrix} 3 \\ -2 \end{pmatrix}$ **25.** $(x_1, x_2) = (-1, 3)$ **27.** $(x_1, x_2, x_3) = (1 + 4r, 1 - 7r/3, r)$

29. System 1: $\mathbf{x} = \begin{pmatrix} 13 \\ -1 - 2t - 6r \\ 10 - 4t + 5s + 2r \\ t \\ s \\ 7 - 3r \\ r \end{pmatrix}$; System 2: No solution

Section 1.7

1. Not a transition matrix   **3.** Not a transition matrix   **5.** A regular transition matrix

7. A transition matrix. Not regular   **9.**
$$\begin{array}{ccc} \text{U} & \text{S} & \text{R} \\ \begin{pmatrix} .4 & .3 & .1 \\ .5 & .5 & .2 \\ .1 & .2 & .7 \end{pmatrix} \end{array}$$
11. 27/50   **13.** $\begin{pmatrix} .29 \\ .437 \\ .273 \end{pmatrix}$

15. Offspring of one dominant and one hybrid parent inherit G from the dominant parent. Half of them inherit g and half inherit G from the hybrid parent. This gives column 1 of the transition matrix. Column 3 is obtained similarly. For column 2, half the offspring of two hybrid parents inherit G from the mother, and half of those also inherit G from the father. Thus the proportion of dominant offspring is $(1/2)(1/2) = 1/4$. The other two numbers in column 2 are obtained similarly.

17. 1/2   **19.** $\begin{pmatrix} 1/4 \\ 1/2 \\ 1/4 \end{pmatrix}$

21. The transition matrix has at least one entry 1 on the main diagonal. It cannot be regular.

23. (a) $\begin{pmatrix} 1 & .5 & .1 \\ 0 & .2 & .4 \\ 0 & .3 & .5 \end{pmatrix}$; **(b)** $\begin{pmatrix} 1 & 0 & 0 \\ 0 & .5 & .2 \\ 0 & .5 & .8 \end{pmatrix}$

25. All nonzero entries are positive. Since more than half the entries in each row and each column are positive, the dot product of any row vector with any column vector has to involve the product of at least one positive component of the row with one positive component of the column vector. Thus the dot product is positive. Consequently, the square of the matrix has all positive entries, so the matrix is regular.

27. $\begin{pmatrix} 1/3 \\ 2/3 \end{pmatrix}$ **29.** $\begin{pmatrix} 5/18 \\ 1/3 \\ 7/18 \end{pmatrix}$ **31.** $\begin{pmatrix} 23/78 \\ 1/6 \\ 7/13 \end{pmatrix}$

33. $(.2020202, .2626263, .2861953, .2491582)^T$

35. $(.2324841, .03184713, .2452229, .2356688, .2547771)^T$

Chapter 2

Section 2.1

1. There are $n - 1$ flops performed on **b** while the first column of A is being fixed up, $n - 2$ flops while the second column of A is being fixed up, and so on. The total number is $(n - 1) + (n - 2) + \cdots + 2 + 1 = n(n - 1)/2$, which has order of magnitude $n^2/2$ for large n.

3. mn if we call each indexed addition a flop   **5.** mn^2 **7.** $2n^3$ **9.** $3n^3$ **11.** $6n^3$

13. $3n^3/2$ **15.** n^3 **17.** $2n$ **19.** w^2n, counting each final division as a flop

21. Table 2.1 is our answer.

23. Yes. The terms of degree less than n^3 in the flop formula will play a role if n is small enough.

25. Yes. The terms of degree less than n^3 in the flop formula will play a role if n is small enough.

Section 2.2

1. It is not significant; no arithmetic operations are involved, just storing of indexed values.

3. $\begin{pmatrix} -18 \\ 5 \end{pmatrix}$ **5.** $\begin{pmatrix} -27 \\ -11 \\ -1 \end{pmatrix}$ **7.** $\begin{pmatrix} 1 \\ -2 \\ 1 \\ 3 \end{pmatrix}$

9. $P = \begin{pmatrix} 0 & 0 & 1 \\ 0 & 1 & 0 \\ 1 & 0 & 0 \end{pmatrix}, L = \begin{pmatrix} 1 & 0 & 0 \\ 3 & 1 & 0 \\ 1 & \frac{1}{3} & 1 \end{pmatrix}, U = \begin{pmatrix} 2 & -1 & 5 \\ 0 & 6 & -23 \\ 0 & 0 & -\frac{1}{3} \end{pmatrix}, \mathbf{x} = \begin{pmatrix} 2 \\ -1 \\ -1 \end{pmatrix}$

11. $L = \begin{pmatrix} 1 & 0 & 0 \\ 3 & 1 & 0 \\ 4 & -10 & 1 \end{pmatrix}, U = \begin{pmatrix} 1 & 2 & -1 \\ 0 & 1 & 5 \\ 0 & 0 & 55 \end{pmatrix}, \mathbf{x} = \begin{pmatrix} -1 \\ 0 \\ 2 \end{pmatrix}$

13. $L = \begin{pmatrix} 1 & 0 & 0 & 0 \\ 2 & 1 & 0 & 0 \\ -1 & 0 & 1 & 0 \\ 4 & 2 & -\frac{3}{2} & 1 \end{pmatrix}, U = \begin{pmatrix} 1 & 2 & -3 & 0 \\ 0 & 1 & 0 & 0 \\ 0 & 0 & -2 & -1 \\ 0 & 0 & 0 & -\frac{1}{2} \end{pmatrix}, \mathbf{x} = \begin{pmatrix} 3 \\ 1 \\ -1 \\ 2 \end{pmatrix}$

15. $\begin{pmatrix} 0 \\ -1 \end{pmatrix}$ **17.** $\begin{pmatrix} -1 \\ 2 \\ 2 \end{pmatrix}$ **19.** $LDU = \begin{pmatrix} 1 & 0 & 0 \\ 0 & 1 & 0 \\ 3 & -4 & 1 \end{pmatrix}\begin{pmatrix} 1 & 0 & 0 \\ 0 & 1 & 0 \\ 0 & 0 & 14 \end{pmatrix}\begin{pmatrix} 1 & 1 & -3 \\ 0 & 1 & 1 \\ 0 & 0 & 1 \end{pmatrix}$

21. $LDU = \begin{pmatrix} 1 & 0 & 0 & 0 \\ 2 & 1 & 0 & 0 \\ -1 & 0 & 1 & 0 \\ 4 & 2 & -1.5 & 1 \end{pmatrix} \begin{pmatrix} 1 & 0 & 0 & 0 \\ 0 & 1 & 0 & 0 \\ 0 & 0 & -2 & 0 \\ 0 & 0 & 0 & -.5 \end{pmatrix} \begin{pmatrix} 1 & 2 & -3 & 0 \\ 0 & 1 & 0 & 0 \\ 0 & 0 & 1 & .5 \\ 0 & 0 & 0 & 1 \end{pmatrix}$

23. Display: $\begin{pmatrix} 1 & 3 & -5 & 2 & 1 \\ 4 & -18 & 30 & 0 & -1 \\ 3 & .1666667 & 9 & -2 & 4.166667 \\ 2 & .2777778 & 1.407407 & -4.185185 & 5.413581 \\ -6 & -1.166667 & .6666667 & -4.141593 & 45.4764 \end{pmatrix}$. For $\mathbf{b}_1$, $\mathbf{x} = \begin{pmatrix} 1 \\ -1 \\ 0 \\ 2 \\ -2 \end{pmatrix}$;

for $\mathbf{b}_2$, $\mathbf{x} = \begin{pmatrix} -5 \\ -8 \\ 9 \\ 11 \\ 4 \end{pmatrix}$; for $\mathbf{b}_3$, $\mathbf{x} = \begin{pmatrix} -8 \\ 9 \\ -11 \\ 3 \\ 7 \end{pmatrix}$

25. $\begin{pmatrix} .4814815 \\ -1.592593 \end{pmatrix}$ **27.** $\begin{pmatrix} .8125 \\ 1.875 \\ -.3125 \end{pmatrix}$ **29.** $\begin{pmatrix} -.09131459 \\ -.08786787 \\ -.4790014 \\ -.01573263 \\ .08439226 \\ .2712737 \end{pmatrix}$

31. The ratios roughly conform on our computer: One solution that requires 100 flops took about $\frac{1}{3}$ of the time to form the L/U display, which requires about 1000/3 flops, etc.

33. The ratios roughly conform on our computer in the sense explained in the answer to Exercise 31.

Section 2.3

1. $x_1 = 0, x_2 = 1$

3. $x_1 = 1, x_2 = .9999$. Yes, it is reasonably accurate. $10x_1 + 1000000x_2 = 1000000$, $-10x_1 + 20x_2 = 10$ is a system that can't be solved without pivoting by a 5-figure computer.

5. $x_1 + 10^{19}x_2 = 10^{19}, -x_1 + 2x_2 = 1$

7. We need to show that $[n(nA)^{-1}]A = I$. It is easy to see that $(rA)B = A(rB) = r(AB)$ for any scalar r and matrices A and B such that AB is defined. Thus we have $[n(nA)^{-1}]A = (nA)^{-1}(nA) = I$.

9. On our computer, $(x_1, x_2) = (-.0001, .0001)$, which is approximately correct.

11. On our computer taking $r = 10^{-14}$ for the roundoff ratio, we obtain $(x_1, x_2) = (-.0001, .0001)$ in single-precision printing, which is approximately correct.

13. System: $10^{-11}x_1 + 10^{10}x_2 = 1, 10^{10}x_1 + 2(10)^{10}x_2 = 1$
(a) $(x_1, x_2) = (-10^{-10}, 10^{-10})$; approximately correct; (b) same answer as part (a)

15. Yes, on our computer, taking default value for roundoff ratio, and using the matrix
$\begin{pmatrix} 10^{-13} & 0 \\ 0 & 10^{-13} \end{pmatrix}$.

17. Solutions: $\begin{pmatrix} -46 \\ 96 \end{pmatrix}$ for $\mathbf{b}$ and $\begin{pmatrix} -36 \\ 78 \end{pmatrix}$ for $\mathbf{c}$. Component differences are at least 10.

19. $\begin{pmatrix} 32 \\ 240 \\ -1500 \\ 1400 \end{pmatrix}$ for **b** and $\begin{pmatrix} 268 \\ -2100 \\ 3720 \\ -1820 \end{pmatrix}$ for **c**. Component differences are at least 236.

21. $H_2^{-1} = \begin{pmatrix} 4 & -6 \\ -6 & 12 \end{pmatrix}$. The inverse does check out OK.

23. Our inverse check for H_4^2 was poor in single precision. Our inverse checks for H_4^4 and H_4^8 were poor even in double precision. We were able to reduce H_4^{16} to the identity, and its inverse check was similar to that for H_4^8.

Chapter 3

Section 3.1

1. $3\sqrt{2}$ **3.** $\sqrt{22}$ **5.** $(1, -3, -4)$ **7.** $(1, 4, 3)$ **9.** $(-9, 6, 19)$ **11.** $(-\frac{8}{5}, -\frac{4}{5}, \frac{12}{5})$
13. $(2/\sqrt{14}, 1/\sqrt{14}, -3/\sqrt{14})$ **15.** 8   **17.** $\cos^{-1}[-3/(2\sqrt{39})]$ **19.** 11
21. $(1, -1, 1)$ (Many other answers are possible.)   **23.** $\cos^{-1}[34/(\sqrt{31}\sqrt{79})]$
25. Perpendicular   **27.** Parallel, opposite direction   **29.** Neither
31. At an angle $\cos^{-1}(\frac{5}{13})$ from north   **33.** **(a)** $50/(\cos\theta)$ lb; **(b)** 50 lb; **(c)** The rope breaks.

Section 3.2

1. **(a)** $(\mathbf{u} + \mathbf{v}) + \mathbf{w} = (u_1 + v_1, \ldots, u_n + v_n) + (w_1, \ldots, w_n)$
$\qquad = (u_1 + v_1 + w_1, \ldots, u_n + v_n + w_n)$
$\qquad = (u_1, \ldots, u_n) + (v_1 + w_1, \ldots, v_n + w_n) = \mathbf{u} + (\mathbf{v} + \mathbf{w})$
(b) $\mathbf{0} + \mathbf{v} = (0, \ldots, 0) + (v_1, \ldots, v_n) = (0 + v_1, \ldots, 0 + v_n) = (v_1, \ldots, v_n) = \mathbf{v}$
(c) $\mathbf{v} + (-\mathbf{v}) = (v_1, \ldots, v_n) + (-v_1, \ldots, -v_n)$
$\qquad = (v_1 - v_1, \ldots, v_n - v_n) = (0, \ldots, 0)$
3. $(4, 16, 8)$ **5.** $(-3, -5, -6)$ **7.** Not a vector space   **9.** Not a vector space
11. Not a vector space
13. Let $\mathbf{u} + \mathbf{v} = \mathbf{u} + \mathbf{w}$. Then $\mathbf{v} = \mathbf{0} + \mathbf{v} = (-\mathbf{u} + \mathbf{u}) + \mathbf{v} = -\mathbf{u} + (\mathbf{u} + \mathbf{v}) = -\mathbf{u} + (\mathbf{u} + \mathbf{w}) = (-\mathbf{u} + \mathbf{u}) + \mathbf{w} = \mathbf{0} + \mathbf{w} = \mathbf{w}$.
15. $r(\mathbf{0}) + r(\mathbf{0}) = r(\mathbf{0} + \mathbf{0}) = r\mathbf{0} = r(\mathbf{0}) + \mathbf{0}$. By **(c)** of Theorem 3.2, we obtain $r(\mathbf{0}) = \mathbf{0}$.
17. If $r\mathbf{v} = \mathbf{0}$ and $r \neq 0$, then $\mathbf{v} = 1\mathbf{v} = [(1/r)r]\mathbf{v} = (1/r)(r\mathbf{v}) = (1/r)\mathbf{0} = \mathbf{0}$ by Theorem 3.2(e).
19. $14\sqrt{10}$ **21.** $(28)^2 = (30 - 2)^2 = 784$
23. $(\mathbf{v} - \mathbf{w}) \cdot (\mathbf{v} + \mathbf{w}) = \mathbf{v} \cdot \mathbf{v} + \mathbf{v} \cdot \mathbf{w} - \mathbf{w} \cdot \mathbf{v} - \mathbf{w} \cdot \mathbf{w} = \mathbf{v} \cdot \mathbf{v} - \mathbf{w} \cdot \mathbf{w} = \|\mathbf{v}\|^2 - \|\mathbf{w}\|^2 = 0$ if and only if $\|\mathbf{v}\| = \|\mathbf{w}\|$.
25. $\|\mathbf{v} - \mathbf{w}\| = \|\mathbf{v} + (-\mathbf{w})\| \leq \|\mathbf{v}\| + \|-\mathbf{w}\| = \|\mathbf{v}\| + |-1|\|\mathbf{w}\| = \|\mathbf{v}\| + \|\mathbf{w}\|$
27. If two adjacent sides are represented by $\mathbf{v}$ and $\mathbf{w}$, the diagonals are represented by $\mathbf{v} + \mathbf{w}$ and $\mathbf{v} - \mathbf{w}$. If $\|\mathbf{v}\| = \|\mathbf{w}\|$, then the diagonals are perpendicular by Exercise 23.
29. $(\|\mathbf{v}\|\mathbf{w} + \|\mathbf{w}\|\mathbf{v}) \cdot (\|\mathbf{v}\|\mathbf{w} - \|\mathbf{w}\|\mathbf{v}) = \|\mathbf{v}\|^2(\mathbf{w} \cdot \mathbf{w}) - \|\mathbf{v}\|\|\mathbf{w}\|(\mathbf{w} \cdot \mathbf{v}) + \|\mathbf{w}\|\|\mathbf{v}\|(\mathbf{v} \cdot \mathbf{w}) - \|\mathbf{w}\|^2(\mathbf{v} \cdot \mathbf{v}) = \|\mathbf{v}\|^2(\mathbf{w} \cdot \mathbf{w}) - \|\mathbf{w}\|^2(\mathbf{v} \cdot \mathbf{v}) = \|\mathbf{v}\|^2\|\mathbf{w}\|^2 - \|\mathbf{w}\|^2\|\mathbf{v}\|^2 = 0$, so the vectors are perpendicular.
31. An inner product space
33. For any vector $\mathbf{v}$ in V, we have $\mathbf{0} \cdot \mathbf{v} = (0 + 0) \cdot \mathbf{v} = \mathbf{0} \cdot \mathbf{v} + \mathbf{0} \cdot \mathbf{v}$ by properties P_1 and P_2. Subtracting $\mathbf{0} \cdot \mathbf{v}$ from both sides, we obtain $0 = \mathbf{0} \cdot \mathbf{v}$.
35. **(a)** Obvious; **(b)** $\|r\mathbf{v}\| = \sqrt{(r\mathbf{v}) \cdot (r\mathbf{v})} = \sqrt{r^2(\mathbf{v} \cdot \mathbf{v})} = |r|\sqrt{\mathbf{v} \cdot \mathbf{v}} = |r|\|\mathbf{v}\|$ for any scalar r in $\mathbb{R}$ and vector $\mathbf{v}$ in V. **(c)** Analogous to the proof of Theorem 3.4. **(d)** Analogous to the proof of Theorem 3.5.

Section 3.3

1. A subspace   **3.** Not a subspace   **5.** Not a subspace   **7.** Not a subspace
9. Not a subspace   **11.** Yes   **13.** Yes   **15.** Yes
17. (a) For any constant c, we have $c = c \sin^2 x + c \cos^2 x$; **(b)** $\cos 2x = (1)\cos^2 x + (-1)\sin^2 x$;
   (c) $\cos 4x = 1 - 2 \sin^2 2x$ so $8 \cos 4x = (8/7)7 + (-16) \sin^2 2x$
19. $\mathbf{b} = 2\mathbf{v}_1 - \mathbf{v}_2 + \mathbf{v}_3$ **21.** Impossible
23. (a) $\mathbf{v}_1$ and $2\mathbf{v}_1 + \mathbf{v}_2$ are obviously in $\mathrm{sp}(\mathbf{v}_1, \mathbf{v}_2)$. Since $\mathbf{v}_2 = (-2)\mathbf{v}_1 + (2\mathbf{v}_1 + \mathbf{v}_2)$, we see that $\mathbf{v}_1$ and
   $\mathbf{v}_2$ are in $\mathrm{sp}(\mathbf{v}_1, 2\mathbf{v}_1 + \mathbf{v}_2)$, so the subspaces are the same.
   (b) $\mathbf{v}_1 + \mathbf{v}_2$ and $\mathbf{v}_1 - \mathbf{v}_2$ are in $\mathrm{sp}(\mathbf{v}_1, \mathbf{v}_2)$, so $\mathrm{sp}(\mathbf{v}_1 + \mathbf{v}_2, \mathbf{v}_1 - \mathbf{v}_2)$ is contained in $\mathrm{sp}(\mathbf{v}_1, \mathbf{v}_2)$. Since
   $\mathbf{v}_1 = (\frac{1}{2})(\mathbf{v}_1 + \mathbf{v}_2) + (\frac{1}{2})(\mathbf{v}_1 - \mathbf{v}_2)$ and $\mathbf{v}_2 = (\frac{1}{2})(\mathbf{v}_1 + \mathbf{v}_2) - (\frac{1}{2})(\mathbf{v}_1 - \mathbf{v}_2)$, we see that $\mathrm{sp}(\mathbf{v}_1, \mathbf{v}_2)$ is
   contained in $\mathrm{sp}(\mathbf{v}_1 + \mathbf{v}_2, \mathbf{v}_1 - \mathbf{v}_2)$.
25. $r\mathbf{v} + s\mathbf{w} = (7r - 9s, -3r + 3s, r, s)$. Checking the first equation, we obtain $4(7r - 9s) +$
   $9(-3r + 3s) - r + 9s = 28r - 36s - 27r + 27s - r + 9s = 0r + 0s = 0$. Substitution in each of
   the other three equations also gives us zero.
27. $\{(1, 1)\}$ **29.** It is the zero subspace.   **31.** $\{(-3, -4, 4, 5)\}$
33. $\{(-1, 1, 0, 1), (2, -5, 1, 0)\}$ **35.** We got 90 percent.   **37.** No   **39.** Impossible

Section 3.4

1. They must be parallel.   **3.** They must be parallel.
5. They must all lie in the same plane containing the origin.   **7.** Dependent; $\{(1, 3)\}$
9. Independent   **11.** Dependent $\{(2, 1), (1, 4)\}$ **13.** Independent   **15.** Independent
17. Dependent; $\{x^2 - 1, x^2 + 1, 4x\}$ **19.** Dependent; $\{1, \sin^2 x\}$ **21.** Independent
23. Dependent
25. The zero vector is in $\mathrm{sp}(\mathbf{v}_1, \mathbf{v}_2, \ldots, \mathbf{v}_k)$ and by hypothesis the expression $\mathbf{0} = 0\mathbf{v}_1 + 0\mathbf{v}_2 + \cdots$
   $+ 0\mathbf{v}_k$ is the only linear combination of the $\mathbf{v}_i$ that yields $\mathbf{0}$. Thus the vectors are independent.
27. A basis   **29.** Not a basis   **31.** A basis
33. The chance that $\mathbf{v}_2$ is parallel to $\mathbf{v}_1$ is essentially zero. The chance that $\mathbf{v}_3$ then lies in the plane
   through the origin containing $\mathbf{v}_1$ and $\mathbf{v}_2$ is essentially zero.
35. Retain $\mathbf{v}_1, \mathbf{v}_2, \mathbf{v}_5$.   **37.** Retain $\mathbf{v}_1, \mathbf{v}_2, \mathbf{v}_3, \mathbf{v}_4, \mathbf{v}_6$.

Section 3.5

1. (a) Basis $\{(-1, 0, 1), (0, -1, 1)\}$; dimension 2; **(b)** $\{(-1, 0, 1), (0, -1, 1), (1, 0, 0)\}$
3. (a) Basis $\{(-1, -1, 2)\}$; dimension 1; **(b)** $\{(-1, -1, 2), (1, 0, 0), (0, 1, 0)\}$
5. (a) 2; **(b)** columns 1 and 2; **(c)** $\{(-4, -1, 0, 8), (12, -13, 8, 0)\}$
7. (a) 3; **(b)** columns 1, 2, and 4; **(c)** $\{(1, -1, 1, 0)\}$
9. (a) 3; **(b)** columns 1, 2, and 3; **(c)** $\{(-5, -2, -2, 6)\}$
11. Rank 3; not invertible   **13.** Rank 3; invertible
15. (a) Interchanging rows keeps the set of row vectors the same, and thus does not change row rank.
   (b) If row i is multiplied by $r \neq 0$, then the original row i can be recovered by multiplying the new
   row i by $(1/r)$, so the row space remains the same. **(c)** If row i is multiplied by r and added to row
   k, then the original row k can be recovered by multiplying row i by $-r$ and adding it to the new
   row k. Thus the row space is not changed.
17. Let $\{\mathbf{b}_1, \mathbf{b}_2, \ldots, \mathbf{b}_n\}$ be a basis for S. Then $\{\mathbf{b}_1, \mathbf{b}_2, \ldots, \mathbf{b}_n\}$ can be expanded, if necessary, to
   a basis of V. Since $\dim(V) = n$, it must be that $\{\mathbf{b}_1, \mathbf{b}_2, \ldots, \mathbf{b}_n\}$ is already a basis for V, so $S = V$.
19. Work with column rank. If $\mathrm{rank}(A) = \mathrm{rank}(A|\mathbf{b})$, then $\mathbf{b}$ lies in the column space of A, so the lin-
   ear system $A\mathbf{x} = \mathbf{b}$ has a solution. (See Theorem 3.8.) Conversely, if $A\mathbf{x} = \mathbf{b}$ has a solution, then $\mathbf{b}$
   lies in the column space of A so $\mathrm{rank}(A) = \mathrm{rank}(A|\mathbf{b})$.
21. By Theorem 3.18, $\mathrm{rank}[A(A^T)] = \mathrm{rank}(A^T)$. But $\mathrm{rank}(A^T) = \mathrm{rank}(A)$ since row rank and column
   rank are the same.

Chapter 4

Section 4.1

1. -15 **3.** 15

5. $\mathbf{b} \cdot \mathbf{p} = b_1(b_2c_3 - b_3c_2) - b_2(b_1c_3 - b_3c_1) + b_3(b_1c_2 - b_2c_1) = b_1b_2c_3 - b_1b_3c_2 - b_2b_1c_3 + b_2b_3c_1 + b_3b_1c_2 - b_3b_2c_1 = 0$. A similar computation shows that $\mathbf{c} \cdot \mathbf{p} = 0$.

7. 120 **9.** -9 **11.** $(a_1b_2 - a_2b_1) = -(b_1a_2 - b_2a_1)$ **13.** $-6\mathbf{i} + 3\mathbf{j} + 5\mathbf{k}$ **15.** $\mathbf{0}$

17. $22\mathbf{i} + 18\mathbf{j} + 2\mathbf{k}$ **19.** 11 **21.** $\sqrt{374}$ **23.** $9\sqrt{2}$ **25.** 16 **27.** $\sqrt{166}/2$ **29.** 24

31. 7 **33.** $\mathbf{a} \cdot (\mathbf{b} \times \mathbf{c}) = -8$, $\mathbf{a} \times (\mathbf{b} \times \mathbf{c}) = 2\mathbf{i} + 2\mathbf{j}$

35. $\mathbf{a} \cdot (\mathbf{b} \times \mathbf{c}) = -27$, $\mathbf{a} \times (\mathbf{b} \times \mathbf{c}) = 24\mathbf{i} + 54\mathbf{j} - 21\mathbf{k}$ **37.** 10 **39.** 6 **41.** $175/6$

43. $71/3$ **45.** Collinear   **47.** Not collinear   **49.** Not coplanar   **51.** Coplanar

53. $\mathbf{a} \times (\mathbf{b} + \mathbf{c}) = \begin{vmatrix} \mathbf{i} & \mathbf{j} & \mathbf{k} \\ a_1 & a_2 & a_3 \\ b_1 + c_1 & b_2 + c_2 & b_3 + c_3 \end{vmatrix}$

$$= \begin{vmatrix} \mathbf{i} & \mathbf{j} & \mathbf{k} \\ a_1 & a_2 & a_3 \\ b_1 & b_2 & b_3 \end{vmatrix} + \begin{vmatrix} \mathbf{i} & \mathbf{j} & \mathbf{k} \\ a_1 & a_2 & a_3 \\ c_1 & c_2 & c_3 \end{vmatrix} = \mathbf{a} \times \mathbf{b} + \mathbf{a} \times \mathbf{c}.$$

Just note that $a_2(b_3 + c_3) = a_2b_3 + a_2c_3$, etc. The demonstration that $(\mathbf{a} + \mathbf{b}) \times \mathbf{c} = \mathbf{a} \times \mathbf{c} + \mathbf{b} \times \mathbf{c}$ is similar.

55. We got 95 percent.

Section 4.2

1. 13 **3.** -21 **5.** 19 **7.** 320 **9.** 0 **11.** -6 **13.** 2 **15.** 4 **17.** 54 **19.** $\frac{1}{2}$

21. If A is diagonal, then AB is obtained from B by multiplying row 1 by a_{11}, row 2 by a_{22}, . . . , row n by a_{nn}. By the scalar multiplication property, we see that $\det(AB) = (a_{11}a_{22} \ldots a_{nn})[\det(B)] = \det(A) \cdot \det(B)$.

23. Expand by minors on the kth row. The minors are the same for the two matrices, so

$$\det(A) = a_{k1}a'_{k1} + a_{k2}a'_{k2} + \cdots + a_{kn}a'_{kn}$$
$$= (b_{k1} + c_{k1})a'_{k1} + (b_{k2} + c_{k2})a'_{k2} + \cdots + (b_{kn} + c_{kn})a'_{kn}$$
$$= b_{k1}a'_{k1} + b_{k2}a'_{k2} + \cdots + b_{kn}a'_{kn} + c_{k1}a'_{k1} + c_{k2}a'_{k2} + \cdots + c_{kn}a'_{kn}$$
$$= \det(B) + \det(C).$$

25. **(a)** The result is obvious for $n = 1$ and $n = 2$. Suppose it is true for $n - 1$. Expand the determinant of an $n \times n$ matrix A by minors across the first row. There are n minors, each of which requires at least $(n - 1)!$ multiplications by our induction assumption. Thus there are at least $n(n - 1)! = n!$ multiplications required.

(b) The answer can't be given without knowing the speed of your computer. For our computer with $n = 50$, we obtained 4.50564×10^{54} years in interpretive BASIC and 8.036651×10^{53} years in compiled BASIC using single precision.

Section 4.3

1. Continuing our list in the text up to only ten items, we may list:

    5. There exists an $n \times n$ matrix B such that $AB = I$.

    6. There exists an $n \times n$ matrix B such that $BA = I$.

    7. $A\mathbf{x} = \mathbf{b}$ has a unique solution for every vector $\mathbf{b}$ in $\mathbb{R}^n$.

    8. Matrix A can be row reduced to the identity matrix.

9. The column rank of A is n.

10. The nullspace of A is $\{0\}$.

3. -74 **5.** -2 **7.** 474 **9.** 420 **11.** -715 **13.** $-2, 5$ **15.** $1, 1 \pm \sqrt{13}$

17. $\text{Det}(A) = (-1)^{rs} \det(R) \cdot \det(S)$

19. $\text{Det}(C^{-1}AC) = \det(C^{-1}) \cdot \det(A) \cdot \det(C) = \det(A) \cdot \det(C^{-1}) \cdot \det(C) = \det(A) \cdot \det(C^{-1}C) = \det(A) \cdot \det(I) = \det(A) \cdot 1 = \det(A)$

21. 13019 for both YUREDUCE and MATCOMP

23. -95090693.71174 for MATCOMP, -95090693.71173999 for YUREDUCE

Section 4.4

1. $\begin{pmatrix} 1/2 & 0 \\ 1/2 & -1 \end{pmatrix}$ **3.** Not invertible   **5.** $\begin{pmatrix} 2/15 & 1/15 & -1/15 \\ -2/15 & 3/5 & 2/5 \\ 1/5 & -1/15 & 1/15 \end{pmatrix}$ **7.** $\mathbf{x} = \begin{pmatrix} 1 \\ 0 \end{pmatrix}$

9. $\mathbf{x} = \begin{pmatrix} 5 \\ -10 \end{pmatrix}$ **11.** $\mathbf{x} = \begin{pmatrix} 1/2 \\ 1/2 \\ -1/2 \end{pmatrix}$ **13.** $\mathbf{x} = \begin{pmatrix} 1/3 \\ 1/3 \\ 0 \end{pmatrix}$ **15.** $x_2 = 10$ **17.** $\begin{pmatrix} 12 & -8 & -7 \\ -11 & 13 & 5 \\ 9 & -6 & -1 \end{pmatrix}$

19. $\begin{pmatrix} -4827 & 114 & -2211 & 2409 \\ -3218 & 76 & -1474 & 1606 \\ 8045 & -190 & 3685 & -4015 \\ 1609 & -38 & 737 & -803 \end{pmatrix}$

Section 4.5

1. $\lambda^2 - 3\lambda + 2$, $\lambda_1 = 1$, $\mathbf{v}_1 = r\begin{pmatrix} -1 \\ 1 \end{pmatrix}$, $\lambda_2 = 2$, $\mathbf{v}_2 = s\begin{pmatrix} 0 \\ 1 \end{pmatrix}$

3. $\lambda^2 - 4\lambda + 3$, $\lambda_1 = 3$, $\mathbf{v}_1 = r\begin{pmatrix} -1 \\ 2 \end{pmatrix}$, $\lambda_2 = 1$, $\mathbf{v}_2 = s\begin{pmatrix} -1 \\ 1 \end{pmatrix}$

5. $\lambda^2 + 1$, no real eigenvalues

7. $-\lambda^3 + 2\lambda^2 + \lambda - 2$, $\lambda_1 = -1$, $\lambda_2 = 1$, $\lambda_3 = 2$, $\mathbf{v}_1 = r\begin{pmatrix} 0 \\ 1 \\ 0 \end{pmatrix}$, $\mathbf{v}_2 = s\begin{pmatrix} 0 \\ -1 \\ 1 \end{pmatrix}$, $\mathbf{v}_3 = t\begin{pmatrix} -1 \\ -1 \\ 1 \end{pmatrix}$

9. $-\lambda^3 + 8\lambda^2 + \lambda - 8$, $\lambda_1 = 1$, $\lambda_2 = 8$, $\lambda_3 = -1$, $\mathbf{v}_1 = r\begin{pmatrix} 0 \\ -1 \\ 1 \end{pmatrix}$, $\mathbf{v}_2 = s\begin{pmatrix} -1 \\ -1 \\ 1 \end{pmatrix}$, $\mathbf{v}_3 = t\begin{pmatrix} 0 \\ 1 \\ 0 \end{pmatrix}$

11. $-\lambda^3 - \lambda^2 + 8\lambda + 12$, $\lambda_1 = \lambda_2 = -2$, $\lambda_3 = 3$, $\mathbf{v}_1 = r\begin{pmatrix} -1 \\ 0 \\ 1 \end{pmatrix}$, $\mathbf{v}_2 = s\begin{pmatrix} 0 \\ 1 \\ 0 \end{pmatrix}$, $\mathbf{v}_3 = t\begin{pmatrix} 0 \\ -1 \\ 1 \end{pmatrix}$

13. $-\lambda^3 + 2\lambda^2 + 4\lambda - 8$, $\lambda_1 = \lambda_2 = 2$, $\lambda_3 = -2$, $\mathbf{v}_1 = r\begin{pmatrix} 0 \\ -1 \\ 1 \end{pmatrix}$, $\mathbf{v}_3 = s\begin{pmatrix} -1 \\ -3 \\ 1 \end{pmatrix}$

15. $-\lambda^3 - 3\lambda^2 + 4$, $\lambda_1 = \lambda_2 = -2$, $\lambda_3 = 1$, $\mathbf{v}_1 = r\begin{pmatrix} 0 \\ 1 \\ 0 \end{pmatrix}$, $\mathbf{v}_3 = s\begin{pmatrix} 3 \\ -1 \\ 3 \end{pmatrix}$

17. The statement is true for $k = 1$. Suppose it is true for $k - 1$. Then $A^k\mathbf{v} = A^{k-1}(A\mathbf{v}) = A^{k-1}(\lambda\mathbf{v}) = \lambda(A^{k-1}\mathbf{v}) = \lambda(\lambda^{k-1}\mathbf{v}) = \lambda^k\mathbf{v}$, so the statement is true for k. Thus by mathematical induction, $A^k = \lambda^k\mathbf{v}$ for all k.

19. The zero vector is in E_λ so E_λ is nonempty. Let $\mathbf{v}_1$ and $\mathbf{v}_2$ be in E_λ. Then $A(\mathbf{v}_1 + \mathbf{v}_2) = A\mathbf{v}_1 + A\mathbf{v}_2 = \lambda\mathbf{v}_1 + \lambda\mathbf{v}_2 = \lambda(\mathbf{v}_1 + \mathbf{v}_2)$ so $\mathbf{v}_1 + \mathbf{v}_2$ is in E_λ. For any scalar r, we have $A(r\mathbf{v}_1) = r(A\mathbf{v}_1) = r(\lambda\mathbf{v}_1) = \lambda(r\mathbf{v}_1)$ so $r\mathbf{v}_1$ is in E_λ. Thus E_λ is nonempty and closed under addition and scalar multiplication, so it is a subspace of $\mathbb{R}^n$.

21. $\lambda_1 = \dfrac{1 + \sqrt{5}}{2}$, $\lambda_2 = \dfrac{1 - \sqrt{5}}{2}$, $\mathbf{v}_1 = \begin{pmatrix} 2 \\ \sqrt{5} - 1 \end{pmatrix}$, $\mathbf{v}_2 = \begin{pmatrix} -2 \\ \sqrt{5} + 1 \end{pmatrix}$

23. **(a)** 0, 1, 2, 5, 12, 29, 70, 169, 408; **(b)** $\begin{pmatrix} 2 & 1 \\ 1 & 0 \end{pmatrix}$; **(c)** 107,578,520,350

25. **(a)** No nonzero vector in the plane is carried into a parallel vector by rotation through 90°.

(b) $\lambda_1 = i$, $\lambda_2 = -i$, $\mathbf{v}_1 = (r + si)\begin{pmatrix} i \\ 1 \end{pmatrix}$, $\mathbf{v}_2 = (p + qi)\begin{pmatrix} -i \\ 1 \end{pmatrix}$

(c) $A^2\mathbf{x} = A(A\mathbf{x})$ corresponds to rotating $\mathbf{x}$ through 180° counterclockwise, which is equivalent to multiplying $\mathbf{x}$ by -1. Thus -1 is an eigenvalue.

(d) A^2 has eigenvalues $\lambda_1 = i^2 = -1$ and $\lambda_2 = (-i)^2 = i^2 = -1$. All nonzero real vectors and all complex vectors are eigenvectors of $A^2 = \begin{pmatrix} -1 & 0 \\ 0 & -1 \end{pmatrix}$.

(e) A^3 has no real eigenvalues, but complex eigenvalues of $\lambda_1 = i^3 = -i$ and $\lambda_2 = (-i)^3 = i$ with eigenvectors of $\mathbf{v}_1 = (r + si)\begin{pmatrix} i \\ 1 \end{pmatrix}$ and $\mathbf{v}_2 = (p + qi)\begin{pmatrix} -i \\ 1 \end{pmatrix}$.

(f) $A^4 = \begin{pmatrix} 1 & 0 \\ 0 & 1 \end{pmatrix}$ has eigenvalues $\lambda_1 = i^4 = 1$ and $\lambda_2 = (-i)^4 = 1$; it corresponds to rotation through 360°. Every nonzero real or complex vector is an eigenvector.

27. $\lambda_1 = -1$, $\lambda_2 = 3$, $\lambda_3 = \lambda_4 = 2$, $\mathbf{v}_1 = r\begin{pmatrix} 1 \\ -1 \\ 1 \\ 1 \end{pmatrix}$, $\mathbf{v}_2 = s\begin{pmatrix} 0 \\ 0 \\ 0 \\ 1 \end{pmatrix}$, $\mathbf{v}_3 = t\begin{pmatrix} 0 \\ -1 \\ 0 \\ 1 \end{pmatrix}$, $\mathbf{v}_4 = p\begin{pmatrix} 0 \\ 1 \\ 0 \\ 0 \end{pmatrix}$

29. $\lambda_1 = 36$, $\lambda_2 = \lambda_3 = 4$, $\lambda_4 = 16$, $\mathbf{v}_1 = r\begin{pmatrix} 0 \\ 0 \\ 1 \\ 1 \end{pmatrix}$, $\mathbf{v}_2 = s\begin{pmatrix} 1 \\ 0 \\ 0 \\ 1 \end{pmatrix}$, $\mathbf{v}_3 = t\begin{pmatrix} 1 \\ 1 \\ 1 \\ 0 \end{pmatrix}$, $\mathbf{v}_4 = p\begin{pmatrix} 0 \\ 1 \\ 0 \\ 0 \end{pmatrix}$

31. $\lambda_1 = -19.20665169352584$; λ_2, $\lambda_3 = 1.603325846762922 \pm 3.319395789764702i$

33. $\lambda_1, \lambda_2 = -12.89594619786273 \pm 13.308655527116964i$; $\lambda_3, \lambda_4 = 40.89594619786273 \pm 17.42594441141775i$

35. If λ is an eigenvalue of A with $\mathbf{v}$ a corresponding eigenvector, then $\lambda + r$ is an eigenvalue of $A + rI$ with $\mathbf{v}$ a corresponding eigenvector. That is, the eigenvalues of $A + rI$ are those of A increased by r, while the corresponding eigenvectors remain the same.

Section 4.6

1. $\lambda_1 = 1$, $\lambda_2 = 3$, $\mathbf{v}_1 = r\begin{pmatrix} 1 \\ 0 \end{pmatrix}$, $\mathbf{v}_2 = s\begin{pmatrix} 1 \\ 1 \end{pmatrix}$, $C = \begin{pmatrix} 1 & 1 \\ 0 & 1 \end{pmatrix}$, $D = \begin{pmatrix} 1 & 0 \\ 0 & 3 \end{pmatrix}$

3. $\lambda_1 = -1$, $\lambda_2 = 3$, $\mathbf{v}_1 = r\begin{pmatrix} -1 \\ 1 \end{pmatrix}$, $\mathbf{v}_2 = s\begin{pmatrix} -2 \\ 1 \end{pmatrix}$, $C = \begin{pmatrix} -1 & -2 \\ 1 & 1 \end{pmatrix}$, $D = \begin{pmatrix} -1 & 0 \\ 0 & 3 \end{pmatrix}$

5. $\lambda_1 = -3$, $\lambda_2 = 1$, $\lambda_3 = 7$, $\mathbf{v}_1 = r\begin{pmatrix} 1 \\ 0 \\ 0 \end{pmatrix}$, $\mathbf{v}_2 = s\begin{pmatrix} 1 \\ 1 \\ 1 \end{pmatrix}$, $\mathbf{v}_3 = t\begin{pmatrix} 1 \\ 1 \\ 0 \end{pmatrix}$, $C = \begin{pmatrix} 1 & 1 & 1 \\ 0 & 1 & 1 \\ 0 & 0 & 1 \end{pmatrix}$, $D = \begin{pmatrix} -3 & 0 & 0 \\ 0 & 1 & 0 \\ 0 & 0 & 7 \end{pmatrix}$

7. $\lambda_1 = -1$, $\lambda_2 = 1$, $\lambda_3 = 2$, $\mathbf{v}_1 = r\begin{pmatrix} -1 \\ 0 \\ 1 \end{pmatrix}$, $\mathbf{v}_2 = s\begin{pmatrix} -1 \\ 0 \\ 3 \end{pmatrix}$, $\mathbf{v}_3 = t\begin{pmatrix} 0 \\ 1 \\ 0 \end{pmatrix}$, $C = \begin{pmatrix} -1 & -1 & 0 \\ 0 & 0 & 1 \\ 1 & 3 & 0 \end{pmatrix}$,

$D = \begin{pmatrix} -1 & 0 & 0 \\ 0 & 1 & 0 \\ 0 & 0 & 2 \end{pmatrix}$

9. Not diagonalizable   **11.** Diagonalizable (symmetric)

13. (a) $I^{-1}PI = P$ so P is similar to itself. **(b)** Let $C^{-1}PC = Q$. Then $(C^{-1})^{-1}QC^{-1} = P$ so Q is similar to P. **(c)** Let $C^{-1}PC = Q$ and $H^{-1}QH = R$. Then $H^{-1}(C^{-1}PC)H = R$, so $(CH)^{-1}P(CH) = R$ and P is similar to R.

15. (a) $A = \begin{pmatrix} .5 & .5 \\ 1 & 0 \end{pmatrix}$; **(b)** neutrally stable; **(c)** $\begin{pmatrix} a_{k+1} \\ a_k \end{pmatrix} = A^k \begin{pmatrix} 1 \\ 0 \end{pmatrix} = (2/3) \begin{pmatrix} 1 \\ 1 \end{pmatrix} + (-1/3)(-1/2)^k \begin{pmatrix} -1 \\ 2 \end{pmatrix}$;

(d) $a_k \approx (2/3)$ for large k.

17. (a) $A = \begin{pmatrix} .5 & .5 \\ 1 & 0 \end{pmatrix}$; **(b)** neutrally stable; **(c)** $\begin{pmatrix} a_{k+1} \\ a_k \end{pmatrix} = A^k \begin{pmatrix} 0 \\ 1 \end{pmatrix} = (1/3) \begin{pmatrix} 1 \\ 1 \end{pmatrix} + (1/3)(-1/2)^k \begin{pmatrix} -1 \\ 2 \end{pmatrix}$;

(d) $a_k \approx (1/3)$ for large k.

19. (a) $A = \begin{pmatrix} 1 & .75 \\ 1 & 0 \end{pmatrix}$; **(b)** unstable; **(c)** $\begin{pmatrix} a_{k+1} \\ a_k \end{pmatrix} = A^k \begin{pmatrix} 1 \\ 0 \end{pmatrix} = (1/4)(3/2)^k \begin{pmatrix} 3 \\ 2 \end{pmatrix} - (1/4)(-1/2)^k \begin{pmatrix} -1 \\ 2 \end{pmatrix}$;

(d) $a_k \approx 3^k/2^{k+1}$ for large k.

21. Not diagonalizable   **23.** Diagonalizable   **25.** Diagonalizable   **27.** Not diagonalizable
29. Diagonalizable

Chapter 5

Section 5.1

1. $(6/5, 8/5)$ **3.** $(1, 0, 0)$, $(0, 2, 0)$ and $(0, 0, 1)$ **5.** $(5/3, 4/3, 1/3)$ **7.** $\begin{pmatrix} 1 & 0 & 0 \\ 0 & 1 & 0 \\ 0 & 0 & 0 \end{pmatrix}$

9. $\begin{pmatrix} 5/7 & 3/7 & 1/7 \\ 3/7 & 5/14 & -3/14 \\ 1/7 & -3/14 & 13/14 \end{pmatrix}$ **11.** $\begin{pmatrix} 1 & 0 & 0 & 0 \\ 0 & 1 & 0 & 0 \\ 0 & 0 & 0 & 0 \\ 0 & 0 & 0 & 1 \end{pmatrix}$

13. Taking $\mathbf{x} = \mathbf{a}$, we have $\mathbf{a} \cdot \mathbf{a} = \|\mathbf{a}\|^2 = 0$ so $\mathbf{a} = \mathbf{0}$.
15. (i) $P^2 = [A(A^TA)^{-1}A^T][A(A^TA)^{-1}A^T] = A(A^TA)^{-1}(A^TA)(A^TA)^{-1}A^T = A(A^TA)^{-1}A^T = P$
(ii) $P^T = [A(A^TA)^{-1}A^T]^T = (A^T)^T[(A^TA)^{-1}]^TA^T = A[(A^TA)^T]^{-1}A^T = A[A^T(A^T)^T]^{-1}A^T$
$= A(A^TA)^{-1}A^T = P$

17. Let $A = P - Q$. By assumption, $A\mathbf{x} = (P - Q)\mathbf{x} = P\mathbf{x} - Q\mathbf{x} = \mathbf{0}$ for all $\mathbf{x}$ in $\mathbb{R}^n$. Thus the dot product of each row vector of A with each vector $\mathbf{x}$ in $\mathbb{R}^n$ is 0. By Exercise 13, this means that each row vector of A is $\mathbf{0}$, so A is the zero matrix. Thus $P = Q$.

19. $\dfrac{29}{31}(1, -2, 3, 1, 4, 0)$ and $\dfrac{26}{31}(1, -2, 3, 1, 4, 0)$

21. $\dfrac{1}{68}(2, 78, 162, 74)$ and $\dfrac{1}{68}(-42, 96, 66, 180)$

23. $\dfrac{1}{680}(-20, -90, -110, 1250, -90)$ and $\dfrac{1}{680}(332, 66, 398, 670, 66)$

Section 5.2

1. (a) $y = \dfrac{116.4}{59} + \dfrac{60.4}{59}x$; **(b)** 7.092 inches   **3. (a)** $p = (.528)e^{.274t}$; **(b)** \$27,271

5. $y = .943 + 2.914x$

7. $y = -.9 + 2.6x$

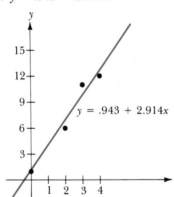

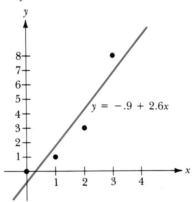

9. $y = .1 - .4x + x^2$

11. $\mathbf{x} = \begin{pmatrix} -1/5 \\ 3/5 \end{pmatrix}$ **13.** $\mathbf{x} = \begin{pmatrix} 0 \\ 2 \\ -1/4 \end{pmatrix}$

15. $y = 1.972881 + 1.023729x$
17. $y = -(1/3) + 1.5x$ where x is in units of 10 people and y is in units of \$100,000.
19. $y = .1 - .4x + x^2$
21. $y = .7587548 + 1.311284x$. A quadratic will fit slightly better. An exponential will not fit as well.

23. $y = .9 + e^{.194x}$ fits a bit better than the computer's fit, which minimizes least-squares using logarithms. The exponential fits best; the linear fit is the worst.

25. $y = 12 - 1.51x$ was our fit. The computer did slightly better. A quadratic fit is the best; an exponential fit is the worst. Even the exponential fits that we guessed, which were better than the computer's attempts using logarithms, were worse than the best linear or quadratic fits.

Section 5.3

1. $\begin{pmatrix} 5/6 & -1/3 & -1/6 \\ -1/3 & 1/3 & -1/3 \\ -1/6 & -1/3 & 5/6 \end{pmatrix}$
3. $\begin{pmatrix} 1/2 & 0 & 7/(10\sqrt{2}) & 1/(10\sqrt{2}) \\ 0 & 1/2 & 1/(10\sqrt{2}) & -7/(10\sqrt{2}) \\ 7/(10\sqrt{2}) & 1/(10\sqrt{2}) & 1/2 & 0 \\ 1/(10\sqrt{2}) & -7/(10\sqrt{2}) & 0 & 1/2 \end{pmatrix}$

5. $\begin{pmatrix} 4/9 & 0 & -2/9 & 4/9 \\ 0 & 4/9 & -4/9 & -2/9 \\ -2/9 & -4/9 & 5/9 & 0 \\ 4/9 & -2/9 & 0 & 5/9 \end{pmatrix}$
7. $\begin{pmatrix} 3/4 & 1/4 & 1/4 & -1/4 \\ 1/4 & 3/4 & -1/4 & 1/4 \\ 1/4 & -1/4 & 3/4 & 1/4 \\ -1/4 & 1/4 & 1/4 & 3/4 \end{pmatrix}$
9. $\begin{pmatrix} 9/4 \\ 12/5 \\ 5 \end{pmatrix}$
11. $\begin{pmatrix} -1 \\ -7 \\ 1 \\ -5 \end{pmatrix}$

13. $\begin{pmatrix} 3/25 & -4/25 & 0 \\ 0 & 0 & 1 \\ 4/25 & 3/25 & 0 \end{pmatrix}$
15. $\begin{pmatrix} 1/4 & -1/4 & 1/4 & 1/4 \\ -1/4 & 1/4 & 1/4 & 1/4 \\ 1/4 & 1/4 & -1/4 & 1/4 \\ 1/4 & 1/4 & 1/4 & -1/4 \end{pmatrix}$

17. Using the preservation of dot product property we have $\|A\mathbf{x}\|^2 = (A\mathbf{x}) \cdot (A\mathbf{x}) = \mathbf{x} \cdot \mathbf{x} = \|\mathbf{x}\|^2$. Thus $\|A\mathbf{x}\| = \|\mathbf{x}\|$.

19. $y = 1.6 + .9x$ **21.** 56.5 gallons

23. Writing $A = \begin{pmatrix} 1 & a_1 \\ 1 & a_2 \\ \vdots & \vdots \\ 1 & a_m \end{pmatrix}$, we have $A^T A = \begin{pmatrix} m & 0 \\ 0 & d \end{pmatrix}$, where $d = \Sigma_{i=1}^m a_i^2$. The least-squares solution

is then

$$\begin{pmatrix} r_0 \\ r_1 \end{pmatrix} = (A^T A)^{-1} A^T \begin{pmatrix} b_1 \\ b_2 \\ \vdots \\ b_m \end{pmatrix} = \begin{pmatrix} 1/m & 0 \\ 0 & 1/d \end{pmatrix} \begin{pmatrix} \Sigma_{i=1}^m b_i \\ \Sigma_{i=1}^m a_i b_i \end{pmatrix} = \begin{pmatrix} (\Sigma_{i=1}^m b_i)/m \\ (\Sigma_{i=1}^m a_i b_i)/d \end{pmatrix}.$$

Thus $r_0 = (\Sigma_{i=1}^m b_i)/m$ and $r_i = (\Sigma_{i=1}^m a_i b_i)/(\Sigma_{i=1}^m a_i^2)$.

Section 5.4

1. $(0, 1, 0)$, $(1/\sqrt{2}, 0, 1/\sqrt{2})$ **3.** $(1/\sqrt{2}, 0, 1/\sqrt{2})$, $(-1/\sqrt{3}, 1/\sqrt{3}, 1/\sqrt{3})$, $(1/\sqrt{6}, 2/\sqrt{6}, -1/\sqrt{6})$

5. $(1/\sqrt{2}, 0, 1/\sqrt{2}, 0)$, $(0, 1, 0, 0)$, $(1/\sqrt{6}, 0, -1/\sqrt{6}, 2/\sqrt{6})$

7. $(1/\sqrt{2}, 0, 1/\sqrt{2}, 0)$, $(-1/\sqrt{6}, 2/\sqrt{6}, 1/\sqrt{6}, 0)$ $(1/\sqrt{3}, 1/\sqrt{3}, -1/\sqrt{3}, 0)$, $(0, 0, 0, 1)$

9. $(-1, -5, 7)$ **11.** $(4, -5, 1, 0, 3)$

13. Since S and T are orthogonal, surely T is a subspace of $S^\perp$. By Theorem 5.7 we know that $\dim(S^\perp) = n - \dim(S) = \dim(T)$. Thus a basis for T must also be a basis for $S^\perp$, so $T = S^\perp$.

15. $4/\sqrt{5}$ **17.** $\sqrt{14}/5$ **19.** $\sqrt{161/27}$ **21.** $\sqrt{10}$

23. Since $\mathbf{0} \cdot \mathbf{w} = 0$ for all $\mathbf{w}$ in S, we see that $S^\perp$ is nonempty. Let $\mathbf{v}_1$ and $\mathbf{v}_2$ be in $S^\perp$. Then $\mathbf{v}_1 \cdot \mathbf{w} = \mathbf{v}_2 \cdot \mathbf{w} = 0$ for all $\mathbf{w}$ in S. Therefore, $(\mathbf{v}_1 + \mathbf{v}_2) \cdot \mathbf{w} = \mathbf{v}_1 \cdot \mathbf{w} + \mathbf{v}_2 \cdot \mathbf{w} = 0 + 0 = 0$ and $(r\mathbf{v}_1) \cdot \mathbf{w} = r(\mathbf{v}_1 \cdot \mathbf{w}) = r0 = 0$ for all $\mathbf{w}$ in S. Thus $S^\perp$ is nonempty and is closed under addition and scalar multiplication, so it is as subspace of $\mathbb{R}^n$.

25. $Q = \begin{pmatrix} 0 & 1/\sqrt{2} \\ 1 & 0 \\ 0 & 1/\sqrt{2} \end{pmatrix}, R = \begin{pmatrix} 1 & 1 \\ 0 & \sqrt{2} \end{pmatrix}$

27. $Q = \begin{pmatrix} 1/\sqrt{2} & -1/\sqrt{3} & 1/\sqrt{6} \\ 0 & 1/\sqrt{3} & 2/\sqrt{6} \\ 1/\sqrt{2} & 1/\sqrt{3} & -1/\sqrt{6} \end{pmatrix}, R = \begin{pmatrix} \sqrt{2} & \sqrt{2} & \sqrt{2} \\ 0 & \sqrt{3} & -1/\sqrt{3} \\ 0 & 0 & 4/\sqrt{6} \end{pmatrix}$

29. $Q = \begin{pmatrix} 1/\sqrt{2} & 0 & 1/\sqrt{6} \\ 0 & 1 & 0 \\ 1/\sqrt{2} & 0 & -1/\sqrt{6} \\ 0 & 0 & 2/\sqrt{6} \end{pmatrix}, R = \begin{pmatrix} \sqrt{2} & \sqrt{2} & 1/\sqrt{2} \\ 0 & 1 & -1 \\ 0 & 0 & \sqrt{6}/2 \end{pmatrix}$ **31.** Our answers checked.

33. $y = 5.476191 - .75x + .2738095x^2$
35. $y = 5.632035 - 1.138889x + .1287879x^2 + .05555556x^3 + .01515152x^4$
37. $y = -5 - 8x + 9x^2 + 0x^3$

Section 5.5

1. $x_1 = 3 - 2t$
$x_2 = -3 + t$

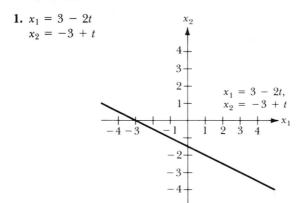

$x_1 = 3 - 2t,$
$x_2 = -3 + t$

3. Clearly d_1 and d_2 are not both zero. If $d_1 \neq 0$ we have $x_1 = c/d_1 + d_2t, x_2 = -d_1t$; and if $d_1 = 0$ we have $x_1 = t, x_2 = c/d_2$.
5. (a) $x_1 = -2 + t, x_2 = 4 - t$; (b) $x_1 = 3 + 3t, x_2 = -1 + 2t, x_3 = 6 + 7t$;
(c) $x_1 = 2 + 3t, x_2 = 0 - 5t, x_3 = 4 + 12t$
7. The lines coincide; all points $(5 - 3t, -1 + t)$ for t in $\mathbb{R}$
9. (a) $(1/2, 3/2)$; (b) $(3/2, -2, 5/2)$; (c) $(-2, 9/2, 17/2)$
11. $(-3/2, -1/2, 15/4)$ **13.** $3x_1 - 2x_2 + 7x_3 = 39$ **15.** $3x_1 - 7x_2 + 3x_3 = 0$
17. $x_1 = -3 + t, x_2 = 2 - 2t, x_3 = 7 + t$ **19.** $(2, 3, 4)$ **21.** $2x_1 + x_2 - x_3 = 2$ **23.** $\sqrt{6}$
25. $12/\sqrt{31}$ **27.** $(1, 4/3, 7/3, 11/3)$ and $(0, 5/3, 5/3, 10/3)$ **29.** $x_1 + 4x_2 + x_3 = 10$
31. $(5, 1)$, a 0-flat   **33.** $(-36, 10, 0) + t(-10, 3, 1)$, a 1-flat
35. $(-28, -12, 0, 0) + t_1(-5, -2, 0, 1) + t_2(7, 3, 1, 0)$, a 2-flat
37. $(-13, 0, -5, 0) + t_1(14, 0, 5, 1) + t_2(-2, 1, 0, 0)$, a 2-flat
39. Expanding by minors on the last column, we obtain a linear equation in $x_1, x_2, \ldots, x_n$ with at least one nonzero coefficient, which is thus the equation a hyperplane in $\mathbb{R}^n$. Substitution of any

point $\mathbf{a}_j$ for x in the last column of the matrix gives a matrix with two equal columns, so
$f(a_{1j}, a_{2j}, \ldots, a_{nj}) = 0$ for $j = 1, 2, \ldots, n$. Thus the hyperplane contains each point $\mathbf{a}_j$.
41. $3/\sqrt{14}$ **43.** 4

Section 5.6

1. $\sqrt{213}$ **3.** $\sqrt{30}$ **5.** 11   **7.** 38   **9.** 2   **11.** $\sqrt{6}/2$ **13.** 2/3
15. Let A be the $n \times 3$ matrix with column vectors $\mathbf{b} - \mathbf{a}$, $\mathbf{c} - \mathbf{a}$, and $\mathbf{d} - \mathbf{a}$. The points are coplanar if and only if $\det(A^T A) = 0$.
17. They are not coplanar.   **19.** **(a)** 0; **(b)** 0

Chapter 6

Section 6.1

1. A linear transformation   **3.** A linear transformation   **5.** A linear transformation
7. Not a linear transformation   **9.** Not a linear transformation   **11.** Not a linear transformation
13. Kernel $= \{\mathbf{0}\}$, one to one, $T^{-1}(u, v) = (u/2 + v/2, -u/2 + v/2)$
15. Kernel $= \{r(-1, 3, 2) \mid r \in \mathbb{R}\}$, not one to one
17. Kernel $= \{r(1, -1, 1) \mid r \in \mathbb{R}\}$, not one to one
19. Kernel $= \{f \in F \mid f(3) = 0\}$, not one to one   **21.** $\mathbb{R}^2$ **23.** $\mathrm{sp}((2, 3, 1), (0, 1, 1))$
25. $\mathbb{R}^2$ **27.** $\mathbb{R}$
29. Let $\mathbf{v}_1$ and $\mathbf{v}_2$ be vectors in V, and let r and s be any scalars. Then

$$
\begin{aligned}
(T_1 + T_2)(r\mathbf{v}_1 + s\mathbf{v}_2) &= T_1(r\mathbf{v}_1 + s\mathbf{v}_2) + T_2(r\mathbf{v}_1 + s\mathbf{v}_2) \\
&= rT_1(\mathbf{v}_1) + sT_1(\mathbf{v}_2) + rT_2(\mathbf{v}_1) + sT_2(\mathbf{v}_2) \\
&= r[T_1(\mathbf{v}_1) + T_2(\mathbf{v}_1)] + s[T_1(\mathbf{v}_2) + T_2(\mathbf{v}_2)] \\
&= r[(T_1 + T_2)(\mathbf{v}_1)] + s[(T_1 + T_2)(\mathbf{v}_2)]
\end{aligned}
$$

so $T_1 + T_2$ is a linear map.
31. Let T_1 and T_2 be in $L(V, W)$. For every vector $\mathbf{v}$ in V, we have $(T_1 + T_2)(\mathbf{v}) = T_1(\mathbf{v}) + T_2(\mathbf{v}) = T_2(\mathbf{v}) + T_1(\mathbf{v}) = (T_2 + T_1)(\mathbf{v})$ so $T_1 + T_2 = T_2 + T_1$. Commutativity of addition in $L(V, W)$ thus follows from commutativity of the addition in W.
   The zero vector is the linear transformation $T_0 \colon V \to W$ such that $T_0(\mathbf{v}) - \mathbf{0}$, the zero vector of W, for all $\mathbf{v}$ in V.
   Proofs of the other characterizing properties of a vector space are similar to the proof of commutativity above. That is, to show that two transformations are equal, simply show that they map each $\mathbf{v}$ in V into the same vector in W.

Section 6.2

1. $(1, -1)$ **3.** $(2, 6, -4)$ **5.** $(2, 1, 3)$ **7.** **(a)** $\begin{pmatrix} 2 & -3 \\ 1 & 1 \end{pmatrix}$; **(b)** $\begin{pmatrix} 3 & -1 \\ -1 & 2 \end{pmatrix}$

9. **(a)** $\begin{pmatrix} 1 & 1 & 1 \\ 1 & -1 & -1 \\ -1 & -1 & 1 \end{pmatrix}$; **(b)** $\begin{pmatrix} 6 & 3 & 5 \\ -2 & -1 & -1 \\ -4 & -3 & 1 \end{pmatrix}$ **11.** **(a)** $\begin{pmatrix} 1 & 1 & 0 \\ 0 & 1 & 1 \\ 1 & 0 & 1 \\ 1 & 1 & 1 \end{pmatrix}$; **(b)** $\begin{pmatrix} 3 & 2 & 1 \\ -1 & -1 & 0 \\ 0 & 0 & -1 \\ 0 & 1 & 1 \end{pmatrix}$

13. Rank 2, nullity 0   **15.** Rank 3, nullity 0   **17.** Rank 3, nullity 0
19. Consider the parallelogram in $\mathbb{R}^2$ representing the sum of two vectors $\mathbf{v}_1$ and $\mathbf{v}_2$ emanating from the *origin*. Reflection in a line through the origin is a rigid motion of the plane corresponding to

turning the plane over with the line acting as axis of rotation. The parallelogram is carried into another having vectors emanating from the origin since the line of reflection passes through the origin. Calling the reflection T, we see that this image parallelogram represents the sum $T(\mathbf{v}_1) + T(\mathbf{v}_2)$ so that $T(\mathbf{v}_1 + \mathbf{v}_2) = T(\mathbf{v}_1) + T(\mathbf{v}_2)$. That is, the vector diagonal of the original parallelogram is carried into the vector diagonal of the image parallelogram. Similarly, the rigid motion T carries $r(\mathbf{v})$ into $r(T(\mathbf{v}))$, so T is a linear transformation.

21. (a) $\begin{pmatrix} 0 & 1 \\ 1 & 0 \end{pmatrix}$; (b) $\begin{pmatrix} 0 & -1 \\ -1 & 0 \end{pmatrix}$ **23.** (a) $\begin{pmatrix} 1 & 0 & 0 \\ 0 & -1 & 0 \\ 0 & 0 & 1 \end{pmatrix}$; (b) $\begin{pmatrix} 1 & 0 & 0 \\ 0 & 1 & 0 \\ 0 & 0 & -1 \end{pmatrix}$

25. The ith basis vector is multiplied by the ith entry on the diagonal.

27. Since A_T is invertible, we know that $A_T\mathbf{a} = A_T\mathbf{b}$ for $\mathbf{a}$ and $\mathbf{b}$ in $\mathbb{R}^n$ if and only if $\mathbf{a} = \mathbf{b}$. Thus $T(\mathbf{a}) = T(\mathbf{b})$ if and only if $\mathbf{a} = \mathbf{b}$ so T is a one-to-one transformation, and T^{-1} is well defined. Let $\mathbf{y}$ be a vector in $\mathbb{R}^n$. Then $T^{-1}(\mathbf{y})$ is the unique vector $\mathbf{x}$ such that $T(\mathbf{x}) = \mathbf{y}$. Then $A_T\mathbf{x} = \mathbf{y}$ so $\mathbf{x} = A_T^{-1}\mathbf{y}$. Thus A_T^{-1} must be the matrix representation of T^{-1} relative to B,B.

29. The linear transformation whose effect on a vector is the same as multiplying the vector on the left by the standard matrix representation of T is T itself.

31. We got 93 percent.   **33.** We got 90 percent.

Section 6.3

1. (a) $\begin{pmatrix} 0 & -1 \\ 1 & 1 \end{pmatrix}$; (b) $\begin{pmatrix} 1 & 1 \\ -1 & 0 \end{pmatrix}$ **3.** (a) $\begin{pmatrix} 0 & 0 & 1 \\ 0 & 1 & 0 \\ 1 & 0 & 0 \end{pmatrix}$; (b) $\begin{pmatrix} 0 & 0 & 1 \\ 0 & 1 & 0 \\ 1 & 0 & 0 \end{pmatrix}$

5. (a) $\begin{pmatrix} 2 & -3 \\ 1 & 1 \end{pmatrix}$; (b) $\begin{pmatrix} 1/2 & 5/2 \\ -3/2 & 5/2 \end{pmatrix}$; (c) $\begin{pmatrix} 1 & 1 \\ 1 & -1 \end{pmatrix}$

7. (a) $\begin{pmatrix} 1 & 1 & 0 \\ 1 & -1 & 0 \\ 1 & 0 & 1 \end{pmatrix}$; (b) $\begin{pmatrix} 0 & 0 & -1 \\ 2 & 1 & 2 \\ 0 & 1 & 0 \end{pmatrix}$; (c) $\begin{pmatrix} 1 & 1 & 0 \\ 1 & 0 & 1 \\ 1 & 1 & 1 \end{pmatrix}$

9. (a) $\begin{pmatrix} 0 & 0 & 0 & 1 \\ 0 & 0 & 1 & 0 \\ 0 & 1 & 0 & 0 \\ 1 & 0 & 0 & 0 \end{pmatrix}$; (b) $\begin{pmatrix} 1 & 1 & 1 & 1 \\ 0 & 0 & 0 & -1 \\ 0 & 0 & -1 & 0 \\ 0 & -1 & 0 & 0 \end{pmatrix}$; (c) $\begin{pmatrix} 1 & 1 & 1 & 1 \\ 1 & 1 & 1 & 0 \\ 1 & 1 & 0 & 0 \\ 1 & 0 & 0 & 0 \end{pmatrix}$

11. Let $\mathbf{u}_1$ and $\mathbf{u}_2$ be in U, and let r and s be scalars. Then

$$(T' \circ T)(r\mathbf{u}_1 + s\mathbf{u}_2) = T'(T(r\mathbf{u}_1 + s\mathbf{u}_2)) = T'[rT(\mathbf{u}_1) + sT(\mathbf{u}_2)]$$
$$= r[T'(T(\mathbf{u}_1))] + s[T'(T(\mathbf{u}_2))]$$
$$= r[(T' \circ T)(\mathbf{u}_1)] + s[(T' \circ T)(\mathbf{u}_2)].$$

Thus $T' \circ T$ is a linear transformation.

13. (a) $T(x_1, x_2, x_3) = (2x_1 + x_2 - 3x_3, x_1 + 3x_2 + 2x_3, 5x_1 + 5x_2 - 4x_3)$; (b) same answer as part (a)

15. Let $A_{T^{-1}}$ be the standard matrix representation of T^{-1}. Now the standard matrix representation of the identity map $\iota: \mathbb{R}^n \to \mathbb{R}^n$ is the identity matrix I. From $T^{-1} \circ T = \iota$ and Theorem 6.10, we obtain $A_{T^{-1}}A_T = I$. Thus $A_{T^{-1}} = (A_T)^{-1}$.

17. (a) $\mathbf{b}_1 = (a, b)$, $\mathbf{b}_2 = (-b, a)$; (b) $\begin{pmatrix} -1 & 0 \\ 0 & 1 \end{pmatrix}$; (c) $\dfrac{1}{a^2 + b^2}\begin{pmatrix} b^2 - a^2 & -2ab \\ -2ab & a^2 - b^2 \end{pmatrix}$

19. (a) We know that $T_A \circ T_B = T_{AB} = T_I$, which is the identity transformation of $\mathbb{R}^n$ into itself.
(b) Suppose that $T_B(\mathbf{v}) = T_B(\mathbf{w})$ for $\mathbf{v}$ and $\mathbf{w}$ in $\mathbb{R}^n$. Then $\mathbf{v} = (T_A \circ T_B)(\mathbf{v}) = T_A(T_B(\mathbf{v})) = T_A(T_B(\mathbf{w})) = (T_A \circ T_B)(\mathbf{w}) = \mathbf{w}$ so T_B is one to one. Also for any $\mathbf{x}$ in $\mathbb{R}^n$, we have $\mathbf{x} = (T_A \circ T_B)(\mathbf{x}) = T_A(T_B(\mathbf{x}))$ so the range of T_A is $\mathbb{R}^n$.

(c) The rank equation for T_A says that $n = \dim(\text{Ker}(T_A)) + \dim(\text{Range}(T_A))$. Since $\dim(\text{Range}(T_A)) = n$ by part (b), it must be that $\dim(\text{Ker}(T_A)) = 0$, so T_A is one to one. A similar argument with the rank equation for T_B shows that T_B has range $\mathbb{R}^n$.

(d) From part (a) we have $T_A \circ T_B = \iota$ while from part (c) we see that T_A^{-1} is well defined. Thus we obtain in succession $T_A \circ T_B = \iota$, $(T_A \circ T_B) \circ T_A = T_A$, $T_A \circ (T_B \circ T_A) = T_A \circ \iota$, $T_A^{-1}T_A \circ (T_B \circ T_A) = T_A^{-1}T_A \circ \iota$, $T_A \circ T_B = \iota$. It follows at once that the standard matrix representation BA of $T_B \circ T_A$ is the identity matrix I.

Section 6.4

1. $\lambda_1 = -1$, $\lambda_2 = 5$, $\mathbf{v}_1 = r(1, 1)$, $\mathbf{v}_2 = r(-1, 1)$, diagonalizable

3. $\lambda_1 = 0$, $\lambda_2 = 1$, $\lambda_3 = 2$, $\mathbf{v}_1 = r(-1, 0, 1)$, $\mathbf{v}_2 = r(0, 1, 0)$, $\mathbf{v}_3 = r(1, 0, 1)$, diagonalizable

5. $\lambda_1 = -2$, $\lambda_2 = 1$, $\lambda_3 = 5$, $\mathbf{v}_1 = r(0, 1, 1)$, $\mathbf{v}_2 = r(-6, 5, 8)$, $\mathbf{v}_3 = r(0, -5, 2)$, diagonalizable

7. $r(x^2 + 2x + 1)$ for any scalar r

9. The zero polynomial only **11.** The zero polynomial only

13. (a) $\lambda = 1$, a or a^2; (b) $\lambda = 1$, a, a^2, or a^3

15. The characteristic polynomial of A is $p(\lambda) = \det(A - \lambda I)$. The characteristic polynomial of D is

$$\begin{aligned} q(\lambda) &= \det(D - \lambda I) \\ &= \det(C^{-1}AC - \lambda I) = \det(C^{-1}AC - \lambda C^{-1}IC) \\ &= \det[C^{-1}(A - \lambda I)C] = \det(C^{-1})\det(C)\det(A - \lambda I) \\ &= \det(C^{-1}C)\det(A - \lambda I) = \det(I)\det(A - \lambda I) = \det(A - \lambda I) \\ &= p(\lambda). \end{aligned}$$

Since A and D have the same characteristic polynomial, they have the same eigenvalues with the same algebraic multiplicities. Since the eigenspaces of A and D are the same, the geometric multiplicities of the equal eigenvalues of A and of D are the same also.

17. Let E_λ consist of the zero vector and all eigenvectors for λ. Then E_λ is nonempty. Let $\mathbf{v}_1$ and $\mathbf{v}_2$ be in E_λ, and let r be any scalar. Then $T(\mathbf{v}_1 + \mathbf{v}_2) = T(\mathbf{v}_1) + T(\mathbf{v}_2) = \lambda\mathbf{v}_1 + \lambda\mathbf{v}_2 = \lambda(\mathbf{v}_1 + \mathbf{v}_2)$, so $\mathbf{v}_1 + \mathbf{v}_2$ is in E_λ. Also $T(r\mathbf{v}_1) = rT(\mathbf{v}_1) = r(\lambda\mathbf{v}_1) = \lambda(r\mathbf{v}_1)$ so $r\mathbf{v}_1$ is in E_λ. Thus E_λ is closed under vector addition and scalar multiplication, so it is a subspace of V.

19. $T(x, y) = \dfrac{1}{a^2 + b^2}\left((b^2 - a^2)x - 2aby, -2abx + (a^2 - b^2)y\right)$

21. $T(x_1, x_2, x_3) = (1/3)(2x_1 - 2x_2 + x_3, -2x_1 - x_2 + 2x_3, x_1 + 2x_2 + 2x_3)$

23. $\lambda_1 = 1$ must be an eigenvalue of algebraic and geometric multiplicity $n - 1$, while $\lambda_2 = -1$ must be an eigenvalue of algebraic and geometric multiplicity 1.

25. $p(\lambda) = (-1)^n(\lambda - \lambda_1)(\lambda - \lambda_2) \cdots (\lambda - \lambda_n)$ has constant term $p(0) = (-1)^{2n}\lambda_1\lambda_2 \cdots \lambda_n$. On the other hand, $p(0) = \det(A - 0I) = \det(A)$.

27. We got 90 percent.

Section 6.5

1. $\begin{pmatrix} -1/\sqrt{2} & 1/\sqrt{2} \\ 1/\sqrt{2} & 1/\sqrt{2} \end{pmatrix}$ **3.** $\begin{pmatrix} 1/4 & -\sqrt{2} & 1/4 \\ \sqrt{2}/4 & 0 & -\sqrt{2}/4 \\ 1/4 & \sqrt{2} & 1/4 \end{pmatrix}$ **5.** $\begin{pmatrix} 1/2 & -1/\sqrt{2} & 1/2 & 0 \\ 1/2 & 0 & -1/2 & -1/\sqrt{2} \\ 1/2 & 0 & -1/2 & 1/\sqrt{2} \\ 1/2 & 1/\sqrt{2} & 1/2 & 0 \end{pmatrix}$

7. Now $C^T = C^{-1}$; thus $A^T = (CDC^{-1})^T = (CDC^T)^T = CD^TC^T = CDC^{-1} = A$, so that A is symmetric.

9. $\lambda_2(\mathbf{v}_1 \cdot \mathbf{v}_2) = \mathbf{v}_1 \cdot \lambda_2\mathbf{v}_2 = \mathbf{v}_1 \cdot A\mathbf{v}_2 = \mathbf{v}_1^T A\mathbf{v}_2$

11. $(1/4)[\|\mathbf{v} + \mathbf{w}\|^2 - \|\mathbf{v} - \mathbf{w}\|^2] = (1/4)[(\mathbf{v} + \mathbf{w}) \cdot (\mathbf{v} + \mathbf{w}) - (\mathbf{v} - \mathbf{w}) \cdot (\mathbf{v} - \mathbf{w})] = (1/4)[\mathbf{v} \cdot \mathbf{v} + 2\mathbf{v} \cdot \mathbf{w} + \mathbf{w} \cdot \mathbf{w} - \mathbf{v} \cdot \mathbf{v} + 2\mathbf{v} \cdot \mathbf{w} - \mathbf{w} \cdot \mathbf{w}] = (1/4)(4\mathbf{v} \cdot \mathbf{w}) = \mathbf{v} \cdot \mathbf{w}$

13. $\begin{pmatrix} 3 & 2 \\ 4 & 3 \end{pmatrix}$ **15.** Not orthogonal **17.** Orthogonal **19.** Orthogonal

21. $T(\mathbf{x}) \cdot T(\mathbf{y}) = (1/\sqrt{2})(-x_1 + x_2, x_1 + x_2) \cdot (1/\sqrt{2})(-y_1 + y_2, y_1 + y_2)$
$= (1/2)[(-x_1 + x_2)(-y_1 + y_2) + (x_1 + x_2)(y_1 + y_2)]$
$= (1/2)(2x_1y_1 + 2x_2y_2) = x_1y_1 + x_2y_2 = \mathbf{x} \cdot \mathbf{y}.$

Section 6.6

1. 16   **3.** 224   **5.** 11   **7.** 330   **9.** $\sqrt{3}$ **11.** $9\sqrt{3}$ **13.** $\sqrt{17}$ **15.** $9\sqrt{17}\pi$

Chapter 7

Section 7.1

1. $U = \begin{pmatrix} 3 & -6 \\ 0 & 1 \end{pmatrix}$, $A = \begin{pmatrix} 3 & -3 \\ -3 & 1 \end{pmatrix}$ **3.** $U = \begin{pmatrix} 1 & -4 & 3 \\ 0 & -1 & -8 \\ 0 & 0 & 0 \end{pmatrix}$, $A = \begin{pmatrix} 1 & -2 & 3/2 \\ -2 & -1 & -4 \\ 3/2 & -4 & 0 \end{pmatrix}$

5. $U = \begin{pmatrix} -2 & 8 \\ 0 & 3 \end{pmatrix}$, $A = \begin{pmatrix} -2 & 4 \\ 4 & 3 \end{pmatrix}$ **7.** $U = \begin{pmatrix} 8 & 5 & -4 \\ 0 & 1 & -2 \\ 0 & 0 & 10 \end{pmatrix}$, $A = \begin{pmatrix} 8 & 5/2 & -2 \\ 5/2 & 1 & -1 \\ -2 & -1 & 10 \end{pmatrix}$

9. $\begin{pmatrix} x_1 \\ x_2 \end{pmatrix} = \begin{pmatrix} 1/\sqrt{2} & 1/\sqrt{2} \\ -1/\sqrt{2} & 1/\sqrt{2} \end{pmatrix}\begin{pmatrix} t_1 \\ t_2 \end{pmatrix}$, $-t_1{}^2 + t_2{}^2$ **11.** $\begin{pmatrix} x_1 \\ x_2 \end{pmatrix} = \begin{pmatrix} 3/\sqrt{10} & -1/\sqrt{10} \\ 1/\sqrt{10} & 3/\sqrt{10} \end{pmatrix}\begin{pmatrix} t_1 \\ t_2 \end{pmatrix}$, $-t_1{}^2 + 9t_2{}^2$

13. $\begin{pmatrix} x_1 \\ x_2 \end{pmatrix} = \begin{pmatrix} 1/\sqrt{2} & 1/\sqrt{2} \\ -1/\sqrt{2} & 1/\sqrt{2} \end{pmatrix}\begin{pmatrix} t_1 \\ t_2 \end{pmatrix}$, $5t_1{}^2 + t_2{}^2$

15. $\begin{pmatrix} x_1 \\ x_2 \\ x_3 \end{pmatrix} = \begin{pmatrix} -1/\sqrt{2} & -1/\sqrt{2} & 1/\sqrt{3} \\ 1/\sqrt{2} & 0 & 1/\sqrt{3} \\ 0 & 1/\sqrt{2} & 1/\sqrt{3} \end{pmatrix}\begin{pmatrix} t_1 \\ t_2 \\ t_3 \end{pmatrix}$, $2t_1{}^2 + 2t_2{}^2 - t_3{}^2$

17. $a + c = k$, $ac = b^2$ **19.** $3.472136t_1{}^2 - 5.472136t_2{}^2$
21. $4.69854t_1{}^2 + 1.323057t_2{}^2 - 4.021597t_3{}^2$
23. $5.5t_1{}^2 - 4t_2{}^2 + 4t_3{}^2 + .5t_4{}^2$

Section 7.2

1. $C = \begin{pmatrix} 1/\sqrt{2} & -1/\sqrt{2} \\ 1/\sqrt{2} & 1/\sqrt{2} \end{pmatrix}$

$t_1{}^2 - t_2{}^2 = 1$

3. $C = \begin{pmatrix} 1/\sqrt{2} & -1/\sqrt{2} \\ 1/\sqrt{2} & 1/\sqrt{2} \end{pmatrix}$

$2t_1{}^2 + 0t_2{}^2 = 4$

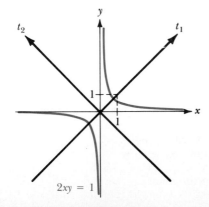

$2xy = 1$

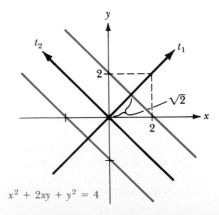

$x^2 + 2xy + y^2 = 4$

5. $C = \begin{pmatrix} 3/\sqrt{10} & -1/\sqrt{10} \\ 1/\sqrt{10} & 3/\sqrt{10} \end{pmatrix}$

$11t_1{}^2 + t_2{}^2 = 4$

7. $C = \begin{pmatrix} 1/\sqrt{5} & -2/\sqrt{5} \\ 2/\sqrt{5} & 1/\sqrt{5} \end{pmatrix}$

$7t_1{}^2 + 2t_2{}^2 = 8$

$10x^2 + 6xy + 2y^2 = 4$

$3x^2 + 4xy + 6y^2 = 8$

9. The symmetric matrix of the quadratic-form portion is $\begin{pmatrix} a & b/2 \\ b/2 & c \end{pmatrix}$. Thus

$$\det(A - \lambda I) = \begin{vmatrix} a - \lambda & b/2 \\ b/2 & c - \lambda \end{vmatrix} = \lambda^2 - (a + c)\lambda + ac - b^2/4.$$

The eigenvalues are given by $\lambda = (1/2)[a + c \pm \sqrt{(a + c)^2 + b^2 - 4ac}$; they are real numbers with the same algebraic sign if $b^2 - 4ac < 0$, with the opposite algebraic sign if $b^2 - 4ac > 0$. One of them is zero if $b^2 - 4ac = 0$. We obtain a (possibly degenerate) ellipse, hyperbola, or parabola accordingly.

11. Let $C^T = \begin{pmatrix} 1 & 0 & 0 \\ 0 & b/r & c/r \\ 0 & -c/r & b/r \end{pmatrix}$, where $r = \sqrt{b^2 + c^2}$. Then C is an orthogonal matrix such that

$\det(C) = 1$ and

$$\begin{pmatrix} t_1 \\ t_2 \\ t_3 \end{pmatrix} = \mathbf{t} = C^T \begin{pmatrix} x \\ y \\ z \end{pmatrix} = \begin{pmatrix} x \\ (by + cz)/r \\ (-cy + bz)/r \end{pmatrix}.$$

Thus

$$\begin{pmatrix} x \\ y \\ z \end{pmatrix} = C\mathbf{t} = \begin{pmatrix} t_1 \\ (bt_2 - ct_3)/r \\ (ct_2 + bt_3)/r \end{pmatrix}$$

represents a rotation of axes that transforms the equation $ax^2 + by + cz = d$ to the form $at_1{}^2 + rt_2 = d$.

13. Hyperboloid of two sheets   **15.** Hyperboloid of one sheet   **17.** Hyperbolic cylinder
19. Hyperboloid of one sheet   **21.** Elliptic cone or hyperboloid of one or two sheets
23. Elliptic cone or hyperboloid of one or two sheets   **25.** Parabolic cylinder or two parallel planes
27. Hyperbolic paraboloid or hyperbolic cylinder

Section 7.3

1. g has a local minimum of -7 at $(0, 0)$.   **3.** g has no local extremum at $(0, 0)$.
5. g has a local maximum of 3 at $(-5, 0)$.   **7.** g has no local extremum at $(0, 0)$.
9. The behavior of g at $(3, 1)$ is not determined.   **11.** g has a local maximum of 4 at $(0, 0, 0)$.
13. g has no local extremum at $(0, 0, 0)$.   **15.** g has no local extremum at $(7, -6, 0)$.
17. $g(x, y) = y^2 + 10$ **19.** $g(x, y) = y^2 + x^3$ **21.** $g(x, y) = x^4 + y^4 + 40$
23. The maximum is .5 at $\pm(1/\sqrt{2})(1, 1)$. The minimum is $-.5$ at $\pm(1/\sqrt{2})(1, -1)$.
25. The maximum is 9 at $\pm(1/\sqrt{10})(-1, 3)$. The minimum is -1 at $\pm(1/\sqrt{10})(3, 1)$.
27. The maximum is 6 at $\pm(1/\sqrt{2})(1, -1)$. The minimum is 0 at $\pm(1/\sqrt{2})(1, 1)$.
29. The maximum is 3 at $\pm(1/\sqrt{3})(1, -1, 1)$. The minimum is 0 at $\pm[1/\sqrt{2a^2 + 2b^2 - 2ab}](a - b, a, b)$.
31. The maximum is 2 at $[\pm 1/\sqrt{2a^2 + 2b^2}](a, b, b, -a)$. The minimum is -2 at $\pm(1/\sqrt{2})(0, -1, 1, 0)$.
33. The local maximum of f is $\lambda_1 a^2$, where λ_1 is the maximum eigenvalue of the symmetric coefficient matrix of the form f; it is assumed at any eigenvector corresponding to λ_1 and of length a. An analogous statement holds for the local minimum of f.
35. g has a local maximum of 5 at $(0, 0, 0)$.   **37.** g has no local extremum at $(0, 0, 0)$.

Section 7.4

1. $w_2 = \begin{pmatrix} 0 \\ -1 \end{pmatrix}$, $w_3 = \begin{pmatrix} 1 \\ -1/3 \end{pmatrix}$, $w_4 = \begin{pmatrix} .75 \\ -1 \end{pmatrix}$

Rayleigh quotients: $-2, 1, 5.2$

Maximum eigenvalue 6, eigenvector $\begin{pmatrix} 1 \\ -1 \end{pmatrix}$

3. $w_2 = \begin{pmatrix} 1 \\ 5/7 \end{pmatrix}$, $w_3 = \begin{pmatrix} 1 \\ 19/29 \end{pmatrix}$, $w_4 = \begin{pmatrix} 1 \\ 65/103 \end{pmatrix}$

Rayleigh quotients: $6, 298/74 \approx 4, 4222/1202 \approx 3.5$

Maximum eigenvalue 3, eigenvector $\begin{pmatrix} 1 \\ 3/5 \end{pmatrix}$

5. $5 \begin{pmatrix} 1/2 & 1/2 \\ 1/2 & 1/2 \end{pmatrix} - \begin{pmatrix} 1/2 & -1/2 \\ -1/2 & 1/2 \end{pmatrix}$ **7.** $2 \begin{pmatrix} 2/3 & 1/3 & 1/3 \\ 1/3 & 1/6 & 1/6 \\ 1/3 & 1/6 & 1/6 \end{pmatrix} - \begin{pmatrix} 1/3 & -1/3 & -1/3 \\ -1/3 & 1/3 & 1/3 \\ -1/3 & 1/3 & 1/3 \end{pmatrix}$

9. $\begin{pmatrix} -1/2 & 1/2 \\ 1/2 & -1/2 \end{pmatrix}$ **11.** $\begin{pmatrix} -1/3 & 1/3 & 1/3 \\ 1/3 & -1/3 & -1/3 \\ 1/3 & -1/3 & -1/3 \end{pmatrix}$

13. $\lambda = 12$, $\mathbf{v} = (-.7058823, 1, -.4117647)$ **15.** $\lambda = 6$, $\mathbf{v} = (-.9032258, 1, -.4193548)$
17. $\lambda = .1882784448992811$, $\mathbf{v} = (1, .2893009, .3203757)$
19. $\lambda_1 = 4.732050807568877$, $\mathbf{v}_1 = (1, -.7320508, 1)$
   $\lambda_2 = 1.267949192431123$, $\mathbf{v}_2 = (.3660254, 1, .3660254)$
   $\lambda_3 = -4$, $\mathbf{v}_3 = (-1, 0, 1)$
21. $\lambda_1 = 16.87586339619508$, $\mathbf{v}_1 = (.9245289, 1, .3678643, .7858846)$
   $\lambda_2 = -15.93189429348535$, $\mathbf{v}_2 = (-.3162426, .6635827, -1, -.004253739)$
   $\lambda_3 = 6.347821447472841$, $\mathbf{v}_3 = (-.5527083, .9894762, .8356429, -1)$
   $\lambda_4 = -.291790550182573$, $\mathbf{v}_4 = (-1, .06924058, .3582734, .9206089)$
23. **(a)** The characteristic polynomial $A - \lambda I = \begin{vmatrix} a - \lambda & b \\ b & c - \lambda \end{vmatrix} = \lambda^2 + (-a - c)\lambda + (ac - b^2)$ has roots

$\lambda = (1/2)[a + c \pm \sqrt{(a + c)^2 - 4(ac - b^2)}] = (1/2)[a + c \pm \sqrt{(a - c)^2 + 4b^2}]$.

(b) If we use part (a), the first row vector of $A - \lambda I$ is

$$(a - \lambda, b) = \left(\frac{1}{2}(a - c \mp \sqrt{(a - c)^2 + 4b^2}), b\right) = (g \mp \sqrt{g^2 + b^2}, b).$$

(c) From part (a), eigenvectors for the matrix are $(-b, g \pm \sqrt{g^2 + b^2}) = (-b, g \pm h)$. Normalizing, we obtain vectors $\dfrac{(-b, g \pm h)}{\sqrt{b^2 + (g \pm h)^2}}$. Using the upper choice of sign and setting $r = \sqrt{b^2 + (g + h)^2}$, we obtain $(-b/r, (g + h)/r)$ as the first column of C. Using the lower choice of sign and setting $s = \sqrt{b^2 + (g - h)^2}$, we obtain $(-b/s, (g - h)/s)$ as the second column of C.

(d) $\det(C) = \dfrac{-b(g - h)}{rs} + \dfrac{b(g + h)}{rs} = \dfrac{2bh}{rs}$; since $h, r, s \geq 0$ we see that the algebraic sign of $\det(C)$ is the same as that of b.

25. $\lambda_1 = -12.00517907692924$, $\lambda_2 = 7.906602974286551$, $\lambda_3 = 17.09857610264269$

27. $\lambda_1 = -5.210618568922174$, $\lambda_2 = 2.856693936892428$, $\lambda_3 = 3.528363748899602$, $\lambda_4 = 7.825560883130143$

29. 5.823349919059785, $-11.91167495952989 \pm 1.357830063519836i$

31. 57.22941613544168, -92.88108454947197, -54.25594801085533, $47.45380821244281 \pm 44.48897425527453i$

Chapter 8

Section 8.1

1. Not a linear transformation   **3.** Not a linear transformation   **5.** A linear transformation

7. A linear transformation

9. A linear transformation

$$A_T = \begin{pmatrix} 4 & 0 & 0 & 0 \\ 0 & 4 & 0 & 0 \\ -18 & 0 & 4 & 0 \\ 12 & -6 & 0 & 4 \end{pmatrix}, \quad A_T \begin{pmatrix} 4 \\ -5 \\ 7 \\ 3 \end{pmatrix} = \begin{pmatrix} 16 \\ -20 \\ -44 \\ 90 \end{pmatrix}$$

so $T(4x^3 - 5x^2 + 7x + 3) = 16x^3 - 20x^2 - 44x + 90$

13. $A_T = \begin{pmatrix} 0 & 0 & 0 & 0 \\ 0 & 14 & 0 & 0 \\ 0 & 0 & -8 & -6 \\ 0 & 0 & 6 & -8 \end{pmatrix}, \quad A_T \begin{pmatrix} 3 \\ -4 \\ 5 \\ -1 \end{pmatrix} = \begin{pmatrix} 0 \\ -56 \\ -34 \\ 38 \end{pmatrix}$

so $T(3e^x - 4e^{3x} + 5 \sin 2x - \cos 2x) = -56e^{3x} - 34 \sin 2x + 38 \cos 2x$

15. (a) If $B = (1, x, x^2, x^3, x^4)$, then D^2 maps B to $(0, 0, 2, 6x, 12x^2)$ so $\{2, 6x, 12x^2\}$ is a basis for the range of D^2; thus rank $(T) = 3$. If $D^2(f) = 0$, then $f = c_1 x + c_2$ and $\{1, x\}$ is a basis for $\text{Ker}(D^2)$; thus null$(T) = 2$ and $5 = \dim(V) = \text{rank}(T) + \text{null}(T) = 3 + 2$.

(b) Using a similar argument we have $\dim(V) = 5$, rank$(T) = 0$, and null $(T) = 5$.

17. If $c_1 \sin x + c_2 \cos x = 0$, then differentiation yields $c_1 \cos x - c_2 \sin x = 0$. Setting $x = 0$, we have $c_2 = 0$ and $c_1 = 0$.

19. If $c_1 e^x + c_2 \sin 2x + c_3 \cos 2x = 0$, then two differentiations yield $c_1 e^x + 2c_2 \cos 2x - 2c_3 \sin 2x = 0$ and $c_1 e^x - 4c_2 \sin 2x - 4c_3 \cos 2x = 0$. Setting $x = 0$ in the three equations, we have the system

$$\begin{aligned} c_1 \qquad\quad + c_3 &= 0 \\ c_1 + 2c_2 \qquad &= 0 \\ c_1 \qquad\quad - 4c_3 &= 0 \end{aligned}$$

whose only solution is $c_1 = c_2 = c_3 = 0$.

21. If $c_1x + c_2x^2 + c_3 \cos x + c_4 \cos x^2 = 0$, then three differentiations yield $c_1 + 2c_2x - c_3 \sin x - 2c_4x \sin x^2 = 0$, $2c_2 - c_3 \cos x - 4c_4x^2 \cos x^2 - 2c_4 \sin x^2 = 0$ and $c_3 \sin x + 8c_4x^3 \sin x^2 - 12c_4x \cos x^2 = 0$. Setting $x = 0$ in the four equations, we have

$$c_3 + c_4 = 0, \qquad c_1 = 0, \qquad 2c_2 - c_3 = 0, \quad \text{and} \quad 0 = 0.$$

One more differentiation at $x = 0$ yields $c_3 - 12c_4 = 0$, and we deduce that $c_1 = c_2 = c_3 = c_4 = 0$.

23. Let $c_1e^{a_1x} + c_2e^{a_2x} + \cdots + c_ne^{a_nx} = 0$. Then $n - 1$ differentiations yield

$$c_1a_1e^{a_1x} + c_2a_2e^{a_2x} + \cdots + c_na_ne^{a_nx} = 0,$$

$$\vdots$$

$$c_1a_1^{n-1}e^{a_1x} + c_2a_2^{n-1}e^{a_2x} + \cdots + c_na_n^{n-1}e^{a_nx} = 0.$$

Setting $x = 0$ in these equations, we obtain the system

$$c_1 \qquad + c_2 + \cdots \qquad + c_n = 0$$
$$a_1c_1 \quad + a_2c_2 + \cdots \quad + a_nc_n = 0$$
$$\vdots$$
$$a_1^{n-1}c_1 + a_2^{n-1}c_2 + \cdots + a_n^{n-1}c_n = 0$$

whose only solution is $c_1 = c_2 = \cdots = c_n = 0$ for distinct a_i.

25. $\{\sin x, \cos x\}$ **27.** $\{e^{2x}, e^{-2x}, \sin 2x, \cos 2x\}$

29. The familiar formulas $\int_a^x (f + g)(t)\, dt = \int_a^x f(t)\, dt + \int_a^x g(t)\, dt$ and $\int_a^x (cf)(t)\, dt = c\int_a^x f(t)\, dt$ show $\int_a^x (\ )\, dt$ to be a linear transformation.

31. $A = \begin{pmatrix} 1 \\ -5 \end{pmatrix}$; $-4x + 20$ **33.** $A = \begin{pmatrix} 4/3 & 0 & 0 \\ 0 & 2 & 0 \\ 0 & 0 & 4 \\ -32/3 & -8 & -8 \end{pmatrix}$; $-4x^3 + 14x^2 - 24$

Section 8.2

1. $(.4, -.3)$ **3.** $(-.7, .8)$ **5.** $T(dx) = -3\, dx$ **7.** $T(dx) = 0$ **9.** $x_2 \approx 1.732$

11. $x_2 \approx .686$ **13.** Orientation reversing   **15.** Orientation reversing

17. $|-7| = 7$ **19.** $\sqrt{13}$

21. (a) Take the paragraph in the text containing Eq. (10), but after Eq. (8) read "decreasing" for "increasing" and "$f'(t) < 0$" for "$f'(t) > 0$." Then put a minus sign before formula (9) and before $\int_c^d$ in Eq. (10).

(b) This follows from the property $-\int_c^d h(t)\, dt = \int_d^c h(t)\, dt$ for an integral.

23. (a) $T(dt) = \begin{pmatrix} f_1'(t_0)\, dt \\ f_2'(t_0)\, dt \\ \vdots \\ f_n'(t_0)\, dt \end{pmatrix}$; **(b)** $A = \begin{pmatrix} f_1'(t_0) \\ f_2'(t_0) \\ \vdots \\ f_n'(t_0) \end{pmatrix}$;

(c) $\sqrt{f_1'(t_0)^2 + f_2'(t_0)^2 + \cdots + f_n'(t_0)^2}$; **(d)** same as (c);

(e) Curve length $= \int_a^b \sqrt{f_1'(t_0)^2 + f_2'(t_0)^2 + \cdots + f_n'(t_0)^2}\, dt$;

(f) $\sqrt{f_1'(t_0)^2 + f_2'(t_0)^2 + \cdots + f_n'(t_0)^2}\, \Delta t$

25. $.02$ **27.** $.05\sqrt{30}$

Section 8.3

1. $(6, -12)$ **3.** $(-24, -16, 48, 48)$ **5.** $\begin{pmatrix} 1/2 & 1/4 & 0 \\ 0 & \pi e^{4\pi} & 4e^{4\pi} \\ \pi & 0 & 2 \end{pmatrix}$ **7.** $\begin{pmatrix} 6\pi \\ 0 \\ 1/\pi \end{pmatrix}$

9. 1; orientation reversing   **11.** 1; orientation reversing   **13.** 3   **15.** $.09\pi$ **17.** .24
19. $\sqrt{26}(.09)\pi$
21. Let $\mathbf{y} = g(\mathbf{x})$ and $\mathbf{x} = f(\mathbf{t})$ so that $\mathbf{y} = g(f(\mathbf{t}))$, and let $f(\mathbf{a}) = \mathbf{b}$. The i,jth entry in the matrix $(g \circ f)'(\mathbf{a})$ is then $\partial y_i / \partial t_j$. By the chain rule of multivariable calculus we have

$$\frac{\partial y_i}{\partial t_j} = \frac{\partial y_i}{\partial x_1}\frac{\partial x_1}{\partial t_j} + \frac{\partial y_i}{\partial x_2}\frac{\partial x_2}{\partial t_j} + \cdots + \frac{\partial y_i}{\partial x_m}\frac{\partial x_m}{\partial t_j} = \left(\frac{\partial y_i}{\partial x_1}, \frac{\partial y_i}{\partial x_2}, \ldots, \frac{\partial y_i}{\partial x_m}\right) \cdot \left(\frac{\partial x_1}{\partial t_j}, \frac{\partial x_2}{\partial t_j}, \ldots, \frac{\partial x_m}{\partial t_j}\right),$$

which is the dot product of the ith row of $g'(\mathbf{b})$ with the jth column of $f'(\mathbf{a})$. Thus $(g \circ f)'(\mathbf{a}) = g'(\mathbf{b})f'(\mathbf{a})$.
23. The volume-change factor for T is the reciprocal of the volume-change factor for T^{-1}. Since the substitution $g(u, v) = ((u + v)/2, (v - u)/2)$ was obtained as the inverse of the transformation $T(x, y) = (x - y, x + y)$, its volume-change factor is the reciprocal of the volume-change factor for T, that is, the reciprocal 1/2 of

$$\begin{vmatrix} 1 & -1 \\ 1 & 1 \end{vmatrix} = 2.$$

25. $\int_{-2}^{1}\int_{-1}^{1} u^2 v^3 (1/2) \, du \, dv$
27. $(1.860805853111703, \pm .7330767879460008), (.2541016883650524, \pm 1.983792411511353)$
29. $(3.077228234333831, 3.386640460128785), (-4.957849742483486, 5.155606081640021)$
31. $(.8518838649739748, -2.173868928754215, -.9867431398353754),$
$(-.3303492803732686, 2.027099071837175, -.8530282645370337)$

Section 8.4

1. $\begin{pmatrix} x_1 \\ x_2 \end{pmatrix} = \begin{pmatrix} 1 & 5 \\ 0 & 1 \end{pmatrix}\begin{pmatrix} k_1 e^{3t} \\ k_2 e^{2t} \end{pmatrix} = \begin{pmatrix} k_1 e^{3t} + 5k_2 e^{2t} \\ k_2 e^{2t} \end{pmatrix}$ **3.** $\begin{pmatrix} x_1 \\ x_2 \end{pmatrix} = \begin{pmatrix} 1 & 1 \\ 1 & -1 \end{pmatrix}\begin{pmatrix} k_1 e^{3t} \\ k_2 e^{-t} \end{pmatrix} = \begin{pmatrix} k_1 e^{3t} + k_2 e^{-t} \\ k_1 e^{3t} - k_2 e^{-t} \end{pmatrix}$

5. $\begin{pmatrix} x_1 \\ x_2 \\ x_3 \end{pmatrix} = \begin{pmatrix} 1 & -1 & 0 \\ -1 & 2 & 1 \\ 2 & 0 & 1 \end{pmatrix}\begin{pmatrix} k_1 e^{-3t} \\ k_2 \\ k_3 e^t \end{pmatrix} = \begin{pmatrix} k_1 e^{-3t} - k_2 \\ -k_1 e^{-3t} + 2k_2 + k_3 e^t \\ 2k_1 e^{-3t} + k_3 e^t \end{pmatrix}$

7. $\begin{pmatrix} x_1 \\ x_2 \\ x_3 \end{pmatrix} = \begin{pmatrix} -5 & -5 & -3 \\ 2 & 2 & 1 \\ 2 & 1 & 1 \end{pmatrix}\begin{pmatrix} k_1 e^{3t} \\ k_2 e^{-t} \\ k_3 e^{2t} \end{pmatrix} = \begin{pmatrix} -5k_1 e^{3t} - 5k_2 e^{-t} - 3k_3 e^{2t} \\ 2k_1 e^{3t} + 2k_2 e^{-t} + k_3 e^{2t} \\ 2k_1 e^{3t} + k_2 e^{-t} + k_3 e^{2t} \end{pmatrix}$

Chapter 9

Section 9.1

1. 19,448   **3.** 1000 lb of blend A and 2500 lb of blend B
5. Maximize $x_1 + 2x_2 + 4x_3$ subject to

$$\begin{pmatrix} 3 & 4 & 1 \\ -2 & -2 & 1 \\ -1 & 1 & 0 \end{pmatrix}\begin{pmatrix} x_1 \\ x_2 \\ x_3 \end{pmatrix} \leq \begin{pmatrix} 12 \\ 4 \\ 1 \end{pmatrix}, \quad \mathbf{x} \geq \mathbf{0}.$$

7. Maximize $-2x_1 - x_2 - x_3 + x_4$ subject to

$$\begin{pmatrix} 1 & 3 & 5 & 0 \\ 1 & -2 & -5 & 1 \end{pmatrix} \begin{pmatrix} x_1 \\ x_2 \\ x_3 \\ x_4 \end{pmatrix} \leq \begin{pmatrix} 10 \\ 5 \end{pmatrix}, \quad \mathbf{x} \geq \mathbf{0}.$$

9. Value 8 at $(2, 1)$ **11.** Value 8 at $(0, 1, 0)$ **13.** Value -24 at $(0, 6, 0)$
15. Value 12 at $(0, 6, 0)$
17. One such problem is to maximize $x_1 + 4x_2$ subject to $x_1 - x_2 \leq 1$, $-x_1 + x_2 \leq 3$, $x_2 \leq 4$, $\mathbf{x} \geq \mathbf{0}$.
19. Value 21 at $(5, 4)$ **21.** Value 30.4 at $(1.6, 0, 7.2)$ **23.** Value -13 at $(0, 0, 2, 15)$

Section 9.2

1.

	x_1	x_2	x_3	y_1	y_2	y_3	q_1	q_2	
y_1	1	2	1	1	0	0	0	0	5
q_1	4	2	-3	0	-1	0	1	0	3
q_2	5	0	-1	0	0	-1	0	1	4
P	-3	2	-5	0			0	0	
	$-9M$	$-2M$	$+4M$		M	M			$-7M$

3.

	x_1	x_2	x_3	y_1	y_2	q_1	q_2	q_3	
q_1	3	-1	0	0	0	1	0	0	5
q_2	0	2	1	0	0	0	1	0	2
y_1	1	4	5	1	0	0	0	0	20
q_3	4	-2	-1	0	-1	0	0	1	3
P	-1	2	-5	0		0	0	0	
	$-7M$	$+M$			M				$-10M$

5. 3.5 oz of food 1 and .75 oz of food 2   **7.** Value 4.5 at $(4.5, 0)$
9. (a) Let x_1 be shipped from W_1 to B_1, x_2 from W_1 to B_2, x_3 from W_2 to B_1, and x_4 from W_2 to B_2.
Maximize $10x_1 + 5x_2 + 4x_3 + 8x_4$ subject to $x_1 + x_2 \leq 30$, $x_3 + x_4 \leq 20$, $x_1 + x_3 = 10$,
$x_2 + x_4 = 40$.
(b)

	x_1	x_2	x_3	x_4	y_1	y_2	q_1	q_2	
y_1	1	1	0	0	1	0	0	0	30
y_2	0	0	1	1	0	1	0	0	20
q_1	1	0	1	0	0	0	1	0	10
q_2	0	1	0	1	0	0	0	1	40
P	10	5	4	8	0	0	0	0	
	$-M$	$-M$	$-M$	$-M$					$-50M$

(c) Value \$270 at $(0, 30, 10, 10)$
11. Value 4.5 at $(4.5, 0)$ **13.** 60 lb of bag 1, 10 lb of bag 2
15. Value 19.85714 at $(2.571429, 0, 2.428572)$ **17.** Value $-35/3$ at $(5/3, 0, 2)$

Section 9.3

1. Minimize $10y_1 + 15y_2$ subject to $y_1 + 3y_2 \geq 3$, $-2y_1 - y_2 \leq 2$.
3. Maximize $6x_1 + 4x_2$ subject to $-2x_1 - x_2 \geq 4$, $3x_1 + 3x_2 \leq 3$.

5. Minimize $8y_1 - y_2 - 2y_3$ subject to $y_1 - y_2 - y_3 \geq 3$, $2y_1 + y_2 - y_3 \geq 4$.

7. Maximize $12x_1 - 12x_2 + 4x_3 - 4x_4$ subject to $x_1 - x_2 \leq 1$, $3x_1 - 3x_2 + x_3 - x_4 \leq 0$, $-x_1 + x_2 - x_3 + x_4 \geq 2$.

9. (a) Minimize $\mathbf{b} \cdot \mathbf{y}$ subject to $A^T\mathbf{y} \geq \mathbf{c}$, $\mathbf{y} \geq \mathbf{0}$.
   (b) Maximize $-\mathbf{b} \cdot \mathbf{x}$ subject to $A^T\mathbf{x} \geq \mathbf{c}$, $\mathbf{x} \geq \mathbf{0}$.
   (c) Minimize $-\mathbf{c} \cdot \mathbf{y}$ subject to $-A\mathbf{y} \geq -\mathbf{b}$ or subject to $A\mathbf{y} \leq \mathbf{b}$, $\mathbf{y} \geq \mathbf{0}$.
   (d) Stating the answer to (c) as a maximum problem, we obtain: Maximize $\mathbf{c} \cdot \mathbf{x}$ subject to $A\mathbf{x} \leq \mathbf{b}$, $\mathbf{x} \geq \mathbf{0}$.

11. 3.5 oz of food 1 and .75 oz of food 2   13. 1000 lb of blend A, 2500 lb of blend B

15. The matrix of data in the tableau will be smallest for the form that contains the fewest ($\geq$ type)-constraints where constants on the righthand sides are positive.

17. Value 4 at (4, 0)   19. Value 34 at (8, 2)

21. The simplex method is a bit easier with the primal problem due to the smaller matrix of data in the tableau.

23. The simplex method is easier with the dual problem due to the smaller matrix of data in the tableau.

25. The simplex method is about the same work with the primal as it is with the dual.

Appendix A

1. Let $P(n)$ be the equation to be proved. Clearly $P(1)$ is true since $1 = \dfrac{1(1+1)(2+1)}{6}$. Assume that $P(k)$ is true. Then

$$
\begin{aligned}
1^2 + 2^2 + 3^2 + \cdots + k^2 + (k+1)^2 &= \frac{k(k+1)(2k+1)}{6} + (k+1)^2 \\
&= \frac{(k+1)(2k^2 + k + 6k + 6)}{6} \\
&= \frac{(k+1)(k+2)(2k+3)}{6}
\end{aligned}
$$

Thus $P(k+1)$ is true, and $P(n)$ holds for all $n \in Z^+$.

3. Let $P(n)$ be the equation to be proved. Clearly $P(1)$ is true. Assume that $P(k)$ is true. Then

$$
\begin{aligned}
1 + 3 + 5 + \cdots + (2k-1) + (2k+1) &= k^2 + 2k + 1 \\
&= (k+1)^2
\end{aligned}
$$

as required. Therefore, $P(n)$ holds for all $n \in Z^+$.

5. Let $P(n)$ be the equation to be proved. Clearly $P(1)$ is true since $a = a(1-r)/(1-r)$. Assume that $P(k)$ is true. Then

$$
\begin{aligned}
a + ar + ar^2 + \cdots + ar^k + ar^{k+1} &= a(1 - r^{k+1})/(1-r) + ar^{k+1} \\
&= a[1 - r^{k+1} + r^{k+1}(1-r)]/(1-r) = a(1 - r^{k+2})/(1-r),
\end{aligned}
$$

which establishes $P(k+1)$. Therefore, $P(n)$ is true for all $n \in Z^+$.

7. The notion of an "interesting property" has not been made precise; it is not well defined. Also, we work in mathematics with two-valued logic: A statement is either true or false, *but not both*. The assertion that not having an interesting property would be an interesting property seems to contradict this two-valued logic. We would be saying that the integer both has and does not have an interesting property.

Index